I0485813

A Brief History of Chemistry

Edited by Paul F. Kisak

Contents

Chapter 1

History of chemistry

Reihen	Gruppe I. — R²O	Gruppe II. — RO	Gruppe III. — R²O³	Gruppe IV. RH⁴ RO²	Gruppe V. RH³ R²O⁵	Gruppe VI. RH² RO³	Gruppe VII. RH R²O⁷	Gruppe VIII. — RO⁴
1	H=1							
2	Li=7	Be=9,4	B=11	C=12	N=14	O=16	F=19	
3	Na=23	Mg=24	Al=27,3	Si=28	P=31	S=32	Cl=35,5	
4	K=39	Ca=40	—=44	Ti=48	V=51	Cr=52	Mn=55	Fe=56, Co=59, Ni=59, Cu=63.
5	(Cu=63)	Zn=65	—=68	—=72	As=75	Se=78	Br=80	
6	Rb=85	Sr=87	?Yt=88	Zr=90	Nb=94	Mo=96	—=100	Ru=104, Rh=104, Pd=106, Ag=108.
7	(Ag=108)	Cd=112	In=113	Sn=118	Sb=122	Te=125	J=127	
8	Cs=133	Ba=137	?Di=138	?Ce=140	—	—	—	— — —
9	(—)							
10	—	—	?Er=178	?La=180	Ta=182	W=184	—	Os=195, Ir=197, Pt=198, Au=199.
11	(Au=199)	Hg=200	Tl=204	Pb=207	Bi=208	—	—	
12	—	—	—	Th=231	—	U=240	—	

The 1871 periodic table constructed by Dmitri Mendeleev. The periodic table is one of the most potent icons in science, lying at the core of chemistry and embodying the most fundamental principles of the field.

The history of chemistry represents a time span from ancient history to the present. By 1000 BC, civilizations used technologies that would eventually form the basis to the various branches of chemistry. Examples include extracting metals from ores, making pottery and glazes, fermenting beer and wine, extracting chemicals from plants for medicine and perfume, rendering fat into soap, making glass, and making alloys like bronze.

The protoscience of chemistry, alchemy, was unsuccessful in explaining the nature of matter and its transformations. However, by performing experiments and recording the results, alchemists set the stage for modern chemistry. The distinction began to emerge when a clear differentiation was made between chemistry and alchemy by Robert Boyle in his work *The Sceptical Chymist* (1661). While both alchemy and chemistry are concerned with matter and its transformations, chemists are seen as applying scientific method to their work.

Chemistry is considered to have become an established science with the work of Antoine Lavoisier, who developed a law of conservation of mass that demanded careful measurement and quantitative observations of chemical phenomena. The **history of chemistry** is intertwined with the history of thermodynamics, especially through the work of Willard Gibbs.[1]

1.1 Ancient history

1.1.1 Early Metallurgy

Main articles: History of ferrous metallurgy and History of metallurgy in the Indian subcontinent

The earliest recorded metal employed by humans seems to be gold which can be found free or "native". Small amounts of natural gold have been found in Spanish caves used during the late Paleolithic period, *c.* 40,000 BC.[2]

Silver, copper, tin and meteoric iron can also be found native, allowing a limited amount of metalworking in ancient cultures.[3] Egyptian weapons made from meteoric iron in about 3000 BC were highly prized as "Daggers from Heaven".[4]

Arguably the first chemical reaction used in a controlled manner was fire. However, for millennia fire was seen simply as a mystical force that could transform one substance into another (burning wood, or boiling water) while producing heat and light. Fire affected many aspects of early societies. These ranged from the simplest facets of everyday life, such as cooking and habitat lighting, to more advanced technologies, such as pottery, bricks, and melting of metals to make tools.

It was fire that led to the discovery of glass and the purification of metals which in turn gave way to the rise of metallurgy. During the early stages of metallurgy, methods of purification of metals were sought, and gold, known in ancient Egypt as early as 2900 BC, became a precious metal.

1.1.2 Bronze Age

Main article: Bronze Age

Certain metals can be recovered from their ores by simply

heating the rocks in a fire: notably tin, lead and (at a higher temperature) copper, a process known as smelting. The first evidence of this extractive metallurgy dates from the 5th and 6th millennium BC, and was found in the archaeological sites of Majdanpek, Yarmovac and Plocnik, all three in Serbia. To date, the earliest copper smelting is found at the Belovode site,[5] these examples include a copper axe from 5500 BC belonging to the Vinča culture.[6] Other signs of early metals are found from the third millennium BC in places like Palmela (Portugal), Los Millares (Spain), and Stonehenge (United Kingdom). However, as often happens with the study of prehistoric times, the ultimate beginnings cannot be clearly defined and new discoveries are continuous and ongoing.

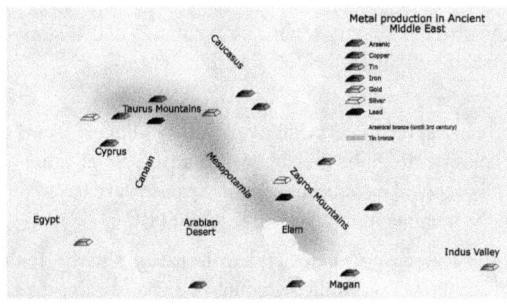

Mining areas of the ancient Middle East. Boxes colors: arsenic is in brown, copper in red, tin in grey, iron in reddish brown, gold in yellow, silver in white and lead in black. Yellow area stands for arsenic bronze, while grey area stands for tin bronze.

These first metals were single ones or as found. By combining copper and tin, a superior metal could be made, an alloy called bronze, a major technological shift which began the Bronze Age about 3500 BC. The Bronze Age was period in human cultural development when the most advanced metalworking (at least in systematic and widespread use) included techniques for smelting copper and tin from naturally occurring outcroppings of copper ores, and then smelting those ores to cast bronze. These naturally occurring ores typically included arsenic as a common impurity. Copper/tin ores are rare, as reflected in the fact that there were no tin bronzes in western Asia before 3000 BC.

After the Bronze Age, the history of metallurgy was marked by armies seeking better weaponry. Countries in Eurasia prospered when they made the superior alloys, which, in turn, made better armor and better weapons. This often determined the outcomes of battles. Significant progress in metallurgy and alchemy was made in ancient India.[7]

1.1.3   Iron Age

Main article: Iron Age

The extraction of iron from its ore into a workable metal is much more difficult than copper or tin. It appears to have been invented by the Hittites in about 1200 BC, beginning the Iron Age. The secret of extracting and working iron was a key factor in the success of the Philistines.[4][8]

In other words, the Iron Age refers to the advent of ferrous metallurgy. Historical developments in ferrous metallurgy can be found in a wide variety of past cultures and civilizations. This includes the ancient and medieval kingdoms and empires of the Middle East and Near East, ancient Iran, ancient Egypt, ancient Nubia, and Anatolia (Turkey), Ancient Nok, Carthage, the Greeks and Romans of ancient Europe, medieval Europe, ancient and medieval China, ancient and medieval India, ancient and medieval Japan, amongst others. Many applications, practices, and devices associated or involved in metallurgy were established in ancient China, such as the innovation of the blast furnace, cast iron, hydraulic-powered trip hammers, and double acting piston bellows.[9][10]

1.1.4   Classical antiquity and atomism

Main article: Atomism

Philosophical attempts to rationalize why different substances have different properties (color, density, smell), exist in different states (gaseous, liquid, and solid), and react in a different manner when exposed to environments, for example to water or fire or temperature changes, led ancient philosophers to postulate the first theories on nature and chemistry. The history of such philosophical theories that relate to chemistry can probably be traced back to every single ancient civilization. The common aspect in all these theories was the attempt to identify a small number of primary classical element that make up all the various substances in nature. Substances like air, water, and soil/earth, energy forms, such as fire and light, and more abstract concepts such as ideas, aether, and heaven, were common in ancient civilizations even in absence of any cross-fertilization; for example in Greek, Indian, Mayan, and ancient Chinese philosophies all considered air, water, earth and fire as primary elements.

Ancient World

Around 420 BC, Empedocles stated that all matter is made up of four elemental substances—earth, fire, air and water. The early theory of atomism can be traced back to ancient Greece and ancient India.[11] Greek atomism dates back

Democritus, Greek philosopher of atomistic school.

Much of the early development of purification methods is described by Pliny the Elder in his Naturalis Historia. He made attempts to explain those methods, as well as making acute observations of the state of many minerals.

1.2 Medieval alchemy

See also: Minima naturalia, a medieval Aristotelian concept analogous to atomism

Seventeenth century alchemical emblem showing the four Classical elements in the corners of the image, alongside the tria prima on the central triangle.

to the Greek philosopher Democritus, who declared that matter is composed of indivisible and indestructible atoms around 380 BC. Leucippus also declared that atoms were the most indivisible part of matter. This coincided with a similar declaration by Indian philosopher Kanada in his Vaisheshika sutras around the same time period.[11] In much the same fashion he discussed the existence of gases. What Kanada declared by sutra, Democritus declared by philosophical musing. Both suffered from a lack of empirical data. Without scientific proof, the existence of atoms was easy to deny. Aristotle opposed the existence of atoms in 330 BC. Earlier, in 380 BC, a Greek text attributed to Polybus argues that the human body is composed of four humours. Around 300 BC, Epicurus postulated a universe of indestructible atoms in which man himself is responsible for achieving a balanced life.

With the goal of explaining Epicurean philosophy to a Roman audience, the Roman poet and philosopher Lucretius[12] wrote *De Rerum Natura* (The Nature of Things)[13] in 50 BC. In the work, Lucretius presents the principles of atomism; the nature of the mind and soul; explanations of sensation and thought; the development of the world and its phenomena; and explains a variety of celestial and terrestrial phenomena.

The elemental system used in Medieval alchemy was developed primarily by the Persian alchemist Jābir ibn Hayyān and rooted in the classical elements of Greek tradition.[14] His system consisted of the four Aristotelian elements of air, earth, fire, and water in addition to two philosophical elements: sulphur, characterizing the principle of combustibility; "the stone which burns", and mercury, characterizing the principle of metallic properties. They were seen by early alchemists as idealized expressions of irreducibile components of the universe[15] and are of larger consideration within philosophical alchemy.

The three metallic principles: sulphur to flammability or combustion, mercury to volatility and stability, and salt to solidity. became the *tria prima* of the Swiss alchemist Paracelsus. He reasoned that Aristotle's four-element theory appeared in bodies as three principles. Paracelsus saw these principles as fundamental and justified them by recourse to the description of how wood burns in fire. Mercury included the cohesive principle, so that when it left in smoke the wood fell apart. Smoke described the volatility (the mercurial principle), the heat-giving flames described flammability (sulphur), and the remnant ash described solidity (salt).[16]

1.2.1 The philosopher's stone

Main article: Alchemy

Alchemy is defined by the Hermetic quest for the

"The Alchemist", by Sir William Douglas, 1855

philosopher's stone, the study of which is steeped in symbolic mysticism, and differs greatly from modern science. Alchemists toiled to make transformations on an esoteric (spiritual) and/or exoteric (practical) level.[17] It was the protoscientific, exoteric aspects of alchemy that contributed heavily to the evolution of chemistry in Greco-Roman Egypt, the Islamic Golden Age, and then in Europe. Alchemy and chemistry share an interest in the composition and properties of matter, and prior to the eighteenth century were not separated into distinct disciplines. The term *chymistry* has been used to describe the blend of alchemy and chemistry that existed before this time.[18]

The earliest Western alchemists, who lived in the first centuries of the common era, invented chemical apparatus. The *bain-marie*, or water bath is named for Mary the Jewess. Her work also gives the first descriptions of the tribikos

and kerotakis.[19] Cleopatra the Alchemist described furnaces and has been credited with the invention of the alembic.[20] Later, the experimental framework established by Jabir ibn Hayyan influenced alchemists as the discipline migrated through the Islamic world, then to Europe in the twelfth century.

During the Renaissance, exoteric alchemy remained popular in the form of Paracelsian iatrochemistry, while spiritual alchemy flourished, realigned to its Platonic, Hermetic, and Gnostic roots. Consequently, the symbolic quest for the philosopher's stone was not superseded by scientific advances, and was still the domain of respected scientists and doctors until the early eighteenth century. Early modern alchemists who are renowned for their scientific contributions include Jan Baptist van Helmont, Robert Boyle, and Isaac Newton.

1.2.2 Problems encountered with alchemy

There were several problems with alchemy, as seen from today's standpoint. There was no systematic naming system for new compounds, and the language was esoteric and vague to the point that the terminologies meant different things to different people. In fact, according to *The Fontana History of Chemistry* (Brock, 1992):

> The language of alchemy soon developed an arcane and secretive technical vocabulary designed to conceal information from the uninitiated. To a large degree, this language is incomprehensible to us today, though it is apparent that readers of Geoffery Chaucer's Canon's Yeoman's Tale or audiences of Ben Jonson's The Alchemist were able to construe it sufficiently to laugh at it.[21]

Chaucer's tale exposed the more fraudulent side of alchemy, especially the manufacture of counterfeit gold from cheap substances. Less than a century earlier, Dante Alighieri also demonstrated an awareness of this fraudulence, causing him to consign all alchemists to the Inferno in his writings. Soon after, in 1317, the Avignon Pope John XXII ordered all alchemists to leave France for making counterfeit money. A law was passed in England in 1403 which made the "multiplication of metals" punishable by death. Despite these and other apparently extreme measures, alchemy did not die. Royalty and privileged classes still sought to discover the philosopher's stone and the elixir of life for themselves.[22]

There was also no agreed-upon scientific method for making experiments reproducible. Indeed, many alchemists included in their methods irrelevant information such as the timing of the tides or the phases of the moon. The esoteric nature and codified vocabulary of alchemy appeared to be more useful in concealing the fact that they could not be sure of very much at all. As early as the 14th century, cracks seemed to grow in the facade of alchemy; and people became sceptical. Clearly, there needed to be a scientific method where experiments can be repeated by other people, and results needed to be reported in a clear language that laid out both what is known and unknown.

1.2.3 Alchemy in the Islamic World

Jābir ibn Hayyān (Geber), a Persian alchemist whose experimental research laid the foundations of chemistry.

Main article: Alchemy and chemistry in medieval Islam

In the Islamic World, the Muslims were translating the works of the ancient Greeks and Egyptians into Arabic and were experimenting with scientific ideas.[23] The development of the modern scientific method was slow and arduous, but an early scientific method for chemistry began emerging among early Muslim chemists, beginning with the 9th century chemist Jābir ibn Hayyān (known as "Geber" in Europe), who is considered as "the father of chemistry".[24][25][26][27] He introduced a systematic and experimental approach to scientific research based in the laboratory, in contrast to the ancient Greek and Egyptian alchemists whose works were largely allegorical and often unintelligble.[28] He also invented and named the alembic (al-anbiq), chemically analyzed many chemical substances,

composed lapidaries, distinguished between alkalis and acids, and manufactured hundreds of drugs.[29] He also refined the theory of five classical elements into the theory of seven alchemical elements after identifying mercury and sulfur as chemical elements.[30]

Among other influential Muslim chemists, Abū al-Rayhān al-Bīrūnī,[31] Avicenna[32] and Al-Kindi refuted the theories of alchemy, particularly the theory of the transmutation of metals; and al-Tusi described a version of the conservation of mass, noting that a body of matter is able to change but is not able to disappear.[33] Rhazes refuted Aristotle's theory of four classical elements for the first time and set up the firm foundations of modern chemistry, using the laboratory in the modern sense, designing and describing more than twenty instruments, many parts of which are still in use today, such as a crucible, cucurbit or retort for distillation, and the head of a still with a delivery tube (ambiq, Latin alembic), and various types of furnace or stove.

For practitioners in Europe, alchemy became an intellectual pursuit after early Arabic alchemy became available through Latin translation, and over time, they improved. Paracelsus (1493–1541), for example, rejected the 4-elemental theory and with only a vague understanding of his chemicals and medicines, formed a hybrid of alchemy and science in what was to be called iatrochemistry. Paracelsus was not perfect in making his experiments truly scientific. For example, as an extension of his theory that new compounds could be made by combining mercury with sulfur, he once made what he thought was "oil of sulfur". This was actually dimethyl ether, which had neither mercury nor sulfur.

1.3 17th and 18th centuries: Early chemistry

See also: Timeline of chemistry and Corpuscularianism

Practical attempts to improve the refining of ores and their extraction to smelt metals was an important source of information for early chemists in the 16th century, among them Georg Agricola (1494–1555), who published his great work *De re metallica* in 1556. His work describes the highly developed and complex processes of mining metal ores, metal extraction and metallurgy of the time. His approach removed the mysticism associated with the subject, creating the practical base upon which others could build. The work describes the many kinds of furnace used to smelt ore, and stimulated interest in minerals and their composition. It is no coincidence that he gives numerous references to the earlier author, Pliny the Elder and his *Naturalis Historia*. Agricola has been described as the "father of metallurgy".[34]

Agricola, author of De re metallica

Workroom, from De re metallica, *1556, Chemical Heritage Foundation*

In 1605, Sir Francis Bacon published *The Proficience and Advancement of Learning*, which contains a description of what would later be known as the scientific method.[35] In 1605, Michal Sedziwój publishes the alchemical treatise *A New Light of Alchemy* which proposed the existence of the "food of life" within air, much later recognized as oxygen. In 1615 Jean Beguin published the *Tyrocinium Chymicum*, an early chemistry textbook, and in it draws the first-ever chemical equation.[36] In 1637 René Descartes publishes *Discours de la méthode*, which contains an outline of the scientific method.

The Dutch chemist Jan Baptist van Helmont's work *Ortus medicinae* was published posthumously in 1648; the book is cited by some as a major transitional work between alchemy and chemistry, and as an important influence on Robert Boyle. The book contains the results of numerous experiments and establishes an early version of the law of conservation of mass. Working during the time just after Paracelsus and iatrochemistry, Jan Baptist van Helmont suggested that there are insubstantial substances other than air and coined a name for them - "gas", from the Greek word *chaos*. In addition to introducing the word "gas" into the vocabulary of scientists, van Helmont conducted several experiments involving gases. Jan Baptist van Helmont is

also remembered today largely for his ideas on spontaneous generation and his 5-year tree experiment, as well as being considered the founder of pneumatic chemistry.

1.3.1 Robert Boyle

Anglo-Irish chemist Robert Boyle (1627–1691) is considered to have refined the modern scientific method for alchemy and to have separated chemistry further from alchemy.[37] Although his research clearly has its roots in the alchemical tradition, Boyle is largely regarded today as the first modern chemist, and therefore one of the founders of modern chemistry, and one of the pioneers of modern experimental scientific method. Although Boyle was not the original discover, he is best known for Boyle's law, which he presented in 1662:[38] the law describes the inversely proportional relationship between the absolute pressure and volume of a gas, if the temperature is kept constant within a closed system.[39][40]

Boyle is also credited for his landmark publication *The Sceptical Chymist* in 1661, which is seen as a cornerstone

Robert Boyle, one of the co-founders of modern chemistry through his use of proper experimentation, which further separated chemistry from alchemy

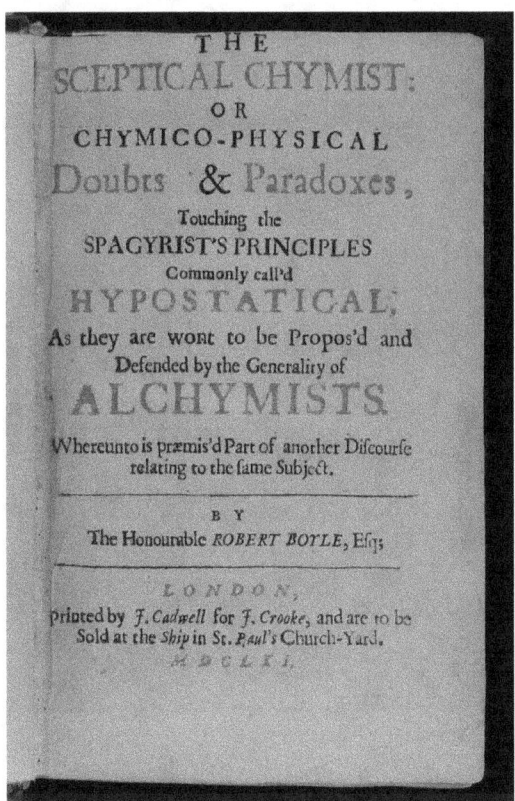

Title page from The sceptical chymist, *1661, Chemical Heritage Foundation*

book in the field of chemistry. In the work, Boyle presents his hypothesis that every phenomenon was the result of collisions of particles in motion. Boyle appealed to chemists to experiment and asserted that experiments denied the limiting of chemical elements to only the classic four: earth, fire, air, and water. He also pleaded that chemistry should cease to be subservient to medicine or to alchemy, and rise to the status of a science. Importantly, he advocated a rigorous approach to scientific experiment: he believed all theories must be proved experimentally before being regarded as true. The work contains some of the earliest modern ideas of atoms, molecules, and chemical reaction, and marks the beginning of the history of modern chemistry.

Boyle also tried to purify chemicals to obtain reproducible reactions. He was a vocal proponent of the mechanical philosophy proposed by René Descartes to explain and quantify the physical properties and interactions of material substances. Boyle was an atomist, but favoured the word *corpuscle* over *atoms*. He commented that the finest division of matter where the properties are retained is at the level of corpuscles. He also performed numerous investigations with an air pump, and noted that the mercury fell as air was pumped out. He also observed that pumping the air out of a container would extinguish a flame and kill small animals placed inside, and well as causing the level of a barometer to

drop. Boyle helped to lay the foundations for the Chemical Revolution with his mechanical corpuscular philosophy.[41] Boyle repeated the tree experiment of van Helmont, and was the first to use indicators which changed colors with acidity.

1.3.2 Development and dismantling of phlogiston

In 1702, German chemist Georg Stahl coined the name "phlogiston" for the substance believed to be released in the process of burning. Around 1735, Swedish chemist Georg Brandt analyzed a dark blue pigment found in copper ore. Brandt demonstrated that the pigment contained a new element, later named cobalt. In 1751, a Swedish chemist and pupil of Stahl's named Axel Fredrik Cronstedt, identified an impurity in copper ore as a separate metallic element, which he named nickel. Cronstedt is one of the founders of modern mineralogy.[42] Cronstedt also discovered the mineral scheelite in 1751, which he named tungsten, meaning "heavy stone" in Swedish.

Joseph Priestley, co-discoverer of the element oxygen, which he called "dephlogisticated air"

In 1754, Scottish chemist Joseph Black isolated carbon dioxide, which he called "fixed air".[43] In 1757, Louis Claude Cadet de Gassicourt, while investigating arsenic compounds, creates Cadet's fuming liquid, later discovered to be cacodyl oxide, considered to be the first synthetic organometallic compound.[44] In 1758, Joseph Black formulated the concept of latent heat to explain the thermochemistry of phase changes.[45] In 1766, English chemist Henry Cavendish isolated hydrogen, which he called "inflammable air". Cavendish discovered hydrogen as a colorless, odourless gas that burns and can form an explosive mixture with air, and published a paper on the production of water by burning inflammable air (that is, hydrogen) in dephlogisticated air (now known to be oxygen), the latter a constituent of atmospheric air (phlogiston theory).

In 1773, Swedish chemist Carl Wilhelm Scheele discovered oxygen, which he called "fire air", but did not immediately publish his achievement.[46] In 1774, English chemist Joseph Priestley independently isolated oxygen in its gaseous state, calling it "dephlogisticated air", and published his work before Scheele.[47][48] During his lifetime, Priestley's considerable scientific reputation rested on his invention of soda water, his writings on electricity, and his discovery of several "airs" (gases), the most famous being what Priestley dubbed "dephlogisticated air" (oxygen). However, Priestley's determination to defend phlogiston theory and to reject what would become the chemical revolution eventually left him isolated within the scientific community.

In 1781, Carl Wilhelm Scheele discovered that a new acid, tungstic acid, could be made from Cronstedt's scheelite (at the time named tungsten). Scheele and Torbern Bergman suggested that it might be possible to obtain a new metal by reducing this acid.[49] In 1783, José and Fausto Elhuyar found an acid made from wolframite that was identical to tungstic acid. Later that year, in Spain, the brothers succeeded in isolating the metal now known as tungsten by reduction of this acid with charcoal, and they are credited with the discovery of the element.[50][51]

1.3.3 Volta and the Voltaic Pile

A voltaic pile on display in the Tempio Voltiano *(the Volta Temple) near Volta's home in Como.*

Italian physicist Alessandro Volta constructed a device for accumulating a large charge by a series of inductions and groundings. He investigated the 1780s discovery "animal electricity" by Luigi Galvani, and found that the electric current was generated from the contact of dissimilar metals, and that the frog leg was only acting as a detector.

Volta demonstrated in 1794 that when two metals and brine-soaked cloth or cardboard are arranged in a circuit they produce an electric current.

In 1800, Volta stacked several pairs of alternating copper (or silver) and zinc discs (electrodes) separated by cloth or cardboard soaked in brine (electrolyte) to increase the electrolyte conductivity.[52] When the top and bottom contacts were connected by a wire, an electric current flowed through the voltaic pile and the connecting wire. Thus, Volta is credited with constructed the first electrical battery to produce electricity. Volta's method of stacking round plates of copper and zinc separated by disks of cardboard moistened with salt solution was termed a voltaic pile.

Thus, Volta is considered to be the founder of the discipline of electrochemistry.[53] A Galvanic cell (or voltaic cell) is an electrochemical cell that derives electrical energy from spontaneous redox reaction taking place within the cell. It generally consists of two different metals connected by a salt bridge, or individual half-cells separated by a porous membrane.

1.3.4 Antoine-Laurent de Lavoisier

Portrait of Monsieur Lavoisier and his wife, *by Jacques-Louis David*

Main articles: Antoine Lavoisier and Chemical Revolution

Although the archives of chemical research draw upon work from ancient Babylonia, Egypt, and especially the Arabs and Persians after Islam, modern chemistry flourished from the time of Antoine-Laurent de Lavoisier, a French chemist who is celebrated as the "father of modern chemistry". Lavoisier demonstrated with careful measurements that transmutation of water to earth was not possible, but that the sediment observed from boiling water came from the container. He burnt phosphorus and sulfur in air, and proved that the products weighed more than the original. Nevertheless, the weight gained was lost from the air. Thus, in 1789, he established the Law of Conservation of Mass, which is also called "Lavoisier's Law."[54]

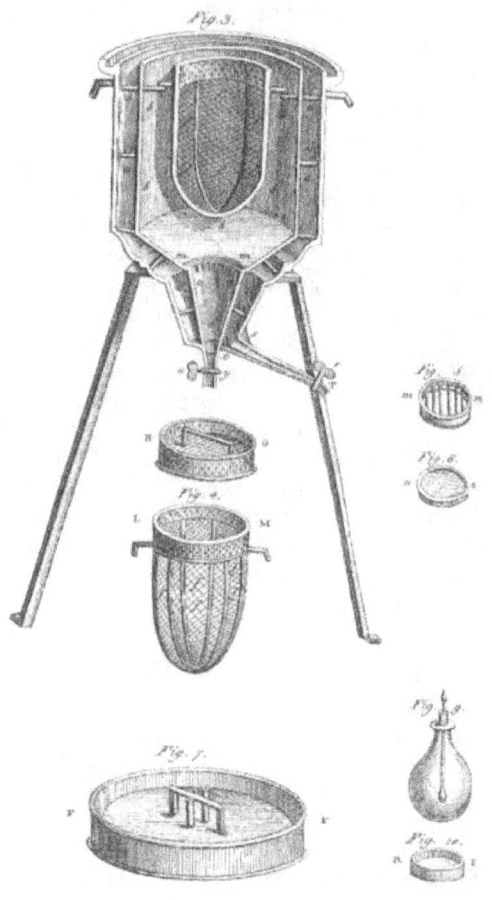

The world's first ice-calorimeter, used in the winter of 1782-83, by Antoine Lavoisier and Pierre-Simon Laplace, to determine the heat involved in various chemical changes; calculations which were based on Joseph Black's prior discovery of latent heat. These experiments mark the foundation of thermochemistry.

Repeating the experiments of Priestley, he demonstrated that air is composed of two parts, one of which combines with metals to form calxes. In *Considérations Générales sur la Nature des Acides* (1778), he demonstrated that the "air" responsible for combustion was also the source of acidity. The next year, he named this portion oxygen (Greek for acid-former), and the other azote (Greek for no life). Lavoisier thus has a claim to the discovery of oxygen along with Priestley and Scheele. He also discovered that the "inflammable air" discovered by Cavendish - which he termed hydrogen (Greek for water-former) - combined with oxygen to produce a dew, as Priestley had reported, which appeared to be water. In *Reflexions sur le Phlogistique* (1783), Lavoisier showed the phlogiston theory of combustion to be inconsistent. Mikhail Lomonosov independently established a tradition of chemistry in Russia in the 18th century. Lomonosov also rejected the phlogiston theory, and anticipated the kinetic theory of gases. Lomonosov regarded heat as a form of motion, and stated the idea of conservation of matter.

Lavoisier worked with Claude Louis Berthollet and others to devise a system of chemical nomenclature which serves as the basis of the modern system of naming chemical compounds. In his *Methods of Chemical Nomenclature* (1787), Lavoisier invented the system of naming and classification still largely in use today, including names such as sulfuric acid, sulfates, and sulfites. In 1785, Berthollet was the first to introduce the use of chlorine gas as a commercial bleach. In the same year he first determined the elemental composition of the gas ammonia. Berthollet first produced a modern bleaching liquid in 1789 by passing chlorine gas through a solution of sodium carbonate - the result was a weak solution of sodium hypochlorite. Another strong chlorine oxidant and bleach which he investigated and was the first to produce, potassium chlorate ($KClO_3$), is known as Berthollet's Salt. Berthollet is also known for his scientific contributions to theory of chemical equilibria via the mechanism of reverse chemical reactions.

Lavoisier's *Traité Élémentaire de Chimie* (Elementary Treatise of Chemistry, 1789) was the first modern chemical textbook, and presented a unified view of new theories of chemistry, contained a clear statement of the Law of Conservation of Mass, and denied the existence of phlogiston. In addition, it contained a list of elements, or substances that could not be broken down further, which included oxygen, nitrogen, hydrogen, phosphorus, mercury, zinc, and sulfur. His list, however, also included light, and caloric, which he believed to be material substances. In the work, Lavoisier underscored the observational basis of his chemistry, stating "I have tried...to arrive at the truth by linking up facts; to suppress as much as possible the use of reasoning, which is often an unreliable instrument which deceives us, in order to follow as much as possible the torch of observation and of experiment." Nevertheless, he believed that the real existence of atoms was philosophically impossible. Lavoisier

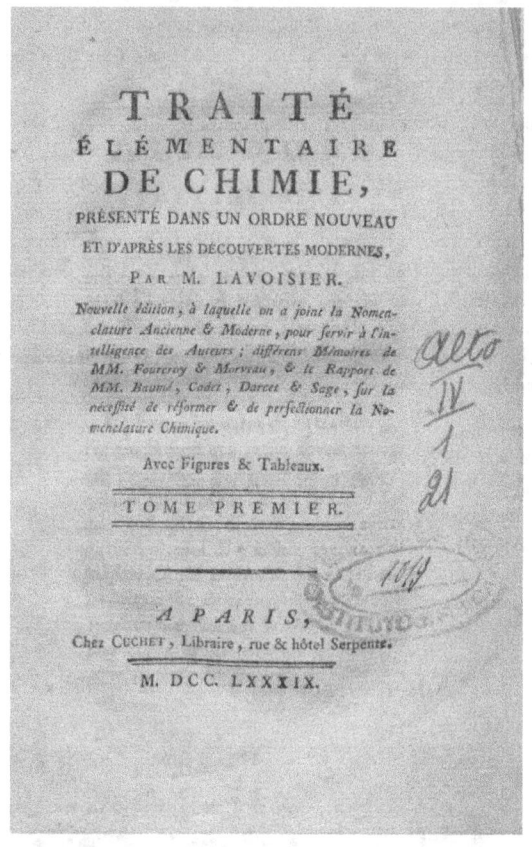

Traité élémentaire de chimie

demonstrated that organisms disassemble and reconstitute atmospheric air in the same manner as a burning body.

With Pierre-Simon Laplace, Lavoisier used a calorimeter to estimate the heat evolved per unit of carbon dioxide produced. They found the same ratio for a flame and animals, indicating that animals produced energy by a type of combustion. Lavoisier believed in the radical theory, believing that radicals, which function as a single group in a chemical reaction, would combine with oxygen in reactions. He believed all acids contained oxygen. He also discovered that diamond is a crystalline form of carbon.

While many of Lavoisier's partners were influential for the advancement of chemistry as a scientific discipline, his wife Marie-Anne Lavoisier was arguably the most influential of them all. Upon their marriage, Mmme. Lavoisier began to study chemistry, English, and drawing in order to help her husband in his work either by translating papers into English, a language which Lavoisier did not know, or by keeping records and drawing the various apparatuses that Lavoisier used in his labs.[55] Through her ability to read and translate articles from Britain for her husband, Lavoisier had access knowledge from many of the chemical advances happening outside of his lab.[55] Furthermore, Mme. Lavoisier kept records of Lavoisier's work and ensured that his works were published.[55] The first sign of Marie-Anne's true potential as a chemist in Lavoisier's lab came when she was translating a book by the scientist Richard Kirwan. While translating, she stumbled upon and corrected multiple errors. When she presented her translation, along with her notes to Lavoisier [55] Her edits and contributions led to Lavoisier's refutation of the theory of phlogiston.

Lavoisier made many fundamental contributions to the science of chemistry. Following Lavoisier's work, chemistry acquired a strict quantitative nature, allowing reliable predictions to be made. The revolution in chemistry which he brought about was a result of a conscious effort to fit all experiments into the framework of a single theory. He established the consistent use of chemical balance, used oxygen to overthrow the phlogiston theory, and developed a new system of chemical nomenclature. Lavoisier was beheaded during the French Revolution.

1.4   19th century

In 1802, French American chemist and industrialist Éleuthère Irénée du Pont, who learned manufacture of gunpowder and explosives under Antoine Lavoisier, founded a gunpowder manufacturer in Delaware known as E. I. du Pont de Nemours and Company. The French Revolution forced his family to move to the United States where du Pont started a gunpowder mill on the Brandywine River in Delaware. Wanting to make the best powder possible, du Pont was vigilant about the quality of the materials he used. For 32 years, du Pont served as president of E. I. du Pont de Nemours and Company, which eventually grew into one of the largest and most successful companies in America.

Throughout the 19th century, chemistry was divided between those who followed the atomic theory of John Dalton and those who did not, such as Wilhelm Ostwald and Ernst Mach.[56] Although such proponents of the atomic theory as Amedeo Avogadro and Ludwig Boltzmann made great advances in explaining the behavior of gases, this dispute was not finally settled until Jean Perrin's experimental investigation of Einstein's atomic explanation of Brownian motion in the first decade of the 20th century.[56]

Well before the dispute had been settled, many had already applied the concept of atomism to chemistry. A major example was the ion theory of Svante Arrhenius which anticipated ideas about atomic substructure that did not fully develop until the 20th century. Michael Faraday was another early worker, whose major contribution

to chemistry was electrochemistry, in which (among other things) a certain quantity of electricity during electrolysis or electrodeposition of metals was shown to be associated with certain quantities of chemical elements, and fixed quantities of the elements therefore with each other, in specific ratios. These findings, like those of Dalton's combining ratios, were early clues to the atomic nature of matter.

1.4.1   John Dalton

John Dalton is remembered for his work on partial pressures in gases, color blindness, and atomic theory

Main articles: John Dalton and Atomic theory

In 1803, English meteorologist and chemist John Dalton proposed Dalton's law, which describes relationship between the components in a mixture of gases and the relative pressure each contributes to that of the overall mixture.[57] Discovered in 1801, this concept is also known as Dalton's law of partial pressures.

Dalton also proposed a modern atomic theory in 1803 which stated that all matter was composed of small indivisible particles termed atoms, atoms of a given element possess unique characteristics and weight, and three types of atoms exist: simple (elements), compound (simple molecules), and complex (complex molecules). In 1808, Dalton first published *New System of Chemical Philosophy* (1808-1827), in which he outlined the first modern scientific description of the atomic theory. This work identified chemical elements as a specific type of atom, therefore rejecting Newton's theory of chemical affinities.

Instead, Dalton inferred proportions of elements in compounds by taking ratios of the weights of reactants, setting the atomic weight of hydrogen to be identically one. Following Jeremias Benjamin Richter (known for introducing the term *stoichiometry*), he proposed that chemical elements combine in integral ratios. This is known as the law of multiple proportions or Dalton's law, and Dalton included a clear description of the law in his *New System of Chemical Philosophy*. The law of multiple proportions is one of the basic laws of stoichiometry used to establish the atomic theory. Despite the importance of the work as the first view of atoms as physically real entities and introduction of a system of chemical symbols, *New System of Chemical Philosophy* devoted almost as much space to the caloric theory as to atomism.

French chemist Joseph Proust proposed the law of definite proportions, which states that elements always combine in small, whole number ratios to form compounds, based on several experiments conducted between 1797 and 1804[58] Along with the law of multiple proportions, the law of definite proportions forms the basis of stoichiometry. The law of definite proportions and constant composition do not prove that atoms exist, but they are difficult to explain without assuming that chemical compounds are formed when atoms combine in constant proportions.

1.4.2   Jöns Jacob Berzelius

Main article: Jöns Jacob Berzelius

A Swedish chemist and disciple of Dalton, Jöns Jacob Berzelius embarked on a systematic program to try to make accurate and precise quantitative measurements and insure the purity of chemicals. Along Lavoisier, Boyle, and Dalton, Berzelius is known as the father of modern chemistry. In 1828 he compiled a table of relative atomic weights, where oxygen was set to 100, and which included all of the elements known at the time. This work provided evidence in favor of Dalton's atomic theory: that inorganic chemical compounds are composed of atoms combined in whole number amounts. He determined the exact elementary constituents of large numbers of compounds. The results strongly confirmed Proust's Law of Definite Proportions. In his weights, he used oxygen as a standard, setting its weight equal to exactly 100. He also measured the weights of 43 elements. In discovering that atomic weights are not integer multiples of the weight of hydrogen, Berzelius also disproved Prout's hypothesis that elements are built up from atoms of hydrogen.

Jöns Jacob Berzelius.

Jöns Jacob Berzelius, the chemist who worked out the modern technique of chemical formula notation and is considered one of the fathers of modern chemistry

Motivated by his extensive atomic weight determinations and in a desire to aid his experiments, he introduced the classical system of chemical symbols and notation with his 1808 publishing of *Lärbok i Kemien*, in which elements are abbreviated by one or two letters to make a distinct abbreviation from their Latin name. This system of chemical notation—in which the elements were given simple written labels, such as O for oxygen, or Fe for iron, with proportions noted by numbers—is the same basic system used today. The only difference is that instead of the subscript number used today (e.g., H_2O), Berzelius used a superscript (H^2O). Berzelius is credited with identifying the chemical elements silicon, selenium, thorium, and cerium. Students working in Berzelius's laboratory also discovered lithium and vanadium.

Berzelius developed the radical theory of chemical combination, which holds that reactions occur as stable groups of atoms called radicals are exchanged between molecules. He believed that salts are compounds of an acid and bases, and discovered that the anions in acids would be attracted to a positive electrode (the anode), whereas the cations in a base would be attracted to a negative electrode (the cathode). Berzelius did not believe in the Vitalism Theory, but instead in a regulative force which produced organization of tissues in an organism. Berzelius is also credited with originating the chemical terms "catalysis", "polymer", "isomer", and "allotrope", although his original definitions differ dramatically from modern usage. For example, he coined the term "polymer" in 1833 to describe organic compounds which shared identical empirical formulas but which differed in overall molecular weight, the larger of the compounds being described as "polymers" of the smallest. By this long superseded, pre-structural definition, glucose ($C_6H_{12}O_6$) was viewed as a polymer of formaldehyde (CH_2O).

1.4.3 New elements and gas laws

Humphry Davy, the discover of several alkali and alkaline earth metals, as well as contributions to the discoveries of the elemental nature of chlorine and iodine.

Main article: Humphry Davy

English chemist Humphry Davy was a pioneer in the field of electrolysis, using Alessandro Volta's voltaic pile to split up common compounds and thus isolate a series of new elements. He went on to electrolyse molten salts and discovered several new metals, especially sodium and potassium, highly reactive elements known as the alkali metals. Potassium, the first metal that was isolated by electrolysis, was discovered in 1807 by Davy, who derived it from caustic potash (KOH). Before the 19th century, no distinction was made between potassium and sodium. Sodium was first isolated by Davy in the same year by passing an electric current through molten sodium hydroxide (NaOH). When Davy heard that Berzelius and Pontin prepared calcium amalgam by electrolyzing lime in mercury, he tried it himself. Davy was successful, and discovered calcium in 1808 by electrolyzing a mixture of lime and mercuric oxide.[59][60] He worked with electrolysis throughout his life and, in 1808, he isolated magnesium, strontium[61] and barium.[62]

Davy also experimented with gases by inhaling them. This experimental procedure nearly proved fatal on several occasions, but led to the discovery of the unusual effects of nitrous oxide, which came to be known as laughing gas. Chlorine was discovered in 1774 by Swedish chemist Carl Wilhelm Scheele, who called it *"dephlogisticated marine acid"* (see phlogiston theory) and mistakenly thought it contained oxygen. Scheele observed several properties of chlorine gas, such as its bleaching effect on litmus, its deadly effect on insects, its yellow-green colour, and the similarity of its smell to that of aqua regia. However, Scheele was unable to publish his findings at the time. In 1810, chlorine was given its current name by Humphry Davy (derived from the Greek word for green), who insisted that chlorine was in fact an element.[63] He also showed that oxygen could not be obtained from the substance known as oxymuriatic acid (HCl solution). This discovery overturned Lavoisier's definition of acids as compounds of oxygen. Davy was a popular lecturer and able experimenter.

Main articles: Joseph Louis Gay-Lussac and Gay-Lussac's law

French chemist Joseph Louis Gay-Lussac shared the interest of Lavoisier and others in the quantitative study of the properties of gases. From his first major program of research in 1801–1802, he concluded that equal volumes of all gases expand equally with the same increase in temperature: this conclusion is usually called "Charles's law", as Gay-Lussac gave credit to Jacques Charles, who had arrived at nearly the same conclusion in the 1780s but had not published it.[64] The law was independently discovered by British natural philosopher John Dalton by 1801, although Dalton's description was less thorough than Gay-Lussac's.[65][66] In 1804 Gay-Lussac made several daring ascents of over 7,000 meters above sea level in hydrogen-filled balloons—a feat not equaled for another 50 years—that allowed him to investigate other aspects of gases. Not only did he gather magnetic measurements at various altitudes, but he also took pressure, temperature, and humidity measurements and samples of air, which he later analyzed chemically.

In 1808 Gay-Lussac announced what was probably his sin-

Joseph Louis Gay-Lussac, who stated that the ratio between the volumes of the reactant gases and the products can be expressed in simple whole numbers.

gle greatest achievement: from his own and others' experiments he deduced that gases at constant temperature and pressure combine in simple numerical proportions by volume, and the resulting product or products—if gases—also bear a simple proportion by volume to the volumes of the reactants. In other words, gases under equal conditions of temperature and pressure react with one another in volume ratios of small whole numbers. This conclusion subsequently became known as "Gay-Lussac's law" or the "Law of Combining Volumes". With his fellow professor at the École Polytechnique, Louis Jacques Thénard, Gay-Lussac also participated in early electrochemical research, investigating the elements discovered by its means. Among other achievements, they decomposed boric acid by using fused potassium, thus discovering the element boron. The two also took part in contemporary debates that modified Lavoisier's definition of acids and furthered his program of analyzing organic compounds for their oxygen and hydrogen content.

The element iodine was discovered by French chemist Bernard Courtois in 1811.[67][68] Courtois gave samples to his friends, Charles Bernard Desormes (1777–1862) and Nicolas Clément (1779–1841), to continue research. He also gave some of the substance to Gay-Lussac and to physicist André-Marie Ampère. On December 6, 1813, Gay-Lussac announced that the new substance was either an element or a compound of oxygen.[69][70][71] It was Gay-

Lussac who suggested the name *"iode"*, from the Greek word ιώδες (iodes) for violet (because of the color of iodine vapor).[67][69] Ampère had given some of his sample to Humphry Davy. Davy did some experiments on the substance and noted its similarity to chlorine.[72] Davy sent a letter dated December 10 to the Royal Society of London stating that he had identified a new element.[73] Arguments erupted between Davy and Gay-Lussac over who identified iodine first, but both scientists acknowledged Courtois as the first to isolate the element.

In 1815, Humphry Davy invented the Davy lamp, which allowed miners within coal mines to work safely in the presence of flammable gases. There had been many mining explosions caused by firedamp or methane often ignited by open flames of the lamps then used by miners. Davy conceived of using an iron gauze to enclose a lamp's flame, and so prevent the methane burning inside the lamp from passing out to the general atmosphere. Although the idea of the safety lamp had already been demonstrated by William Reid Clanny and by the then unknown (but later very famous) engineer George Stephenson, Davy's use of wire gauze to prevent the spread of flame was used by many other inventors in their later designs. There was some discussion as to whether Davy had discovered the principles behind his lamp without the help of the work of Smithson Tennant, but it was generally agreed that the work of both men had been independent. Davy refused to patent the lamp, and its invention led to him being awarded the Rumford medal in 1816.[74]

Main articles: Amedeo Avogadro and Avogadro's law

After Dalton published his atomic theory in 1808, certain of his central ideas were soon adopted by most chemists. However, uncertainty persisted for half a century about how atomic theory was to be configured and applied to concrete situations; chemists in different countries developed several different incompatible atomistic systems. A paper that suggested a way out of this difficult situation was published as early as 1811 by the Italian physicist Amedeo Avogadro (1776-1856), who hypothesized that equal volumes of gases at the same temperature and pressure contain equal numbers of molecules, from which it followed that relative molecular weights of any two gases are the same as the ratio of the densities of the two gases under the same conditions of temperature and pressure. Avogadro also reasoned that simple gases were not formed of solitary atoms but were instead compound molecules of two or more atoms. Thus Avogadro was able to overcome the difficulty that Dalton and others had encountered when Gay-Lussac reported that above 100 °C the volume of water vapor was twice the volume of the oxygen used to form it. According to Avogadro, the molecule of oxygen had split into two atoms in the course of forming water vapor.

Amedeo Avogadro, who postulated that, under controlled conditions of temperature and pressure, equal volumes of gases contain an equal number of molecules. This is known as Avogadro's law.

Avogadro's hypothesis was neglected for half a century after it was first published. Many reasons for this neglect have been cited, including some theoretical problems, such as Jöns Jacob Berzelius's "dualism", which asserted that compounds are held together by the attraction of positive and negative electrical charges, making it inconceivable that a molecule composed of two electrically similar atoms—as in oxygen—could exist. An additional barrier to acceptance was the fact that many chemists were reluctant to adopt physical methods (such as vapour-density determinations) to solve their problems. By mid-century, however, some leading figures had begun to view the chaotic multiplicity of competing systems of atomic weights and molecular formulas as intolerable. Moreover, purely chemical evidence began to mount that suggested Avogadro's approach might be right after all. During the 1850s, younger chemists, such as Alexander Williamson in England, Charles Gerhardt and Charles-Adolphe Wurtz in France, and August Kekulé in Germany, began to advocate reforming theoretical chemistry to make it consistent with Avogadrian theory.

1.4.4 Wöhler and the vitalism debate

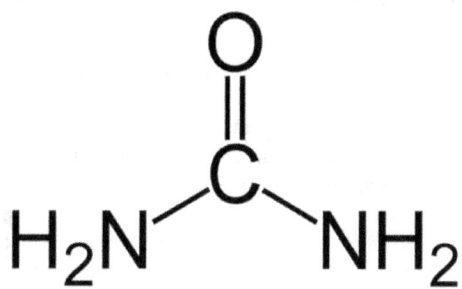

Structural formula of urea

Main articles: Vitalism, Friedrich Wöhler and Wöhler synthesis

In 1825, Friedrich Wöhler and Justus von Liebig performed the first confirmed discovery and explanation of isomers, earlier named by Berzelius. Working with cyanic acid and fulminic acid, they correctly deduced that isomerism was caused by differing arrangements of atoms within a molecular structure. In 1827, William Prout classified biomolecules into their modern groupings: carbohydrates, proteins and lipids. After the nature of combustion was settled, another dispute, about vitalism and the essential distinction between organic and inorganic substances, began. The vitalism question was revolutionized in 1828 when Friedrich Wöhler synthesized urea, thereby establishing that organic compounds could be produced from inorganic starting materials and disproving the theory of vitalism. Never before had an organic compound been synthesized from inorganic material.

This opened a new research field in chemistry, and by the end of the 19th century, scientists were able to synthesize hundreds of organic compounds. The most important among them are mauve, magenta, and other synthetic dyes, as well as the widely used drug aspirin. The discovery of the artificial synthesis of urea contributed greatly to the theory of isomerism, as the empirical chemical formulas for urea and ammonium cyanate are identical (see Wöhler synthesis). In 1832, Friedrich Wöhler and Justus von Liebig discovered and explained functional groups and radicals in relation to organic chemistry, as well as first synthesizing benzaldehyde. Liebig, a German chemist, made major contributions to agricultural and biological chemistry, and worked on the organization of organic chemistry. Liebig is considered the "father of the fertilizer industry" for his discovery of nitrogen as an essential plant nutrient, and his formulation of the Law of the Minimum which described the effect of individual nutrients on crops.

1.4.5   mid-1800s

In 1840, Germain Hess proposed Hess's law, an early statement of the law of conservation of energy, which establishes that energy changes in a chemical process depend only on the states of the starting and product materials and not on the specific pathway taken between the two states. In 1847, Hermann Kolbe obtained acetic acid from completely inorganic sources, further disproving vitalism. In 1848, William Thomson, 1st Baron Kelvin (commonly known as Lord Kelvin) established the concept of absolute zero, the temperature at which all molecular motion ceases. In 1849, Louis Pasteur discovered that the racemic form of tartaric acid is a mixture of the levorotatory and dextrotatory forms, thus clarifying the nature of optical rotation and advancing the field of stereochemistry.[75] In 1852, August Beer proposed Beer's law, which explains the relationship between the composition of a mixture and the amount of light it will absorb. Based partly on earlier work by Pierre Bouguer and Johann Heinrich Lambert, it established the analytical technique known as spectrophotometry.[76] In 1855, Benjamin Silliman, Jr. pioneered methods of petroleum cracking, which made the entire modern petrochemical industry possible.[77]

Main articles: Stanislao Cannizzaro and Karlsruhe Congress

Formulas of acetic acid given by August Kekulé in 1861.

Avogadro's hypothesis began to gain broad appeal among chemists only after his compatriot and fellow scientist Stanislao Cannizzaro demonstrated its value in 1858, two years after Avogadro's death. Cannizzaro's chemical interests had originally centered on natural products and on reactions of aromatic compounds; in 1853 he discovered that when benzaldehyde is treated with concentrated base, both benzoic acid and benzyl alcohol are produced—a phenomenon known today as the Cannizzaro reaction. In his 1858 pamphlet, Cannizzaro showed that a complete return to the ideas of Avogadro could be used to construct a consistent and robust theoretical structure that fit nearly all of the available empirical evidence. For instance, he pointed to evidence that suggested that not all elementary gases consist of two atoms per molecule—some were monatomic, most were diatomic, and a few were even more complex.

Another point of contention had been the formulas for compounds of the alkali metals (such as sodium) and the alkaline earth metals (such as calcium), which, in view of their striking chemical analogies, most chemists had wanted to assign to the same formula type. Cannizzaro argued that placing these metals in different categories had the beneficial result of eliminating certain anomalies when using

their physical properties to deduce atomic weights. Unfortunately, Cannizzaro's pamphlet was published initially only in Italian and had little immediate impact. The real breakthrough came with an international chemical congress held in the German town of Karlsruhe in September 1860, at which most of the leading European chemists were present. The Karlsruhe Congress had been arranged by Kekulé, Wurtz, and a few others who shared Cannizzaro's sense of the direction chemistry should go. Speaking in French (as everyone there did), Cannizzaro's eloquence and logic made an indelible impression on the assembled body. Moreover, his friend Angelo Pavesi distributed Cannizzaro's pamphlet to attendees at the end of the meeting; more than one chemist later wrote of the decisive impression the reading of this document provided. For instance, Lothar Meyer later wrote that on reading Cannizzaro's paper, "The scales seemed to fall from my eyes."[78] Cannizzaro thus played a crucial role in winning the battle for reform. The system advocated by him, and soon thereafter adopted by most leading chemists, is substantially identical to what is still used today.

1.4.6 Perkin, Crookes, and Nobel

In 1856, Sir William Henry Perkin, age 18, given a challenge by his professor, August Wilhelm von Hofmann, sought to synthesize quinine, the anti-malaria drug, from coal tar. In one attempt, Perkin oxidized aniline using potassium dichromate, whose toluidine impurities reacted with the aniline and yielded a black solid—suggesting a "failed" organic synthesis. Cleaning the flask with alcohol, Perkin noticed purple portions of the solution: a byproduct of the attempt was the first synthetic dye, known as mauveine or Perkin's mauve. Perkin's discovery is the foundation of the dye synthesis industry, one of the earliest successful chemical industries.

German chemist August Kekulé von Stradonitz's most important single contribution was his structural theory of organic composition, outlined in two articles published in 1857 and 1858 and treated in great detail in the pages of his extraordinarily popular *Lehrbuch der organischen Chemie* ("Textbook of Organic Chemistry"), the first installment of which appeared in 1859 and gradually extended to four volumes. Kekulé argued that tetravalent carbon atoms - that is, carbon forming exactly four chemical bonds - could link together to form what he called a "carbon chain" or a "carbon skeleton," to which other atoms with other valences (such as hydrogen, oxygen, nitrogen, and chlorine) could join. He was convinced that it was possible for the chemist to specify this detailed molecular architecture for at least the simpler organic compounds known in his day. Kekulé was not the only chemist to make such claims in this era. The Scottish chemist Archibald Scott Couper published a substantially similar theory nearly simultaneously, and the Russian chemist Aleksandr Butlerov did much to clarify and expand structure theory. However, it was predominantly Kekulé's ideas that prevailed in the chemical community.

British chemist and physicist William Crookes is noted for his cathode ray studies, fundamental in the development of atomic physics. His researches on electrical discharges through a rarefied gas led him to observe the dark space around the cathode, now called the Crookes dark space. He demonstrated that cathode rays travel in straight lines and produce phosphorescence and heat when they strike certain materials. A pioneer of vacuum tubes, Crookes invented the Crookes tube - an early experimental discharge tube, with partial vacuum with which he studied the behavior of cathode rays. With the introduction of spectrum analysis by Robert Bunsen and Gustav Kirchhoff (1859-1860), Crookes applied the new technique to the study of selenium compounds. Bunsen and Kirchoff had previously used spectroscopy as a means of chemical analysis to discover caesium and rubidium. In 1861, Crookes used this process to discover thallium in some seleniferous deposits. He continued work on that new element, isolated it, stud-

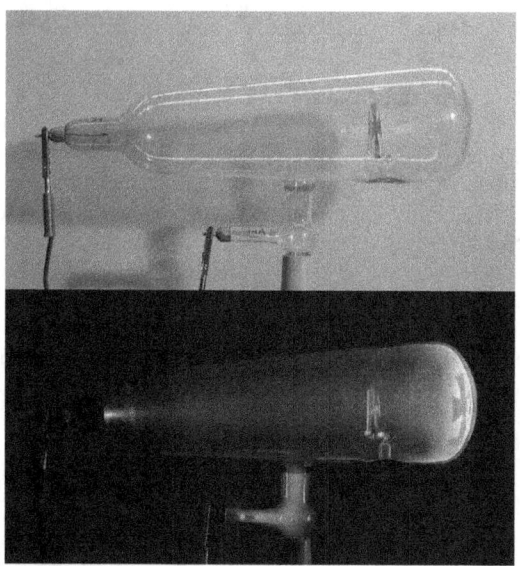

A Crookes tube (2 views): light and dark. Electrons travel in straight lines from the cathode (left), as evidenced by the shadow cast from the Maltese cross on the fluorescence of the righthand end. The anode is at the bottom wire.

ied its properties, and in 1873 determined its atomic weight. During his studies of thallium, Crookes discovered the principle of the Crookes radiometer, a device that converts light radiation into rotary motion. The principle of this radiometer has found numerous applications in the development of sensitive measuring instruments.

In 1862, Alexander Parkes exhibited Parkesine, one of the earliest synthetic polymers, at the International Exhibition in London. This discovery formed the foundation of the modern plastics industry. In 1864, Cato Maximilian Guldberg and Peter Waage, building on Claude Louis Berthollet's ideas, proposed the law of mass action. In 1865, Johann Josef Loschmidt determined the exact number of molecules in a mole, later named Avogadro's number.

In 1865, August Kekulé, based partially on the work of Loschmidt and others, established the structure of benzene as a six carbon ring with alternating single and double bonds. Kekulé's novel proposal for benzene's cyclic structure was much contested but was never replaced by a superior theory. This theory provided the scientific basis for the dramatic expansion of the German chemical industry in the last third of the 19th century. Today, the large majority of known organic compounds are aromatic, and all of them contain at least one hexagonal benzene ring of the sort that Kekulé advocated. Kekulé is also famous for having clarified the nature of aromatic compounds, which are compounds based on the benzene molecule. In 1865, Adolf von Baeyer began work on indigo dye, a milestone in modern industrial organic chemistry which revolutionized the dye industry.

Swedish chemist and inventor Alfred Nobel found that when nitroglycerin was incorporated in an absorbent inert substance like *kieselguhr* (diatomaceous earth) it became safer and more convenient to handle, and this mixture he patented in 1867 as dynamite. Nobel later on combined nitroglycerin with various nitrocellulose compounds, similar to collodion, but settled on a more efficient recipe combining another nitrate explosive, and obtained a transparent, jelly-like substance, which was a more powerful explosive than dynamite. Gelignite, or blasting gelatin, as it was named, was patented in 1876; and was followed by a host of similar combinations, modified by the addition of potassium nitrate and various other substances.

1.4.7   Mendeleev's periodic table

Main articles: Dmitri Mendeleev, Periodic table and History of the periodic table

An important breakthrough in making sense of the list of

Dmitri Mendeleev, responsible for organizing the known chemical elements in a periodic table.

known chemical elements (as well as in understanding the internal structure of atoms) was Dmitri Mendeleev's development of the first modern periodic table, or the periodic classification of the elements. Mendeleev, a Russian chemist, felt that there was some type of order to the elements and he spent more than thirteen years of his life collecting data and assembling the concept, initially with the idea of resolving some of the disorder in the field for his students. Mendeleev found that, when all the known chemical elements were arranged in order of increasing atomic weight, the resulting table displayed a recurring pattern, or periodicity, of properties within groups of elements.

Mendeleev's law allowed him to build up a systematic periodic table of all the 66 elements then known based on atomic mass, which he published in *Principles of Chemistry* in 1869. His first Periodic Table was compiled on the basis of arranging the elements in ascending order of atomic weight and grouping them by similarity of properties.

Mendeleev had such faith in the validity of the periodic law that he proposed changes to the generally accepted values for the atomic weight of a few elements and, in his version of the periodic table of 1871, predicted the locations within the table of unknown elements together with their properties. He even predicted the likely properties of three yet-to-be-discovered elements, which he called ekaboron (Eb), ekaaluminium (Ea), and ekasilicon (Es), which proved to be good predictors of the properties of scandium, gallium, and germanium, respectively, which each fill the spot in the periodic table assigned by Mendeleev.

At first the periodic system did not raise interest among chemists. However, with the discovery of the predicted elements, notably gallium in 1875, scandium in 1879, and germanium in 1886, it began to win wide acceptance. The subsequent proof of many of his predictions within his lifetime brought fame to Mendeleev as the founder of the periodic law. This organization surpassed earlier attempts at classification by Alexandre-Émile Béguyer de Chancourtois, who published the telluric helix, an early, three-dimensional version of the periodic table of the elements in 1862, John Newlands, who proposed the law of octaves (a precursor to the periodic law) in 1864, and Lothar Meyer, who developed an early version of the periodic table with 28 elements organized by valence in 1864. Mendeleev's table did not include any of the noble gases, however, which had not yet been discovered. Gradually the periodic law and table became the framework for a great part of chemical theory. By the time Mendeleyev died in 1907, he enjoyed international recognition and had received distinctions and awards from many countries.

In 1873, Jacobus Henricus van 't Hoff and Joseph Achille Le Bel, working independently, developed a model of chemical bonding that explained the chirality experiments of Pasteur and provided a physical cause for optical activity in chiral compounds.[79] van 't Hoff's publication, called V*oorstel tot Uitbreiding der Tegenwoordige in de Scheikunde gebruikte Structuurformules in de Ruimte*, etc. (Proposal for the development of 3-dimensional chemical structural formulae) and consisting of twelve pages text and one page diagrams, gave the impetus to the development of stereochemistry. The concept of the "asymmetrical carbon atom", dealt with in this publication, supplied an explanation of the occurrence of numerous isomers, inexplicable by means of the then current structural formulae. At the same time he pointed out the existence of relationship between optical activity and the presence of an asymmetrical carbon atom.

1.4.8 Josiah Willard Gibbs

Main articles: Josiah Willard Gibbs and Statistical mechanics
 American mathematical physicist J. Willard Gibbs's work

J. Willard Gibbs formulated a concept of thermodynamic equilibrium of a system in terms of energy and entropy. He also did extensive work on chemical equilibrium, and equilibria between phases.

on the applications of thermodynamics was instrumental in transforming physical chemistry into a rigorous deductive science. During the years from 1876 to 1878, Gibbs worked on the principles of thermodynamics, applying them to the complex processes involved in chemical reactions. He discovered the concept of chemical potential, or the "fuel" that makes chemical reactions work. In 1876 he published his most famous contribution, "On the Equilibrium of Heterogeneous Substances", a compilation of his work on thermodynamics and physical chemistry which laid out the concept of free energy to explain the physical basis of chemical equilibria.[80] In these essays were the beginnings of Gibbs' theories of phases of matter: he considered each state of matter a phase, and each substance a component. Gibbs took all of the variables involved in a chemical reaction - temperature, pressure, energy, volume, and entropy -

and included them in one simple equation known as Gibbs' phase rule.

Within this paper was perhaps his most outstanding contribution, the introduction of the concept free energy, now universally called Gibbs free energy in his honor. The Gibbs free energy relates the tendency of a physical or chemical system to simultaneously lower its energy and increase its disorder, or entropy, in a spontaneous natural process. Gibbs's approach allows a researcher to calculate the change in free energy in the process, such as in a chemical reaction, and how fast it will happen. Since virtually all chemical processes and many physical ones involve such changes, his work has significantly impacted both the theoretical and experiential aspects of these sciences. In 1877, Ludwig Boltzmann established statistical derivations of many important physical and chemical concepts, including entropy, and distributions of molecular velocities in the gas phase.[81] Together with Boltzmann and James Clerk Maxwell, Gibbs created a new branch of theoretical physics called statistical mechanics (a term that he coined), explaining the laws of thermodynamics as consequences of the statistical properties of large ensembles of particles. Gibbs also worked on the application of Maxwell's equations to problems in physical optics. Gibbs's derivation of the phenomenological laws of thermodynamics from the statistical properties of systems with many particles was presented in his highly influential textbook *Elementary Principles in Statistical Mechanics*, published in 1902, a year before his death. In that work, Gibbs reviewed the relationship between the laws of thermodynamics and statistical theory of molecular motions. The overshooting of the original function by partial sums of Fourier series at points of discontinuity is known as the Gibbs phenomenon.

1.4.9 Late 19th century

German engineer Carl von Linde's invention of a continuous process of liquefying gases in large quantities formed a basis for the modern technology of refrigeration and provided both impetus and means for conducting scientific research at low temperatures and very high vacuums. He developed a methyl ether refrigerator (1874) and an ammonia refrigerator (1876). Though other refrigeration units had been developed earlier, Linde's were the first to be designed with the aim of precise calculations of efficiency. In 1895 he set up a large-scale plant for the production of liquid air. Six years later he developed a method for separating pure liquid oxygen from liquid air that resulted in widespread industrial conversion to processes utilizing oxygen (e.g., in steel manufacture).

In 1883, Svante Arrhenius developed an ion theory to explain conductivity in electrolytes.[82] In 1884, Jacobus Henricus van 't Hoff published *Études de Dynamique chimique* (Studies in Dynamic Chemistry), a seminal study on chemical kinetics.[83] In this work, van 't Hoff entered for the first time the field of physical chemistry. Of great importance was his development of the general thermodynamic relationship between the heat of conversion and the displacement of the equilibrium as a result of temperature variation. At constant volume, the equilibrium in a system will tend to shift in such a direction as to oppose the temperature change which is imposed upon the system. Thus, lowering the temperature results in heat development while increasing the temperature results in heat absorption. This principle of mobile equilibrium was subsequently (1885) put in a general form by Henry Louis Le Chatelier, who extended the principle to include compensation, by change of volume, for imposed pressure changes. The van 't Hoff-Le Chatelier principle, or simply Le Chatelier's principle, explains the response of dynamic chemical equilibria to external stresses.[84]

In 1884, Hermann Emil Fischer proposed the structure of purine, a key structure in many biomolecules, which he later synthesized in 1898. He also began work on the chemistry of glucose and related sugars.[85] In 1885, Eugene Goldstein named the cathode ray, later discovered to be composed of electrons, and the canal ray, later discovered to be positive hydrogen ions that had been stripped of their electrons in a cathode ray tube; these would later be named protons.[86] The year 1885 also saw the publishing of J. H. van 't Hoff's *L'Équilibre chimique dans les Systèmes gazeux ou dissous à l'État dilué* (Chemical equilibria in gaseous systems or strongly diluted solutions), which dealt with this theory of dilute solutions. Here he demonstrated that the "osmotic pressure" in solutions which are sufficiently dilute is proportionate to the concentration and the absolute temperature so that this pressure can be represented by a formula which only deviates from the formula for gas pressure by a coefficient **i**. He also determined the value of i by various methods, for example by means of the vapor pressure and François-Marie Raoult's results on the lowering of the freezing point. Thus van 't Hoff was able to prove that thermodynamic laws are not only valid for gases, but also for dilute solutions. His pressure laws, given general validity by the electrolytic dissociation theory of Arrhenius (1884-1887) - the first foreigner who came to work with him in Amsterdam (1888) - are considered the most comprehensive and important in the realm of natural sciences. In 1893, Alfred Werner discovered the octahedral structure of cobalt complexes, thus establishing the field of coordination chemistry.[87]

Ramsay's discovery of the noble gases

Main articles: William Ramsay and Noble gas

The most celebrated discoveries of Scottish chemist William Ramsay were made in inorganic chemistry. Ramsay was intrigued by the British physicist John Strutt, 3rd Baron Rayleigh's 1892 discovery that the atomic weight of nitrogen found in chemical compounds was lower than that of nitrogen found in the atmosphere. He ascribed this discrepancy to a light gas included in chemical compounds of nitrogen, while Ramsay suspected a hitherto undiscovered heavy gas in atmospheric nitrogen. Using two different methods to remove all known gases from air, Ramsay and Lord Rayleigh were able to announce in 1894 that they had found a monatomic, chemically inert gaseous element that constituted nearly 1 percent of the atmosphere; they named it argon.

The following year, Ramsay liberated another inert gas from a mineral called cleveite; this proved to be helium, previously known only in the solar spectrum. In his book *The Gases of the Atmosphere* (1896), Ramsay showed that the positions of helium and argon in the periodic table of elements indicated that at least three more noble gases might exist. In 1898 Ramsay and the British chemist Morris W. Travers isolated these elements—called neon, krypton, and xenon—from air brought to a liquid state at low temperature and high pressure. Sir William Ramsay worked with Frederick Soddy to demonstrate, in 1903, that alpha particles (helium nuclei) were continually produced during the radioactive decay of a sample of radium. Ramsay was awarded the 1904 Nobel Prize for Chemistry in recognition of "services in the discovery of the inert gaseous elements in air, and his determination of their place in the periodic system."

In 1897, J. J. Thomson discovered the electron using the cathode ray tube. In 1898, Wilhelm Wien demonstrated that canal rays (streams of positive ions) can be deflected by magnetic fields, and that the amount of deflection is proportional to the mass-to-charge ratio. This discovery would lead to the analytical technique known as mass spectrometry.[88]

Marie and Pierre Curie

Main articles: Marie Curie, Pierre Curie and Henri Becquerel

Marie Skłodowska-Curie was a Polish-born French physicist and chemist who is famous for her pioneering research on radioactivity. She and her husband are considered to have laid the cornerstone of the nuclear age with their re-

Marie Curie, a pioneer in the field of radioactivity and the first twice-honored Nobel laureate (and still the only one in two different sciences)

Pierre Curie, known for his work on radioactivity as well as on ferromagnetism, paramagnetism, and diamagnetism; notably Curie's law and Curie point.

search on radioactivity. Marie was fascinated with the work of Henri Becquerel, a French physicist who discovered in 1896 that uranium casts off rays similar to the X-rays discovered by Wilhelm Röntgen. Marie Curie began studying uranium in late 1897 and theorized, according to a 1904 article she wrote for Century magazine, "that the emission of rays by the compounds of uranium is a property of the metal itself—that it is an atomic property of the element uranium independent of its chemical or physical state." Curie took Becquerel's work a few steps further, conducting her own experiments on uranium rays. She discovered that the rays remained constant, no matter the condition or form of the uranium. The rays, she theorized, came from the element's atomic structure. This revolutionary idea created the field of atomic physics and the Curies coined the word *radioactivity* to describe the phenomena.

Pierre and Marie further explored radioactivity by working to separate the substances in uranium ores and then using the electrometer to make radiation measurements to 'trace' the minute amount of unknown radioactive element among the fractions that resulted. Working with the mineral pitchblende, the pair discovered a new radioactive element in 1898. They named the element polonium, after Marie's native country of Poland. On December 21, 1898, the Curies detected the presence of another radioactive ma-

terial in the pitchblende. They presented this finding to the French Academy of Sciences on December 26, proposing that the new element be called radium. The Curies then went to work isolating polonium and radium from naturally occurring compounds to prove that they were new elements. In 1902, the Curies announced that they had produced a decigram of pure radium, demonstrating its existence as a unique chemical element. While it took three years for them to isolate radium, they were never able to isolate polonium. Along with the discovery of two new elements and finding techniques for isolating radioactive isotopes, Curie oversaw the world's first studies into the treatment of neoplasms, using radioactive isotopes. With Henri Becquerel and her husband, Pierre Curie, she was awarded the 1903 Nobel Prize for Physics. She was the sole winner of the 1911 Nobel Prize for Chemistry. She was the first woman to win a Nobel Prize, and she is the only woman to win the award in two different fields.

While working with Marie to extract pure substances from ores, an undertaking that really required industrial resources but that they achieved in relatively primitive conditions, Pierre himself concentrated on the physical study (including luminous and chemical effects) of the new radiations. Through the action of magnetic fields on the rays given out by the radium, he proved the existence of particles electrically positive, negative, and neutral; these Ernest Rutherford was afterward to call alpha, beta, and gamma rays. Pierre then studied these radiations by calorimetry and also observed the physiological effects of radium, thus opening the way to radium therapy. Among Pierre Curie's discoveries were that ferromagnetic substances exhibited a critical temperature transition, above which the substances lost their ferromagnetic behavior - this is known as the "Curie point." He was elected to the Academy of Sciences (1905), having in 1903 jointly with Marie received the Royal Society's prestigious Davy Medal and jointly with her and Becquerel the Nobel Prize for Physics. He was run over by a carriage in the rue Dauphine in Paris in 1906 and died instantly. His complete works were published in 1908.

Ernest Rutherford

Ernest Rutherford, discoverer of the nucleus and considered the father of nuclear physics

New Zealand-born chemist and physicist Ernest Rutherford is considered to be "the father of nuclear physics." Rutherford is best known for devising the names alpha, beta, and gamma to classify various forms of radioactive "rays" which were poorly understood at his time (alpha and beta rays

are particle beams, while gamma rays are a form of high-energy electromagnetic radiation). Rutherford deflected alpha rays with both electric and magnetic fields in 1903. Working with Frederick Soddy, Rutherford explained that radioactivity is due to the transmutation of elements, now known to involve nuclear reactions.

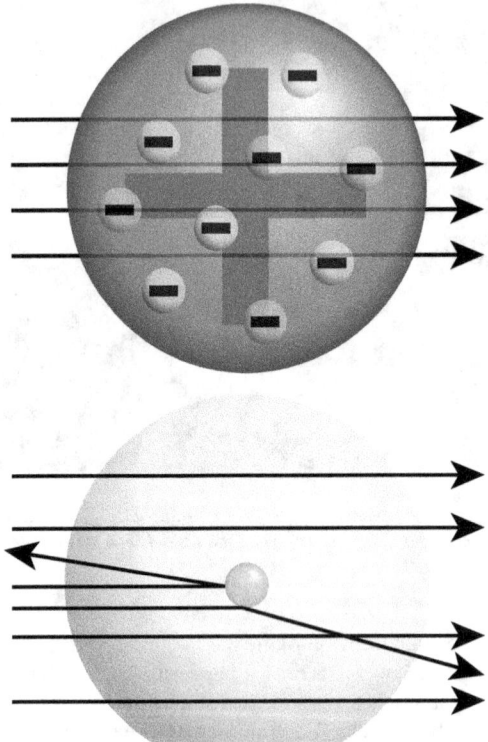

Top: Predicted results based on the then-accepted plum pudding model of the atom. Bottom: Observed results. Rutherford disproved the plum pudding model and concluded that the positive charge of the atom must be concentrated in a small, central nucleus.

He also observed that the intensity of radioactivity of a radioactive element decreases over a unique and regular amount of time until a point of stability, and he named the halving time the "half-life." In 1901 and 1902 he worked with Frederick Soddy to prove that atoms of one radioactive element would spontaneously turn into another, by expelling a piece of the atom at high velocity. In 1906 at the University of Manchester, Rutherford oversaw an experiment conducted by his students Hans Geiger (known for the Geiger counter) and Ernest Marsden. In the Geiger–Marsden experiment, a beam of alpha particles, generated by the radioactive decay of radon, was directed normally onto a sheet of very thin gold foil in an evacuated chamber. Under the prevailing plum pudding model, the alpha particles should all have passed through the foil and hit the detector screen, or have been deflected by, at most, a few degrees.

However, the actual results surprised Rutherford. Although many of the alpha particles did pass through as expected, many others were deflected at small angles while others were reflected back to the alpha source. They observed that a very small percentage of particles were deflected through angles much larger than 90 degrees. The gold foil experiment showed large deflections for a small fraction of incident particles. Rutherford realized that, because some of the alpha particles were deflected or reflected, the atom had a concentrated centre of positive charge and of relatively large mass - Rutherford later termed this positive center the "atomic nucleus". The alpha particles had either hit the positive centre directly or passed by it close enough to be affected by its positive charge. Since many other particles passed through the gold foil, the positive centre would have to be a relatively small size compared to the rest of the atom - meaning that the atom is mostly open space. From his results, Rutherford developed a model of the atom that was similar to the solar system, known as Rutherford model. Like planets, electrons orbited a central, sun-like nucleus. For his work with radiation and the atomic nucleus, Rutherford received the 1908 Nobel Prize in Chemistry.

1.5 20th century

The first Solvay Conference was held in Brussels in 1911 and was considered a turning point in the world of physics and chemistry.

In 1903, Mikhail Tsvet invented chromatography, an important analytic technique. In 1904, Hantaro Nagaoka proposed an early nuclear model of the atom, where electrons orbit a dense massive nucleus. In 1905, Fritz Haber and Carl Bosch developed the Haber process for making ammonia, a milestone in industrial chemistry with deep consequences in agriculture. The Haber process, or Haber-Bosch process, combined nitrogen and hydrogen to form ammonia in industrial quantities for production of fertilizer and munitions. The food production for half the world's current population depends on this method for producing fertilizer. Haber, along with Max Born, proposed the Born–Haber cycle as a method for evaluating the lattice energy of an ionic solid. Haber has also been described as the "father of chemical warfare" for his work developing and deploying chlorine and other poisonous gases during World War I.

Robert A. Millikan, who is best known for measuring the charge on the electron, won the Nobel Prize in Physics in 1923.

In 1905, Albert Einstein explained Brownian motion in a way that definitively proved atomic theory. Leo Baekeland invented bakelite, one of the first commercially successful plastics. In 1909, American physicist Robert Andrews Millikan - who had studied in Europe under Walther Nernst and Max Planck - measured the charge of individual electrons with unprecedented accuracy through the oil drop experiment, in which he measured the electric charges on tiny falling water (and later oil) droplets. His study established that any particular droplet's electrical charge is a multiple of a definite, fundamental value — the electron's charge — and thus a confirmation that all electrons have the same charge and mass. Beginning in 1912, he spent several years investigating and finally proving Albert Einstein's proposed linear relationship between energy and frequency, and providing the first direct photoelectric support for Planck's constant. In 1923 Millikan was awarded the Nobel Prize for Physics.

In 1909, S. P. L. Sørensen invented the pH concept and develops methods for measuring acidity. In 1911, Antonius Van den Broek proposed the idea that the elements on the periodic table are more properly organized by positive nuclear charge rather than atomic weight. In 1911, the first Solvay Conference was held in Brussels, bringing together most of the most prominent scientists of the day. In 1912, William Henry Bragg and William Lawrence Bragg proposed Bragg's law and established the field of X-ray crystallography, an important tool for elucidating the crystal structure of substances. In 1912, Peter Debye develops the concept of molecular dipole to describe asymmetric charge distribution in some molecules.

1.5.1 Niels Bohr

Niels Bohr, the developer of the Bohr model of the atom, and a leading founder of quantum mechanics

Main articles: Niels Bohr and Bohr model

In 1913, Niels Bohr, a Danish physicist, introduced the concepts of quantum mechanics to atomic structure by proposing what is now known as the Bohr model of the atom, where electrons exist only in strictly defined circular orbits around the nucleus similar to rungs on a ladder. The Bohr Model is a planetary model in which the negatively charged electrons orbit a small, positively charged nucleus similar to the planets orbiting the Sun (except that the orbits are not planar) - the gravitational force of the solar system is mathematically akin to the attractive Coulomb (electrical) force between the positively charged nucleus and the negatively charged electrons.

In the Bohr model, however, electrons orbit the nucleus in orbits that have a set size and energy - the energy levels are said to be *quantized*, which means that only certain orbits with certain radii are allowed; orbits in between simply don't exist. The energy of the orbit is related to its size - that is, the lowest energy is found in the smallest orbit. Bohr also postulated that electromagnetic radiation is absorbed or emitted when an electron moves from one orbit to another. Because only certain electron orbits are permitted, the emission of light accompanying a jump of an electron from an excited energy state to ground state produces a unique emission spectrum for each element.

Niels Bohr also worked on the principle of complementarity, which states that an electron can be interpreted in two mutually exclusive and valid ways. Electrons can be interpreted as wave or particle models. His hypothesis was that an incoming particle would strike the nucleus and create an excited compound nucleus. The formed the basis of his liquid drop model and later provided a theory base for the explanation of nuclear fission.

In 1913, Henry Moseley, working from Van den Broek's earlier idea, introduces concept of atomic number to fix inadequacies of Mendeleev's periodic table, which had been based on atomic weight. The peak of Frederick Soddy's career in radiochemistry was in 1913 with his formulation of the concept of isotopes, which stated that certain elements exist in two or more forms which have different atomic weights but which are indistinguishable chemically. He is remembered for proving the existence of isotopes of certain radioactive elements, and is also credited, along with others, with the discovery of the element protactinium in 1917. In 1913, J. J. Thomson expanded on the work of Wien by showing that charged subatomic particles can be separated by their mass-to-charge ratio, a technique known as mass spectrometry.

1.5.2 Gilbert N. Lewis

Main article: Gilbert N. Lewis

American physical chemist Gilbert N. Lewis laid the foundation of valence bond theory; he was instrumental in developing a bonding theory based on the number of electrons in the outermost "valence" shell of the atom. In 1902, while Lewis was trying to explain valence to his students, he depicted atoms as constructed of a concentric series of cubes with electrons at each corner. This "cubic atom" explained the eight groups in the periodic table and represented his idea that chemical bonds are formed by electron transference to give each atom a complete set of eight outer electrons (an "octet").

Lewis's theory of chemical bonding continued to evolve and, in 1916, he published his seminal article "The Atom of the Molecule", which suggested that a chemical bond is a pair of electrons shared by two atoms. Lewis's model equated the classical chemical bond with the sharing of a pair of electrons between the two bonded atoms. Lewis introduced the "electron dot diagrams" in this paper to symbolize the electronic structures of atoms and molecules. Now known as Lewis structures, they are discussed in virtually every introductory chemistry book.

Shortly after publication of his 1916 paper, Lewis became involved with military research. He did not return to the subject of chemical bonding until 1923, when he masterfully summarized his model in a short monograph entitled Valence and the Structure of Atoms and Molecules. His renewal of interest in this subject was largely stimulated by the activities of the American chemist and General Electric researcher Irving Langmuir, who between 1919 and 1921 popularized and elaborated Lewis's model. Langmuir subsequently introduced the term *covalent bond*. In 1921, Otto Stern and Walther Gerlach establish concept of quantum mechanical spin in subatomic particles.

For cases where no sharing was involved, Lewis in 1923 developed the electron pair theory of acids and base: Lewis redefined an acid as any atom or molecule with an incomplete octet that was thus capable of accepting electrons from another atom; bases were, of course, electron donors. His theory is known as the concept of Lewis acids and bases. In 1923, G. N. Lewis and Merle Randall published *Thermodynamics and the Free Energy of Chemical Substances*, first modern treatise on chemical thermodynamics.

The 1920s saw a rapid adoption and application of Lewis's model of the electron-pair bond in the fields of organic and coordination chemistry. In organic chemistry, this was primarily due to the efforts of the British chemists Arthur Lapworth, Robert Robinson, Thomas Lowry, and Christopher Ingold; while in coordination chemistry, Lewis's bonding model was promoted through the efforts of the American chemist Maurice Huggins and the British chemist Nevil Sidgwick.

1.5.3 Quantum mechanics

Main articles: Louis de Broglie, Wolfgang Pauli, Erwin Schrödinger and Werner Heisenberg

In 1924, French quantum physicist Louis de Broglie published his thesis, in which he introduced a revolutionary theory of electron waves based on wave–particle duality in his thesis. In his time, the wave and particle interpretations of light and matter were seen as being at odds with one another, but de Broglie suggested that these seemingly different characteristics were instead the same behavior observed from different perspectives — that particles can behave like waves, and waves (radiation) can behave like particles. Broglie's proposal offered an explanation of the restriction motion of electrons within the atom. The first publications of Broglie's idea of "matter waves" had drawn little attention from other physicists, but a copy of his doctoral thesis chanced to reach Einstein, whose response was enthusiastic. Einstein stressed the importance of Broglie's work both explicitly and by building further on it.

In 1925, Austrian-born physicist Wolfgang Pauli developed the Pauli exclusion principle, which states that no two electrons around a single nucleus in an atom can occupy the same quantum state simultaneously, as described by four quantum numbers. Pauli made major contributions to quantum mechanics and quantum field theory - he was awarded the 1945 Nobel Prize for Physics for his discovery of the Pauli exclusion principle - as well as solid-state physics, and he successfully hypothesized the existence of the neutrino. In addition to his original work, he wrote masterful syntheses of several areas of physical theory that are considered classics of scientific literature.

$$ H(t) \mid \psi(t) \rangle = i\hbar \frac{d}{dt} \mid \psi(t) \rangle $$

The Schrödinger equation

In 1926 at the age of 39, Austrian theoretical physicist Erwin Schrödinger produced the papers that gave the foundations of quantum wave mechanics. In those papers he described his partial differential equation that is the basic equation of quantum mechanics and bears the same relation to the mechanics of the atom as Newton's equations of motion bear to planetary astronomy. Adopting a proposal made by Louis de Broglie in 1924 that particles of matter have a dual nature and in some situations act like waves, Schrödinger introduced a theory describing the behaviour of such a system by a wave equation that is now known as the Schrödinger equation. The solutions to Schrödinger's equation, unlike the solutions to Newton's equations, are wave functions that can only be related to the probable occurrence of physical events. The readily visualized sequence of events of the planetary orbits of Newton is, in quantum mechanics, replaced by the more abstract notion of probability. (This aspect of the quantum theory made Schrödinger and several other physicists profoundly unhappy, and he devoted much of his later life to formulating philosophical objections to the generally accepted interpretation of the theory that he had done so much to create.)

German theoretical physicist Werner Heisenberg was one of the key creators of quantum mechanics. In 1925, Heisenberg discovered a way to formulate quantum mechanics in terms of matrices. For that discovery, he was awarded the Nobel Prize for Physics for 1932. In 1927 he published his uncertainty principle, upon which he built his philosophy and for which he is best known. Heisenberg was able to demonstrate that if you were studying an electron in an atom you could say where it was (the electron's location) or where it was going (the electron's velocity), but it was impossible to express both at the same time. He also made important contributions to the theories of the hydrodynamics of turbulent flows, the atomic nucleus, ferromagnetism, cosmic rays, and subatomic particles, and he was instrumental in planning the first West German nuclear reactor at Karlsruhe, together with a research reactor in Munich, in 1957. Considerable controversy surrounds his work on atomic research during World War II.

1.5.4   Quantum chemistry

Main article: Quantum chemistry

Some view the birth of quantum chemistry in the discovery of the Schrödinger equation and its application to the hydrogen atom in 1926. However, the 1927 article of Walter Heitler and Fritz London[89] is often recognised as the first milestone in the history of quantum chemistry. This is the first application of quantum mechanics to the diatomic hydrogen molecule, and thus to the phenomenon of the chemical bond. In the following years much progress was accomplished by Edward Teller, Robert S. Mulliken, Max Born, J. Robert Oppenheimer, Linus Pauling, Erich Hückel, Douglas Hartree, Vladimir Aleksandrovich Fock, to cite a few.

Still, skepticism remained as to the general power of quantum mechanics applied to complex chemical systems. The situation around 1930 is described by Paul Dirac:[90]

> The underlying physical laws necessary for

the mathematical theory of a large part of physics and the whole of chemistry are thus completely known, and the difficulty is only that the exact application of these laws leads to equations much too complicated to be soluble. It therefore becomes desirable that approximate practical methods of applying quantum mechanics should be developed, which can lead to an explanation of the main features of complex atomic systems without too much computation.

Hence the quantum mechanical methods developed in the 1930s and 1940s are often referred to as theoretical molecular or atomic physics to underline the fact that they were more the application of quantum mechanics to chemistry and spectroscopy than answers to chemically relevant questions. In 1951, a milestone article in quantum chemistry is the seminal paper of Clemens C. J. Roothaan on Roothaan equations.[91] It opened the avenue to the solution of the self-consistent field equations for small molecules like hydrogen or nitrogen. Those computations were performed with the help of tables of integrals which were computed on the most advanced computers of the time.

In the 1940s many physicists turned from molecular or atomic physics to nuclear physics (like J. Robert Oppenheimer or Edward Teller). Glenn T. Seaborg was an American nuclear chemist best known for his work on isolating and identifying transuranium elements (those heavier than uranium). He shared the 1951 Nobel Prize for Chemistry with Edwin Mattison McMillan for their independent discoveries of transuranium elements. Seaborgium was named in his honour, making him the only person, along Albert Einstein, for whom a chemical element was named during his lifetime.

1.5.5   Molecular biology and biochemistry

Main articles: History of molecular biology and History of biochemistry

By the mid 20th century, in principle, the integration of physics and chemistry was extensive, with chemical properties explained as the result of the electronic structure of the atom; Linus Pauling's book on *The Nature of the Chemical Bond* used the principles of quantum mechanics to deduce bond angles in ever-more complicated molecules. However, though some principles deduced from quantum mechanics were able to predict qualitatively some chemical features for biologically relevant molecules, they were, till the end of the 20th century, more a collection of rules, observations, and recipes than rigorous ab initio quantitative methods.

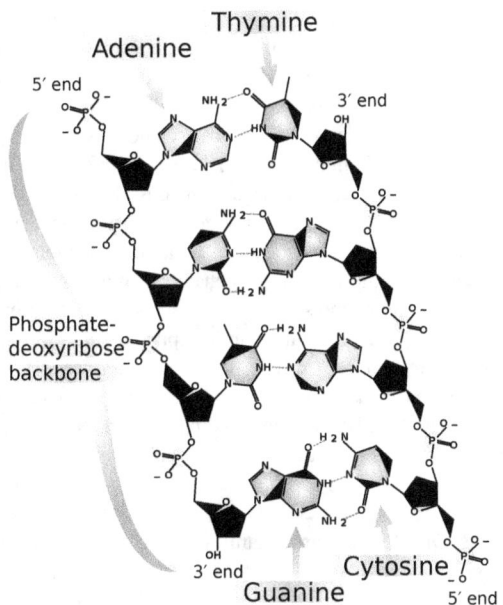

Diagrammatic representation of some key structural features of DNA

This heuristic approach triumphed in 1953 when James Watson and Francis Crick deduced the double helical structure of DNA by constructing models constrained by and informed by the knowledge of the chemistry of the constituent parts and the X-ray diffraction patterns obtained by Rosalind Franklin.[92] This discovery lead to an explosion of research into the biochemistry of life.

In the same year, the Miller–Urey experiment demonstrated that basic constituents of protein, simple amino acids, could themselves be built up from simpler molecules in a simulation of primordial processes on Earth. Though many questions remain about the true nature of the origin of life, this was the first attempt by chemists to study hypothetical processes in the laboratory under controlled conditions.

In 1983 Kary Mullis devised a method for the in-vitro amplification of DNA, known as the polymerase chain reaction (PCR), which revolutionized the chemical processes used in the laboratory to manipulate it. PCR could be used to synthesize specific pieces of DNA and made possible the sequencing of DNA of organisms, which culminated in the huge human genome project.

An important piece in the double helix puzzle was solved by one of Pauling's students Matthew Meselson and Frank Stahl, the result of their collaboration (Meselson–Stahl experiment) has been called as "the most beautiful experiment in biology".

They used a centrifugation technique that sorted molecules according to differences in weight. Because nitrogen atoms are a component of DNA, they were labelled and therefore tracked in replication in bacteria.

1.5.6   Late 20th century

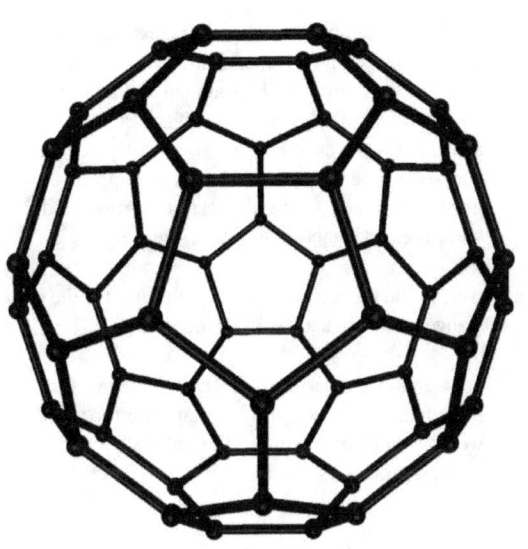

Buckminsterfullerene, C_{60}

In 1970, John Pople developed the Gaussian program greatly easing computational chemistry calculations.[93] In 1971, Yves Chauvin offered an explanation of the reaction mechanism of olefin metathesis reactions.[94] In 1975, Karl Barry Sharpless and his group discovered a stereoselective oxidation reactions including Sharpless epoxidation,[95][96] Sharpless asymmetric dihydroxylation,[97][98][99] and Sharpless oxyamination.[100][101][102] In 1985, Harold Kroto, Robert Curl and Richard Smalley discovered fullerenes, a class of large carbon molecules superficially resembling the geodesic dome designed by architect R. Buckminster Fuller.[103] In 1991, Sumio Iijima used electron microscopy to discover a type of cylindrical fullerene known as a carbon nanotube, though earlier work had been done in the field as early as 1951. This material is an important component in the field of nanotechnology.[104] In 1994, Robert A. Holton and his group achieved the first total synthesis of Taxol.[105][106][107] In 1995, Eric Cornell and Carl Wieman produced the first Bose–Einstein condensate, a substance that displays quantum mechanical properties on the macroscopic scale.[108]

1.6 Mathematics and chemistry

Classically, before the 20th century, chemistry was defined as the science of the nature of matter and its transformations. It was therefore clearly distinct from physics which was not concerned with such dramatic transformation of matter. Moreover, in contrast to physics, chemistry was not using much of mathematics. Even some were particularly reluctant to use mathematics within chemistry. For example, Auguste Comte wrote in 1830:

> Every attempt to employ mathematical methods in the study of chemical questions must be considered profoundly irrational and contrary to the spirit of chemistry.... if mathematical analysis should ever hold a prominent place in chemistry -- an aberration which is happily almost impossible -- it would occasion a rapid and widespread degeneration of that science.

However, in the second part of the 19th century, the situation changed and August Kekulé wrote in 1867:

> I rather expect that we shall someday find a mathematico-mechanical explanation for what we now call atoms which will render an account of their properties.

1.7 Scope of chemistry

After the discovery by Rutherford and Bohr of the atomic structure in 1912, and by Marie and Pierre Curie of radioactivity, scientists had to change their viewpoint on the nature of matter. The experience acquired by chemists was no longer pertinent to the study of the whole nature of matter but only to aspects related to the electron cloud surrounding the atomic nuclei and the movement of the latter in the electric field induced by the former (see Born–Oppenheimer approximation). The range of chemistry was thus restricted to the nature of matter around us in conditions which are not too far (or exceptionally far) from standard conditions for temperature and pressure and in cases where the exposure to radiation is not too different from the natural microwave, visible or UV radiations on Earth. Chemistry was therefore re-defined as the science of matter that deals with the composition, structure, and properties of substances and with the transformations that they undergo.

However the meaning of matter used here relates explicitly to substances made of atoms and molecules, disregarding the matter within the atomic nuclei and its nuclear reaction or matter within highly ionized plasmas. This does not mean that chemistry is never involved with plasma or nuclear sciences or even bosonic fields nowadays, since areas such as Quantum Chemistry and Nuclear Chemistry are currently well developed and formally recognized subfields of study under the Chemical sciences (Chemistry), but what is now formally recognized as subject of study under the Chemistry category as a science is always based on the use of concepts that describe or explain phenomena either from matter or to matter in the atomic or molecular scale, including the study of the behavior of many molecules as an aggregate or the study of the effects of a single proton on a single atom, but excluding phenomena that deal with different (more "exotic") types of matter (e.g. Bose–Einstein condensate, Higgs boson, dark matter, naked singularity, etc.) and excluding principles that refer to intrinsic abstract laws of nature in which their concepts can be formulated completely without a precise formal molecular or atomic paradigmatic view (e.g. Quantum Chromodynamics, Quantum Electrodynamics, String Theory, parts of Cosmology (see Cosmochemistry), certain areas of Nuclear Physics (see Nuclear Chemistry), etc.). Nevertheless, the field of chemistry is still, on our human scale, very broad and the claim that *chemistry is everywhere* is accurate.

1.7.1 Chemical industry

Main article: Chemical industry

The later part of the nineteenth century saw a huge increase in the exploitation of petroleum extracted from the earth for the production of a host of chemicals and largely replaced the use of whale oil, coal tar and naval stores used previously. Large-scale production and refinement of petroleum provided feedstocks for liquid fuels such as gasoline and diesel, solvents, lubricants, asphalt, waxes, and for the production of many of the common materials of the modern world, such as synthetic fibers, plastics, paints, detergents, pharmaceuticals, adhesives and ammonia as fertilizer and for other uses. Many of these required new catalysts and the utilization of chemical engineering for their cost-effective production.

In the mid-twentieth century, control of the electronic structure of semiconductor materials was made precise by the creation of large ingots of extremely pure single crystals of silicon and germanium. Accurate control of their chemical composition by doping with other elements made the production of the solid state transistor in 1951 and made possible the production of tiny integrated circuits for use in electronic devices, especially computers.

1.8 See also

1.8.1 Histories and timelines

- Atomic theory
- Cupellation
- History of chromatography
- History of electrochemistry
- History of the molecule
- History of molecular biology
- History of physics
- History of science and technology
- History of the periodic table
- History of thermodynamics
- History of energy
- History of molecular theory
- History of materials science
- List of years in science
- Nobel Prize in chemistry
- Timeline of atomic and subatomic physics
- Timeline of chemical elements discoveries
- Timeline of chemistry
- Timeline of materials technology
- Timeline of thermodynamics, statistical mechanics, and random processes
- The Chemical History of a Candle
- The Mystery of Matter: Search for the Elements (PBS film)

1.8.2 Notable chemists

listed chronologically:

- List of chemists
- Robert Boyle, 1627–1691
- Joseph Black, 1728–1799
- Joseph Priestley, 1733–1804

- Carl Wilhelm Scheele, 1742–1786
- Antoine Lavoisier, 1743–1794
- Alessandro Volta, 1745–1827
- Jacques Charles, 1746–1823
- Claude Louis Berthollet, 1748–1822
- Amedeo Avogadro, 1776-1856
- Joseph-Louis Gay-Lussac, 1778–1850
- Humphry Davy, 1778–1829
- Jöns Jakob Berzelius, inventor of modern chemical notation, 1779–1848
- Justus von Liebig, 1803–1873
- Louis Pasteur, 1822–1895
- Stanislao Cannizzaro, 1826–1910
- Friedrich August Kekulé von Stradonitz, 1829–1896
- Dmitri Mendeleev, 1834–1907
- Josiah Willard Gibbs, 1839–1903
- J. H. van 't Hoff, 1852–1911
- William Ramsay, 1852–1916
- Svante Arrhenius, 1859–1927
- Walther Nernst, 1864–1941
- Marie Curie, 1867–1934
- Gilbert N. Lewis, 1875–1946
- Otto Hahn, 1879–1968
- Irving Langmuir, 1881–1957
- Linus Pauling, 1901–1994
- Glenn T. Seaborg, 1912–1999
- Robert Burns Woodward, 1917-1979
- Frederick Sanger, 1918-2013
- Rudolph A. Marcus, 1923-
- Elias James Corey, 1928-
- Harold Kroto, 1939-
- Peter Atkins, 1940-
- Richard Smalley, 1943–2005

1.9   Notes

[1] Selected Classic Papers from the History of Chemistry

[2] "History of Gold". Gold Digest. Retrieved 2007-02-04.

[3] Photos, E., 'The Question of Meteorictic versus Smelted Nickel-Rich Iron: Archaeological Evidence and Experimental Results' *World Archaeology* Vol. 20, No. 3, Archaeometallurgy (February 1989), pp. 403–421. Online version accessed on 2010-02-08.

[4] W. Keller (1963) *The Bible as History*, p. 156 ISBN 0-340-00312-X

[5] Radivojević, Miljana; Rehren, Thilo; Pernicka, Ernst; Šljivar, Dušan; Brauns, Michael; Borić, Dušan (2010). "On the origins of extractive metallurgy: New evidence from Europe". *Journal of Archaeological Science* **37** (11): 2775. doi:10.1016/j.jas.2010.06.012.

[6] Neolithic Vinca was a metallurgical culture Stonepages from news sources November 2007

[7] Will Durant wrote in *The Story of Civilization I: Our Oriental Heritage*:

> "Something has been said about the chemical excellence of cast iron in ancient India, and about the high industrial development of the Gupta times, when India was looked to, even by Imperial Rome, as the most skilled of the nations in such chemical industries as dyeing, tanning, soap-making, glass and cement... By the sixth century the Hindus were far ahead of Europe in industrial chemistry; they were masters of calcinations, distillation, sublimation, steaming, fixation, the production of light without heat, the mixing of anesthetic and soporific powders, and the preparation of metallic salts, compounds and alloys. The tempering of steel was brought in ancient India to a perfection unknown in Europe till our own times; King Porus is said to have selected, as a specially valuable gift from Alexander, not gold or silver, but thirty pounds of steel. The Moslems took much of this Hindu chemical science and industry to the Near East and Europe; the secret of manufacturing "Damascus" blades, for example, was taken by the Arabs from the Persians, and by the Persians from India."

[8] B. W. Anderson (1975) *The Living World of the Old Testament*, p. 154, ISBN 0-582-48598-3

[9] R. F. Tylecote (1992) A History of Metallurgy ISBN 0-901462-88-8

[10] Temple, Robert K.G. (2007). *The Genius of China: 3,000 Years of Science, Discovery, and Invention* (3rd edition). London: André Deutsch. pp. 44–56. ISBN 978-0-233-00202-6.

[11] Will Durant (1935), *Our Oriental Heritage*:

> "Two systems of Hindu thought propound physical theories suggestively similar to those of Greece. Kanada, founder of the Vaisheshika philosophy, held that the world was composed of atoms as many in kind as the various elements. The Jains more nearly approximated to Democritus by teaching that all atoms were of the same kind, producing different effects by diverse modes of combinations. Kanada believed light and heat to be varieties of the same substance; Udayana taught that all heat comes from the sun; and Vachaspati, like Newton, interpreted light as composed of minute particles emitted by substances and striking the eye."

[12] Simpson, David (29 June 2005). "Lucretius (c. 99 - c. 55 BCE)". *The Internet History of Philosophy*. Retrieved 2007-01-09.

[13] Lucretius (50 BCE). "de Rerum Natura (On the Nature of Things)". *The Internet Classics Archive*. Massachusetts Institute of Technology. Retrieved 2007-01-09. Check date values in: |date= (help)

[14] Norris, John A. (2006). "The Mineral Exhalation Theory of Metallogenesis in Pre-Modern Mineral Science". *Ambix* **53**: 43. doi:10.1179/174582306X93183.

[15] Clulee, Nicholas H. (1988). *John Dee's Natural Philosophy*. Routledge. p. 97. ISBN 978-0-415-00625-5.

[16] Strathern, 2000. Page 79.

[17] Holmyard, E.J. (1957). *Alchemy*. New York: Dover, 1990. pp. 15, 16.

[18] William Royall Newman. *Atoms and Alchemy: Chymistry and the experimental origins of the scientific revolution*. University of Chicago Press, 2006. p.xi

[19] Holmyard, E.J. (1957). *Alchemy*. New York: Dover, 1990. pp. 48, 49.

[20] Stanton J. Linden. *The alchemy reader: from Hermes Trismegistus to Isaac Newton* Cambridge University Press. 2003. p.44

[21] Brock, William H. (1992). *The Fontana History of Chemistry*. London, England: Fontana Press. pp. 32–33. ISBN 0-00-686173-3.

[22] Brock, William H. (1992). *The Fontana History of Chemistry*. London, England: Fontana Press. ISBN 0-00-686173-3.

[23] The History of Ancient Chemistry

[24] Derewenda, Zygmunt S.; Derewenda, ZS (2007). "On wine, chirality and crystallography". *Acta Crystallographica Section A: Foundations of Crystallography* **64** (Pt 1): 246–258 [247]. Bibcode:2008AcCrA..64..246D. doi:10.1107/S0108767307054293. PMID 18156689.

[25] John Warren (2005). "War and the Cultural Heritage of Iraq: a sadly mismanaged affair", *Third World Quarterly*, Volume 26, Issue 4 & 5, p. 815-830.

[26] Dr. A. Zahoor (1997), JABIR IBN HAIYAN (Jabir), University of Indonesia

[27] Paul Vallely, How Islamic inventors changed the world, *The Independent*

[28] Kraus, Paul, Jâbir ibn Hayyân, *Contribution à l'histoire des idées scientifiques dans l'Islam. I. Le corpus des écrits jâbiriens. II. Jâbir et la science grecque,*. Cairo (1942-1943). Repr. By Fuat Sezgin, (Natural Sciences in Islam. 67-68), Frankfurt. 2002:

"To form an idea of the historical place of Jabir's alchemy and to tackle the problem of its sources, it is advisable to compare it with what remains to us of the alchemical literature in the Greek language. One knows in which miserable state this literature reached us. Collected by Byzantine scientists from the tenth century, the corpus of the Greek alchemists is a cluster of incoherent fragments, going back to all the times since the third century until the end of the Middle Ages."

"The efforts of Berthelot and Ruelle to put a little order in this mass of literature led only to poor results, and the later researchers, among them in particular Mrs. Hammer-Jensen, Tannery, Lagercrantz, von Lippmann, Reitzenstein, Ruska, Bidez, Festugiere and others, could make clear only few points of detail...

The study of the Greek alchemists is not very encouraging. An even surface examination of the Greek texts shows that a very small part only was organized according to true experiments of laboratory: even the supposedly technical writings, in the state where we find them today, are unintelligible nonsense which refuses any interpretation.

It is different with Jabir's alchemy. The relatively clear description of the processes and the alchemical apparatuses, the methodical classification of the substances, mark an experimental spirit which is extremely far away from the weird and odd esotericism of the Greek texts. The theory on which Jabir supports his operations is one of clearness and of an impressive unity. More than with the other Arab authors, one notes with him a balance between theoretical teaching and practical teaching, between the *'ilm* and the *'amal*. In vain one would seek in the Greek texts a work as systematic as that which is presented for example in the *Book of Seventy*."

(cf. Ahmad Y Hassan. "A Critical Reassessment of the Geber Problem: Part Three". Retrieved 2008-08-09.)

[29] Will Durant (1980). *The Age of Faith (The Story of Civilization, Volume 4)*, p. 162-186. Simon & Schuster. ISBN 0-671-01200-2.

[30] Strathern, Paul. (2000), *Mendeleyev's Dream – the Quest for the Elements*, New York: Berkley Books

[31] Marmura, Michael E.; Nasr, Seyyed Hossein (1965). "*An Introduction to Islamic Cosmological Doctrines. Conceptions of Nature and Methods Used for Its Study by the Ikhwan Al-Safa'an, Al-Biruni, and Ibn Sina* by Seyyed Hossein Nasr". *Speculum* **40** (4): 744–746. doi:10.2307/2851429. JSTOR 2851429.

[32] Robert Briffault (1938). *The Making of Humanity*, p. 196-197.

[33] Alakbarov, Farid (2001). "A 13th-Century Darwin? Tusi's Views on Evolution". *Azerbaijan International* **9**: 2.

[34] Karl Alfred von Zittel (1901) *History of Geology and Palaeontology*, p. 15

[35] Asarnow, Herman (2005-08-08). "Sir Francis Bacon: Empiricism". *An Image-Oriented Introduction to Backgrounds for English Renaissance Literature*. University of Portland. Retrieved 2007-02-22.

[36] Crosland, M.P. (1959). "The use of diagrams as chemical 'equations' in the lectures of William Cullen and Joseph Black." *Annals of Science*, Vol 15, No. 2, June

[37] Robert Boyle

[38] Acott, Chris (1999). "The diving "Law-ers": A brief resume of their lives.". *South Pacific Underwater Medicine Society journal* **29** (1). ISSN 0813-1988. OCLC 16986801. Retrieved 17 April 2009.

[39] Levine, Ira. N (1978). "Physical Chemistry" University of Brooklyn: McGraw-Hill

[40] Levine, Ira. N. (1978), p12 gives the original definition.

[41] Ursula Klein (July 2007). "Styles of Experimentation and Alchemical Matter Theory in the Scientific Revolution". *Metascience* (Springer) **16** (2): 247–256 [247]. doi:10.1007/s11016-007-9095-8. ISSN 1467-9981.

[42] Nordisk familjebok – Cronstedt: "*den moderna mineralogiens och geognosiens grundläggare*" = "*the modern mineralogy's and geognosie's founder*"

[43] Cooper, Alan (1999). "Joseph Black". *History of Glasgow University Chemistry Department*. University of Glasgow Department of Chemistry. Archived from the original on 2006-04-10. Retrieved 2006-02-23.

[44] Seyferth, Dietmar (2001). "Cadet's Fuming Arsenical Liquid and the Cacodyl Compounds of Bunsen". *Organometallics* **20** (8): 1488–1498. doi:10.1021/om0101947.

[45] Partington, J.R. (1989). *A Short History of Chemistry*. Dover Publications, Inc. ISBN 0-486-65977-1.

[46] Kuhn, 53–60; Schofield (2004), 112–13. The difficulty in precisely defining the time and place of the "discovery" of oxygen, within the context of the developing chemical revolution, is one of Thomas Kuhn's central illustrations of the gradual nature of paradigm shifts in *The Structure of Scientific Revolutions*.

[47] "Joseph Priestley". *Chemical Achievers: The Human Face of Chemical Sciences*. Chemical Heritage Foundation. 2005. Retrieved 2007-02-22.

[48] "Carl Wilhelm Scheele". *History of Gas Chemistry*. Center for Microscale Gas Chemistry, Creighton University. 2005-09-11. Retrieved 2007-02-23.

[49] Saunders, Nigel (2004). *Tungsten and the Elements of Groups 3 to 7 (The Periodic Table)*. Chicago: Heinemann Library. ISBN 1-4034-3518-9.

[50] "ITIA Newsletter" (PDF). International Tungsten Industry Association. June 2005. Retrieved 2008-06-18.

[51] "ITIA Newsletter" (PDF). International Tungsten Industry Association. December 2005. Retrieved 2008-06-18.

[52] Mottelay, Paul Fleury (2008). *Bibliographical History of Electricity and Magnetism* (Reprint of 1892 ed.). Read Books. p. 247. ISBN 1-4437-2844-6.

[53] "Inventor Alessandro Volta Biography". *The Great Idea Finder*. The Great Idea Finder. 2005. Retrieved 2007-02-23.

[54] Lavoisier, Antoine (1743-1794) -- from Eric Weisstein's World of Scientific Biography, ScienceWorld

[55] http://www.humantouchofchemistry.com/marieanne-lavoisier.htm

[56] Pullman, Bernard (2004). *The Atom in the History of Human Thought*. Reisinger, Axel. USA: Oxford University Press Inc. ISBN 0-19-511447-7.

[57] "John Dalton". *Chemical Achievers: The Human Face of Chemical Sciences*. Chemical Heritage Foundation. 2005. Retrieved 2007-02-22.

[58] "Proust, Joseph Louis (1754-1826)". *100 Distinguished Chemists*. European Association for Chemical and Molecular Science. 2005. Retrieved 2007-02-23.

[59] Enghag, P. (2004). "11. Sodium and Potassium". *Encyclopedia of the elements*. Wiley-VCH Weinheim. ISBN 3-527-30666-8.

[60] Davy, Humphry (1808). "On some new Phenomena of Chemical Changes produced by Electricity, particularly the Decomposition of the fixed Alkalies, and the Exhibition of the new Substances, which constitute their Bases". *Philosophical Transactions of the Royal Society of London* (Royal Society of London.) **98** (0): 1–45. doi:10.1098/rstl.1808.0001.

[61] Weeks, Mary Elvira (1933). "XII. Other Elements Isolated with the Aid of Potassium and Sodium: Beryllium, Boron, Silicon and Aluminum". *The Discovery of the Elements*. Easton, Pennsylvania: Journal of Chemical Education. ISBN 0-7661-3872-0.

[62] Robert E. Krebs (2006). *The history and use of our earth's chemical elements: a reference guide*. Greenwood Publishing Group. p. 80. ISBN 0-313-33438-2.

[63] Sir Humphry Davy (1811). "On a Combination of Oxymuriatic Gas and Oxygene Gas". *Philosophical Transactions of the Royal Society* **101** (0): 155–162. doi:10.1098/rstl.1811.0008.

[64] Gay-Lussac, J. L. (L'An X – 1802), "Recherches sur la dilatation des gaz et des vapeurs" [Researches on the expansion of gases and vapors], *Annales de chimie* **43**: 137–175 Check date values in: |date= (help). English translation (extract). On page 157, Gay-Lussac mentions the unpublished findings of Charles: "*Avant d'aller plus loin, je dois prévenir que quoique j'eusse reconnu un grand nombre de fois que les gaz oxigène, azote, hydrogène et acide carbonique, et l'air atmosphérique se dilatent également depuis 0° jusqu'a 80°, le cit. Charles avait remarqué depuis 15 ans la même propriété dans ces gaz ; mais n'avant jamais publié ses résultats, c'est par le plus grand hasard que je les ai connus.*" (Before going further, I should inform [you] that although I had recognized many times that the gases oxygen, nitrogen, hydrogen, and carbonic acid [i.e., carbon dioxide], and atmospheric air also expand from 0° to 80°, citizen Charles had noticed 15 years ago the same property in these gases; but having never published his results, it is by the merest chance that I knew of them.)

[65] J. Dalton (1802) "Essay IV. On the expansion of elastic fluids by heat," *Memoirs of the Literary and Philosophical Society of Manchester*, vol. 5, pt. 2, pages 595-602.

[66] http://www.chemistryexplained.com/Fe-Ge/Gay-Lussac-Joseph-Louis.html

[67] Courtois, Bernard (1813). "Découverte d'une substance nouvelle dans le Vareck". *Annales de chimie* **88**: 304. In French, seaweed that had been washed onto the shore was called "varec", "varech", or "vareck", whence the English word "wrack". Later, "varec" also referred to the ashes of such seaweed: The ashes were used as a source of iodine and salts of sodium and potassium.

[68] Swain, Patricia A. (2005). "Bernard Courtois (1777–1838) famed for discovering iodine (1811), and his life in Paris from 1798" (PDF). *Bulletin for the History of Chemistry* **30** (2): 103.

[69] Gay-Lussac, J. (1813). "Sur un nouvel acide formé avec la substance découverte par M. Courtois". *Annales de chimie* **88**: 311.

[70] Gay-Lussac, J. (1813). "Sur la combinaison de l'iode avec d'oxigène". *Annales de chimie* **88**: 319.

[71] Gay-Lussac, J. (1814). "Mémoire sur l'iode". *Annales de chimie* **91**: 5.

[72] Davy, H. (1813). "Sur la nouvelle substance découverte par M. Courtois, dans le sel de Vareck". *Annales de chimie* **88**: 322.

[73] Davy, Humphry (January 1, 1814). "Some Experiments and Observations on a New Substance Which Becomes a Violet Coloured Gas by Heat". *Phil. Trans. R. Soc. Lond.* **104**: 74. doi:10.1098/rstl.1814.0007.

[74] David Knight, 'Davy, Sir Humphry, baronet (1778–1829)', Oxford Dictionary of National Biography, Oxford University Press, 2004 accessed 6 April 2008

[75] "History of Chirality". Stheno Corporation. 2006. Archived from the original on 2007-03-07. Retrieved 2007-03-12.

[76] "Lambert-Beer Law". Sigrist-Photometer AG. 2007-03-07. Retrieved 2007-03-12.

[77] "Benjamin Silliman, Jr. (1816–1885)". *Picture History*. Picture History LLC. 2003. Retrieved 2007-03-24.

[78] Moore, F. J. (1931). *A History of Chemistry*. McGraw-Hill. pp. 182–1184. ISBN 0-07-148855-3. (2nd edition)

[79] "Jacobus Henricus van't Hoff". *Chemical Achievers: The Human Face of Chemical Sciences*. Chemical Heritage Foundation. 2005. Retrieved 2007-02-22.

[80] O'Connor, J. J.; Robertson, E.F. (1997). "Josiah Willard Gibbs". *MacTutor*. School of Mathematics and Statistics University of St Andrews, Scotland. Retrieved 2007-03-24.

[81] Weisstein, Eric W. (1996). "Boltzmann, Ludwig (1844–1906)". *Eric Weisstein's World of Scientific Biography*. Wolfram Research Products. Retrieved 2007-03-24.

[82] "Svante August Arrhenius". *Chemical Achievers: The Human Face of Chemical Sciences*. Chemical Heritage Foundation. 2005. Retrieved 2007-02-22.

[83] "Jacobus H. van 't Hoff: The Nobel Prize in Chemistry 1901". *Nobel Lectures, Chemistry 1901–1921*. Elsevier Publishing Company. 1966. Retrieved 2007-02-28.

[84] "Henry Louis Le Châtelier". *World of Scientific Discovery*. Thomson Gale. 2005. Retrieved 2007-03-24.

[85] "Emil Fischer: The Nobel Prize in Chemistry 1902". *Nobel Lectures, Chemistry 1901–1921*. Elsevier Publishing Company. 1966. Retrieved 2007-02-28.

[86] "History of Chemistry". *Intensive General Chemistry*. Columbia University Department of Chemistry Undergraduate Program. Retrieved 2007-03-24.

[87] "Alfred Werner: The Nobel Prize in Chemistry 1913". *Nobel Lectures, Chemistry 1901–1921*. Elsevier Publishing Company. 1966. Retrieved 2007-03-24.

[88] "Alfred Werner: The Nobel Prize in Physics 1911". *Nobel Lectures, Physics 1901–1921*. Elsevier Publishing Company. 1967. Retrieved 2007-03-24.

[89] W. Heitler and F. London, *Wechselwirkung neutraler Atome und Homöopolare Bindung nach der Quantenmechanik*, Z. Physik, 44, 455 (1927).

[90] P.A.M. Dirac, *Quantum Mechanics of Many-Electron Systems*, Proc. R. Soc. London, A 123, 714 (1929).

[91] C.C.J. Roothaan, *A Study of Two-Center Integrals Useful in Calculations on Molecular Structure*, J. Chem. Phys., 19, 1445 (1951).

[92] Watson, J. and Crick, F., "Molecular Structure of Nucleic Acids" *Nature, April 25, 1953, p 737–8*

[93] W. J. Hehre, W. A. Lathan, R. Ditchfield, M. D. Newton, and J. A. Pople, Gaussian 70 (Quantum Chemistry Program Exchange, Program No. 237, 1970).

[94] *Catalyse de transformation des oléfines par les complexes du tungstène. II. Télomérisation des oléfines cycliques en présence d'oléfines acycliques* Die Makromolekulare Chemie Volume 141, Issue 1, Date: 9 February **1971**, Pages: 161–176 Par Jean-Louis Hérisson, Yves Chauvin doi:10.1002/macp.1971.021410112

[95] Katsuki, T.; Sharpless, K. B. *J. Am. Chem. Soc.* **1980**, *102*, 5974. (doi:10.1021/ja00538a077)

[96] Hill, J. G.; Sharpless, K. B.; Exon, C. M.; Regenye, R. *Org. Syn.*, Coll. Vol. 7, p.461 (1990); Vol. 63, p.66 (1985). (Article)

[97] Jacobsen, E. N.; Marko, I.; Mungall, W. S.; Schroeder, G.; Sharpless, K. B. *J. Am. Chem. Soc.* **1988**, *110*, 1968. (doi:10.1021/ja00214a053)

[98] Kolb, H. C.; Van Nieuwenhze, M. S.; Sharpless, K. B. *Chem. Rev.* **1994**, *94*, 2483–2547. (Review) (doi:10.1021/cr00032a009)

[99] Gonzalez, J.; Aurigemma, C.; Truesdale, L. *Org. Syn.*, Coll. Vol. 10, p.603 (2004); Vol. 79, p.93 (2002). (Article)

[100] Sharpless, K. B.; Patrick, D. W.; Truesdale, L. K.; Biller, S. A. *J. Am. Chem. Soc.* **1975**, *97*, 2305. (doi:10.1021/ja00841a071)

[101] Herranz, E.; Biller, S. A.; Sharpless, K. B. *J. Am. Chem. Soc.* **1978**, *100*, 3596–3598. (doi:10.1021/ja00479a051)

[102] Herranz, E.; Sharpless, K. B. *Org. Syn.*, Coll. Vol. 7, p.375 (1990); Vol. 61, p.85 (1983). (Article)

[103] "The Nobel Prize in Chemistry 1996". *Nobelprize.org*. The Nobel Foundation. Retrieved 2007-02-28.

[104] "Benjamin Franklin Medal awarded to Dr. Sumio Iijima, Director of the Research Center for Advanced Carbon Materials, AIST". National Institute of Advanced Industrial Science and Technology. 2002. Retrieved 2007-03-27.

[105] *First total synthesis of taxol 1.* Functionalization of the B ring Robert A. Holton, Carmen Somoza, Hyeong Baik Kim, Feng Liang, Ronald J. Biediger, P. Douglas Boatman, Mitsuru Shindo, Chase C. Smith, Soekchan Kim, et al.; J. Am. Chem. Soc.; **1994**; 116(4); 1597–1598. DOI Abstract

[106] *First total synthesis of taxol. 2.* Completion of the C and D rings Robert A. Holton, Hyeong Baik Kim, Carmen Somoza, Feng Liang, Ronald J. Biediger, P. Douglas Boatman, Mitsuru Shindo, Chase C. Smith, Soekchan Kim, and et al. J. Am. Chem. Soc.; **1994**; 116(4) pp 1599–1600 DOI Abstract

[107] *A synthesis of taxusin* Robert A. Holton, R. R. Juo, Hyeong B. Kim, Andrew D. Williams, Shinya Harusawa, Richard E. Lowenthal, Sadamu Yogai J. Am. Chem. Soc.; **1988**; 110(19); 6558–6560. Abstract

[108] "Cornell and Wieman Share 2001 Nobel Prize in Physics". *NIST News Release.* National Institute of Standards and Technology. 2001. Retrieved 2007-03-27.

1.10 References

- Selected classic papers from the history of chemistry

- Biographies of Chemists

- Eric R. Scerri, The Periodic Table: Its Story and Its Significance, Oxford University Press, 2006.

1.11 Further reading

- Servos, John W., *Physical chemistry from Ostwald to Pauling : the making of a science in America*, Princeton, N.J. : Princeton University Press, 1990. ISBN 0-691-08566-8

Documentaries

- BBC (2010). *Chemistry: A Volatile History.*

1.12 External links

- ChemisLab - Chemists of the Past

- SHAC: Society for the History of Alchemy and Chemistry

Chapter 2

History of thermodynamics

The 1698 **Savery Engine** *– the world's first commercially-useful steam engine: built by Thomas Savery*

The **history of thermodynamics** is a fundamental strand in the history of physics, the history of chemistry, and the history of science in general. Owing to the relevance of thermodynamics in much of science and technology, its history is finely woven with the developments of classical mechanics, quantum mechanics, magnetism, and chemical kinetics, to more distant applied fields such as meteorology, information theory, and biology (physiology), and to technological developments such as the steam engine, internal combustion engine, cryogenics and electricity generation. The development of thermodynamics both drove and was driven by atomic theory. It also, albeit in a sub-

tle manner, motivated new directions in probability and statistics; see, for example, the timeline of thermodynamics.

2.1 History

See also: Timeline of thermodynamics

2.1.1 Contributions from ancient and medieval times

See also: History of heat and Vacuum

The ancients viewed heat as that related to fire. In 3000 BC, the ancient Egyptians viewed heat as related to origin mythologies.[1] In the Western philosophical tradition, after much debate about the primal element among earlier pre-Socratic philosophers, Empedocles proposed a four-element theory, in which all substances derive from earth, water, air, and fire. The Empedoclean element of fire is perhaps the principal ancestor of later concepts such as phlogiston and caloric. Around 500 BC, the Greek philosopher Heraclitus became famous as the "flux and fire" philosopher for his proverbial utterance: "All things are flowing." Heraclitus argued that the three principal elements in nature were fire, earth, and water.

Atomism is a central part of today's relationship between thermodynamics and statistical mechanics. Ancient thinkers such as Leucippus and Democritus, and later the Epicureans, by advancing atomism, laid the foundations for the later atomic theory. Until experimental proof of atoms was later provided in the 20th century, the atomic theory was driven largely by philosophical considerations and scientific intuition.

The 5th century BC, Greek philosopher Parmenides, in his only known work, a poem conventionally titled *On Nature*,

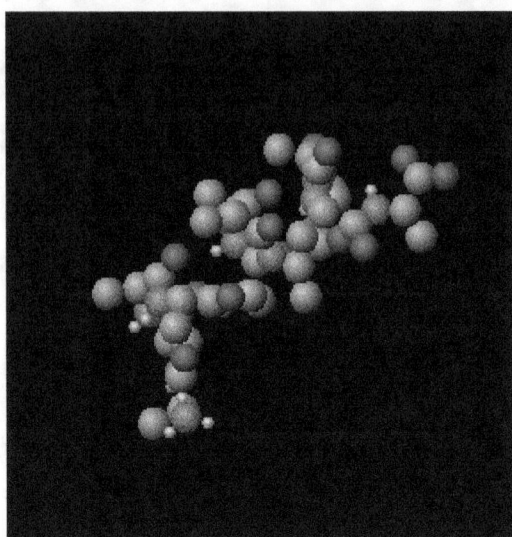

Heating a body, such as a segment of protein alpha helix (above), tends to cause its atoms to vibrate more, and to expand or change phase, if heating is continued; an axiom of nature noted by Herman Boerhaave in the in 1700s.

uses verbal reasoning to postulate that a void, essentially what is now known as a vacuum, in nature could not occur. This view was supported by the arguments of Aristotle, but was criticized by Leucippus and Hero of Alexandria. From antiquity to the Middle Ages various arguments were put forward to prove or disapprove the existence of a vacuum and several attempts were made to construct a vacuum but all proved unsuccessful.

The European scientists Cornelius Drebbel, Robert Fludd, Galileo Galilei and Santorio Santorio in the 16th and 17th centuries were able to gauge the relative "coldness" or "hotness" of air, using a rudimentary air thermometer (or thermoscope). This may have been influenced by an earlier device which could expand and contract the air constructed by Philo of Byzantium and Hero of Alexandria.

Around 1600, the English philosopher and scientist Francis Bacon surmised: "Heat itself, its essence and quiddity is motion and nothing else." In 1643, Galileo Galilei, while generally accepting the 'sucking' explanation of *horror vacui* proposed by Aristotle, believed that nature's vacuum-abhorrence is limited. Pumps operating in mines had already proven that nature would only fill a vacuum with water up to a height of ~30 feet. Knowing this curious fact, Galileo encouraged his former pupil Evangelista Torricelli to investigate these supposed limitations. Torricelli did not believe that vacuum-abhorrence (*Horror vacui*) in the sense of Aristotle's 'sucking' perspective, was responsible for raising the water. Rather, he reasoned, it was the result of the pressure exerted on the liquid by the surrounding air.

To prove this theory, he filled a long glass tube (sealed at one end) with mercury and upended it into a dish also containing mercury. Only a portion of the tube emptied (as shown adjacent); ~30 inches of the liquid remained. As the mercury emptied, and a partial vacuum was created at the top of the tube. The gravitational force on the heavy element Mercury prevented it from filling the vacuum.

2.1.2   Transition from chemistry to thermo-chemistry

See also: History of chemistry

The theory of phlogiston arose in the 17th century, late

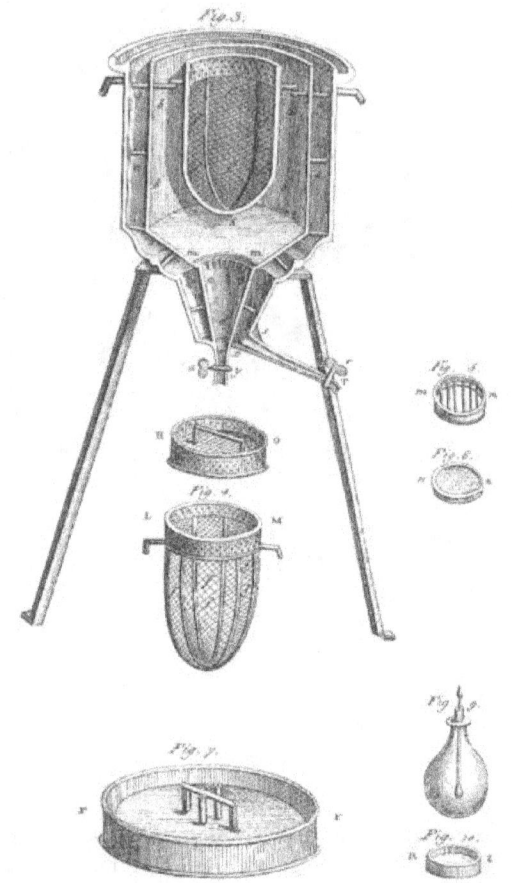

*The world's first **ice-calorimeter**, used in the winter of 1782-83, by Antoine Lavoisier and Pierre-Simon Laplace, to determine the heat evolved in various chemical changes; calculations which were based on Joseph Black's prior discovery of latent heat. These experiments mark the foundation of thermochemistry.*

in the period of alchemy. Its replacement by caloric theory in the 18th century is one of the historical markers of the transition from alchemy to chemistry. Phlogiston was a hypothetical substance that was presumed to be liberated from combustible substances during burning, and from metals during the process of rusting. Caloric, like phlogiston, was also presumed to be the "substance" of heat that would flow from a hotter body to a cooler body, thus warming it.

The first substantial experimental challenges to caloric theory arose in Rumford's 1798 work, when he showed that boring cast iron cannons produced great amounts of heat which he ascribed to friction, and his work was among the first to undermine the caloric theory. The development of the steam engine also focused attention on calorimetry and the amount of heat produced from different types of coal. The first quantitative research on the heat changes during chemical reactions was initiated by Lavoisier using an ice calorimeter following research by Joseph Black on the latent heat of water.

More quantitative studies by James Prescott Joule in 1843 onwards provided soundly reproducible phenomena, and helped to place the subject of thermodynamics on a solid footing. William Thomson, for example, was still trying to explain Joule's observations within a caloric framework as late as 1850. The utility and explanatory power of kinetic theory, however, soon started to displace caloric and it was largely obsolete by the end of the 19th century. Joseph Black and Lavoisier made important contributions in the precise measurement of heat changes using the calorimeter, a subject which became known as thermochemistry.

2.1.3 Phenomenological thermodynamics

- Boyle's law (1662)

- Charles's law was first published by Joseph Louis Gay-Lussac in 1802, but he referenced unpublished work by Jacques Charles from around 1787. The relationship had been anticipated by the work of Guillaume Amontons in 1702.

- Gay-Lussac's law (1802)

2.1.4 Birth of thermodynamics as science

At its origins, thermodynamics was the study of engines. A precursor of the engine was designed by the German scientist Otto von Guericke who, in 1650, designed and built the world's first vacuum pump and created the world's first ever vacuum known as the Magdeburg hemispheres. He was driven to make a vacuum in order to disprove Aristotle's long-held supposition that 'Nature abhors a vacuum'.

Shortly thereafter, Irish physicist and chemist Robert Boyle had learned of Guericke's designs and in 1656, in coordina-

Robert Boyle. 1627-1691

tion with English scientist Robert Hooke, built an air pump. Using this pump, Boyle and Hooke noticed the pressure-volume correlation: P.V=constant. In that time, air was assumed to be a system of motionless particles, and not interpreted as a system of moving molecules. The concept of thermal motion came two centuries later. Therefore Boyle's publication in 1660 speaks about a mechanical concept: the air spring.[2] Later, after the invention of the thermometer, the property temperature could be quantified. This tool gave Gay-Lussac the opportunity to derive his law, which led shortly later to the ideal gas law. But, already before the establishment of the ideal gas law, an associate of Boyle's named Denis Papin built in 1679 a bone digester, which is a closed vessel with a tightly fitting lid that confines steam until a high pressure is generated.

Later designs implemented a steam release valve to keep the machine from exploding. By watching the valve rhythmically move up and down, Papin conceived of the idea of a piston and cylinder engine. He did not however follow through with his design. Nevertheless, in 1697, based on Papin's designs, engineer Thomas Savery built the first engine. Although these early engines were crude and inefficient, they attracted the attention of the leading scientists of the time. One such scientist was Sadi Carnot, the "father of thermodynamics", who in 1824 published *Reflections on the Motive Power of Fire*, a discourse on heat, power, and engine efficiency. This marks the start of thermodynamics as a modern science.

A Watt steam engine, the steam engine that propelled the Industrial Revolution in Britain and the world

Hence, prior to 1698 and the invention of the Savery Engine, horses were used to power pulleys, attached to buckets, which lifted water out of flooded salt mines in England. In the years to follow, more variations of steam engines were built, such as the Newcomen Engine, and later the Watt Engine. In time, these early engines would eventually be utilized in place of horses. Thus, each engine began to be associated with a certain amount of "horse power" depending upon how many horses it had replaced. The main problem with these first engines was that they were slow and clumsy, converting less than 2% of the input fuel into useful work. In other words, large quantities of coal (or wood) had to be burned to yield only a small fraction of work output. Hence the need for a new science of engine dynamics was born.

Most cite Sadi Carnot's 1824 book *Reflections on the Motive Power of Fire* as the starting point for thermodynamics as a modern science. Carnot defined "motive power" to be the expression of the *useful effect* that a motor is capable of producing. Herein, Carnot introduced us to the first modern day definition of "work": *weight lifted through a height*. The desire to understand, via formulation, this *useful effect* in relation to "work" is at the core of all modern day thermodynamics.

In 1843, James Joule experimentally found the mechanical equivalent of heat. In 1845, Joule reported his best-known experiment, involving the use of a falling weight to spin a paddle-wheel in a barrel of water, which allowed him to estimate a mechanical equivalent of heat of 819 ft·lbf/Btu (4.41 J/cal). This led to the theory of conservation of energy and explained why heat can do work.

In 1850, the famed mathematical physicist Rudolf Clausius defined the term entropy *S* to be the heat lost or turned into waste, stemming from the Greek word *entrepein* meaning

Sadi Carnot (1796-1832): the "father" of thermodynamics

to turn.

The name "thermodynamics", however, did not arrive until 1854, when the British mathematician and physicist William Thomson (Lord Kelvin) coined the term *thermodynamics* in his paper *On the Dynamical Theory of Heat*.[3]

In association with Clausius, in 1871, the Scottish mathematician and physicist James Clerk Maxwell formulated a new branch of thermodynamics called *Statistical Thermodynamics*, which functions to analyze large numbers of particles at equilibrium, i.e., systems where no changes are occurring, such that only their average properties as temperature *T*, pressure *P*, and volume *V* become important.

Soon thereafter, in 1875, the Austrian physicist Ludwig Boltzmann formulated a precise connection between entropy *S* and molecular motion:

$$S = k \log W$$

being defined in terms of the number of possible states [W] such motion could occupy, where k is the Boltzmann's constant.

The following year, 1876, was a seminal point in the development of human thought. During this essential period, chemical engineer Willard Gibbs, the first person in America to be awarded a PhD in engineering (Yale), published an obscure 300-page paper titled: *On the Equilibrium of Heterogeneous Substances*, wherein he formulated one grand equality, the Gibbs free energy equation, which suggested a measure of the amount of "useful work" attainable in reacting systems. Gibbs also originated the concept we now know as enthalpy H, calling it "a heat function for constant pressure".[4] The modern word *enthalpy* would be coined many years later by Heike Kamerlingh Onnes,[5] who based it on the Greek word *enthalpein* meaning *to warm*.

Building on these foundations, those as Lars Onsager, Erwin Schrödinger, and Ilya Prigogine, and others, functioned to bring these engine "concepts" into the thoroughfare of almost every modern-day branch of science.

2.1.5 Kinetic theory

Main article: Kinetic theory

The idea that heat is a form of motion is perhaps an ancient one and is certainly discussed by Francis Bacon in 1620 in his *Novum Organum*. The first written scientific reflection on the microscopic nature of heat is probably to be found in a work by Mikhail Lomonosov, in which he wrote:

> "(..) movement should not be denied based on the fact it is not seen. Who would deny that the leaves of trees move when rustled by a wind, despite it being unobservable from large distances? Just as in this case motion remains hidden due to perspective, it remains hidden in warm bodies due to the extremely small sizes of the moving particles. In both cases, the viewing angle is so small that neither the object nor their movement can be seen."

During the same years, Daniel Bernoulli published his book *Hydrodynamics* (1738), in which he derived an equation for the pressure of a gas considering the collisions of its atoms with the walls of a container. He proves that this pressure is two thirds the average kinetic energy of the gas in a unit volume. Bernoulli's ideas, however, made little impact on the dominant caloric culture. Bernoulli made a connection with Gottfried Leibniz's *vis viva* principle, an early formulation of the principle of conservation of energy, and the two theories became intimately entwined throughout their history. Though Benjamin Thompson suggested that heat was a form of motion as a result of his experiments in 1798, no attempt was made to reconcile theoretical and experimental

approaches, and it is unlikely that he was thinking of the *vis viva* principle.

John Herapath later independently formulated a kinetic theory in 1820, but mistakenly associated temperature with momentum rather than *vis viva* or kinetic energy. His work ultimately failed peer review and was neglected. John James Waterston in 1843 provided a largely accurate account, again independently, but his work received the same reception, failing peer review even from someone as well-disposed to the kinetic principle as Davy.

Further progress in kinetic theory started only in the middle of the 19th century, with the works of Rudolf Clausius, James Clerk Maxwell, and Ludwig Boltzmann. In his 1857 work *On the nature of the motion called heat*, Clausius for the first time clearly states that heat is the average kinetic energy of molecules. This interested Maxwell, who in 1859 derived the momentum distribution later named after him. Boltzmann subsequently generalized his distribution for the case of gases in external fields.

Boltzmann is perhaps the most significant contributor to kinetic theory, as he introduced many of the fundamental concepts in the theory. Besides the Maxwell–Boltzmann distribution mentioned above, he also associated the kinetic energy of particles with their degrees of freedom. The Boltzmann equation for the distribution function of a gas in non-equilibrium states is still the most effective equation for studying transport phenomena in gases and metals. By introducing the concept of thermodynamic probability as the number of microstates corresponding to the current macrostate, he showed that its logarithm is proportional to entropy.

2.2 Branches of

The following list gives a rough outline as to when the major branches of thermodynamics came into inception:

- Thermochemistry - 1780s

- Classical thermodynamics - 1824

- Chemical thermodynamics - 1876

- Statistical mechanics - c. 1880s

- Equilibrium thermodynamics

- Engineering thermodynamics

- Chemical engineering thermodynamics - c. 1940s

- Non-equilibrium thermodynamics - 1941

- Small systems thermodynamics - 1960s

- Biological thermodynamics - 1957

- Ecosystem thermodynamics - 1959

- Relativistic thermodynamics - 1965

- Quantum thermodynamics - 1968

- Black hole thermodynamics - c. 1970s

- Geological thermodynamics - c. 1970s

- Biological evolution thermodynamics - 1978

- Geochemical thermodynamics - c. 1980s

- Atmospheric thermodynamics - c. 1980s

- Natural systems thermodynamics - 1990s

- Supramolecular thermodynamics - 1990s

- Earthquake thermodynamics - 2000

- Drug-receptor thermodynamics - 2001

- Pharmaceutical systems thermodynamics – 2002

Ideas from thermodynamics have also been applied in other fields, for example:

- Thermoeconomics - c. 1970s

2.3   Entropy and the second law

Main article: History of entropy

Even though he was working with the caloric theory, Sadi Carnot in 1824 suggested that some of the caloric available for generating useful work is lost in any real process. In March 1851, while grappling to come to terms with the work of James Prescott Joule, Lord Kelvin started to speculate that there was an inevitable loss of useful heat in all processes. The idea was framed even more dramatically by Hermann von Helmholtz in 1854, giving birth to the spectre of the heat death of the universe.

In 1854, William John Macquorn Rankine started to make use in calculation of what he called his *thermodynamic function*. This has subsequently been shown to be identical to the concept of entropy formulated by Rudolf Clausius in 1865. Clausius used the concept to develop his classic statement of the second law of thermodynamics the same year.

2.4   Heat transfer

Main article: Heat transfer

The phenomenon of heat conduction is immediately grasped in everyday life. In 1701, Sir Isaac Newton published his law of cooling. However, in the 17th century, it came to be believed that all materials had an identical conductivity and that differences in sensation arose from their different heat capacities.

Suggestions that this might not be the case came from the new science of electricity in which it was easily apparent that some materials were good electrical conductors while others were effective insulators. Jan Ingen-Housz in 1785-9 made some of the earliest measurements, as did Benjamin Thompson during the same period.

The fact that warm air rises and the importance of the phenomenon to meteorology was first realised by Edmund Halley in 1686. Sir John Leslie observed that the cooling effect of a stream of air increased with its speed, in 1804.

Carl Wilhelm Scheele distinguished heat transfer by thermal radiation (radiant heat) from that by convection and conduction in 1777. In 1791, Pierre Prévost showed that all bodies radiate heat, no matter how hot or cold they are. In 1804, Leslie observed that a matte black surface radiates heat more effectively than a polished surface, suggesting the importance of black body radiation. Though it had become to be suspected even from Scheele's work, in 1831 Macedonio Melloni demonstrated that black body radiation could be reflected, refracted and polarised in the same way as light.

James Clerk Maxwell's 1862 insight that both light and radiant heat were forms of electromagnetic wave led to the start of the quantitative analysis of thermal radiation. In 1879, Jožef Stefan observed that the total radiant flux from a blackbody is proportional to the fourth power of its temperature and stated the Stefan–Boltzmann law. The law was derived theoretically by Ludwig Boltzmann in 1884.

2.5   Cryogenics

In 1702 Guillaume Amontons introduced the concept of absolute zero based on observations of gases. In 1810, Sir John Leslie froze water to ice artificially. The idea of absolute zero was generalised in 1848 by Lord Kelvin. In 1906, Walther Nernst stated the third law of thermodynamics.

2.6 See also

- Conservation of energy: Historical development

- History of Chemistry

- History of Physics

- Maxwell's thermodynamic surface

- Timeline of thermodynamics, statistical mechanics, and random processes

- Thermodynamics

- Timeline of heat engine technology

- Timeline of low-temperature technology

2.7 References

[1] J. Gwyn Griffiths (1955). "The Orders of Gods in Greece and Egypt (According to Herodotus)". *The Journal of Hellenic Studies* **75**: 21–23. doi:10.2307/629164. JSTOR 629164.

[2] New Experiments physico-mechanicall, Touching the Spring of the Air and its Effects (1660).

[3] Thomson, W. (1854). "On the Dynamical Theory of Heat". *Transactions of the Royal Society of Edinburgh* **21** (part I): 123. doi:10.1017/s0080456800032014. |chapter= ignored (help) reprinted in Sir William Thomson, LL.D. D.C.L., F.R.S. (1882). *Mathematical and Physical Papers* **1**. London, Cambridge: C.J. Clay, M.A. & Son, Cambridge University Press. p. 232. Hence Thermo-dynamics falls naturally into two Divisions, of which the subjects are respectively, *the relation of heat to the forces acting between contiguous parts of bodies, and the relation of heat to electrical agency.*

[4] Laidler, Keith (1995). *The World of Physical Chemistry.* Oxford University Press. p. 110.

[5] Howard, Irmgard (2002). "H Is for Enthalpy, Thanks to Heike Kamerlingh Onnes and Alfred W. Porter". *Journal of Chemical Education* (ACS Publications) **79** (6): 697. Bibcode:2002JChEd..79..697H. doi:10.1021/ed079p697.

2.8 Further reading

- Cardwell, D.S.L. (1971). *From Watt to Clausius: The Rise of Thermodynamics in the Early Industrial Age.* London: Heinemann. ISBN 0-435-54150-1.

- Leff, H.S. & Rex, A.F. (eds) (1990). *Maxwell's Demon: Entropy, Information and Computing.* Bristol: Adam Hilger. ISBN 0-7503-0057-4.

2.9 External links

- History of Statistical Mechanics and Thermodynamics - Timeline (1575 to 1980) @ Hyperjeff.net

- History of Thermodynamics - University of Waterloo

- Thermodynamic History Notes - WolframScience.com

- Brief History of Thermodynamics - Berkeley [PDF]

- History of Thermodynamics - ThermodynamicStudy.net

- Historical Background of Thermodynamics - Carnegie-Mellon University

- History of Thermodynamics - In Pictures

Chapter 3

Ferrous metallurgy

Bloomery smelting during the Middle Ages.

Ferrous metallurgy involves processes and alloys based on iron. It began far back in prehistory. The earliest surviving iron artifacts, from the 4th millennium BC in Egypt,[1] were made from meteoritic iron-nickel.[2] By the end of the 2nd millennium BC iron was being produced from iron ores from South of the Saharan Africa to China.[3][4] The use of wrought iron (worked iron) was known by the 1st millen-

nium BC. During the medieval period, means were found in Europe of producing wrought iron from cast iron (in this context known as pig iron) using finery forges. For all these processes, charcoal was required as fuel.

Steel (with a carbon content between pig iron and wrought iron) was first produced in antiquity as an alloy. Its process of production, Wootz, was exported before the 4th century BC to ancient China, Africa, the Middle East and Europe. Archaeological evidence of cast iron appears in 5th century BC China.[5] New methods of producing it by carburizing bars of iron in the cementation process were devised in the 17th century. During the Industrial Revolution, new methods of producing bar iron without charcoal were devised and these were later applied to produce steel. In the late 1850s, Henry Bessemer invented a new steelmaking process, that involved blowing air through molten pig iron to burn off carbon, and so to produce mild steel. This and other 19th-century and later processes have displace the use of wrought iron. Today, wrought iron is no longer produced.

3.1 Meteoric iron

Iron was extracted from iron-nickel meteorites, which comprise about 6% of all meteorites that fall on the earth. That source can often be identified with certainty because of the unique crystalline features ("Widmanstatten figures") of that material, which are preserved when the metal is worked cold or at low temperature. Those artifacts include, for example, a bead from the 5th millennium BC found in Iran[2] and spear tips and ornaments from Ancient Egypt and Sumer around 4000 BC.[6] Meteoric iron has been identified also in a Chinese axe head from the middle of the 2nd millennium BC.

These early uses appear to have been largely ceremonial or ornamental. Meteoritic iron is very rare, and the metal was probably very expensive, perhaps more expensive than gold. The early Hittites are known to have bartered iron (meteoritic or smelted) for silver, at a rate of 40 times the iron's

Willamette Meteorite, the sixth largest in the world, is an iron-nickel meteorite

Iron meteorites consist overwhelmingly of nickel-iron alloys. The metal taken from these meteorites is known as meteoric iron and was one of the earliest sources of usable iron available to humans.

weight, with Assyria.[7]

Meteoric iron was also fashioned into tools in the Arctic, about the year 1000, when the Thule people of Greenland began making harpoons, knives, ulos and other edged tools from pieces of the Cape York meteorite. Typically pea-size bits of metal were cold-hammered into disks and fitted to a bone handle.[2] These artifacts were also used as trade goods with other Arctic peoples: tools made from the Cape York

meteorite have been found in archaeological sites more than 1,000 miles (1,600 km) distant. When the American polar explorer Robert Peary shipped the largest piece of the meteorite to the American Museum of Natural History in New York City in 1897, it still weighed over 33 tons. Another example of a late use of meteoritic iron is an adze from around 1000 AD found in Sweden.[2]

Media related to Objects made from meteoritic iron at Wikimedia Commons

3.2 Native iron

Native iron in the metallic state occurs rarely as small inclusions in certain basalt rocks. Besides meteoritic iron, Thule people of Greenland have used native iron from the Disko region.[2]

3.3 Iron smelting and the Iron Age

Iron smelting—the extraction of usable metal from oxidized iron ores—is more difficult than tin and copper smelting. While these metals and their alloys can be cold-worked or melted in relatively simple furnaces (such as the kilns used for pottery) and cast into molds, smelted iron requires hot-working and can be melted only in specially designed furnaces. Thus it is not surprising that humans only mastered the technology of smelted iron after several millennia of bronze metallurgy.

The place and time for the discovery of iron smelting is not known, partly because of the difficulty of distinguishing metal extracted from nickel-containing ores from hot-worked meteoritic iron.[2] The archaeological evidence seems to point to the Middle East area, during the Bronze Age in the 3rd millennium BC. However iron artifacts remained a rarity until the 12th century BC.

The Iron Age is conventionally defined by the widespread replacement of bronze weapons and tools with those of steel.[8] That transition happened at different times in different places, as the technology spread.. Mesopotamia was fully into the Iron Age by 900 BC. Although Egypt produced iron artifacts, bronze remained dominant until its conquest by Assyria in 663 BC. The Iron Age began in Central Europe about 500 BC, and in India and China between 1200 and 500 BC.[9] Around 500 BC, the Nubians who had learned from the Assyrians the use of iron and were expelled from Egypt, became major manufacturers and exporters of iron.[10]

3.3.1 Ancient Near East

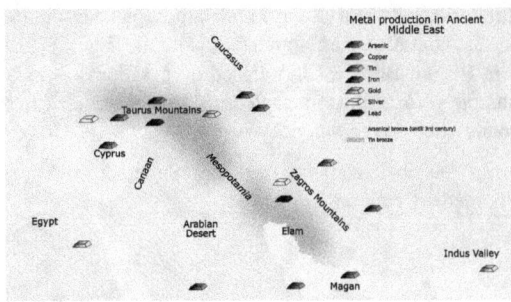

Mining areas of the ancient Middle East. Boxes colors: arsenic is in brown, copper in red, tin in grey, iron in reddish brown, gold in yellow, silver in white and lead in black. Yellow area stands for arsenic bronze, while grey area stands for tin bronze

One of the earliest smelted iron artifacts, a dagger with an iron blade found in a Hattic tomb in Anatolia, dated from 2500 BC.[11] About 1500 BC, increasing numbers of non-meteoritic, smelted iron objects appeared in Mesopotamia, Anatolia, and Egypt.[2] Nineteen iron objects were found in the tomb of Egyptian ruler Tutankhamun, died in 1323 BC, including an iron dagger with a golden hilt, an Eye of Horus, the mummy's head-stand and sixteen models of an artisan's tools.[12] An Ancient Egyptian sword bearing the name of pharaoh Merneptah as well as a battle axe with an iron blade and gold-decorated bronze shaft were both found in the excavation of Ugarit.[11]

Although iron objects dating from the Bronze Age have been found across the Eastern Mediterranean, bronzework appears to have greatly predominated during this period.[13] By the 12th century BC, iron smelting and forging, of weapons and tools, was common from Sub-Saharan Africa through India. As the technology spread, iron came to replace bronze as the dominant metal used for tools and weapons across the Eastern Mediterranean (the Levant, Cyprus, Greece, Crete, Anatolia, and Egypt).[8]

Iron was originally smelted in bloomeries, furnaces where bellows were used to force air through a pile of iron ore and burning charcoal. The carbon monoxide produced by the charcoal reduced the iron oxide from the ore to metallic iron. The bloomery, however, was not hot enough to melt the iron, so the metal collected in the bottom of the furnace as a spongy mass, or *bloom*. Workers filled the bloom's pores with ash and slag. Then they reheated the bloom to soften the iron and melt the slag, and then repeatedly beat and folded it to force out the molten slag. This laborious, time-consuming process produced wrought iron, a malleable but fairly soft alloy.

Concurrent with the transition from bronze to iron was the discovery of carburization, the process of adding carbon to wrought iron. While the iron bloom contained some carbon, the subsequent hot-working oxidized most of it. Smiths in the Middle East discovered that wrought iron could be turned into a much harder product by heating the finished piece in a bed of charcoal, and then quenching it in water or oil. This procedure turned the outer layers of the piece into steel, an alloy of iron and iron carbides, with an inner core of less brittle iron.

Theories on the origin of iron smelting

The development of iron smelting was traditionally attributed to the Hittites of Anatolia of the Late Bronze Age.[14] It was believed that they maintained a monopoly on iron working, and that their empire had been based on that advantage. According to that theory, the ancient Sea Peoples, who invaded the Eastern Mediterranean and destroyed the Hittite empire at the end of the Late Bronze Age, were responsible for spreading the knowledge through that region. This theory is no longer held in the mainstream of scholarship,[14] since there is no archaeological evidence of the alleged Hittite monopoly. While there are some iron objects from Bronze Age Anatolia, the number is comparable to iron objects found in Egypt and other places of the same time period, and only a small number of those objects were weapons.[13]

A more recent theory claims that the development of iron technology was driven by the disruption of the copper and tin trade routes, due to the collapse of the empires at the end of the Late Bronze Age.[14] These metals, especially tin, were not widely available and metal workers had to transport them over long distances, whereas iron ores were widely available. However, no known archaeological evidence suggests a shortage of bronze or tin in the Early Iron Age.[15] Bronze objects remained abundant, and these objects have the same percentage of tin as those from the Late Bronze Age.

3.3.2 Indian Sub-Continent

The History of metallurgy in the Indian subcontinent began in the 2nd millennium BC. Archaeological sites in Gangetic plains have yielded iron implements dated between 1800 – 1200 BC.[16] By the early 13th century BC, iron smelting was practiced on a large scale in India.[16] In Southern India (present day Mysore) iron was in use 12th to 11th centuries BC.[17] The technology of iron metallurgy advanced in the politically stable Maurya period.[18] and during a period of peaceful settlements in the 1st millennium BC.[17]

Iron artifacts such as spikes, knives, daggers, arrow-heads, bowls, spoons, saucepans, axes, chisels, tongs, door fittings etc., dated from 600 to 200 BC, have been discov-

The Iron pillar of Delhi

Dagger and its scabbard, India, 17th–18th century. Blade: Damascus steel inlaid with gold; hilt: jade; scabbard: steel with engraved, chased and gilded decoration

ered at several archaeological sites of India.[9] The Greek historian Herodotus wrote the first western account of the use of iron in India.[9] The Indian mythological texts, the Upanishads, have mentions of weaving, pottery, and metallurgy as well.[19] The Romans had high regard for the excellence of steel from India in the time of the Gupta Empire.[20]

Perhaps as early as 500 BC, although certainly by 200 AD, high quality steel was produced in southern India by the crucible technique. In this system, high-purity wrought iron, charcoal, and glass were mixed in a crucible and heated until the iron melted and absorbed the carbon.[21] Iron chain was used in Indian suspension bridges as early as the 4th century.[22]

Wootz steel was produced in India and Sri Lanka from around 300 BC.[21] Wootz steel is famous from Classical Antiquity for its durability and ability to hold an edge. When asked by King Porus to select a gift, Alexander is said to have chosen, over gold or silver, thirty pounds of steel.[20] Wootz steel was originally a complex alloy with iron as its main component together with various trace elements. Recent studies have suggested that its qualities may have been due to the formation of carbon nanotubes in the metal.[23] According to Will Durant, the technology passed to the Persians and from them to Arabs who spread it through the Middle East.[20] In the 16th century, the Dutch carried the technology from South India to Europe, where it was mass-produced.[24]

Steel was produced in Sri Lanka from 300 BC[21] by furnaces blown by the monsoon winds. The furnaces were dug into the crests of hills, and the wind was diverted into the air vents by long trenches. This arrangement created a zone of high pressure at the entrance, and a zone of low pressure at the top of the furnace. The flow is believed to have allowed higher temperatures than bellows-driven furnaces could produce, resulting in better-quality iron.[25][26][27] Steel made in Sri Lanka was traded extensively within the region and in the Islamic world.
See also Steel#Wootz steel and Damascus steel

One of the world's foremost metallurgical curiosities is an

iron pillar located in the Qutb complex, Delhi. The pillar is made of wrought iron (98% Fe), is almost seven meters high and weighs more than six tonnes.[28] The pillar was erected by Chandragupta II Vikramaditya and has withstood 1,600 years of exposure to heavy rains with relatively little corrosion.

3.3.3 Iron Age Europe

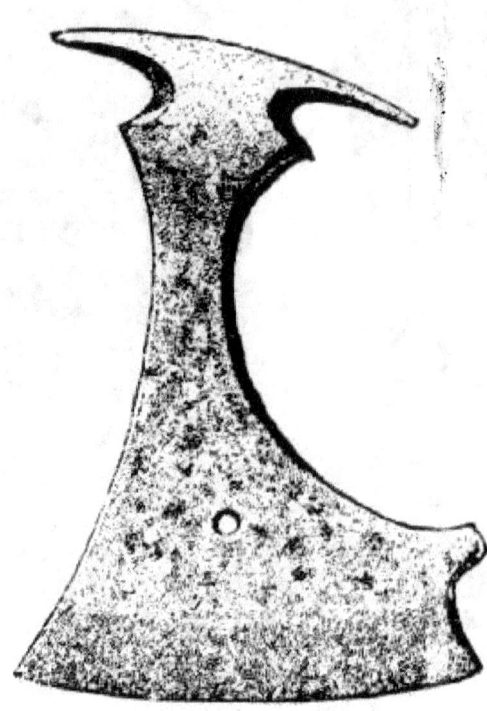

Axe made of iron, dating from Swedish Iron Age, found at Gotland, Sweden

Iron working was introduced to Greece in the late 10th century BC.[29] The earliest marks of Iron Age in Central Europe are artifacts from the Hallstatt C culture (8th century BC). Throughout the 7th to 6th centuries BC, iron artifacts remained luxury items reserved for an elite. This changed dramatically shortly after 500 BC with the rise of the La Tène culture, from which time iron metallurgy also became common in Northern Europe and Britain. The spread of ironworking in Central and Western Europe is associated with Celtic expansion. By the 1st century BC, Noric steel was famous for its quality and sought-after by the Roman military.

The annual iron output of the Roman Empire is estimated at 84,750 t,[30] while the similarly populous Han China produced around 5,000 t.[31]

3.3.4 China

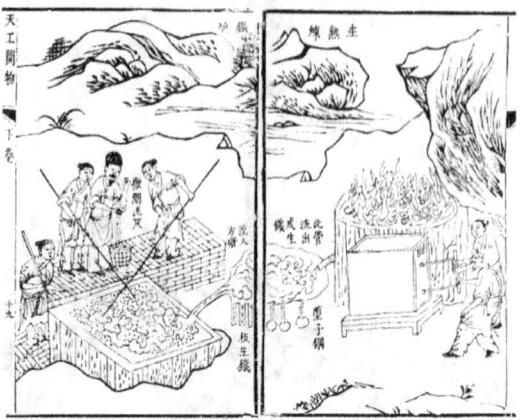

The process of fining iron ore to make wrought iron from pig iron, with the right illustration displaying men working a blast furnace, from the Tiangong Kaiwu *encyclopedia, 1637*

Historians debate whether bloomery-based ironworking ever spread to China from the Middle East. One theory suggests that metallurgy was introduced through Central Asia.[32] The earliest cast iron artifacts, dating to 5th century BC, were discovered by archaeologists in what is now modern Luhe County, Jiangsu in China. Cast iron was used in ancient China for warfare, agriculture, and architecture.[5] Around 500 BC, metalworkers in the southern state of Wu achieved a temperature of 1130 °C. At this temperature, iron combines with 4.3% carbon and melts. The liquid iron can be cast into molds, a method far less laborious than individually forging each piece of iron from a bloom. This technology would be known in Europe from early medieval times on.[33]

Cast iron is rather brittle and unsuitable for striking implements. It can, however, be *decarburized* to steel or wrought iron by heating it in air for several days. In China, these iron working methods spread northward, and by 300 BC, iron was the material of choice throughout China for most tools and weapons. A mass grave in Hebei province, dated to the early 3rd century BC, contains several soldiers buried with their weapons and other equipment. The artifacts recovered from this grave are variously made of wrought iron, cast iron, malleabilized cast iron, and quench-hardened steel, with only a few, probably ornamental, bronze weapons.

During the Han Dynasty (202 BC–220 AD), the government established ironworking as a state monopoly (repealed during the latter half of the dynasty and returned to private entrepreneurship) and built a series of large blast furnaces in Henan province, each capable of producing several tons of iron per day. By this time, Chinese metallurgists had discovered how to fine molten pig iron, stirring it

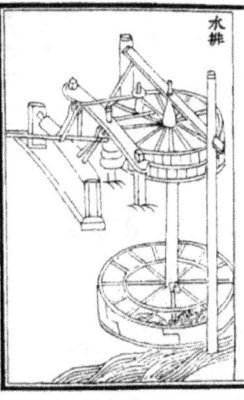

An illustration of furnace bellows operated by waterwheels, from the Nong Shu, *by Wang Zhen, 1313 AD, during the Yuan Dynasty in China*

in the open air until it lost its carbon and could be hammered (wrought). (In modern Mandarin-Chinese, this process is now called *chao*, literally, stir frying.) By the 1st century BC, Chinese metallurgists had found that wrought iron and cast iron could be melted together to yield an alloy of intermediate carbon content, that is, steel.[34][35][36] According to legend, the sword of Liu Bang, the first Han emperor, was made in this fashion. Some texts of the era mention "harmonizing the hard and the soft" in the context of ironworking; the phrase may refer to this process. The ancient city of Wan (Nanyang) from the Han period forward was a major center of the iron and steel industry.[37] Along with their original methods of forging steel, the Chinese had also adopted the production methods of creating Wootz steel, an idea imported from India to China by the 5th century AD.[38] During Han Dynasty, the Chinese were also the first to apply hydraulic power (i.e. a waterwheel) in working the bellows of the blast furnace. This was recorded in the year 31 AD, as an innovation of the engineer Du Shi, Prefect of Nanyang.[39] Although Du Shi was the first to apply water power to bellows in metallurgy, the first drawn and printed illustration of its operation with water power appeared in 1313 AD, in the Yuan Dynasty era text called the *Nong Shu*.[40] In the 11th century, there is evidence of the production of steel in Song China using two techniques: a "berganesque" method that produced inferior, heterogeneous steel and a precursor to the modern Bessemer process that utilized partial decarbonization via repeated forging under a cold blast.[41] By the 11th century, there was a large amount of deforestation in China due to the iron industry's demands for charcoal.[42] By this time however, the Chinese had learned to use bituminous coke to replace charcoal, and with this switch in resources many acres of prime timberland in China were spared.[42] The change of fuel resources from charcoal to coal was pioneered in Roman

Britain by the 2nd century AD, although it was also practiced in the Germanic Rhineland at the time.[43]

See also: Early Japanese Ironworking

3.3.5 Indigenous South of the Saharan Africa

Iron Age finds in East and Southern Africa, corresponding to the early 1st millennium AD Bantu expansion

Main article: Iron metallurgy in Africa

Inhabitants at Termit, in eastern Niger became the first iron smelting people in West Africa around 1500 BC.[44] Iron and copper working spread southward through the continent, reaching the Cape around AD 200.[45][46] The widespread use of iron revolutionized the Bantu-speaking farming communities who adopted it, driving out and absorbing the rock tool using hunter-gatherer societies they encountered as they expanded to farm wider areas of savanna. The technologically superior Bantu-speakers spread across southern Africa and became wealthy and powerful, producing iron for tools and weapons in large, industrial quantities.[45]

In the region of the Aïr Mountains in Niger there are signs of independent copper smelting between 2500–1500 BC. The process was not in a developed state, indicating smelting was not foreign. It became mature about the 1500 BC.[47]

Similarly, smelting in bloomery-type furnaces in West Africa and forging for tools appear in the Nok culture in Africa by 500 BC.[48][49][50] The earliest records of bloomery-type furnaces in East Africa are discoveries of smelted iron and carbon in Nubia and Axum that date back between 1,000-500 BCE.[51][52] Particularly in Meroe, there are known to have been ancient bloomeries that produced metal tools for the Nubians and Kushites and produced surplus for their economy.

In the regions of Tanzania inhabited by the Haya people, carbon dating has shown that blast furnaces were as old as 2000 years, whereas steel of this calibre did not appear in Europe until several centuries later.[53]

3.3.6   Medieval Islamic world

Iron technology was further advanced by several inventions in medieval Islam, during the so-called Islamic Golden Age. These included a variety of water-powered and wind-powered industrial mills for metal production, including geared gristmills and forges. By the 11th century, every province throughout the Muslim world had these industrial mills in operation, from Islamic Spain and North Africa in the west to the Middle East and Central Asia in the east.[54] There are also 10th-century references to cast iron, as well as archeological evidence of blast furnaces being used in the Ayyubid and Mamluk empires from the 11th century, thus suggesting a diffusion of Chinese metal technology to the Islamic world.[55]

Geared gristmills[56] were invented by Muslim engineers, and were used for crushing metallic ores before extraction. Gristmills in the Islamic world were often made from both watermills and windmills. In order to adapt water wheels for gristmilling purposes, cams were used for raising and releasing trip hammers.[57] The first forge driven by a hydropowered water mill rather than manual labour was invented in the 12th century Islamic Spain.[58][58]

One of the most famous steels produced in the medieval Near East was Damascus steel used for swordmaking, and mostly produced in Damascus, Syria, in the period from 900 to 1750. This was produced using the crucible steel method, based on the earlier Indian wootz steel. This process was adopted in the Middle East using locally produced steels. The exact process remains unknown, but it allowed carbides to precipitate out as micro particles arranged in sheets or bands within the body of a blade. Carbides are far harder than the surrounding low carbon steel, so swordsmiths could produce an edge that cut hard materials with the precipitated carbides, while the bands of softer steel let the sword as a whole remain tough and flexible. A team of researchers based at the Technical University of Dresden that uses X-rays and electron microscopy to exam-

ine Damascus steel discovered the presence of cementite nanowires[59] and carbon nanotubes.[60] Peter Paufler, a member of the Dresden team, says that these nanostructures give Damascus steel its distinctive properties[61] and are a result of the forging process.[61][62]

3.4   Medieval and Early Modern Europe

There was no fundamental change in the technology of iron production in Europe for many centuries. European metal workers continued to produce iron in bloomeries. However, the Medieval period brought two developments—the use of water power in the bloomery process in various places (outlined above), and the first European production in cast iron.

3.4.1   Powered bloomeries

Main article: Bloomery

Sometime in the medieval period, water power was applied to the bloomery process. It is possible that this was at the Cistercian Abbey of Clairvaux as early as 1135, but it was certainly in use in early 13th century France and Sweden.[63] In England, the first clear documentary evidence for this is the accounts of a forge of the Bishop of Durham, near Bedburn in 1408,[64] but that was certainly not the first such ironworks. In the Furness district of England, powered bloomeries were in use into the beginning of the 18th century, and near Garstang until about 1770.

The Catalan Forge was a variety of powered bloomery. Bloomeries with hot blast were used in upstate New York in the mid-19th century.

3.4.2   Blast furnace

Main article: blast furnace

Cast iron development lagged in Europe, as the smelters could only achieve temperatures of about 1000 C; or perhaps they did not want hotter temperatures, as they were seeking to produce blooms as a precursor of wrought iron, not cast iron. Through a good portion of the Middle Ages, in Western Europe, iron was still being made by the working of iron blooms into wrought iron. Some of the earliest casting of iron in Europe occurred in Sweden, in two sites, Lapphyttan and Vinarhyttan, between 1150 and 1350. Some scholars have speculated the practice followed the Mongols across Russia to these sites, but there is no clear

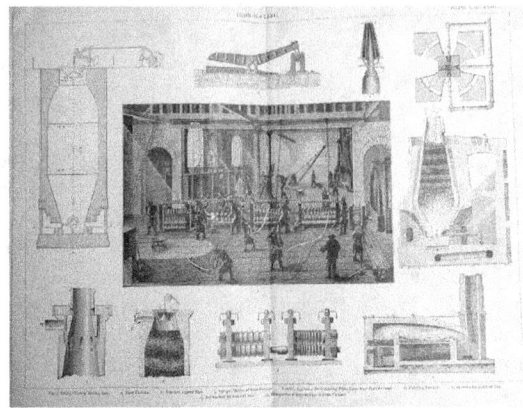

Ironmaking described in "The Popular Encyclopedia" vol. VII, published 1894

proof of this hypothesis, and it would certainly not explain the pre-Mongol datings of many of these iron-production centres. In any event, by the late 14th century, a market for cast iron goods began to form, as a demand developed for cast iron cannonballs.

3.4.3 Osmond process

Iron from furnaces such as Lapphyttan was refined into wrought iron by the osmond process. The pig iron from the furnace was melted in front of a blast of air and the droplets caught on a staff (which was spun). This formed a ball of iron, known as an osmond. This was probably a traded commodity by c. 1200.

3.4.4 Finery process

An alternative method of decarburising pig iron was the finery process, which seems to have been devised in the region around Namur in the 15th century. By the end of that century, this Walloon process spread to the *Pay de Bray* on the eastern boundary of Normandy, and then to England, where it became the main method of making wrought iron by 1600. It was introduced to Sweden by Louis de Geer in the early 17th century and was used to make the oregrounds iron favoured by English steelmakers.

A variation on this was the German process. This became the main method of producing bar iron in Sweden.

3.4.5 Cementation steel

In the early 17th century, ironworkers in Western Europe had developed the cementation process for carburiz-

ing wrought iron. Wrought iron bars and charcoal were packed into stone boxes, then held at a red heat for up to a week. During this time, carbon diffused into the iron, producing a product called *cement steel* or *blister steel*. One of the earliest places where this was used in England was at Coalbrookdale, where Sir Basil Brooke had two cementation furnaces (recently excavated). For a time in the 1610s, he owned a patent on the process, but had to surrender this in 1619. He probably used Forest of Dean iron as his raw material, but it was soon found that oregrounds iron was more suitable. The quality of the steel could be improved by faggoting, producing the so-called shear steel.

3.4.6 Crucible steel

In the 1740s, Benjamin Huntsman found a means of melting blister steel, made by the cementation process, in crucibles. The resulting crucible steel, usually cast in ingots, was more homogeneous than blister steel.

3.5 Transition to coke in England

3.5.1 Beginnings

Early iron smelting used charcoal as both the heat source and the reducing agent. By the 18th century, the availability of wood for making charcoal was limiting the expansion of iron production, so that England became increasingly dependent for a considerable part of the iron required by its industry, on Sweden (from the mid-17th century) and then from about 1725 also on Russia.

Smelting with coal (or its derivative coke) was a long sought objective. The production of pig iron with coke was probably achieved by Dud Dudley in the 1620s, and with a mixed fuel made from coal and wood again in the 1670s. However this was probably only a technological rather than a commercial success. Shadrach Fox may have smelted iron with coke at Coalbrookdale in Shropshire in the 1690s, but only to make cannonballs and other cast iron products such as shells. However, in the peace after the Nine Years War, there was no demand for these.[65][66]

3.5.2 Abraham Darby and his successors

Main article: Abraham Darby I

In 1707, Abraham Darby patented a method of making cast iron pots. His pots were thinner and hence cheaper than those of his rivals. Needing a larger supply of pig iron he leased the blast furnace at Coalbrookdale in 1709. There, he made iron using coke, thus establishing the first successful business in Europe to do so. His products were all of cast iron, though his immediate successors attempted (with little commercial success) to fine this to bar iron.[67]

Bar iron thus continued normally to be made with charcoal pig iron until the mid-1750s. In 1755 Abraham Darby II (with partners) opened a new coke-using furnace at Horsehay in Shropshire, and this was followed by others. These supplied coke pig iron to finery forges of the traditional kind for the production of bar iron. The reason for the delay remains controversial.[68]

3.5.3 New forge processes

Schematic drawing of a puddling furnace

It was only after this that economically viable means of converting pig iron to bar iron began to be devised. A process known as potting and stamping was devised in the 1760s and improved in the 1770s, and seems to have been widely adopted in the West Midlands from about 1785. However, this was largely replaced by Henry Cort's puddling process, patented in 1784, but probably only made to work with grey pig iron in about 1790. These processes permitted the great expansion in the production of iron that constitutes the Industrial Revolution for the iron industry.[69]

In the early 19th century, Hall discovered that the addition of iron oxide to the charge of the puddling furnace caused a violent reaction, in which the pig iron was decarburised, this became known as 'wet puddling'. It was also found possible to produce steel by stopping the puddling process before decarburisation was complete.

3.6 Hot blast

Main article: Hot blast

The efficiency of the blast furnace was improved by the change to hot blast, patented by James Beaumont Neilson in Scotland in 1828. This further reduced production costs. Within a few decades, the practice was to have a 'stove' as large as the furnace next to it into which the waste gas (containing CO) from the furnace was directed and burnt. The resultant heat was used to preheat the air blown into the furnace.[70]

3.7 Industrial steelmaking

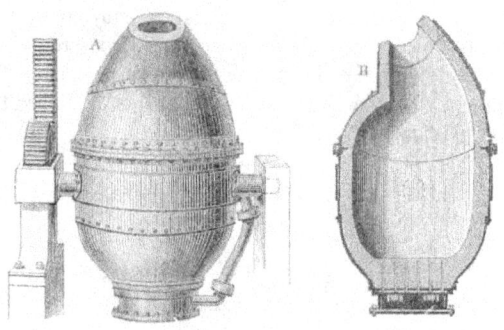

Schematic drawing of a Bessemer converter

Apart from some production of puddled steel, English steel continued to be made by the cementation process, sometimes followed by remelting to produce crucible steel. These were batch-based processes whose raw material was bar iron, particularly Swedish oregrounds iron.

The problem of mass-producing cheap steel was solved in 1855 by Henry Bessemer, with the introduction of the Bessemer converter at his steelworks in Sheffield, England. (An early converter can still be seen at the city's Kelham Island Museum). In the Bessemer process, molten pig iron from the blast furnace was charged into a large crucible, and then air was blown through the molten iron from below, igniting the dissolved carbon from the coke. As the carbon burned off, the melting point of the mixture increased, but the heat from the burning carbon provided the extra energy needed to keep the mixture molten. After the carbon content in the melt had dropped to the desired level, the air draft was cut off: a typical Bessemer converter could convert a 25-ton batch of pig iron to steel in half an hour.

Finally, the basic oxygen process was introduced at the

Voest-Alpine works in 1952; a modification of the basic Bessemer process, it lances oxygen from above the steel (instead of bubbling air from below), reducing the amount of nitrogen uptake into the steel. The basic oxygen process is used in all modern steelworks; the last Bessemer converter in the U.S. was retired in 1968. Furthermore, the last three decades have seen a massive increase in the mini-mill business, where scrap steel only is melted with an electric arc furnace. These mills only produced bar products at first, but have since expanded into flat and heavy products, once the exclusive domain of the integrated steelworks.

Until these 19th-century developments, steel was an expensive commodity and only used for a limited number of purposes where a particularly hard or flexible metal was needed, as in the cutting edges of tools and springs. The widespread availability of inexpensive steel powered the Second Industrial Revolution and modern society as we know it. Mild steel ultimately replaced wrought iron for almost all purposes, and wrought iron is no longer commercially produced. With minor exceptions, alloy steels only began to be made in the late 19th century. Stainless steel was developed on the eve of the First World War and was not widely used until the 1920s.

3.8   See also

- Damascus Steel
- Wootz steel
- History of steelmaking
- Iron Age
- Nok culture
- Non-ferrous extractive metallurgy
- Roman metallurgy

3.9   Notes

[1] Rehren, T et al. (2013). "5,000 years old Egyptian iron beads made from hammered meteoritic iron". *Journal of Archaeological Science* **40**: 4785–4792. doi:10.1016/j.jas.2013.06.002.

[2] Photos, E. (1989). "The Question of Meteoritic versus Smelted Nickel-Rich Iron: Archaeological Evidence and Experimental Results". *World Archaeology* **20** (3): 403–421. doi:10.1080/00438243.1989.9980081.

[3] Miller, Duncan E.; Der Merwe, N.J. Van (1994). "Early Metal Working in South of the Saharan Africa". *Journal of African History* **35**: 1–36. doi:10.1017/s0021853700025949.

[4] Stuiver, Minze; Der Merwe, N.J. Van (1968). "Radiocarbon Chronology of the Iron Age in South of the Saharan Africa". *Current Anthropology*.

[5] Donald B. Wagner (1993). *Iron and Steel in Ancient China*. BRILL. pp. 335–340. ISBN 978-90-04-09632-5.

[6] R. F. Tylecote, *A History of Metallurgy* (2nd edn, 1992), 3

[7] Klass R Veenhof; Jesper Eidem (2008). *Mesopotamia: The Old Assyrian Period: The Old Assyrian Period*. *Orbis Biblicus et Orientalis*. German: Vandenhoeck & Ruprecht GmbH & Co KG. p. 84. ISBN 3525534523. Retrieved 4 November 2013.

[8] Waldbaum, Jane C. *From Bronze to Iron*. Göteburg: Paul Astöms Förlag (1978): 56–58.

[9] Marco Ceccarelli (2000). *International Symposium on History of Machines and Mechanisms: Proceedings HMM Symposium*. Springer. ISBN 0-7923-6372-8. pp 218

[10] Collins, Rober O. and Burns, James M. The History of Sub-Saharan Africa. New York:Cambridge University Press, p. 37. ISBN 978-0-521-68708-9.

[11] Richard Cowen () *The Age of Iron* Chapter 5 in a series of essays on Geology, History, and People prepares for a course of the University of California at Davis. Online version accessed on 2010-02-11.

[12] *The Tomb of Tut-Ankh-Amen: Discovered by the Late Earl of Carnarvon and Howard Carter, Volume 3*

[13] Waldbaum 1978: 23.

[14] Muhly, James D. 'Metalworking/Mining in the Levant' pp. 174-183 in *Near Eastern Archaeology* ed. Suzanne Richard (2003), pp. 179-180.

[15] Muhly 2003:180.

[16] Tewari, Rakesh (2003). "The origins of iron-working in India: new evidence from the Central Ganga Plain and the Eastern Vindhyas" (PDF). *Antiquity* **77** (297): 536–544.

[17] I. M. Drakonoff (1991). *Early Antiquity*. University of Chicago Press. ISBN 0-226-14465-8. pp 372

[18] J. F. Richards et al. (2005).*The New Cambridge History of India*. Cambridge University Press. ISBN 0-521-36424-8. pp 64

[19] Patrick Olivelle (1998). *Upanisads*. Oxford University Press. ISBN 0-19-283576-9. pp xxix

[20] Will Durant (), *The Story of Civilization I: Our Oriental Heritage*

[21] G. Juleff (1996). "An ancient wind powered iron smelting technology in Sri Lanka". *Nature* **379** (3): 60–63. Bibcode:1996Natur.379...60J. doi:10.1038/379060a0.

[22] Suspension bridge. (2007). In Encyclopædia Britannica. Retrieved April 5, 2007, from Encyclopædia Britannica Online

[23] Sanderson, Katharine (2006-11-15). "Sharpest cut from nanotube sword: Carbon nanotech may have given swords of Damascus their edge". Nature. Retrieved 2006-11-17.

[24] Roy Porter (2003). *The Cambridge History of Science*. Cambridge University Press. ISBN 0-521-57199-5. pp 684

[25] Juleff, G. (1996). "An ancient wind powered iron smelting technology in Sri Lanka". *Nature* **379** (3): 60–63. Bibcode:1996Natur.379...60J. doi:10.1038/379060a0.

[26] http://www.fluent.com/about/news/newsletters/04v13i1/a27.htm

[27] Simulation of air flows through a Sri Lankan wind driven furnace, submitted to J. Arch. Sci, 2003.

[28] R. Balasubramaniam (2002), *Delhi Iron Pillar: New Insights*. Aryan Books International, Delhi ISBN 81-7305-223-9.

[29] Riederer, Josef; Wartke, Ralf-B.: "Iron", Cancik, Hubert; Schneider, Helmuth (eds.): Brill's New Pauly, Brill 2009

[30] Craddock, Paul T. (2008): "Mining and Metallurgy", in: Oleson, John Peter (ed.): *The Oxford Handbook of Engineering and Technology in the Classical World*, Oxford University Press, ISBN 978-0-19-518731-1, p. 108

[31] Wagner, Donald B.: "The State and the Iron Industry in Han China", NIAS Publishing, Copenhagen 2001, ISBN 87-87062-77-1, p. 73

[32] Pigott, Vincent C. (1999). *The Archaeometallurgy of the Asian Old World*. Philadelphia: University of Pennsylvania Museum of Archaeology and Anthropology. ISBN 0-924171-34-0, p. 8.

[33] Giannichedda, Enrico (2007): "Metal production in Late Antiquity", in *Technology in Transition AD 300-650* L. Lavan E.Zanini & A. Sarantis Brill, eds., Leiden; p. 200

[34] Needham, Volume 4, Part 3, 197.

[35] Needham, Volume 4, Part 3, 277.

[36] Needham, Volume 4, Part 3, 563 g

[37] Needham, Volume 4, Part 3, 86.

[38] Needham, Volume 4, Part 1, 282.

[39] Needham, Volume 4, Part 2, 370

[40] Needham, Volume 4, Part 2, 371.

[41] Hartwell, Robert (1966). "Markets, Technology and the Structure of Enterprise in the Development of the Eleventh Century Chinese Iron and Steel Industry". *Journal of Economic History* **26**: 53–54.

[42] Ebrey, 158.

[43] Smith, A. H. V. (1997). "Provenance of Coals from Roman Sites in England and Wales". *Britannia* **28**: 297–324 (322–324). doi:10.2307/526770.

[44] Iron in Africa: Revisiting the History – Unesco (2002)

[45] Miller, Duncan E.; Der Merwe, N.J. Van (1994). "Early Metal Working in Sub Saharan Africa". *Journal of African History* **35**: 1–36. doi:10.1017/s0021853700025949.

[46] Stuiver, Minze; Der Merwe, N.J. Van (1968). "Radiocarbon Chronology of the Iron Age in Sub-Saharan Africa". *Current Anthropology*.

[47] Ehret, Christopher (2002). The Civilizations of Africa. Charlottesville: University of Virginia, pp. 136, 137 ISBN 0-8139-2085-X.

[48] Miller, Duncan E.; Der Merwe, N.J. Van (1994). "Early Metal Working in South of the Saharan Africa'". *Journal of African History* **35**: 1–36. doi:10.1017/s0021853700025949.

[49] Stuiver, Minze; Der Merwe, N.J. Van (1968). "Radiocarbon Chronology of the Iron Age in South of the Saharan Africa'". *Current Anthropology*.

[50] Tylecote 1975 (see below)

[51] A History of Sub-Saharan Africa

[52] The Nubian Past

[53] "Africa's Ancient Steelmakers". *Time magazine*. 1978-09-25. Retrieved 2007-09-21.

[54] Adam Robert Lucas (2005), "Industrial Milling in the Ancient and Medieval Worlds: A Survey of the Evidence for an Industrial Revolution in Medieval Europe", *Technology and Culture* **46** (1): 1-30 [10-1 & 27]

[55] R. L. Miller (October 1988). "Ahmad Y. Al-Hassan and Donald R. Hill, *Islamic technology: an illustrated history*". *Medical History* **32** (4): 466–7. doi:10.1017/s0025727300048602.

[56] Donald Routledge Hill (1996), "Engineering", p. 781, in (Rashed & Morelon 1996, pp. 751–95)

[57] Donald Routledge Hill, "Mechanical Engineering in the Medieval Near East", *Scientific American*, May 1991, p. 64-69. (cf. Donald Routledge Hill, Mechanical Engineering)

[58] Adam Lucas (2006), *Wind, Water, Work: Ancient and Medieval Milling Technology*, p. 65. BRILL, ISBN 90-04-14649-0.

[59] Kochmann, W.; Reibold M., Goldberg R., Hauffe W., Levin A. A., Meyer D. C., Stephan T., Müller H., Belger A., Paufler P. (2004). "Nanowires in ancient Damascus steel". *Journal of Alloys and Compounds* **372**: L15–L19. doi:10.1016/j.jallcom.2003.10.005. ISSN 0925-8388.
Levin, A. A.; Meyer D. C., Reibold M., Kochmann W., Pätzke N., Paufler P. (2005). "Microstructure of a genuine

Damascus sabre" (PDF). *Crystal Research and Technology* **40** (9): 905–916. doi:10.1002/crat.200410456.

[60] Reibold, M.; Levin A. A., Kochmann W., Pätzke N., Meyer D. C. (16 November 2006). "Materials: Carbon nanotubes in an ancient Damascus sabre". *Nature* **444** (7117): 286. Bibcode:2006Natur.444..286R. doi:10.1038/444286a. PMID 17108950.

[61] Legendary Swords' Sharpness, Strength From Nanotubes, Study Says

[62] Sanderson, Katharine (2006-11-15). "Sharpest cut from nanotube sword: Carbon nanotech may have given swords of Damascus their edge". Nature (journal). Retrieved 2006-11-17.

[63] Lucas, A. R. (2005). "Industrial milling in the ancient and Medieval Worlds". *Technology and Culture* **46**: 19.

[64] R. F. Tylecote, *A History of Metallurgy*, 76.

[65] King, P. W. (2002). "Dud Dudley's contribution to metallurgy". *Historical Metallurgy* **36** (1): 43–53.

[66] King, P. W. (2001). "Sir Clement Clerke and the adoption of coal in metallurgy". *Trans. Newcomen Soc.* **73** (1): 33–52. doi:10.1179/tns.2001.002.

[67] A. Raistrick, *A dynasty of Ironfounders* (1953; 1989); N. Cox, 'Imagination and innovation of an industrial pioneer: The first Abraham Darby' *Industrial Archaeology Review* 12(2) (1990), 127-144.

[68] A. Raistrick, *Dynasty*; C. K. Hyde, *Technological change and the British iron industry 1700–1870* (Princeton, 1977), 37-41; P. W. King, 'The Iron Trade in England and Wales 1500–1815' (Ph.D. thesis, Wolverhampton University, 2003), 128-41.

[69] G. R. Morton and N. Mutton, 'The transition to Cort's puddling process' *Journal of Iron and Steel Institute* 205(7) (1967), 722-8; R. A. Mott (ed. P. Singer), *Henry Cort: The great finer: creator of puddled iron* (1983); P. W. King, 'Iron Trade', 185-93.

[70] A. Birch, *Economic History of the British Iron and Steel Industry* , 181-9; C. K. Hyde, *Technological Change and the British iron industry* (Princeton 1977), 146-59.

3.10 References

- Ebrey, Walthall, Palais, (2006). *East Asia: A Cultural, Social, and Political History*. Boston: Houghton Mifflin Company.

- Knowles, Anne Kelly. (2013) *Mastering Iron: The Struggle to Modernize an American Industry, 1800–1868* (University of Chicago Press) 334 pages

- Needham, Joseph (1986). *Science and Civilization in China: Volume 4, Part 2*; Needham, Joseph (1986). *Science and Civilization in China: Volume 4, Part 3*.

- Pleiner, R. (2000) *Iron in Archaeology. The European Bloomery Smelters*, Praha, Archeologický Ústav Av Cr.

- Wagner, Donald (1996). *Iron and Steel in Ancient China*. Leiden: E. J. Brill.

- Woods, Michael and Mary B. Woods (2000). *Ancient Machines: From Wedges to Waterwheels. Minneapolis*: Twenty-First Century Books.

Chapter 4

History of metallurgy in South Asia

Coin of Samudragupta (c. 350—375) with Garuda pillar. British Museum.

The iron pillar of Delhi (375—413).

The **history of metallurgy in South Asia** began prior to the 3rd millennium BCE and continued well into the British Raj.[1] Metals and related concepts were mentioned in various early Vedic age texts. The Rigveda already uses the Sanskrit term **Ayas** (metal). The Indian cultural and commercial contacts with the Near East and the Greco-Roman world enabled an exchange of metallurgic sciences.[2] With the advent of the Mughals, India's Mughal Empire (established: April 21, 1526—ended: September 21, 1857) further improved the established tradition of metallurgy and metal working in India.[3]

The imperial policies of the British Raj led to stagnation of metallurgy in India as the British regulated mining and metallurgy—used in India previously by its rulers to build armies and resist England during various wars.[4]

4.1   Overview

Recent excavations in Middle Ganga Valley done by archaeologist Rakesh Tewari show iron working in India begun as early as 2800 BCE.[5] Archaeological sites in India, such as Malhar, Dadupur, Raja Nala Ka Tila and Lahuradewa in the state of Uttar Pradesh show iron implements in the period between 1800 BCE - 1200 BCE. Sahi (1979: 366) concluded that by the early 13th century BCE, iron smelting was definitely practiced on a bigger scale in India, suggesting that the date the technology's inception may well be placed as early as the 26th century BCE.[6]

Bronze Chola Statue of Nataraja *at the Metropolitan Museum of Art, New York City.*

The Black and Red Ware culture was another early Iron Age archaeological culture of the northern Indian subcontinent. It is dated to roughly the 12th – 9th centuries BCE, and associated with the Rigvedic Vedic civilization. It extended from the upper Gangetic plain in Uttar Pradesh to the eastern Vindhya range and West Bengal.

Perhaps as early as 300 BCE, although certainly by 200 CE, high quality steel was being produced in southern India by what Europeans would later call the crucible technique. In this system, high-purity wrought iron, charcoal, and glass were mixed in crucibles and heated until the iron melted and absorbed the carbon. The resulting high-carbon steel, called *fūlāḏ* ???? in Arabic and *wootz* by later Europeans, was exported throughout much of Asia and Europe.

Will Durant wrote in *The Story of Civilization I: Our Oriental Heritage*:

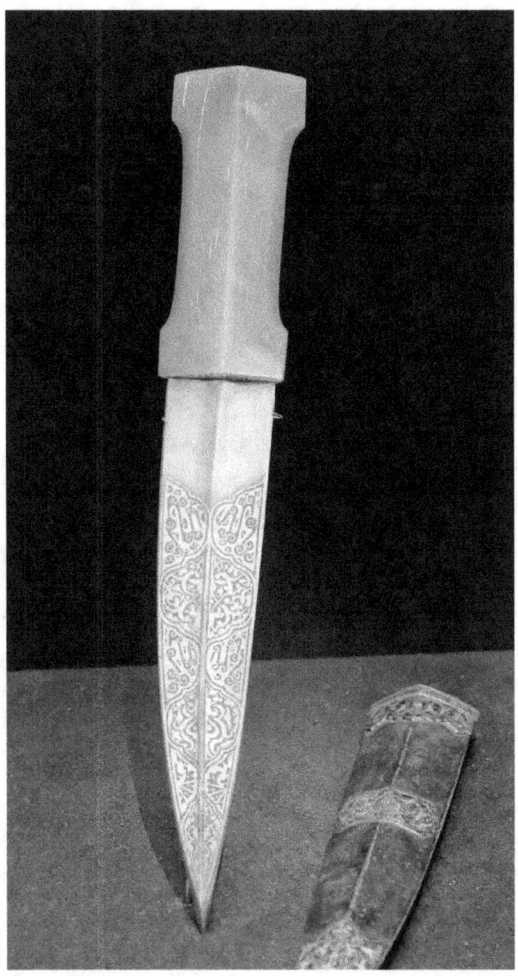

Dagger and its scabbard, India, 17th—18th century. Blade: Damascus steel inlaid with gold; hilt: jade; scabbard: steel with engraved, chased and gilded decoration.

> "Something has been said about the chemical excellence of cast iron in ancient India, and about the high industrial development of the Gupta times, when India was looked to, even by Imperial Rome, as the most skilled of the nations in such chemical industries as dyeing, tanning, soap-making, glass and cement... By the sixth century the Hindus were far ahead of Europe in industrial chemistry; they were masters of calcinations, distillation, sublimation, steaming, fixation, the production of light without heat, the mixing of anesthetic and soporific powders, and the preparation of metallic salts, compounds and alloys. The tempering of steel was brought in ancient India to a perfection unknown in Europe till our own times; King Porus is said to have selected, as a specially valuable gift for Alexander, not gold or silver, but thirty pounds of steel. The Moslems took much of this Hindu chemical science and industry to the Near East and Europe; the secret of manufacturing "Damascus" blades, for example, was taken by the Arabs from the Persians, and by the Persians from India."

4.2 Hindu, Buddhist, Jain and other texts

The Sanskrit term Ayas means metal and can refer to bronze, copper or iron.

Akbarnama—written in August 12, 1602—depicts the defeat of Baz Bahadur of Malwa by the Mughal troops in 1561. The Mughals extensively improved metal weapons and armor used by the armies of India.

4.2.1   Rigveda

The Rig Veda refers to ayas, and also states that the Dasyus had Ayas (RV 2.20.8). In RV 4.2.17, "the gods [are] smelting like copper/metal ore the human generations".

The references to Ayas in the Rig Veda probably refer to bronze or copper rather than to iron.[7] However, D. K. Chakrabarti (1992) argued: "It should be clear that any controversy regarding the meaning of ayas in the Rgveda or the problem of the Rgvedic familiarity or unfamiliarity with iron is pointless. There is no positive evidence either way. It can mean both copper-bronze and iron and, strictly on the basis of the contexts, there is no reason to choose between the two."

Hyder Ali (c. 1722-1782)—the ruler of the Kingdom of Mysore until 1782—developed military rockets using metal cylinders to contain the combustion powder.

4.2.2   Arthasastra

The Arthasastra lays down the role of the Director of Metals, the Director of Forest Produce and the Director of Mining.[8] It is the duty of the Director of Metals to establish factories for different metals. The Director of Mines is responsible for the inspection of mines. The Arthasastra also refers to counterfeit coins.[8]

4.2.3   Other texts

There are many references to Ayas in the early Indian texts.[9]

The Atharva Veda and the Satapatha Brahmana refer to *krsna ayas* ("black metal"), which could be iron (but possibly also iron ore and iron items not made of smelted iron). There is also some controversy if the term syamayas ("black metal) refers to iron or not. In later texts the term refers to iron. In earlier texts, it could possibly also refer to darker-than-copper bronze, an alloy of copper and tin.[10][11] Copper can also become black by heating it.[12] Oxidation with the use of sulphides can produce the same effect.[12][13]

The Yajurveda seems to know iron.[8] In the Taittiriya Samhita are references to ayas and at least one reference to smiths.[8] The Satapatha Brahmana 6.1.3.5 refers to the smelting of metallic ore.[14] In the Manu Smriti (6.71), the following analogy is found: "For as the impurities of metallic ores, melted in the blast (of a furnace), are consumed, even so the taints of the organs are destroyed through the suppression of the breath." Metal was also used in agriculture, and the Buddhist text Suttanipata has the following analogy: "for as a ploughshare that has got hot during the day when thrown into the water splashes, hisses and smokes in volumes..."[8]

In the Charaka Samhita an analogy occurs that probably refers to the lost wax technique.[14] The Silpasastras (the Manasara, the Manasollasa (Abhilashitartha-Chintamani) and the Uttarabhaga of Silparatna) describe the lost wax technique in detail.[14]

The Silappadikaram says that copper-smiths were in Puhar and in Madura.[14] According to the History of the Han Dynasty by Ban Gu, Kashmir and "Tien-chu" were rich in metals.[14]

An influential Indian metallurgist and alchemist was Nagarjuna (born 931). He wrote the treatise *Rasaratnakara* that deals with preparations of *rasa* (mercury) compounds. It gives a survey of the status of metallurgy and alchemy in the land. Extraction of metals such as silver, gold, tin and copper from their ores and their purification were also mentioned in the treatise. The Rasa Ratnasamuccaya describes the extraction and use of copper.[15]

4.3 Archaeology

Chakrabarti (1976) has identified six early iron-using centres in India: Baluchistan, the Northwest, the Indo-Gangetic divide and the upper Gangetic valley, eastern India, Malwa and Berar in central India and the megalithic south India.[8] The central Indian region seems to be the earliest iron-using centre.[16]

According to Tewari, iron using and iron "was prevalent in the Central Ganga Plain and the Eastern Vindhyas from the early 2nd millennium BC."[17]

The earliest evidence for smelted iron in India dates to 2300 to 1000 BCE.[18] These early findings also occur in places like the Deccan and the earliest evidence for smelted iron occurs in Central India, not in north-western India.[19] Moreover, the dates for iron in India are earlier than in those of Central Asia, and according to some scholars (e.g. Koshelenko 1986) the dates for smelted iron may actually be much earlier in India than in Central Asia and Iran.[20]

The Iron Age did however not necessary imply a major social transformation, and Gregory Possehl wrote that "the Iron Age is more of a continuation of the past then a break with it".[21]

Archaeological data suggests that India was "an independent and early centre of iron technology."[22] According to Shaffer, the "nature and context of the iron objects involved [of the BRW culture] are very different from early iron objects found in Southwest Asia."[23] In Central Asia, the development of iron technology was not necessarily connected with Indo-Iranian migrations either.[24]

J.M. Kenoyer (1995) also remarks that there is a "long break in tin acquisition" necessary for the production of "tin bronzes" in the Indus Valley region, suggesting a lack of contact with Baluchistan and northern Afghanistan, or the lack of migrants from the north-west who could have procured tin.

4.3.1 Indus Valley Civilization

The copper-bronze metallurgy in the Harappan civilization was widespread and had a high variety and quality.[25] The early use of iron may have developed from the practice of copper-smelting.[26] Iron ore and iron items have been unearthed in eight Indus Valley sites, some of them dating to before 2600 BCE.[27] There remains the possibility that some of these items were made of smelted iron, and the term "krsna ayas" might possibly also refer to these iron items, even if they are not made of smelted iron.

Lothali copper is unusually pure, lacking the arsenic typically used by coppersmiths across the rest of the Indus valley. Workers mixed tin with copper for the manufacture of celts, arrowheads, fishhooks, chisels, bangles, rings, drills and spearheads, although weapon manufacturing was minor. They also employed advanced metallurgy in following the *cire perdue* technique of casting, and used more than one-piece moulds for casting birds and animals.[28] They also invented new tools such as curved saws and twisted drills unknown to other civilizations at the time.[29]

4.4 Metals

4.4.1 Brass

Brass was used in Lothal and Atranjikhera in the 3rd and 2nd millennium BCE.[30] Brass and probably zinc was also found at Taxila in 4th to 3rd century BCE contexts.[31]

4.4.2 Copper

Copper technology may date back to the 4th millennium BCE in the Himalaya region.[15] It is the first element to be discovered in metallurgy, Copper and its alloys were also used to create copper-bronze images such as Buddhas or Hindu/Mahayana Buddhist deities.[14] Xuanzang also noted that there were copper-bronze Buddha images in Magadha.[14] In Varanasi, each stage of the image manufacturing process is handled by a specialist.[32]

Other metal objects made by Indian artisans include lamps.[33] Copper was also a component in the razors for the tonsure ceremony.[14]

One of the most important sources of history in South Asia are the royal records of grants engraved on copper-plate grants (tamra-shasan or tamra-patra). Because copper does not rust or decay, they can survive indefinitely. Collections of archaeological texts from the copper-plates and rock-inscriptions have been compiled and published by the Archaeological Survey of India during the past century. The earliest known copper-plate known as the Sohgaura copper-plate is a Maurya record that mentions famine relief efforts. It is one of the very few pre-Ashoka Brahmi inscriptions in India.

4.4.3 Gold and silver

The deepest gold mines of the Ancient world were found in the Maski region in Karnataka.[34] There were ancient silver mines in northwest India. Dated to the middle of the 1st millennium BCE. gold and silver were also used for making utensils for the royal family and nobilities.the royal family wore costly fabrics so it may be assumed that gold and silver were beaten into thin fibres and embroidered or woven into fabrics or dress.

4.4.4 Iron

See also: Iron pillar of Delhi

 Recent excavations in Middle Ganges Valley show iron working in India may have begun as early as 1800 BCE.[35] In the 5th century BCE, the Greek historian Herodotus observed that "Indian and the Persian army used arrows tipped with iron."[36] Ancient Romans used armour and cutlery made of Indian iron. Pliny the Elder also mentioned Indian iron.[36] Muhammad al-Idrisi wrote the Hindus excelled in the manufacture of iron, and that it would be impossible to find anything to surpass the edge from Hindwani steel.[37] Quintus Curtius wrote about an Indian present of steel to Alexander.[38] *Ferrum indicum* appeared in the list of articles subject to duty under Marcus Aurelius and Commodus.[8] Indian Wootz steel was held in high regard

The iron pillar of Delhi

in Europe, and Indian iron was often considered to be the best.[39]

Wootz and steel

Main articles: Wootz steel and Damascus steel

The first form of crucible steel was wootz, developed in India some time around 300 BCE. In its production the iron was mixed with glass and then slowly heated and then cooled. As the mixture cooled the glass would bond to impurities in the steel and then float to the surface, leaving the steel considerably more pure. Carbon could enter the iron by diffusing in through the porous walls of the crucibles. Carbon dioxide would not react with the iron, but the small amounts of carbon monoxide could, adding carbon to the mix with some level of control. Wootz was widely exported throughout the Middle East, where it was combined with a local production technique around 1000 CE to produce Damascus steel, famed throughout the world.[40] Wootz derives from the Kannada term for steel *ukku*.[41] Indian wootz steel was the first high quality steel that was produced.

Henry Yule quoted the 12th-century Arab Edrizi who wrote: "The Hindus excel in the manufacture of iron, and in the preparations of those ingredients along with which it is fused to obtain that kind of soft iron which is usually styled Indian steel (Hindiah). They also have workshops wherein are forged the most famous sabres in the world. ...It is not possible to find anything to surpass the edge that you get from Indian steel (al-hadid al-Hindi).[36]

As early as the 17th century, Europeans knew of India's ability to make crucible steel from reports brought back by travelers who had observed the process at several places in southern India. Several attempts were made to import the process, but failed because the exact technique remained a mystery. Studies of wootz were made in an attempt to understand its secrets, including a major effort by the famous scientist, Michael Faraday, son of a blacksmith. Working with a local cutlery manufacturer he wrongly concluded that it was the addition of aluminium oxide and silica from the glass that gave wootz its unique properties.

After the Indian rebellion of 1857, many Indian wootz steel swords were destroyed by order of the British authorities.[36] Metal working suffered a decline during the British Empire, but steel production was revived in India by Jamsetji Tata.

4.4.5 Zinc

Zinc was extracted in India as early as 10th century BCE and finds detail process in 5th century BCE Arthasastra . Zinc production may have begun in India, and ancient northwestern India is the earliest known civilization that produced zinc on an industrial scale.[42] The distillation technique was developed around 1200 CE at Zawar in Rajasthan.[30]

In the 17th century, China exported Zinc to Europe under the name of totamu or tutenag. The term tutenag may derive from the South Indian term *Tutthanagaa* (zinc).[43] In 1597, Libavius, a metallurgist in England received some quantity of Zinc metal and named it as Indian/Malabar lead.[44] In 1738, William Champion is credited with patenting in Britain a process to extract zinc from calamine in a smelter, a technology that bore a strong resemblance to and was probably inspired by the process used in the Zawar zinc mines in Rajasthan.[36] His first patent was rejected by the patent court on grounds of plagiarising the technology common in India. However, he was granted the patent on his second submission of patent approval. Postlewayt's Universal Dictionary of 1751 still wasn't aware of how Zinc was produced.[31]

The Arthasastra describes the production of zinc.[45] The Rasaratnakara by Nagarjuna describes the production of brass and zinc.[46] There are references of medicinal uses of zinc in the Charaka Samhita (300 BCE). The Rasaratna Samuchaya (800 CE) explains the existence of two types of

ores for zinc metal, one of which is ideal for metal extraction while the other is used for medicinal purpose.[47] It also describes two methods of zinc distillation.[31]

4.5 Early History (—200 BCE)

Recent excavations in Middle Ganges Valley conducted by archaeologist Rakesh Tewari show iron working in India may have begun as early as 1800 BCE.[35] Archaeological sites in India, such as Malhar, Dadupur, Raja Nala Ka Tila and Lahuradewa in the state of Uttar Pradesh show iron implements in the period between 1800 BCE-1200 BCE.[35] Sahi (1979: 366) concluded that by the early 13th century BCE, iron smelting was definitely practiced on a bigger scale in India, suggesting that the date the technology's early period may well be placed as early as the 16th century BCE.[35]

Some of the early iron objects found in India are dated to 1400 BCE by employing the method of radio carbon dating.[48] Spikes, knives, daggers, arrow-heads, bowls, spoons, saucepans, axes, chisels, tongs, door fittings etc. ranging from 600 BCE—200 BCE have been discovered from several archaeological sites.[48] In Southern India (present day Mysore) iron appeared as early as the 12th or 11th century BCE.[49] These developments were too early for any significant close contact with the northwest of the country.[49]

The earliest available Bronze age swords of copper discovered from the Harappan sites in Pakistan date back to 2300 BCE.[50] Swords have been recovered in archaeological findings throughout the Ganges-Jamuna Doab region of India, consisting of bronze but more commonly copper.[50] Diverse specimens have been discovered in Fatehgarh, where there are several varieties of hilt.[50] These swords have been variously dated to periods between 1700-1400 BCE, but were probably used more extensively during the opening centuries of the 1st millennium BCE.[50]

The beginning of the 1st millennium BCE saw extensive developments in iron metallurgy in India.[49] Technological advancement and mastery of iron metallurgy was achieved during this period of peaceful settlements.[49] The years between 322—185 BCE saw several advancements being made to the technology involved in metallurgy during the politically stable Maurya period (322—185 BCE).[51] Greek historian Herodotus (431—425 BCE) wrote the first western account of the use of iron in India.[48]

Perhaps as early as 300 BCE—although certainly by 200 CE—high quality steel was being produced in southern India by what Europeans would later call the crucible technique.[52] In this system, high-purity wrought iron, charcoal, and glass were mixed in a crucible and heated until the iron melted and absorbed the carbon.[52] The first crucible steel was the wootz steel that originated in India before the beginning of the common era.[53] Wootz steel was widely exported and traded throughout ancient Europe, China, the Arab world, and became particularly famous in the Middle East, where it became known as Damascus steel. Archaeological evidence suggests that this manufacturing process was already in existence in South India well before the Christian era.[54][55]

Zinc mines of Zawar, near Udaipur, Rajasthan, were active during 400 BCE.[56] There are references of medicinal uses of zinc in the Charaka Samhita (300 BCE).[56] The Rasaratna Samuccaya (800 CE) explains the existence of two types of ores for zinc metal, one of which is ideal for metal extraction while the other is used for medicinal purpose.[56] The *Periplus Maris Erythraei* mentions weapons of Indian iron and steel being exported from India to Greece.[57]

4.6 Early Common Era—Early Modern Era

The world's first iron pillar was the Iron pillar of Delhi—erected at the times of Chandragupta II Vikramaditya (375–413).[58] The swords manufactured in Indian workshops find written mention in the works of Muhammad al-Idrisi (flourished 1154).[59] Indian Blades made of Damascus steel found their way into Persia.[57] European scholars—during the 14th century—studied Indian casting and metallurgy technology.[60]

Indian metallurgy under the Mughal emperor Akbar (reign: 1556-1605) produced excellent small firearms.[61] Gommans (2002) holds that Mughal handguns were stronger and more accurate than their European counterparts.[62]

Srivastava & Alam (2008) comment on Indian coinage of the Mughal Empire (established: April 21, 1526 - ended: September 21, 1857) during Akbar's regime:[63]

Akbar reformed Mughal currency to make it one of the best known of its time. The new regime possessed a fully functioning trimetallic (silver, copper, and gold) currency, with an open minting system in which anyone willing to pay the minting charges could bring metal or old or foreign coin to the mint and have it struck. All monetary exchanges were, however, expressed in copper coins in Akbar's time. In the 17th century, following the silver influx from the New World, silver rupee with new fractional denominations replaced the copper coin as a common medium of circulation. Akbar's aim

was to establish a uniform coinage throughout his empire; some coins of the old regime and regional kingdoms also continued.

Statues of *Nataraja* and *Vishnu* were cast during the reign of the imperial Chola dynasty (200-1279) in the 9th century.[60] The casting could involve a mixture of five metals: copper, zinc, tin, gold, and silver.[60]

Considered one of the most remarkable feats in metallurgy, the Seamless celestial globe was manufactured in Kashmir by Ali Kashmiri ibn Luqman in 998 AH (1589-90 CE) borrowing from an earlier Hindu tradition, and twenty other such globes were later produced in Lahore and Kashmir during the Mughal Empire.[64] Before they were rediscovered in the 1980s, it was believed by modern metallurgists to be technically impossible to produce metal globes without any seams, even with modern technology.[64] These Mughal metallurgists pioneered the method of lost-wax casting in order to produce these globes.[64]

4.7 Colonial British Era—Republic of India

In the *The New Cambridge History of India: Science, Technology and Medicine in Colonial India*, scholar David Arnold examines the effect of the British Raj in Indian mining and metallurgy:[4]

With the partial exception of coal, foreign competition, aided by the absence of tariff barriers and lack of technological innovation, held back the development of mining and metal-working technology in India until the early 20th century. The relatively crude, labour-intensive nature of surviving mining techniques contributed to the false impression that India was poorly endowed with mineral resources or that they were inaccessible or otherwise difficult and unremunerative to work. But the fate of mining and metallurgy was affected by political as well as by economic and technological considerations.

The British were aware of the part metal-working had played in supporting indigenous powers in the past through the production of arms and ammunition, and, just as they introduced an Arms Act in 1878 to restrict Indian access to firearms, so they sought to limit India's ability to mine and work metals that might sustain it in future wars and rebellions. This was especially the case with Rajasthan, a region rich in metals. In the 1820s James Tod identified the 'mines of Mewar' as one of the means that had enabled its masters 'so long to struggle against superior power, and to raise those magnificent structures which would do honour to the most potent kingdoms of the west'. Indian skill in the difficult art of casting brass cannon had made Indian artillery a formidable adversary from the reign of Akbar to the Maratha and Sikh wars 300 years later. But by the early 19th century most of the mines in Rajasthan had been abandoned: the caste of miners was 'extinct'.

During the Company period, as military opponents were eliminated and princely states extinguished, so was the local capacity to mine and work metals steadily eroded. As late as the Rebellion of 1857, the mining of lead for ammunition at Ajmer was perceived as a threat the British would no longer countenance and the mines were closed down.

The first iron-cased and metal-cylinder rockets which the non-Indian world discovered was invented before the rule of Tipu Sultan, ruler of the South Indian Kingdom of Mysore, and his father Hyder Ali. They subsequently used them successfully in the 1780s.[65] He successfully used these iron-cased rockets against the larger forces of the British East India Company during the Anglo-Mysore Wars.[65]

4.8 Further reading

- Agarwal, D.P. 2000. Ancient Metal Technology and Archaeology of South Asia. New Delhi: Aryan Books International. ISBN 81-7305-177-1

- Biswas, Arun Kumar. 1994. Minerals and Metals in Ancient India. Vol. 1 Archaeological Evidence. New Delhi: D. K. Printworld (P) Ltd.

- Dilip K. Chakrabarti. The Early use of Iron In India. 1992. New Delhi: The Oxford University Press.

- Chakrabarti D.K. (1996a). Copper and its Alloys in Ancient India. Delhi: Munshiram Manoharlal Publishers Private Limited

- Mukherjee, M. 1978 Metalcraftsmen of India, Calcutta

- Rakesh Tewari, 2003, The origins of iron-working in India: new evidence from the Central Ganga Plain and the Eastern Vindhyas

- Srinivasan, Sharda and Srinivasa Rangnathan. 2004. India's Legendary Wootz Steel. Bangalore: Tata Steel.

- Tripathi, Vibha (Ed.). 1998. Archaeometallurgy in India. Delhi: Sharada Publishing House.

- Tripathi, Vibha. 2001. The Age of Iron in India. New Delhi: Aryan Books International.

- Allchin, F.R. (1979), *South Asian Archaeology 1975: Papers from the Third International Conference of the Association of South Asian Archaeologists in Western Europe, Held in Paris* edited by J.E.van Lohuizen-de Leeuw, Brill Academic Publishers, ISBN 90-04-05996-2.

- Arnold, David (2004), *The New Cambridge History of India: Science, Technology and Medicine in Colonial India*, Cambridge University Press, ISBN 0-521-56319-4.

- Balasubramaniam, R. (2002), *Delhi Iron Pillar: New Insights*, Indian Institute of Advanced Studies, ISBN 81-7305-223-9.

- Ceccarelli, Marco (2000), *International Symposium on History of Machines and Mechanisms: Proceedings HMM Symposium*, Springer, ISBN 0-7923-6372-8.

- Craddock, P.T. etc. (1983). "Zinc production in medieval India", *World Archaeology*, **15** (2), Industrial Archaeology.

- Drakonoff, I. M. (1991), *Early Antiquity*, University of Chicago Press, ISBN 0-226-14465-8.

- Edgerton etc. (2002), *Indian and Oriental Arms and Armour*, Courier Dover Publications, ISBN 0-486-42229-1.

- Gommans, Jos J. L. (2002), *Mughal Warfare: Indian Frontiers and Highroads to Empire, 1500-1700*, Routledge, ISBN 0-415-23989-3.

- Juleff, G. (1996), "An ancient wind powered iron smelting technology in Sri Lanka", *Nature*, **379** (3): 60–63.

- Mondal, Biswanath (2004), *Proceedings of the National Conference on Investment Casting: NCIC 2003*, Allied Publishers, ISBN 81-7764-659-1.

- Prasad, P. C. (2003), *Foreign Trade and Commerce in Ancient India*, Abhinav Publications, ISBN 81-7017-053-2.

- Richards, J. F. etc. (2005), *The New Cambridge History of India*, Cambridge University Press, ISBN 0-521-36424-8.

- Savage-Smith, Emilie (1985), *Islamicate Celestial Globes: Their History, Construction, and Use*, Smithsonian Institution Press.

- Srinivasan, S. & Ranganathan, S., *Wootz Steel: An Advanced Material of the Ancient World*, Indian Institute of Science.

- Srinivasan, S. (1994), *Wootz crucible steel: a newly discovered production site in South India*, Institute of Archaeology, University College London, **5**: 49-61.

- Srinivasan, S. and Griffiths, D., *South Indian wootz: evidence for high-carbon steel from crucibles from a newly identified site and preliminary comparisons with related finds*, Material Issues in Art and Archaeology-V, Materials Research Society Symposium Proceedings Series, Vol. 462.

- Srivastava, A.L. & Alam, Muzaffar (2008), *India*, Encyclopædia Britannica.

- Tewari, Rakesh (2003), "The origins of Iron Working in India: New evidence from the Central Ganga plain and the Eastern Vindhyas", *Antiquity*, **77**: 536-544.

- P. Yule–A. Hauptmann–M. Hughes. 1989 [1992]. The Copper Hoards of the Indian Subcontinent: Preliminaries for an Interpretation, *Jahrbuch des Römisch-Germanischen Zentralmuseums Mainz* 36, 193–275, ISSN 0076-2741 = http://archiv.ub. uni-heidelberg.de/savifadok/volltexte/2009/509/

4.9   See also

- Rasayana

- Damascus steel

- Wootz steel

- Crucible steel

- Iron pillar of Delhi

- Copper-plate grant

- Iron Age India, Iron Age

- Indian coinage

- Science and technology in ancient India

4.10   References

[1] See *Tewari (2003)* and *Arnold, 100-101*.

[2] For Near East see *Edgerton, 56* and *Prasad, chapter IX*. Greco-Roman world: *Mondal, 2-3*.

[3] Gommans (2002)

[4] Arnold, 100-101

[5] e.g. R. Tewari 2003

[6] Origins of Iron Ore

[7] (e.g. Frawley 1991)

[8] Chakrabarti 1992

[9] A review of literary references to Ayas in the early Indian texts can be found in Chakrabarti 1996 and Chakrabarti 1992.

[10] (Sethna 1992: 235)

[11] Agarwal, Vishal (2003), "A Reply to Michael Witzel's 'Ein Fremdling im Rgveda'" (PDF), *Journal of Indo-European Studies* **31** (1-2): 107–185

[12] Kazanas, Nicholas: Addendum to The AIT and Scholarship

[13] In AV 11.3.7. Lohita (red copper) is compared with blood, and syama (swarthy metal) with flesh (maam-sa). This could be an analogy that describes how black metal (flesh) is produced by red metal (blood). Kazanas, Nicholas: Addendum to The AIT and Scholarship

[14] Chakrabarti 1996

[15] Copper Technology in the Central Himalayas Goes Back to 2000BC

[16] e.g., Cf. Chakrabarti 1992; Erdosy 1995

[17] Rakesh Tewari 2003

[18] (see Bryant 2001: 246-248)

[19] (Bryant 2001: 246)

[20] (see Bryant 2001: 247)

[21] cited in Bryant 2001

[22] Rakesh Tewari 2003; Chakrabarti 1976, 1992:171; Tripathi, Vibha. 2001; Erdosy 1995

[23] Shaffer 1989, cited in Chakrabarti 1992:171

[24] H. P. Francfort, Fouilles de Shortugai, Recherches sur L'Asie Centrale Protohistorique Paris: Diffusion de Boccard, 1989, p. 450

[25] Jim Shaffer 1992 "The Indus Valley, Baluchistan and Helmand Traditions: Neolithic Through Bronze Age." In Chronologies in Old World Archaeology. Second Edition. R.W. Ehrich, (Ed.). Chicago: University of Chicago Press. I:441-464, II:425-446., cited in Possehl 1992

[26] Gregory Possehl, The Indus Civilization, 2002:94

[27] (see Bryant 2001: 246-248, 339)

[28] S. R. Rao, *Lothal* (ASI, 1985), pp. 42

[29] S. R. Rao, *Lothal* (ASI, 1985), pp. 41-42

[30] The Bill of Contentions

[31] Craddock *et al.* 1983

[32] Chakrabarti 1996, with reference to Mukherjee, M. 1978

[33] http://www.chennaionline.com/artscene/craftpalace/history/lamps.asp

[34] They date to the middle of the 1st millennium BCE. Srinivasan, Sharda and Srinivasa Rangnathan. 2004

[35] Tewari (2003)

[36] Srinivasan, Sharda and Srinivasa Rangnathan. 2004

[37] Srinivasan, Sharda and Srinivasa Rangnathan. 2004; W. Egerton, Indian and Oriental Armour, London (1896).

[38] J.M. Heath 1839, quoted by Chakrabarti 1992; G. N. Pant, Indian Arms and Armour, Vol. I and II, National Museum, New Delhi (1980)

[39] e.g. James Stodart 1818, Robert Hadfield, quoted by Chakrabarti 1992:3-6, 119; Robert Hadfield, Sinhalese iron and steel of ancient origin, Journal of the Iron and Steel Institute, 85 (1912).

[40] C. S. Smith, A History of Metallography, University Press, Chicago (1960); Juleff 1996; Srinivasan, Sharda and Srinivasa Rangnathan 2004

[41] http://metalrg.iisc.ernet.in/~{}wootz/heritage/WOOTZ.htm

[42] Craddock *et al.* 1983. (The earliest evidence for the production of zinc comes from India. Srinivasan, Sharda and Srinivasa Rangnathan. 2004)

[43] India Was the First to Smelt Zinc by Distillation Process

[44] Arun Kumar Biswas, Zinc and related alloys

[45] TKS Book Series

[46] ; Srinivasan, Sharda and Srinivasa Rangnathan. 2004

[47]

[48] Ceccarelli, 218

[49] Drakonoff, 372

[50] Allchin, 111-114

[51] Richards etc., 64

[52] Juleff 1996

[53] Srinivasan & Ranganathan

[54] Srinivasan 1994

[55] Srinivasan & Griffiths

[56] Craddock (1983)

[57] Prasad, chapter IX

[58] Balasubramaniam, R. (2002)

[59] Edgerton, 56

[60] Mondal, 2-3

[61] Gommans, 154

[62] Gommans, 155

[63] Srivastava & Alam (2008)

[64] Savage-Smith (1985)

[65] "Hyder Ali, prince of Mysore, developed war rockets with an important change: the use of metal cylinders to contain the combustion powder. Although the hammered soft iron he used was crude, the bursting strength of the container of black powder was much higher than the earlier paper construction. Thus a greater internal pressure was possible, with a resultant greater thrust of the propulsive jet. The rocket body was lashed with leather thongs to a long bamboo stick. Range was perhaps up to three-quarters of a mile (more than a kilometre). Although individually these rockets were not accurate, dispersion error became less important when large numbers were fired rapidly in mass attacks. They were particularly effective against cavalry and were hurled into the air, after lighting, or skimmed along the hard dry ground. Hyder Ali's son, Tippu Sultan, continued to develop and expand the use of rocket weapons, reportedly increasing the number of rocket troops from 1,200 to a corps of 5,000. In battles at Seringapatam in 1792 and 1799 these rockets were used with considerable effect against the British." - Encyclopædia Britannica (2008), *rocket and missile.*

• Edwin Bryant (2001). *The Quest for the Origins of Vedic Culture: The Indo-Aryan Migration Debate.* Oxford University Press. ISBN 0-19-516947-6.

• Craddock, P.T. *et al.*, Zinc production in medieval India, World Archaeology, vol.15, no.2, Industrial Archaeology, 1983

• G. Juleff, "An ancient wind powered iron smeting technology in Sri Lanka", *Nature* **379** (3), 60-63 (January 1996)

• Erdosy, George: 1995; "The Prelude to urbanization", in The Archaeology of the Early Historic South Asia: The Emergence of cities and states. Allchin, F. R. *et al.* (eds.), Cambridge 1995.

• Frawley, David (1995). Gods, Sages and Kings. 1991.Lotus Press, Twin Lakes, Wisconsin ISBN 0-910261-37-7

- Kenoyer, J.M. (1995). Interaction Systems, Specialized crafts and Culture Change. In: Indo-Aryans of Ancient South Asia. Ed. George Erdosy.. ISBN 3110144476

- Sethna, K.D. 1992. The Problem of Aryan Origins. New Delhi: Aditya Prakashan. ISBN 81-85179-67-0

- S. R. Rao, Lothal (published by the Director General, Archaeological Survey of India, 1985)

- Shaffer, Jim. Mathura: A protohistoric Perspective in D.M. Srinivasan (ed.), Mathura, the Cultural Heritage, 1989, pp. 171–180. Delhi.

- J.D. Verhoeven, A.H. Pendray, and W.E. Dauksch. (1998). *The Key Role of Impurities in Ancient Damascus Steel Blades.* Journal of Metals. 50(9). pp. 58–64.

- Lynn Willies *et al.* 1984, Ancient Zinc and Lead Mining in Rajasthan, India. World Archaeology, Vol.16, No. 2, Mines and Quarries.

4.11 Terminology for ayas

4.11.1 Other terms

- Prastarika: metal trader

- Sulbhadhatusastra: science of metals

- panchaloha, sarva loha: the five base metals (tin, lead, iron, copper, silver)

4.12 External links

- The origins of Iron-working in India

- Copper Technology in the Central Himalayas Dates Back to 2000 BCE

- TKS Metallurgy Bibliography

- Zinc production in Ancient India

- Wootz steel: an advanced material of the ancient world

- Indian heritage in metallurgy

- Zinc and related alloys

- Smelting of Zinc by Distillation Process

- Iron In Kumaun Goes Back To First Millennium BC D.P. Agrawal and Manikant Shah

Chapter 5

Atomism

Atomism (from Greek ἄτομον, *atomon*, i.e. "uncuttable", "indivisible"[1][2][3]) is a natural philosophy that developed in several ancient traditions. The atomists theorized that nature consists of two fundamental principles: *atom* and *void*. Unlike their modern scientific namesake in atomic theory, philosophical atoms come in an infinite variety of shapes and sizes, each indestructible, immutable and surrounded by a void where they collide with the others or hook together forming a cluster. Clusters of different shapes, arrangements, and positions give rise to the various macroscopic substances in the world.[4][5]

References to the concept of atomism and its atoms are found in ancient India and ancient Greece. In India the Jain,[6][7] Ajivika and Carvaka schools of atomism may date back to the 4th century BCE.[8] The Nyaya and Vaisheshika schools later developed theories on how atoms combined into more complex objects.[9] In the West, atomism emerged in the 5th century BCE with Leucippus and Democritus.[10] Whether Indian culture influenced Greek or vice versa or whether both evolved independently is a matter of dispute.[11]

The particles of chemical matter for which chemists and other natural philosophers of the early 19th century found experimental evidence were thought to be indivisible, and therefore were given the name "atom", long used by the atomist philosophy.

However, in the 20th century, the "atoms" of the chemists were found to be composed of even smaller entities: electrons, neutrons, and protons, and further experiments showed that protons and neutrons are made of quarks. Although the connection to historical atomism is at best tenuous, elementary particles have thus become a modern analog of philosophical atoms, despite the misnomer in chemistry.

5.1 Reductionism

Philosophical atomism is a reductive argument: not only that everything is composed of atoms and void, but that nothing they compose really exists: the only things that really exist are atoms ricocheting off each other mechanistically in an otherwise empty void. Atomism stands in contrast to a substance theory wherein a prime material continuum remains qualitatively invariant under division (for example, the ratio of the four classical elements would be the same in any portion of a homogeneous material).

Indian Buddhists, such as Dharmakirti and others, also developed distinctive theories of atomism, for example, involving momentary (instantaneous) atoms, that flash in and out of existence (Kalapas).

5.2 Greek atomism

In the 5th century BC, Leucippus and his pupil Democritus proposed that all matter was composed of small indivisible particles called atoms, in order to reconcile two conflicting schools of thought on the nature of reality. On one side was Heraclitus, who believed that the nature of all existence is change. On the other side was Parmenides, who believed instead that all change is illusion.

Parmenides denied the existence of motion, change and void. He believed all existence to be a single, all-encompassing and unchanging mass (a concept known as monism), and that change and motion were mere illusions. This conclusion, as well as the reasoning that led to it, may indeed seem baffling to the modern empirical mind, but Parmenides explicitly rejected sensory experience as the path to an understanding of the universe, and instead used purely abstract reasoning. Firstly, he believed there is no such thing as void, equating it with non-being (i.e. "if the void *is*, then it is not nothing; therefore it is not the void"). This in turn meant that motion is impossible, because there is no void to move into.[12] [13] He also wrote all that *is* must be an indivisible unity, for if it were manifold, then there would have to be a void that could divide it (and he did not believe the void exists). Finally, he stated that the all encompassing Unity is unchanging, for the Unity already encompasses all that is and can be.[14]

Democritus accepted most of Parmenides' arguments, except for the idea that change is an illusion. He believed change was real, and if it was not then at least the illusion had to be explained. He thus supported the concept of void, and stated that the universe is made up of many Parmenidean entities that move around in the void.[12] The void is infinite and provides the space in which the atoms can pack or scatter differently. The different possible packings and scatterings within the void make up the shifting outlines and bulk of the objects that organisms feel, see, eat, hear, smell, and taste. While organisms may feel hot or cold, hot and cold actually have no real existence. They are simply sensations produced in organisms by the different packings and scatterings of the atoms in the void that compose the object that organisms sense as being "hot" or "cold".

The work of Democritus only survives in secondhand reports, some of which are unreliable or conflicting. Much of the best evidence of Democritus' theory of atomism is reported by Aristotle in his discussions of Democritus' and Plato's contrasting views on the types of indivisibles composing the natural world.[15]

5.2.1   Geometry and atoms

Plato (c. 427 — c. 347 BC), were he familiar with the atomism of Democritus, would have objected to its mechanistic materialism. He argued that atoms just crashing into other atoms could never produce the beauty and form of the world. In Plato's *Timaeus*, (28B – 29A) the character of Timeaus insisted that the cosmos was not eternal but was created, although its creator framed it after an eternal, unchanging model.

One part of that creation were the four simple bodies of fire, air, water, and earth. But Plato did not consider these corpuscles to be the most basic level of reality, for in his view they were made up of an unchanging level of reality, which was mathematical. These simple bodies were geometric solids, the faces of which were, in turn, made up of triangles. The square faces of the cube were each made up of four isosceles right-angled triangles and the triangular faces of the tetrahedron, octahedron, and icosahedron were each made up of six right-angled triangles.

He postulated the geometric structure of the simple bodies of the four elements as summarized in the table to the right. The cube, with its flat base and stability, was assigned to earth; the tetrahedron was assigned to fire because its penetrating points and sharp edges made it mobile. The points and edges of the octahedron and icosahedron were blunter and so these less mobile bodies were assigned to air and water. Since the simple bodies could be decomposed into triangles, and the triangles reassembled into atoms of different elements, Plato's model offered a plausible account of changes among the primary substances.[16][17]

5.2.2   The rejection of atoms

Sometime before 330 BC Aristotle asserted that the elements of fire, air, earth, and water were not made of atoms, but were continuous. Aristotle considered the existence of a void, which was required by atomic theories, to violate physical principles. Change took place not by the rearrangement of atoms to make new structures, but by transformation of matter from what it was in potential to a new actuality. A piece of wet clay, when acted upon by a potter, takes on its potential to be an actual drinking mug. Aristotle has often been criticized for rejecting atomism, but in ancient Greece the atomic theories of Democritus remained "pure speculations, incapable of being put to any experimental test. Granted that atomism was, in the long run, to prove far more fruitful than any qualitative theory of matter, in the short run the theory that Aristotle proposed must have seemed in some respects more promising".[18][19]

5.2.3   Later ancient atomism

Epicurus (341–270) studied atomism with Nausiphanes who had been a student of Democritus. Although Epicurus was certain of the existence of atoms and the void, he was less sure we could adequately explain specific natural phenomena such as earthquakes, lightning, comets, or the phases of the Moon (Lloyd 1973, 25–6). Few of Epicurus's writings survive and those that do reflect his interest in applying Democritus's theories to assist people in taking responsibility for themselves and for their own happiness— since he held there are no gods around that can help them. He understood gods' role as moral ideals.

His ideas are also represented in the works of his follower Lucretius, who wrote *On the Nature of Things*. This scientific work in poetic form illustrates several segments of Epicurean theory on how the universe came into its current stage and it shows that the phenomena we perceive are actually composite forms. The atoms and the void are eternal and in constant motion. Atomic collisions create objects, which are still composed of the same eternal atoms whose motion for a while is incorporated into the created entity. Human sensations and meteorological phenomena are also explained by Lucretius in terms of atomic motion.

5.2.4 Atomism and ethics

Some later philosophers attributed the idea that man created gods; the gods did not create man to Democritus. For example, Sextus Empiricus noted:

> Some people think that we arrived at the idea of gods from the remarkable things that happen in the world. Democritus ... says that the people of ancient times were frightened by happenings in the heavens such as thunder, lightning, ..., and thought that they were caused by gods.[20]

Three hundred years after Epicurus, Lucretius in his epic poem *On the Nature of Things* would depict him as the hero who crushed the monster Religion through educating the people in what was possible in the atoms and what was *not* possible in the atoms. However, Epicurus expressed a non-aggressive attitude characterized by his statement: "The man who best knows how to meet external threats makes into one family all the creatures he can; and those he can not, he at any rate does not treat as aliens; and where he finds even this impossible, he avoids all dealings, and, so far as is advantageous, excludes them from his life."

5.3 Indian atomism

The Indian atomistic position, like many movements in Indian Philosophy and Mathematics, starts with an argument from Linguistics. The Vedic etymologist and grammarian Yaska (c. 7th century BC) in his Nirukta, in dealing with models for how linguistic structures get to have their meanings, takes the atomistic position that words are the "primary" carrier of meaning – i.e. words have a preferred ontological status in defining meaning. This position was to be the subject of a fierce debate in the Indian tradition from the early Christian era till the 18th century, involving different philosophers from the Nyaya, Mimamsa and Buddhist schools.

In the pratishakhya text (c. 2nd century BCE), the gist of the controversy was stated cryptically in the sutra form as "saMhitA pada-prakr^tiH".[21] According to the atomist view, the words (*pada*) would be the primary elements (*prakrti*) out of which the sentence is constructed, while the holistic view considers the sentence as the primary entity, originally "given" in its context of utterance, and the words are arrived at only through analysis and abstraction.[22]

These two positions came to be called *a-kShaNDa-pakSha* (indivisibility or sentence-holism), a position developed later by Bhartrihari (c. 500 AD), vs. *kShaNDa-pakSha* (atomism), a position adopted by the Mimamsa and Nyaya schools (Note: *kShanDa* = fragmented; "a-kShanDa" = whole).

Between the 5th and 3rd centuries BC, the atom (anu or aṇor) is mentioned in the Bhagavad Gita (Chapter 8, Verse 9):

*kaviṁ purāṇam anuśāsitāram **aṇor** aṇīyāṁsam anusmared yaḥ sarvasya dhātāram acintya-rūpam āditya-varṇaṁ tamasaḥ parastāt*

One meditates on the omniscient, primordial, the controller, smaller than the atom, yet the maintainer of everything; whose form is inconceivable, resplendent like the sun and totally transcendental to material nature

The ancient *shāshvata-vāda* doctrine of eternalism, which held that elements are eternal, is also suggestive of a possible starting point for atomism (Gangopadhyaya 1981).

The atomist position had transcended language into epistemology by the time that Nyaya–Vaisesika, Buddhist and Jaina theology were developing mature philosophical positions.

Will Durant wrote in *Our Oriental Heritage*:

> "Two systems of Indian thought propound physical theories suggestively similar to those of Greece. Kanada, founder of the Vaisheshika philosophy, held that the world was composed of atoms as many in kind as the various elements. The Jains more nearly approximated to Democritus by teaching that all atoms were of the same kind, producing different effects by diverse modes of combinations. Kanada believed light and heat to be varieties of the same substance; Udayana taught that all heat comes from the sun; and Vachaspati, like Newton, interpreted light as composed of minute particles emitted by substances and striking the eye."

Indian atomism in the Middle Ages was still mostly philosophical and/or religious in intent, though it was also scientific. Because the "infallible Vedas", the oldest Hindu texts, do not mention atoms (though they do mention elements), atomism was not orthodox in many schools of Hindu philosophy, although accommodationist interpretations or assumptions of lost text justified the use of atomism for non-orthodox schools of Hindu thought. The Buddhist and Jaina schools, however, were more willing to accept the ideas of atomism.

5.3.1 Nyaya–Vaisesika school

Main articles: Nyaya and Vaisesika

The Nyaya–Vaisesika school developed one of the earliest forms of atomism; scholars date the Nyaya and Vaisesika texts from the 6th to 1st centuries BC. Like the Buddhist atomists, the Vaisesika had a pseudo-Aristotelian theory of atomism. They posited the four elemental atom types, but in Vaisesika physics atoms had 24 different possible qualities, divided between general extensive properties and specific (intensive) properties. Like the Jaina school, the Nyaya–Vaisesika atomists had elaborate theories of how atoms combine. In both Jaina and Vaisesika atomism, atoms first combine in pairs (dyads), and then group into trios of pairs (triads), which are the smallest visible units of matter.[23]

5.3.2 Buddhist school

Main article: Buddhist atomism

The Buddhist atomists had very qualitative, Aristotelian-style atomic theory. According to ancient Buddhist atomism, which probably began developing before the 4th century BC, there are four kinds of atoms, corresponding to the standard elements. Each of these elements has a specific property, such as solidity or motion, and performs a specific function in mixtures, such as providing support or causing growth. Like the Hindu Jains, the Buddhists were able to integrate a theory of atomism with their theological presuppositions. Later Indian Buddhist philosophers, such as Dharmakirti and Dignāga, considered atoms to be point-sized, durationless, and made of energy.

5.3.3 Jaina school

Further information: Jain cosmology, Dravya (Jainism) and Karma in Jainism

The most elaborate and well-preserved Indian theory of atomism comes from the philosophy of the Jaina school, dating back to at least the 6th century BC. Some of the Jain texts that refer to matter and atoms are Pancastikayasara, Kalpasutra, Tattvarthasutra and Pannavana Suttam. The Jains envisioned the world as consisting wholly of atoms, except for souls. Paramāṇus or atoms were considered as the basic building blocks of all matter. Their concept of atoms was very similar to classical atomism, differing primarily in the specific properties of atoms. Each atom, according to Jain philosophy, has one kind of taste, one smell, one color, and two kinds of touch, though it is unclear what was meant by "kind of touch". Atoms can exist in one of two states: subtle, in which case they can fit in infinitesimally small spaces, and gross, in which case they have extension and occupy a finite space. Certain characteristics of Paramāṇu correspond with that sub-atomic particles. For example Paramāṇu is characterized by continuous motion either in a straight line or in case of attractions from other Paramāṇus, it follows a curved path. This corresponds with the description of orbit of electrons across the Nucleus. Ultimate particles are also described as particles with positive (Snigdha i.e. smooth charge) and negative (Rūksa – rough) charges that provide them the binding force. Although atoms are made of the same basic substance, they can combine based on their eternal properties to produce any of six "aggregates", which seem to correspond with the Greek concept of "elements": earth, water, shadow, sense objects, karmic matter, and unfit matter. To the Jains, karma was real, but was a naturalistic, mechanistic phenomenon caused by buildups of subtle karmic matter within the soul. They also had detailed theories of how atoms could combine, react, vibrate, move, and perform other actions, all of which were thoroughly deterministic.

5.3.4 Time as explained in *Srimad-Bhagavatam*

"The material manifestation's ultimate particle, which is indivisible and not formed into a body, is called the atom (*parama-aṇuh*). It exists always as an invisible identity, even after the dissolution of all forms. The material body is but a combination of such atoms, but it is misunderstood by the common man.

Atoms are the ultimate state of the manifest universe. When they stay in their own forms without forming different bodies, they are called the unlimited oneness. There are certainly different bodies in physical forms, but the atoms themselves form the complete manifestation.

One can estimate time by measuring the movement of the atomic combination of bodies. Time is the potency of the almighty Personality of Godhead, Hari, who controls all physical movement although He is not visible in the physical world.

The division of gross time is calculated as follows: two atoms make one double atom, and three double atoms make one hexatom. This hexatom is visible in the sunshine which enters through the holes of a window screen. One can clearly see that the hexatom goes up towards the sky.

The atom is described as an invisible particle, but when six such atoms combine together, they are called a *trasareṇu*, and this is visible in the sunshine pouring through the holes of a window screen."

— Excerpted from *Srimad-Bhagavatam* (*SB* 3.11.1–5) by A. C. Bhaktivedanta Swami Prabhupada, courtesy of the Bhaktivedanta Book Trust International (prabhupadabooks.com)

5.4 Islamic atomism

See also: Early Islamic philosophy: Atomism and Alchemy and chemistry in medieval Islam

Atomistic philosophies are found very early in Islamic philosophy and was influenced by earlier Greek and to some extent Indian philosophy.[24][25] Like both the Greek and Indian versions, Islamic atomism was a charged topic that had the potential for conflict with the prevalent religious orthodoxy, but it was instead more often favoured by orthodox Islamic theologians. It was such a fertile and flexible idea that, as in Greece and India, it flourished in some leading schools of Islamic thought.

5.4.1 Asharite atomism

See also: Ash'ari

The most successful form of Islamic atomism was in the Asharite school of Islamic theology, most notably in the work of the theologian al-Ghazali (1058–1111). In Asharite atomism, atoms are the only perpetual, material things in existence, and all else in the world is "accidental" meaning something that lasts for only an instant. Nothing accidental can be the cause of anything else, except perception, as it exists for a moment. Contingent events are not subject to natural physical causes, but are the direct result of God's constant intervention, without which nothing could happen. Thus nature is completely dependent on God, which meshes with other Asharite Islamic ideas on causation, or the lack thereof (Gardet 2001). Al-Ghazali also used the theory to support his theory of occasionalism. In a sense, the Asharite theory of atomism has far more in common with Indian atomism than it does with Greek atomism.[26]

5.4.2 Averroism

See also: Averroism

Other traditions in Islam rejected the atomism of the Asharites and expounded on many Greek texts, especially those of Aristotle. An active school of philosophers in Spain, including the noted commentator Averroes (AD 1126–1198) explicitly rejected the thought of al-Ghazali and turned to an extensive evaluation of the thought of Aristotle. Averroes commented in detail on most of the works of Aristotle and his commentaries did much to guide the interpretation of Aristotle in later Jewish and Christian scholastic thought.

5.5 Medieval European speculations

While Aristotelian philosophy eclipsed the importance of the atomists in late Roman and medieval Europe, their work was still preserved and exposited through commentaries on the works of Aristotle. In the 2nd century, Galen (AD 129–216) presented extensive discussions of the Greek atomists, especially Epicurus, in his Aristotle commentaries. According to historian of atomism Joshua Gregory, there was no serious work done with atomism from the time of Galen until Gassendi and Descartes resurrected it in the 17th century; "the gap between these two 'modern naturalists' and the ancient Atomists marked "the exile of the atom" and "it is universally admitted that the Middle Ages had abandoned Atomism, and virtually lost it."

However, although the ancient atomists' works were unavailable, Scholastic thinkers still had Aristotle's critiques of atomism. In the medieval universities there were expressions of atomism. For example, in the 14th century Nicholas of Autrecourt considered that matter, space, and time were all made up of indivisible atoms, points, and instants and that all generation and corruption took place by the rearrangement of material atoms. The similarities of his ideas with those of al-Ghazali suggest that Nicholas may have been familiar with Ghazali's work, perhaps through Averroes' refutation of it (Marmara, 1973–74).

5.5.1 Scholastic *minima naturalia*

Main article: Minima naturalia

Although the atomism of Epicurus had fallen out of favor in the centuries of Scholasticism, a related Aristotelian concept, that of *minima naturalia* (natural minima) received extensive consideration. *Minima naturalia* were theorized by Aristotle as the smallest parts into which a homogeneous natural substance (e.g., flesh, bone, or wood) could be divided and still retain its essential character. Speculation on *minima naturalia* provided philosophical background for the mechanistic philosophy of early modern thinkers like Descartes, and for the alchemical works of Geber and Daniel Sennert, who in turn influenced the corpuscularian alchemist Robert Boyle, one of the founders of modern chemistry.[27][28]

Unlike the atomism of Leucippus, Democritus, and Epicurus, and also unlike the later atomic theory of John Dalton, the Aristotelian natural minimum was not conceptualized as physically indivisible--"atomic" in the contemporary sense. Instead, the concept was rooted in Aristotle's hylomorphic worldview, which held that every physical thing is a compound of matter (Greek *hyle*) and an immaterial substantial form (Greek *morphe*) that imparts its essential nature and structure. For instance, a rubber ball for a hylomorphist like Aristotle would be rubber (matter) structured by spherical shape (form).

Aristotle's intuition was that there is some smallest size beyond which matter could no longer be structured as flesh, or bone, or wood, or some other such organic substance that for Aristotle, living before the microscope, could be considered homogeneous. For instance, if flesh were divided beyond its natural minimum, what would be left might be a large amount of the element water, and smaller amounts of the other elements. But whatever water or other elements were left, they would no longer have the "nature" of flesh: in hylomorphic terms, they would no longer be matter structured by the form of flesh; instead the remaining water, e.g., would be matter structured by the form of water, not the form of flesh. This is suggestive of modern chemistry, in which, e.g., a bar of gold can be continually divided until one has a single atom of gold, but further division yields only subatomic particles (electrons, quarks, etc.) which are no longer "gold." However, the parallel is not exact: *minima naturalia* are not a direct anticipation of modern chemical and physical concepts.

A chief theme in late Roman and Scholastic commentary on this concept is reconciling *minima naturalia* with the general Aristotelian principle of infinite divisibility. Commentators like John Philoponus and Thomas Aquinas reconciled these aspects of Aristotle's thought by distinguishing between mathematical and "natural" divisibility. With few exceptions, much of the curriculum in the universities of Europe was based on such Aristotelianism for most of the Middle Ages (Kargon 1966). Scholasticism was standard science in the time of Isaac Newton, but in the 17th century, a renewed interest in Epicurean atomism and corpuscularianism as a hybrid or an alternative to Aristotelian physics had begun to mount outside the classroom.

5.6 Atomic renaissance

The main figures in the rebirth of atomism were René Descartes, Pierre Gassendi, and Robert Boyle, as well as other notable figures.

One of the first groups of atomists in England was a cadre of amateur scientists known as the Northumberland circle, led by Henry Percy (1585–1632), the 9th Earl of Northumberland. Although they published little of account, they helped to disseminate atomistic ideas among the burgeoning scientific culture of England, and may have been particularly influential to Francis Bacon, who became an atomist around 1605, though he later rejected some of the claims of atomism. Though they revived the classical form of atomism, this group was among the scientific avant-garde: the Northumberland circle contained nearly half of the confirmed Copernicans prior to 1610 (the year of Galileo's The Starry Messenger). Other influential atomists of late 16th and early 17th centuries include Giordano Bruno, Thomas Hobbes (who also changed his stance on atomism late in his career), and Thomas Hariot. A number of different atomistic theories were blossoming in France at this time, as well (Clericuzio 2000).

Galileo Galilei (1564–1642) was an advocate of atomism in his 1612, *Discourse on Floating Bodies* (Redondi 1969). In The Assayer, Galileo offered a more complete physical system based on a corpuscular theory of matter, in which all phenomena—with the exception of sound—are produced by "matter in motion". Galileo identified some basic problems with Aristotelian physics through his experiments. He utilized a theory of atomism as a partial replacement, but he was never unequivocally committed to it. For example, his experiments with falling bodies and inclined planes led him to the concepts of circular inertial motion and accelerating free-fall. The current Aristotelian theories of impetus and terrestrial motion were inadequate to explain these. While atomism did not explain the law of fall either, it was a more promising framework in which to develop an explanation because motion was conserved in ancient atomism (unlike Aristotelian physics).

René Descartes' (1596–1650) "mechanical" philosophy of corpuscularism had much in common with atomism, and is considered, in some senses, to be a different version of it. Descartes thought everything physical in the universe to be made of tiny *vortices* of matter. Like the ancient atomists, Descartes claimed that sensations, such as taste or temperature, are caused by the shape and size of tiny pieces of matter. The main difference between atomism and Descartes' concept was the existence of the void. For him, there could be no vacuum, and all matter was constantly swirling to prevent a void as corpuscles moved through other matter. Another key distinction between Descartes' view and classical atomism is the mind/body duality of Descartes, which allowed for an independent realm of existence for thought, soul, and most importantly, God. Gassendi's concept was closer to classical atomism, but with no atheistic overtone.

Pierre Gassendi (1592–1655) was a Catholic priest from France who was also an avid natural philosopher. He was particularly intrigued by the Greek atomists, so he set out to "purify" atomism from its heretical and atheistic philosophical conclusions (Dijksterhuis 1969). Gassendi formulated his atomistic conception of mechanical philosophy partly in response to Descartes; he particularly opposed Descartes' reductionist view that only purely mechanical explanations of physics are valid, as well as the application of geometry to the whole of physics (Clericuzio 2000).

5.6.1 Corpuscularianism

Main article: Corpuscularianism

Corpuscularianism is similar to atomism, except that where atoms were supposed to be indivisible, corpuscles could in principle be divided. In this manner, for example, it was theorized that mercury could penetrate into metals and modify their inner structure, a step on the way towards transmutative production of gold. Corpuscularianism was associated by its leading proponents with the idea that some of the properties that objects appear to have are artifacts of the perceiving mind: 'secondary' qualities as distinguished from 'primary' qualities.[29] Not all corpuscularianism made use of the primary-secondary quality distinction, however. An influential tradition in medieval and early modern alchemy argued that chemical analysis revealed the existence of robust corpuscles that retained their identity in chemical compounds (to use the modern term). William R. Newman has dubbed this approach to matter theory "chymical atomism," and has argued for its significance to both the mechanical philosophy and to the chemical atomism that emerged in the early 19th century.[30] Corpuscularianism stayed a dominant theory over the next several hundred years and retained its links with alchemy in the work of scientists such as Robert Boyle and Isaac Newton in the 17th century.[31][32] It was used by Newton, for instance, in his development of the corpuscular theory of light. The form that came to be accepted by most English scientists after Robert Boyle (1627–1692) was an amalgam of the systems of Descartes and Gassendi. In The Sceptical Chymist (1661), Boyle demonstrates problems that arise from chemistry, and offers up atomism as a possible explanation. The unifying principle that would eventually lead to the acceptance of a hybrid corpuscular–atomism was mechanical philosophy, which became widely accepted by physical sciences.

5.7 Atomic theory

Main article: Atomic theory

By the late 18th century, the useful practices of engineering and technology began to influence philosophical explanations for the composition of matter. Those who speculated on the ultimate nature of matter began to verify their "thought experiments" with some repeatable demonstrations, when they could.

Roger Boscovich provided the first general mathematical theory of atomism, based on the ideas of Newton and Leibniz but transforming them so as to provide a programme for atomic physics.[33]

In 1808, John Dalton assimilated the known experimental work of many people to summarize the empirical evidence on the composition of matter. He noticed that distilled water everywhere analyzed to the same elements, hydrogen and oxygen. Similarly, other purified substances decomposed to the same elements in the same proportions by weight.

> Therefore we may conclude that the ultimate particles of all homogeneous bodies are perfectly alike in weight, figure, etc. In other words, every particle of water is like every other particle of water; every particle of hydrogen is like every other particle of hydrogen, etc.

Furthermore, he concluded that there was a unique atom for each element, using Lavoisier's definition of an element as a substance that could not be analyzed into something simpler. Thus, Dalton concluded the following.

> Chemical analysis and synthesis go no farther than to the separation of particles one from another, and to their reunion. No new creation or destruction of matter is within the reach of chemical agency. We might as well attempt to introduce a new planet into the solar system, or to annihilate one already in existence, as to create or destroy a particle of hydrogen. All the changes we can produce, consist in separating particles that are in a state of cohesion or combination, and joining those that were previously at a distance.

And then he proceeded to give a list of relative weights in the compositions of several common compounds, summarizing:

> 1st. That water is a binary compound of hydrogen and oxygen, and the relative weights of the two elementary atoms are as 1:7, nearly;

2nd. That ammonia is a binary compound of hydrogen and azote nitrogen, and the relative weights of the two atoms are as 1:5, nearly...

Dalton concluded that the fixed proportions of elements by weight suggested that the atoms of one element combined with only a limited number of atoms of the other elements to form the substances that he listed.

5.7.1 Atomic theory controversy

Dalton's atomic theory remained controversial throughout the 19th century.[34] Whilst the Law of definite proportion was accepted, the hypothesis that this was due to atoms was not so widely accepted. For example in 1826 when Sir Humphry Davy presented Dalton the Royal Medal from the Royal Society, Davy said that the theory only became useful when the atomic conjecture was ignored.[35] Sir Benjamin Collins Brodie in 1866 published the first part of his Calculus of Chemical Operations[36] as a non-atomic alternative to the Atomic Theory. He described atomic theory as a 'Thoroughly materialistic bit of joiners work'.[37] Alexander Williamson used his Presidential Address to the London Chemical Society in 1869[38] to defend the Atomic Theory against its critics and doubters. This in turn led to further meetings at which the positivists again attacked the supposition that there were atoms. The matter was finally resolved in Dalton's favour in the early 20th century with the rise of atomic physics.

5.8 See also

- Becoming (philosophy)
- History of chemistry
- Infinite divisibility
- Ontological pluralism
- Physical ontology

5.9 Notes

[1] ἄτομον. Liddell, Henry George; Scott, Robert; *A Greek–English Lexicon* at the Perseus Project

[2] "atom". Online Etymology Dictionary.

[3] The term 'atomism' is recorded in English since 1670–80 (*Random House Webster's Unabridged Dictionary*, 2001, "atomism").

[4] Aristotle, *Metaphysics* I, 4, 985^b 10–15.

[5] Berryman, Sylvia, "Ancient Atomism", *The Stanford Encyclopedia of Philosophy* (Fall 2008 Edition), Edward N. Zalta (ed.), http://plato.stanford.edu/archives/fall2008/entries/atomism-ancient/

[6] Gangopadhyaya, Mrinalkanti (1981). *Indian Atomism: History and Sources*. Atlantic Highlands, New Jersey: Humanities Press. ISBN 0-391-02177-X. OCLC 10916778.

[7] Iannone, A. Pablo (2001). *Dictionary of World Philosophy*. Routledge. pp. 83,356. ISBN 0-415-17995-5. OCLC 44541769.

[8] Thomas McEvilley, The Shape of Ancient Thought: Comparative Studies in Greek and Indian Philosophies ISBN 1-58115-203-5, Allwarth Press, 2002, p. 317-321.

[9] Richard King, Indian philosophy: an introduction to Hindu and Buddhist thought, , Edinburgh University Press, 1999, ISBN 0-7486-0954-7, pp. 105-107.

[10] The atomists, Leucippus and Democritus: fragments, a text and translation with a commentary by C.C.W. Taylor, University of Toronto Press Incorporated 1999, ISBN 0-8020-4390-9, pp. 157-158.

[11] Teresi, Dick (2003). *Lost Discoveries: The Ancient Roots of Modern Science*. Simon & Schuster. pp. 213–214. ISBN 0-7432-4379-X.

[12] Andrew G. van Melsen (1952). *From Atomos to Atom*. Mineola, N.Y.: Dover Publications. ISBN 0486495841.

[13] Bertrand Russel (1946). *History of Western Philosophy*. London: Routledge. p. 75. ISBN 0415325056.

[14] Andrew G. van Melsen. (1952). *From Atomos to Atom: The History and Concept of the Atom*. Dover Phoenix Editions. ISBN 0-486-49584-1

[15] Berryman, Sylvia, "Democritus", *The Stanford Encyclopedia of Philosophy* (Fall 2008 Edition), Edward N. Zalta (ed.), http://plato.stanford.edu/archives/fall2008/entries/democritus

[16] Lloyd, Geoffrey (1970). *Early Greek Science: Thales to Aristotle*. London; New York: Chatto and Windus; W. W. Norton & Company. pp. 74–77. ISBN 0-393-00583-6.

[17] Cornford, Francis Macdonald (1957). *Plato's Cosmology: The* Timaeus *of Plato*. New York: Liberal Arts Press. pp. 210–239. ISBN 0-87220-386-7.

[18] Lloyd, Geoffrey (1968). *Aristotle: The Growth and Structure of his Thought*. Cambridge: Cambridge University Press. p. 165. ISBN 0-521-09456-9.

[19] Lloyd, Geoffrey (1970). *Early Greek Science: Thales to Aristotle*. London; New York: Chatto and Windus; W. W. Norton & Company. pp. 108–109. ISBN 0-393-00583-6.

[20] Taylor, C. C. W. (1999). *The Atomists, Leucippus and Democritus: a text and translation with commentary by C. C. W. Taylor*. Toronto; Buffalo: University of Toronto Press. ISBN 0-8020-4390-9.

[21] Bimal Krishna Matilal (1990). *The word and the world: India's contribution to the study of language*. Oxford. Yaska is dealt with in Chapter 3. ISBN 0-19-562515-3.

[22] McEvilley (2002), 317–320

[23] Teresi, Dick (2003). *Lost Discoveries: The Ancient Roots of Modern Science*. Simon & Schuster. pp. 213–214. ISBN 0-7432-4379-X.

[24] Saeed, Abdullah (2006). *Islamic Thought: An Introduction*. Routledge. p. 95. ISBN 978-0415364096.

[25] Michael Marmura (1976). "God and his creation:Two medieval Islamic views". In R. M. Savory. *Introduction to Islamic Civilization*. Cambridge University Press. p. 49.

[26] Shlomo Pines (1986). *Studies in Arabic versions of Greek texts and in mediaeval science* **2**. Brill Publishers. pp. 355–6. ISBN 965-223-626-8.

[27] John Emery Murdoch; Christoph Herbert Lüthy; William Royall Newman (1 January 2001). "The Medieval and Renaissance Tradition of Minima Naturalia". *Late Medieval and Early Modern Corpuscular Matter Theories*. BRILL. pp. 91–133. ISBN 90-04-11516-1.

[28] Alan Chalmers (4 June 2009). *The Scientist's Atom and the Philosopher's Stone: How Science Succeeded and Philosophy Failed to Gain Knowledge of Atoms*. Springer. pp. 75–96. ISBN 978-90-481-2362-9.

[29] The Mechanical Philosophy - Early modern 'atomism' ("corpuscularianism" as it was known)

[30] William R. Newman, "The Significance of 'Chymical Atomism'," in Edith Sylla and W. R. Newman, eds., *Evidence and Interpretation: Studies on Early Science and Medicine in Honor of John E. Murdoch* (Leiden: Brill, 2009), pp. 248-264 and Newman, *Atoms and Alchemy: Chymistry and the Experimental Origins of the Scientific Revolution* (Chicago: University of Chicago Press, 2006)

[31] Levere, Trevor, H. (2001). *Transforming Matter – A History of Chemistry for Alchemy to the Buckyball*. The Johns Hopkins University Press. ISBN 0-8018-6610-3.

[32] Corpuscularianism - Philosophical Dictionary

[33] Lancelot Law Whyte Essay on Atomism, 1961, p 54.

[34] Brock(ed), W.H. (1967). *The Atomic Debates*. Leicester University Press. p. 1.

[35] Davy(ed), J. *Collected Works of Sir Humphrey Davy*. Bart. p. 93 vol 8.

[36] Brodie, Sir Benjamin Collins (1866). *Philosophical Transactions of the Royal Society*. pp. 781–859 vol I56.

[37] Brock(ed), W.H. (1967). *The Atomic Debates*. Leicester University Press. p. 12.

[38] Brock(ed), W.H. (1967). *The Atomic Debates*. Leicester University Press. p. 15.

5.10   References

- Clericuzio, Antonio. *Elements, Principles, and Corpuscles; a study of atomism and chemistry in the seventeenth century*. Dordrecht; Boston: Kluwer Academic Publishers, 2000.

- Cornford, Francis MacDonald. *Plato's Cosmology: The* Timaeus *of Plato*. New York: Liberal Arts Press, 1957.

- Dijksterhuis, E. *The Mechanization of the World Picture*. Trans. by C. Dikshoorn. New York: Oxford University Press, 1969. ISBN 0-691-02396-4

- Firth, Raymond. *Religion: A Humanist Interpretation*. Routledge, 1996. ISBN 0-415-12897-8.

- Gangopadhyaya, Mrinalkanti. *Indian Atomism: history and sources*. Atlantic Highlands, New Jersey: Humanities Press, 1981. ISBN 0-391-02177-X

- Gardet, L. "djuz'" in *Encyclopaedia of Islam CD-ROM Edition, v. 1.1*. Leiden: Brill, 2001.

- Gregory, Joshua C. *A Short History of Atomism*. London: A. and C. Black, Ltd, 1981.

- Kargon, Robert Hugh. *Atomism in England from Hariot to Newton*. Oxford: Clarendon Press, 1966.

- Lloyd, G. E. R. *Aristotle: The Growth and Structure of his Thought*. Cambridge: Cambridge University Press, 1968. ISBN 0-521-09456-9

- Lloyd, G. E. R. *Greek Science After Aristotle*. New York: W. W. Norton, 1973. ISBN 0-393-00780-4

- Marmara, Michael E. "Causation in Islamic Thought." *Dictionary of the History of Ideas*. New York: Charles Scribner's Sons, 1973-74. online at the of Virginia Electronic Text Center.

- McEvilley, Thomas (2002). *The Shape of Ancient Thought: Comparative Studies in Greek and Indian Philosophies*. New York: Allworth Communications Inc. ISBN 1-58115-203-5.

- Redondi, Pietro. *Galileo Heretic*. Translated by Raymond Rosenthal. Princeton, NJ: Princeton University Press, 1987. ISBN 0-691-02426-X

5.11   External links

- The dictionary definition of atomism at Wiktionary

- *Dictionary of the History of Ideas*: Atomism: Antiquity to the Seventeenth Century

- *Dictionary of the History of Ideas*: Atomism in the Seventeenth Century

- Jonathan Schaffer, "Is There a Fundamental Level?" *Nous* 37 (2003): 498–517. Article by a philosopher who opposes atomism

- Article on traditional Greek atomism

- Atomism from the 17th to the 20th Century at Stanford Encyclopedia of Philosophy

Chapter 6

Minima naturalia

Minima naturalia ("natural minima")[n 1] were theorized by Aristotle as the smallest parts into which a homogeneous natural substance (e.g., flesh, bone, or wood) could be divided and still retain its essential character. In this context, "nature" means formal nature. Thus, "natural minimum" may be taken to mean "formal minimum": the minimum amount of matter necessary to instantiate a certain form.

Speculation on *minima naturalia* in late Antiquity, in the Islamic world, and by Scholastic and Renaissance thinkers in Europe provided a conceptual bridge between the atomism of ancient Greece and the mechanistic philosophy of early modern thinkers like Descartes, which in turn provided a background for the rigorously mathematical and experimental atomism of modern science.[1][2]

6.1 Aristotle's initial suggestion

According to Aristotle, the Pre-Socratic Greek philosopher Anaxagoras had taught that every thing, and every portion of a thing, contains within itself an infinite number of like and unlike parts. For example, Anaxagoras maintained that there must be blackness as well as whiteness in snow; how, otherwise, could it be turned into dark water? Aristotle criticized Anaxagoras' theory on multiple grounds, among them the following:[1][3]

- Animals and plants cannot be infinitely small according to Aristotle; thus the relatively homogeneous substances of which they are composed (e.g., bone and flesh in animals, or wood in plants) could not be infinitely small, either, but must have a smallest determinate size—i.e., a natural minimum.

- On Anaxagoras' argument in which all things contain all others infinitely, water could be drawn from flesh, then flesh from that water, and water from that flesh, and so on. However, as above, because there is a smallest determinate size beyond which a further divided substance would no longer be flesh, any further cycle of such drawings out would be impossible.

- Moreover, "[s]ince every body must diminish in size when something is taken from it, and flesh is quantitatively definite in respect both of greatness and smallness, it is clear that from the minimum quantity of flesh no body can be separated out; for the flesh left would be less than the minimum of flesh."[3]

Unlike the atomism of Leucippus, Democritus, and Epicurus, and also unlike the later atomic theory of John Dalton, the Aristotelian natural minimum was not conceptualized as physically indivisible--"atomic" in the contemporary sense. Instead, the concept was rooted in Aristotle's hylomorphic worldview, which held that every physical thing is a compound of matter (Greek *hyle*) and a substantial form (Greek *morphe*) that imparts its essential nature and structure. For instance, a rubber ball for a hylomorphist like Aristotle would be rubber (matter) structured by spherical shape (form).

Aristotle's intuition was that there is some smallest size beyond which matter could no longer be structured as flesh, or bone, or wood, or some other such organic substance that (for Aristotle, living before the microscope) could be considered homogeneous. For instance, if flesh were divided beyond its natural minimum, what would remain might be some elemental water, and smaller amounts of the other elements (e.g., earth) with which water was thought to mix to form flesh. But whatever was left, the water (or earth, etc.), would no longer have the formal "nature" of flesh in particular – the remaining matter would have the form of water (or earth, etc.) rather than the substantial form of flesh.

This is suggestive of modern chemistry, in which, e.g., a bar of gold can be continually divided until one has a single atom of gold, but further division of that atom of gold yields only subatomic particles (electrons, quarks, etc.) which are no longer the chemical element gold. Just as water alone is not flesh, electrons alone are not gold. Although suggestive, the parallel is not exact: the Aristotelian concept of the natural minimum of a substance is not a direct anticipation of the modern concept of an atom of a certain chemical element.

6.2 Scholastic elaboration

Aristotle's brief comments on *minima naturalia* in the *Physics* and *Meteorology* prompted further speculations by later philosophers. The idea was taken up by John Philoponus and Simplicius of Cilicia in late Antiquity and by the Islamic Aristotelian Averroes (Ibn Rushd).

Minima naturalia were discussed by Scholastic and Renaissance thinkers including Roger Bacon, Albertus Magnus, Thomas Aquinas, Giles of Rome, Siger of Brabant, Boethius of Dacia, Richard of Middleton, Duns Scotus, John of Jandun, William of Ockham, William Alnwick, Walter Bury, Adam de Wodeham, Jean Buridan, Gregory of Rimini, John Dumbleton, Nicole Oresme, John Marsilius Inguen,[n 2] John Wycliffe, Albert of Saxony, Facinus de Ast, Peter Alboinis of Mantua, Paul of Venice, Gaetano of Thiene, Alessandro Achillini, Luis Coronel, Juan de Celaya, Domingo de Soto, Didacus de Astudillo, Ludovicus Buccaferrea, Francisco de Toledo, and Benedict Pereira.[1] Of this list, the most influential Scholastic thinkers on *minima naturalia* were Duns Scotus and Gregory of Rimini.[1]

A chief theme in later commentary is reconciling *minima naturalia* with the general Aristotelian principle of infinite divisibility.[2] Commentators like Philoponus and Aquinas reconciled these aspects of Aristotle's thought by distinguishing between mathematical and "natural" divisibility. For example, in his commentary on Aristotle's *Physics*, Aquinas writes of natural minima that, "although a body, considered mathematically, is divisible to infinity, the natural body is not divisible to infinity. For in a mathematical body nothing but quantity is considered. And in this there is nothing repugnant to division to infinity. But in a natural body the form also is considered, which form requires a determinate quantity and also other accidents. Whence it is not possible for quantity to be found in the species of flesh except as determined within some termini."[4]

6.3 Influence on corpuscularianism

In the early modern period, Aristotelian hylomorphism fell out of favor with the rise of the "mechanical philosophy" of thinkers like Descartes and John Locke, who were more sympathetic to the ancient Greek atomism of Democritus than to the natural minima of Aristotle. However, the concept of *minima naturalia* continued to shape philosophical thinking even among these mechanistic philosophers in the transitional centuries between the Aristotelianism of the medieval Scholastics and the worked-out atomic theory of modern scientists like Dalton.

The mechanist Pierre Gassendi discussed *minima naturalia* in the course of expounding his opposition to Scholastic Aristotelianism, and his own attempted reconciliation between the atomism of Epicurus and the Catholic faith. Aristotle's *mininma naturalia* became "corpuscles" in the alchemical works of Geber and Daniel Sennert, who in turn influenced the corpuscularian alchemist Robert Boyle, one of the founders of modern chemistry. Boyle occasionally referred to his postulated corpuscles as *minima naturalia*.[2]

6.4 Notes

[1] *Minima naturalia* is the conventional Latin translation of Greek ελάχιστα ("elachista," singular ελάχιστον, "elachiston"), which means "minima."

[2] Not to be confused with Marsilius of Inghen[1]

6.5 References

[1] John Emery Murdoch; Christoph Herbert Lüthy; William Royall Newman (1 January 2001). "The Medieval and Renaissance Tradition of Minima Naturalia". *Late Medieval and Early Modern Corpuscular Matter Theories*. BRILL. pp. 91–133. ISBN 90-04-11516-1.

[2] Alan Chalmers (4 June 2009). *The Scientist's Atom and the Philosopher's Stone: How Science Succeeded and Philosophy Failed to Gain Knowledge of Atoms*. Springer. pp. 75–96. ISBN 978-90-481-2362-9.

[3] Aristotle, *Physics* 1.4, 187b14–21.

[4] Thomas Aquinas. *In octo libros Physicorum expositio*. Sed dicendum quod licet corpus, mathematice acceptum, sit divisibile in infinitum, corpus tamen naturale non est divisibile in infinitum. In corpore enim mathematico non consideratur nisi quantitas, in qua nihil invenitur divisioni in infinitum repugnans; sed in corpore naturali consideratur forma naturalis, quae requirit determinatam quantitatem sicut et alia accidentia. Unde non potest inveniri quantitas in specie carnis nisi infra aliquos terminos determinata.

Chapter 7

Alchemy

"Alchemist" redirects here. For other uses, see Alchemist (disambiguation) and Alchemy (disambiguation).

Alchemy is an influential tradition whose practitioners

The Emerald Tablet, a key text of Western Alchemy, in a 17th-century edition.

have, from antiquity, claimed it to be the precursor to profound powers. As described by Paul-Jacques Malouin in

The Encyclopedia of Diderot, it is the chemistry of the subtlest kind which allows one to observe extraordinary chemical operations at a more rapid pace – operations that require a long time for nature to produce.[1] Definitions of the objectives of alchemy are varied but historically have typically included one or more of the following goals: the creation of the fabled philosopher's stone; the ability to transmute base metals into the noble metals (gold or silver); and development of an elixir of life, which would confer youth and longevity.

Though alchemy played a significant role in the development of early modern science,[2] it differs significantly from modern science in its inclusion of Hermetic principles and practices related to mythology, magic, religion, and spirituality. It is recognized as a protoscience that contributed to the development of modern chemistry and medicine. Alchemists developed a structure of basic laboratory techniques, theory, terminology, and experimental method, some of which are still in use today. However, alchemists predated modern foundations of chemistry, such as scientific skepticism, atomic theory, the modern understanding of a chemical element and a chemical substance, the periodic table and conservation of mass and stoichiometry. Instead, they believed in four elements, and cryptic symbolism and mysticism was an integral part of alchemical work.

7.1 Overview

The ostensible goals of alchemy are often given as the transmutation of common metals into gold (known as chrysopoeia), the creation of a panacea, and the discovery of a universal solvent.[3] However, these only highlight certain aspects of alchemy. Alchemists have historically rewritten and evolved their explanation of alchemy, so it is difficult to define it simply.[4] H.J. Sheppard gives the following as a comprehensive summary:

> Alchemy is the art of liberating parts of the

Cosmos from temporal existence and achieving perfection which, for metals is gold, and for man, longevity, then immortality and, finally, redemption. Material perfection was sought through the action of a preparation (Philosopher's Stone for metals; Elixir of Life for humans), while spiritual ennoblement resulted from some form of inner revelation or other enlightenment (Gnosis, for example, in Hellenistic and western practices).[5]

Modern discussions of alchemy are generally split into an examination of its exoteric practical applications and its esoteric aspects. The former is pursued by historians of the physical sciences who have examined the subject in terms of protochemistry, medicine, and charlatanism. The latter interests psychologists, spiritual and new age communities, hermetic philosophers, and historians of esotericism.[6] The subject has also made an ongoing impact on literature and the arts. Despite the modern split, numerous sources stress an integration of esoteric and exoteric approaches to alchemy. Holmyard, when writing on exoteric aspects, states that they cannot be properly appreciated if the esoteric is not always kept in mind.[7] The prototype for this model can be found in Bolos of Mendes's 3rd-century BCE work *Physika kai Mystika* ("On Physical and Mystical Matters").[8] Marie-Louise von Franz tells us the double approach of Western alchemy was set from the start, when Greek philosophy was mixed with Egyptian and Mesopotamian technology. The technological, operative approach, which she calls extraverted, and the mystic, contemplative, psychological one, which she calls introverted, are not mutually exclusive but complementary since meditation requires practice in the real world and vice versa.[9]

7.1.1   Relation to chemistry

Main article: History of chemistry
Practical applications of alchemy produced a wide range of contributions to medicine and the physical sciences. The alchemist Robert Boyle[10] is credited as being the father of chemistry. Paracelsian iatrochemistry emphasized the medicinal application of alchemy (continued in plant alchemy, or spagyric).[11] Studies of alchemy also influenced Isaac Newton's theory of gravity.[12] Academic historical research supports that the alchemists were searching for a material substance using physical methods.[13]

Alchemists made contributions to the "chemical" industries of the day—ore testing and refining, metalworking, production of gunpowder, ink, dyes, paints, cosmetics, leather tanning, ceramics, glass manufacture, preparation of extracts, liquors, and so on. Alchemists contributed distillation to Western Europe. The attempts of alchemists to arrange information on substances, so as to clarify and anticipate

Scientific apparatus in the alchemist's workshop, 1580.

the products of their chemical reactions, resulted in early conceptions of chemical elements and the first rudimentary periodic tables. They learned how to extract metals from ores, and how to compose many types of inorganic acids and bases.

During the 17th century, practical alchemy started to disappear in favor of its younger offshoot chemistry,[14] as it was renamed by Robert Boyle, the "father of modern chemistry".[15] In his book, *The Skeptical Chymist*, Boyle attacked Paracelsus and the natural philosophy of Aristotle, which was taught at universities. However, Boyle's biographers, in their emphasis that he laid the foundations of modern chemistry, neglect how steadily he clung to the scholastic sciences and to alchemy, in theory, practice and doctrine.[16] The decline of alchemy continued in the 18th century with the birth of modern chemistry, which provided a more precise and reliable framework within a new view of the universe based on rational materialism.

7.1.2   Relation to Hermeticism

In the eyes of a variety of esoteric and Hermetic practitioners, the heart of alchemy is spiritual. Transmutation of lead into gold is presented as an analogy for personal transmutation, purification, and perfection.[8] This approach is often termed 'spiritual', 'esoteric', or 'internal' alchemy.

Early alchemists, such as Zosimos of Panopolis (c. AD 300), highlight the spiritual nature of the alchemical quest, symbolic of a religious regeneration of the human soul.[17] This approach continued in the Middle Ages, as metaphysical aspects, substances, physical states, and material processes were used as metaphors for spiritual entities, spiritual states, and, ultimately, transformation. In this sense, the literal meanings of 'Alchemical Formulas' were a blind, hiding their true spiritual philosophy. Practitioners and patrons such as Melchior Cibinensis and Pope Innocent

VIII existed within the ranks of the church, while Martin Luther applauded alchemy for its consistency with Christian teachings.[18] Both the transmutation of common metals into gold and the universal panacea symbolized evolution from an imperfect, diseased, corruptible, and ephemeral state toward a perfect, healthy, incorruptible, and everlasting state, so the philosopher's stone then represented a mystic key that would make this evolution possible. Applied to the alchemist himself, the twin goal symbolized his evolution from ignorance to enlightenment, and the stone represented a hidden spiritual truth or power that would lead to that goal. In texts that are written according to this view, the cryptic alchemical symbols, diagrams, and textual imagery of late alchemical works typically contain multiple layers of meanings, allegories, and references to other equally cryptic works; and must be laboriously decoded to discover their true meaning.

In his 1766 *Alchemical Catechism*, Théodore Henri de Tschudi denotes that the usage of the metals was a symbol:

> Q. When the Philosophers speak of gold and silver, from which they extract their matter, are we to suppose that they refer to the vulgar gold and silver?
> A. By no means; vulgar silver and gold are dead, while those of the Philosophers are full of life.[1]

1. ^ Théodore Henri de Tschudi. Hermetic Catechism in his *L'Etoile Flamboyant ou la Société des Franc-Maçons considerée sous tous les aspects*. 1766. (A.E. Waite translation as found in *The Hermetic and Alchemical Writings of Paracelsus*.)

During the renaissance, alchemy broke into more distinct schools placing spiritual alchemists in high contrast with those working with literal metals and chemicals.[19] While most spiritual alchemists also incorporate elements of exotericism, examples of a purely spiritual alchemy can be traced back as far as the 16th century, when Jacob Boehme used alchemical terminology in strictly mystical writings.[20] Another example can be found in the work of Heinrich Khunrath (1560–1605) who viewed the process of transmutation as occurring within the alchemist's spirit.[19]

The recent work of Lawrence M. Principe and William R. Newman, rejects the 'spiritual interpretation' of alchemy, especially as applied to medieval, 16th- and 17th-century alchemy, showing that it arose predominantly as a product of the Victorian occult revival.[21] There is evidence to support that some classical alchemical sources were adulterated during this time to give greater weight to the spiritual aspects of alchemy.[22][23] Despite this, other scholars such as Calian and Tilton reject this view as entirely historically in-

accurate, drawing examples of historical spiritual alchemy from Boehme, Isaac Newton, and Michael Maier.[24]

7.2 Etymology

Main article: Chemistry (etymology)

The word alchemy was borrowed from Old French *alquemie*, *alkimie*, taken from Medieval Latin *alchymia*, and which is in turn borrowed from Arabic *al-kīmiyā'* (الكيمياء) 'philosopher's stone'. The Arabic word is borrowed from Late Greek *chēmeía* (χημεία), *chēmía* (χημία)[25] 'black magic' with the agglutination of the Arabic definite article *al-* (ال).[26] This ancient Greek word was derived from[27] the early Greek name for Egypt, *Chēmia* (Χημία), based on the Egyptian name for Egypt, *kēme* (hieroglyphic *khmi*, lit. 'black earth', as opposed to red desert sand).[26]

The Medieval Latin form was influenced by Greek *chymeia* (χυμεία) meaning 'mixture' and referring to pharmaceutical chemistry.[28]

7.3 History

Alchemy covers several philosophical traditions spanning some four millennia and three continents. These traditions' general penchant for cryptic and symbolic language makes it hard to trace their mutual influences and "genetic" relationships. One can distinguish at least three major strands, which appear to be largely independent, at least in their earlier stages: Chinese alchemy, centered in China and its zone of cultural influence; Indian alchemy, centered on the Indian subcontinent; and Western alchemy, which occurred around the Mediterranean and whose center has shifted over the millennia from Greco-Roman Egypt, to the Islamic world, and finally medieval Europe. Chinese alchemy was closely connected to Taoism and Indian alchemy with the Dharmic faiths, whereas Western alchemy developed its own philosophical system that was largely independent of, but influenced by, various Western religions. It is still an open question whether these three strands share a common origin, or to what extent they influenced each other.

7.3.1 Alchemy in Greco-Roman Egypt

The start of Western alchemy may generally be traced to Hellenistic Egypt, where the city of Alexandria was a center of alchemical knowledge, and retained its pre-eminence through most of the Greek and Roman periods.[29] Here, elements of technology, religion, mythology, and Hellenistic

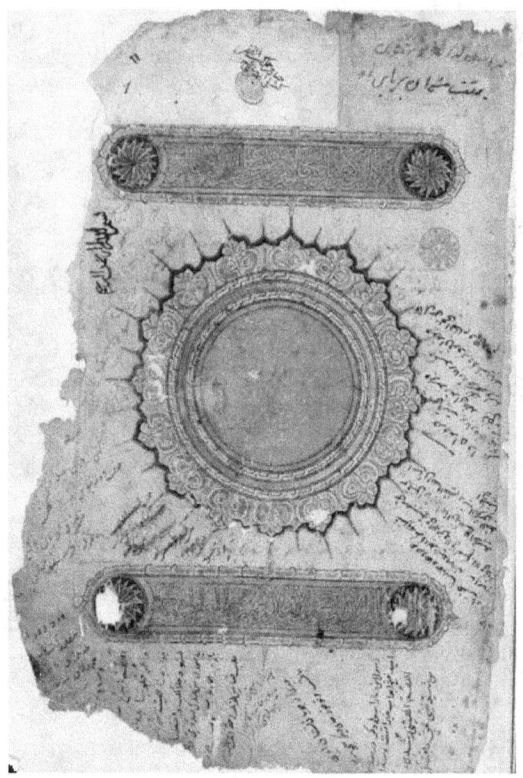

Kimiya-yi sa'ādat (The Alchemy of Happiness) – a text on Islamic philosophy and spiritual alchemy by Al-Ghazālī (1058–1111).

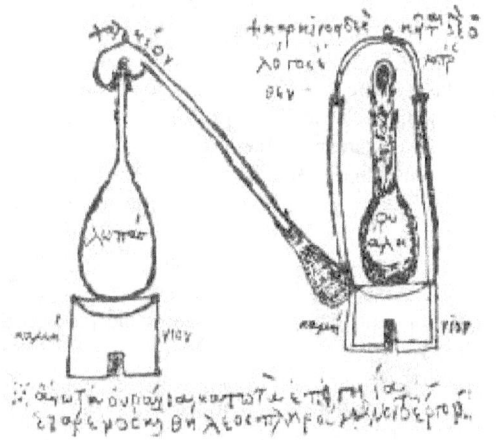

Ambix, cucurbit and retort of Zosimos, from Marcelin Berthelot, Collection des anciens alchimistes grecs (3 vol., Paris, 1887–1888).

philosophy, each with their own much longer histories, combined to form the earliest known records of alchemy in the West. Zosimos of Panopolis wrote the oldest known

books on alchemy, while Mary the Jewess is credited as being the first non-fictitious Western alchemist. They wrote in Greek and lived in Egypt under Roman rule.

Mythology – Zosimos of Panopolis asserted that alchemy dated back to Pharaonic Egypt where it was the domain of the priestly class, though there is little to no evidence for his assertion.[30] Alchemical writers used Classical figures from Greek, Roman, and Egyptian mythology to illuminate their works and allegorize alchemical transmutation.[31] These included the pantheon of gods related to the Classical planets, Isis, Osiris, Jason, and many others.

The central figure in the mythology of alchemy is Hermes Trismegistus (or Thrice-Great Hermes). His name is derived from the god Thoth and his Greek counterpart Hermes. Hermes and his caduceus or serpent-staff, were among alchemy's principal symbols. According to Clement of Alexandria, he wrote what were called the "forty-two books of Hermes", covering all fields of knowledge.[32] The *Hermetica* of Thrice-Great Hermes is generally understood to form the basis for Western alchemical philosophy and practice, called the hermetic philosophy by its early practitioners. These writings were collected in the first centuries of the common era.

Technology – The dawn of Western alchemy is sometimes associated with that of metallurgy, extending back to 3500 BCE.[33] Many writings were lost when the emperor Diocletian ordered the burning of alchemical books[34] after suppressing a revolt in Alexandria (292 CE). Few original Egyptian documents on alchemy have survived, most notable among them the Stockholm papyrus and the Leyden papyrus X. Dating from 300 to 500 CE, they contained recipes for dyeing and making artificial gemstones, cleaning and fabricating pearls, and manufacturing of imitation gold and silver.[35] These writings lack the mystical, philosophical elements of alchemy, but do contain the works of Bolus of Mendes (or Pseudo-Democritus), which aligned these recipes with theoretical knowledge of astrology and the classical elements.[36] Between the time of Bolus and Zosimos, the change took place that transformed this metallurgy into a Hermetic art.[37]

Philosophy – Alexandria acted as a melting pot for philosophies of Pythagoreanism, Platonism, Stoicism and Gnosticism which formed the origin of alchemy's character.[36] An important example of alchemy's roots in Greek philosophy, originated by Empedocles and developed by Aristotle, was that all things in the universe were formed from only four elements: earth, air, water, and fire. According to Aristotle, each element had a sphere to which it belonged and to which it would return if left undisturbed.[38] The four elements of the Greek were mostly qualitative aspects of matter, not quantitative, as our modern elements are; "...True alchemy never regarded earth, air, water, and fire as corporeal or chemical substances in the present-day sense of the word. The four elements are simply the primary, and most general, qualities by means of which the amorphous and purely quantitative substance of all bodies first reveals itself in differentiated form."[39] The Roman emperor Caligula is said "to have instituted experiments for producing gold out of orpiment (arsenic sulfide)."[40] Later alchemists extensively developed the mystical aspects of this concept.

Alchemy coexisted alongside emerging Christianity. Lactantius believed Hermes Trismegistus had prophesied its birth. Augustine (354–430 CE) later affirmed this, but also condemned Trismegistus for idolatry.[41] Examples of Pagan, Christian, and Jewish alchemists can be found during this period.

Most of the Greco-Roman alchemists preceding Zosimos are known only by pseudonyms, such as Moses, Isis, Cleopatra, Democritus, and Ostanes. Others authors such as Komarios, and Chymes, we only know through fragments of text. After 400 CE, Greek alchemical writers occupied themselves solely in commenting on the works of these predecessors.[42] By the middle of the 7th century alchemy was almost an entirely mystical discipline.[43] It was at that time that Khalid Ibn Yazid sparked its migration from Alexandria to the Islamic world, facilitating the translation and preservation of Greek alchemical texts in the 8th and 9th centuries.[44]

7.3.2 Alchemy in the Islamic world

Main article: Alchemy and chemistry in medieval Islam

After the fall of the Roman Empire, the focus of alchemical development moved to the Islamic World. Much more is known about Islamic alchemy because it was better documented: indeed, most of the earlier writings that have come down through the years were preserved as Arabic translations.[45] The word *alchemy* itself was derived from the Arabic word *al-kīmiyā'* (الكيمياء). The early Islamic world was a melting pot for alchemy. Platonic and Aristotelian thought, which had already been somewhat appropriated into hermetical science, continued to be assimilated during the late 7th and early 8th centuries through Syriac translations and scholarship.

In the late 8th century, Jābir ibn Hayyān (known as "Geber" in Europe) introduced a new approach to alchemy, based on scientific methodology and controlled experimentation in the laboratory, in contrast to the ancient Greek and Egyptian alchemists whose works were often allegorical and unintelligible, with very little concern for laboratory work.[46] Jabir is thus "considered by many to be the father of chemistry",[47] albeit others reserve that title for Robert Boyle or Antoine Lavoisier. The science historian, Paul

Jabir ibn Hayyan (Geber), considered the "father of chemistry", introduced a scientific and experimental approach to alchemy.

Kraus, wrote:

To form an idea of the historical place of Jabir's alchemy and to tackle the problem of its sources, it is advisable to compare it with what remains to us of the alchemical literature in the Greek language. One knows in which miserable state this literature reached us. Collected by Byzantine scientists from the tenth century, the corpus of the Greek alchemists is a cluster of incoherent fragments, going back to all the times since the third century until the end of the Middle Ages.

The efforts of Berthelot and Ruelle to put a little order in this mass of literature led only to poor results, and the later researchers, among them in particular Mrs. Hammer-Jensen, Tannery, Lagercrantz, von Lippmann, Reitzenstein, Ruska, Bidez, Festugiere and others, could make clear only few points of detail ....

The study of the Greek alchemists is not very encouraging. An even surface examination of the Greek texts shows that a very small part only was organized according to true experiments of laboratory: even the supposedly technical writings, in the state where we find them today, are unintelligible nonsense which refuses any interpretation.

It is different with Jabir's alchemy. The relatively clear description of the processes and the alchemical apparati, the methodical classification of the substances, mark an experimental spirit which is extremely far away from the weird and odd esotericism of the Greek texts. The theory on which Jabir supports his operations is one of clearness and of an impressive unity. More than with the other Arab authors, one notes with him a balance between theoretical teaching and practical teaching, between the *'ilm* and the *'amal*. In vain one would seek in the Greek texts a work as systematic as that which is presented, for example, in the *Book of Seventy*.[46]

Jabir himself clearly recognized and proclaimed the importance of experimentation:

> The first essential in chemistry is that thou shouldest perform practical work and conduct experiments,
>
> for he who performs not practical work nor makes experiments will never attain to the least degree of mastery.[48]

Early Islamic chemists such as Jabir Ibn Hayyan (جابر بن حيان in Arabic, Geberus in Latin; usually rendered in English as Geber), Al-Kindi (Alkindus) and Muhammad ibn Zakarīya Rāzi (Rasis or Rhazes in Latin) contributed a number of key chemical discoveries, such as the muriatic (hydrochloric acid), sulfuric and nitric acids, and more. The discovery that aqua regia, a mixture of nitric and hydrochloric acids, could dissolve the noblest metal, gold, was to fuel the imagination of alchemists for the next millennium.

Islamic philosophers also made great contributions to alchemical hermeticism. The most influential author in this regard was arguably Jabir. Jabir's ultimate goal was *Takwin*, the artificial creation of life in the alchemical laboratory, up to, and including, human life. He analyzed each Aristotelian element in terms of four basic qualities of *hotness*, *coldness*, *dryness*, and *moistness*.[49] According to Jabir, in each metal two of these qualities were interior and two were exterior. For example, lead was externally cold and dry, while gold was hot and moist. Thus, Jabir theorized, by rearranging the qualities of one metal, a different metal would result.[49] By this reasoning, the search for the philosopher's stone was introduced to Western alchemy. Jabir developed an elaborate numerology whereby the root letters of a substance's name in Arabic, when treated with various transformations, held correspondences to the element's physical properties.

The elemental system used in medieval alchemy also originated with Jabir. His original system consisted of seven elements, which included the five classical elements (aether, air, earth, fire, and water) in addition to two chemical elements representing the metals: sulphur, "the stone which burns", which characterized the principle of combustibility, and mercury, which contained the idealized principle of metallic properties. Shortly thereafter, this evolved into eight elements, with the Arabic concept of the three metallic principles: sulphur giving flammability or combustion, mercury giving volatility and stability, and salt giving solidity.[50] The atomic theory of corpuscularianism, where all physical bodies possess an inner and outer layer of minute particles or corpuscles, also has its origins in the work of Jabir.[51]

From the 9th to 14th centuries, alchemical theories faced criticism from a variety of practical Muslim chemists, including Alkindus,[52] Abū al-Rayhān al-Bīrūnī,[53] Avicenna[54] and Ibn Khaldun. In particular, they wrote refutations against the idea of the transmutation of metals.

7.3.3 Alchemy in medieval Europe

Painting by Joseph Wright of Derby, 1771.

The introduction of alchemy to Latin Europe occurred on 11 February 1144, with the completion of Robert of Chester's translation of the Arabic *Book of the Composition of Alchemy*. Although European craftsmen and technicians preexisted, Robert notes in his preface that alchemy was unknown in Latin Europe at the time of his writing. The translation of Arabic texts concerning numerous disciplines including alchemy flourished in 12th-century Toledo, Spain, through contributors like Gerard of Cremona and Adelard of Bath.[55] Translations of the time included the Turba Philosophorum, and the works of Avicenna and al-Razi. These brought with them many new words to the European vocabulary for which there was no previous Latin equivalent. Alcohol, carboy, elixir, and athanor are examples.[56]

Meanwhile, theologian contemporaries of the translators made strides towards the reconciliation of faith and experimental rationalism, thereby priming Europe for the influx of alchemical thought. Saint Anselm (1033–1109) put forth the opinion that faith and rationalism were compatible and encouraged rationalism in a Christian context.

Peter Abelard (1079–1142) followed Anselm's work, laying down the foundation for acceptance of Aristotelian thought before the first works of Aristotle had reached the West. And later, Robert Grosseteste (1170–1253) used Abelard's methods of analysis and added the use of observation, experimentation, and conclusions when conducting scientific investigations. Grosseteste also did much work to reconcile Platonic and Aristotelian thinking.[57]

Through much of the 12th and 13th centuries, alchemical knowledge in Europe remained centered on translations, and new Latin contributions were not made. The efforts of the translators were succeeded by that of the encyclopaedists. Albertus Magnus and Roger Bacon are the most notable of these.[58] Their works explained and summarized the newly imported alchemical knowledge in Aristotelian terms. There is little to suggest that Albertus Magnus (1193–1280), a Dominican, was himself an alchemist. In his authentic works such as the Book of Minerals, he observed and commented on the operations and theories of alchemical authorities like Hermes and Democritus, and unnamed alchemists of his time. Albertus critically compared these to the writings of Aristotle and Avicenna, where they concerned the transmutation of metals. From the time shortly after his death through to the 15th century, twenty-eight or more alchemical tracts were misattributed to him, a common practice giving rise to his reputation as an accomplished alchemist.[59] Likewise, alchemical texts have been attributed to Albert's student Thomas Aquinas (1225–1274).

Roger Bacon (1214–1294) was an Oxford Franciscan who studied a wide variety of topics including optics, languages and medicine. After studying the Pseudo-Aristotelian Secretum Secretorum around 1247, he dramatically shifted his studies towards a vision of a universal science which included alchemy and astrology. Bacon maintained that Albertus Magnus' ignorance of the fundamentals of alchemy prevented a complete picture of wisdom. While alchemy was not more important to him than any of the other sciences, and he did not produce symbolic allegorical works, Bacon's contributions advanced alchemy's connections to soteriology and Christian theology. Bacon's writings demonstrated an integration of morality, salvation, alchemy, and the prolongation of life. His correspondence with Pope Clement IV highlighted this integration, calling attention to the importance of alchemy to the papacy.[60] Like the Greeks before him, Bacon acknowledged the division of alchemy into the practical and theoretical. He noted that the theoretical lay outside the scope of Aristotle, the natural philosophers, and all Latin writers of his time. The practical, however, confirmed the theoretical thought experiment, and Bacon advocated its uses in natural science and medicine.[61]

Soon after Bacon, the influential work of Pseudo-Geber (sometimes identified as Paul of Taranto) appeared. His Summa Perfectionis remained a staple summary of alchemical practice and theory through the medieval and renaissance periods. It was notable for its inclusion of practical chemical operations alongside sulphur-mercury theory, and the unusual clarity with which they were described.[62] By the end of the 13th century, alchemy had developed into a fairly structured system of belief. Adepts believed in the macrocosm-microcosm theories of Hermes, that is to say, they believed that processes that affect minerals and other substances could have an effect on the human body (for example, if one could learn the secret of purifying gold, one could use the technique to purify the human soul). They believed in the four elements and the four qualities as described above, and they had a strong tradition of cloaking their written ideas in a labyrinth of coded jargon set with traps to mislead the uninitiated. Finally, the alchemists practiced their art: they actively experimented with chemicals and made observations and theories about how the universe operated. Their entire philosophy revolved around their belief that man's soul was divided within himself after the fall of Adam. By purifying the two parts of man's soul, man could be reunited with God.[63]

In the 14th century, alchemy became more accessible to Europeans outside the confines of Latin speaking churchmen and scholars. Alchemical discourse shifted from scholarly philosophical debate to an exposed social commentary on the alchemists themselves.[64] Dante, Piers Plowman, and Chaucer all painted unflattering pictures of alchemists as thieves and liars. Pope John XXII's 1317 edict, Spondent quas non exhibent forbade the false promises of transmutation made by pseudo-alchemists.[65] In 1403, Henry IV of England banned the practice of multiplying metals (although it was possible to buy a licence to attempt to make gold alchemically, and a number were granted by Henry VI and Edward IV[66]). These critiques and regulations centered more around pseudo-alchemical charlatanism than the actual study of alchemy, which continued with an increasingly Christian tone. The 14th century saw the Christian imagery of death and resurrection employed in the alchemical texts of Petrus Bonus, John of Rupescissa and in works written in the name of Raymond Lull and Arnold of Villanova.[67]

Nicolas Flamel is a well known alchemist, but a good example of pseudepigraphy, the practice of giving your works the name of someone else, usually more famous. Though the historical Flamel existed, the writings and legends assigned to him only appeared in 1612.[68][69] Flamel was not a religious scholar as were many of his predecessors, and his entire interest in the subject revolved around the pursuit of the philosopher's stone. His work spends a great deal of time describing the processes and reactions, but never actually gives the formula for carrying out the transmutations. Most of 'his' work was aimed at gathering alchemical knowledge that had existed before him, especially as regarded the philosopher's stone.[70]

Through the late Middle Ages (1300–1500) alchemists were much like Flamel: they concentrated on looking for the philosophers' stone. Bernard Trevisan and George Ripley made similar contributions in the 14th and 15th centuries. Their cryptic allusions and symbolism led to wide variations in interpretation of the art.

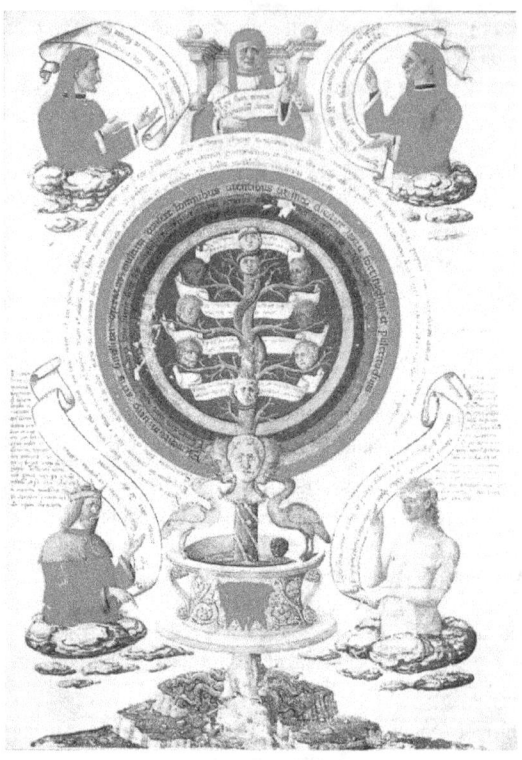

Page from alchemic treatise of Ramon Llull, 16th century.

7.3.4 Alchemy in the Renaissance and modern age

Further information: Renaissance magic and natural magic During the Renaissance, Hermetic and Platonic foundations were restored to European alchemy. The dawn of medical, pharmaceutical, occult, and entrepreneurial branches of alchemy followed.

In the late 15th century, Marsilo Ficino translated the Corpus Hermeticum and the works of Plato into Latin. These were previously unavailable to Europeans who for the first time had a full picture of the alchemical theory that Bacon had declared absent. Renaissance Humanism and Renaissance Neoplatonism guided alchemists away from physics to refocus on mankind as the alchemical vessel.

Esoteric systems developed that blended alchemy into a broader occult Hermeticism, fusing it with magic, astrology, and Christian cabala.[71][72] A key figure in this development was German Heinrich Cornelius Agrippa (1486–1535), who received his Hermetic education in Italy in the schools of the humanists. In his *De Occulta Philosophia*, he attempted to merge Kabbalah, Hermetism, and alchemy. He was instrumental in spreading this new blend of Hermeticism outside the borders of Italy.[73][74]

Philippus Aureolus Paracelsus, (Theophrastus Bombastus von Hohenheim, 1493–1541) cast alchemy into a new form, rejecting some of Agrippa's occultism and moving away from chrysopoeia. Paracelsus pioneered the use of chemicals and minerals in medicine and wrote, "Many have said of Alchemy, that it is for the making of gold and silver. For me such is not the aim, but to consider only what virtue and power may lie in medicines."[75]

His hermetical views were that sickness and health in the body relied on the harmony of man the microcosm and Nature the macrocosm. He took an approach different from those before him, using this analogy not in the manner of soul-purification but in the manner that humans must have certain balances of minerals in their bodies, and that certain illnesses of the body had chemical remedies that could cure them.[76] Paracelsian practical alchemy, especially herbal medicine and plant remedies has since been named spagyrics (a synonym for alchemy from the Greek words meaning *to separate* and *to join together*, based on the Latin alchemic maxim: *solve et coagula*).[77] Iatrochemistry also refers to the pharmaceutical applications of alchemy championed by Paracelsus.

John Dee (13 July 1527 – December, 1608) followed Agrippa's occult tradition. Though better known for angel summoning, divination, and his role as astrologer, cryptographer, and consultant to Queen Elizabeth I, Dee's alchemical *Monas Hieroglyphica*, written in 1564 was his most popular and influential work. His writing portrayed alchemy as a sort of terrestrial astronomy in line with the Hermetic axiom *As above so below*.[78] During the 17th century, a short-lived "supernatural" interpretation of alchemy became popular, including support by fellows of the Royal Society: Robert Boyle and Elias Ashmole. Proponents of the supernatural interpretation of alchemy believed that the philosopher's stone might be used to summon and communicate with angels.[79]

"Alchemist Sędziwój" (1566–1636) by Jan Matejko, 1867.

Entrepreneurial opportunities were not uncommon for the alchemists of Renaissance Europe. Alchemists were contracted by the elite for practical purposes related to mining, medical services, and the production of chemicals, medicines, metals, and gemstones.[80] Rudolf II, Holy Roman Emperor, in the late 16th century, famously received and sponsored various alchemists at his court in Prague, including Dee and his associate Edward Kelley. King James IV of Scotland,[81] Julius, Duke of Brunswick-Lüneburg, Henry V, Duke of Brunswick-Lüneburg, Augustus, Elector of Saxony, Julius Echter von Mespelbrunn, and Maurice, Landgrave of Hesse-Kassel all contracted alchemists.[82] John's son Arthur Dee worked as a court physician to Michael I of Russia and Charles I of England but also compiled the alchemical book *Fasciculus Chemicus*.

Though most of these appointments were legitimate, the trend of pseudo-alchemical fraud continued through the Renaissance. *Betrüger* would use sleight of hand, or claims of secret knowledge to make money or secure patronage. Legitimate mystical and medical alchemists such as Michael Maier and Heinrich Khunrath wrote about fraudulent transmutations, distinguishing themselves from the con artists.[83] False alchemists were sometimes prosecuted for fraud.

The terms "chemia" and "alchemia" were used as synonyms in the early modern period, and the differences between alchemy, chemistry and small-scale assaying and metallurgy were not as neat as in the present day. There were impor-

tant overlaps between practitioners, and trying to classify them into alchemists, chemists and craftsmen is anachronistic. For example, Tycho Brahe (1546–1601), an alchemist better known for his astronomical and astrological investigations, had a laboratory built at his Uraniborg observatory/research institute. Michael Sendivogius (*Michał Sędziwój*, 1566–1636), a Polish alchemist, philosopher, medical doctor and pioneer of chemistry wrote mystical works but is also credited with distilling oxygen in a lab sometime around 1600. Sendivogious taught his technique to Cornelius Drebbel who, in 1621, applied this in a submarine. Isaac Newton devoted considerably more of his writing to the study of alchemy (see Isaac Newton's occult studies) than he did to either optics or physics. Other early modern alchemists who were eminent in their other studies include Robert Boyle, and Jan Baptist van Helmont. Their Hermetism complemented rather than precluded their practical achievements in medicine and science.

7.3.5   The decline of European alchemy

Robert Boyle

The decline of European alchemy was brought about by the rise of modern science with its emphasis on rigorous quantitative experimentation and its disdain for "ancient wisdom". Although the seeds of these events were planted as early as the 17th century, alchemy still flourished for some two hundred years, and in fact may have reached its apogee in the 18th century. As late as 1781 James Price claimed to have produced a powder that could transmute mercury into silver or gold. Early modern European alchemy continued to exhibit a diversity of theories, practices, and purposes: "Scholastic and anti-Aristotelian, Paracelsian and anti-Paracelsian, Hermetic, Neoplatonic, mechanistic, vitalistic, and more—plus virtually every combination and compromise thereof."[84]

Robert Boyle (1627–1691) pioneered the scientific method in chemical investigations. He assumed nothing in his experiments and compiled every piece of relevant data. Boyle would note the place in which the experiment was carried out, the wind characteristics, the position of the Sun and Moon, and the barometer reading, all just in case they proved to be relevant.[85] This approach eventually led to the founding of modern chemistry in the 18th and 19th centuries, based on revolutionary discoveries of Lavoisier and John Dalton.

Beginning around 1720, a rigid distinction was drawn between "alchemy" and "chemistry" for the first time.[21][86] By the 1740s, "alchemy" was now restricted to the realm of gold making, leading to the popular belief that alchemists were charlatans, and the tradition itself nothing more than a fraud.[84][86] In order to protect the developing science of modern chemistry from the negative censure of which alchemy was being subjected, academic writers during the scientific Enlightenment attempted, for the sake of survival, to separate and divorce the "new" chemistry from the "old" practices of alchemy. This move was mostly successful, and the consequences of this continued into the 19th and 20th centuries, and even to the present day.[87]

During the occult revival of the early 19th century, alchemy received new attention as an occult science.[88][89] The esoteric or occultist school, which arose during the 19th century, held (and continues to hold) the view that the substances and operations mentioned in alchemical literature are to be interpreted in a spiritual sense, and it downplays the role of the alchemy as a practical tradition or protoscience.[21][90][91] This interpretation further forwarded the view that alchemy is an art primarily concerned with spiritual enlightenment or illumination, as opposed to the physical manipulation of apparatus and chemicals, and claims that the obscure language of the alchemical texts were an allegorical guise for spiritual, moral or mystical processes.[91]

In the 19th-century revival of alchemy, the two most seminal figures were Mary Anne Atwood and Ethan Allen Hitchcock, who independently published similar works regarding spiritual alchemy. Both forwarded a completely esoteric view of alchemy, as Atwood claimed: "No modern art or chemistry, notwithstanding all its surreptitious claims, has any thing in common with Alchemy."[92][93] At-

wood's work influenced subsequent authors of the occult revival including Eliphas Levi, Arthur Edward Waite, and Rudolf Steiner. Hitchcock, in his *Remarks Upon Alchymists* (1855) attempted to make a case for his spiritual interpretation with his claim that the alchemists wrote about a spiritual discipline under a materialistic guise in order to avoid accusations of blasphemy from the church and state. In 1845, Baron Carl Reichenbach, published his studies on Odic force, a concept with some similarities to alchemy, but his research did not enter the mainstream of scientific discussion.[94]

7.3.6   Indian alchemy

Main article: Rasayana
See also: History of metallurgy in the Indian subcontinent

According to the Encyclopædia Britannica, the Vedas describe a connection between eternal life and gold. The use of Mercury for alchemy is first documented in the 3rd - 4th century CE Artha-śāstra. Buddhist texts from the 2nd to 5th centuries CE mention the transmutation of base metals to gold. Greek alchemy may have been introduced to Ancient India through the invasions of Alexander the Great in 325 BCE, and kingdoms that were culturally influenced by the Greeks like Gandhāra, although hard evidence for this is lacking.[95]

The 11th-century Persian chemist and physician Abū Rayhān Bīrūnī, who visited Gujarat as part of the court of Mahmud of Ghazni, reported that they

> have a science similar to alchemy which is quite peculiar to them, which in Sanskrit is called Rasayāna and in Persian Rasavātam. It means the art of obtaining/manipulating *Rasa*: nectar, mercury, and juice. This art was restricted to certain operations, metals, drugs, compounds, and medicines, many of which have mercury as their core element. Its principles restored the health of those who were ill beyond hope and gave back youth to fading old age.

The goals of alchemy in India included the creation of a divine body (Sanskrit *divya-deham*) and immortality while still embodied (Sanskrit *jīvan-mukti*). Sanskrit alchemical texts include much material on the manipulation of mercury and sulphur, that are homologized with the semen of the god Śiva and the menstrual blood of the goddess Devī.

Some early alchemical writings seem to have their origins in the Kaula tantric schools associated to the teachings of the personality of Matsyendranath. Other early writings are found in the Jaina medical treatise *Kalyāṇakārakam* of Ugrāditya, written in South India in the early 9th century.[96]

Two famous early Indian alchemical authors were Nāgārjuna Siddha and Nityanātha Siddha. Nāgārjuna Siddha was a Buddhist monk. His book, *Rasendramangalam*, is an example of Indian alchemy and medicine. Nityanātha Siddha wrote *Rasaratnākara*, also a highly influential work. In Sanskrit, *rasa* translates to "mercury", and Nāgārjuna Siddha was said to have developed a method of converting mercury into gold.[97]

Reliable scholarship on Indian alchemy has been advanced in a major way by the publication of *The Alchemical Body* by David Gordon White.[98] Trustworthy scholarship on Indian alchemy must now take the findings of this work into account.

An important modern bibliography on Indian alchemical studies has also been provided by David Gordon White at Oxford Bibliographies Online.[99]

Representative works in Sanskrit

The contents of the following thirty-nine Sanskrit alchemical treatises have been analysed in detail in G. Jan Meulenbeld's *History of Indian Medical Literature*.:[100]

- Ānandakanda
- Āyurvedaprakāśa
- Gorakṣasaṃhitā
- Kākacaṇḍeśvarīmatatantra
- Kākacaṇḍīśvarakalpatantra
- Kūpīpakvarasanirmāṇavijñāna
- Pāradasaṃhitā
- Rasabhaiṣajyakalpanāvijñāna
- Rasādhyāya
- Rasahṛdayatantra
- Rasajalanidhi
- Rasakāmadhenu
- Rasakaumudī
- Rasamañjarī
- Rasamitra
- Rasāmṛta
- Rasapaddhati

- Rasapradīpa
- Rasaprakāśasudhākara
- Rasarājalakṣmī
- Rasaratnadīpikā
- Rasaratnākara
- Rasaratnasamuccaya
- Rasārṇava
- Rasārṇavakalpa
- Rasasaṃketakalikā
- Rasasāra
- Rasataraṅgiṇī
- Rasāyanasāra
- Rasayogasāgara
- Rasayogaśataka
- Rasendracintāmaṇi
- Rasendracūḍāmaṇi
- Rasendramaṅgala
- Rasendrapurāṇa
- Rasendrasambhava
- Rasendrasārasaṅgraha
- Rasoddhāratantra (or Rasasaṃhitā)
- Rasopaniṣad

The discussion of these works in HIML gives a summary of the contents of each work, their special features, and where possible the evidence concerning their dating. Chapter 13 of HIML, *Various works on rasaśāstra and ratnaśāstra* (or *Various works on alchemy and gems*) gives brief details of a further 655 (six hundred and fifty-five) treatises. In some cases Meulenbeld gives notes on the contents and authorship of these works; in other cases references are made only to the unpublished manuscripts of these titles.

A great deal remains to be discovered about Indian alchemical literature. The content of the Sanskrit alchemical corpus has not yet (2014) been adequately integrated into the wider general history of alchemy.

Taoist Alchemists often use this alternate version of the Taijitu.

7.3.7 Chinese alchemy

Main article: Chinese alchemy

Whereas European alchemy eventually centered on the transmutation of base metals into noble metals, Chinese alchemy had a more obvious connection to medicine. The philosopher's stone of European alchemists can be compared to the Grand Elixir of Immortality sought by Chinese alchemists. However, in the hermetic view, these two goals were not unconnected, and the philosopher's stone was often equated with the universal panacea; therefore, the two traditions may have had more in common than initially appears.

Black powder may have been an important invention of Chinese alchemists. As previously stated above, Chinese alchemy was more related to medicine. It is said that the Chinese invented gunpowder while trying to find a potion for eternal life. Described in 9th-century texts and used in fireworks in China by the 10th century, it was used in cannons by 1290. From China, the use of gunpowder spread to Japan, the Mongols, the Muslim world, and Europe. Gunpowder was used by the Mongols against the Hungarians in 1241, and in Europe by the 14th century.

Chinese alchemy was closely connected to Taoist forms of traditional Chinese medicine, such as Acupuncture and Moxibustion, and to martial arts such as Tai Chi Chuan and Kung Fu (although some Tai Chi schools believe that their art derives from the philosophical or hygienic branches of Taoism, not Alchemical). In fact, in the early Song dynasty, followers of this Taoist idea (chiefly the elite and upper class) would ingest mercuric sulfide, which, though

tolerable in low levels, led many to suicide. Thinking that this consequential death would lead to freedom and access to the Taoist heavens, the ensuing deaths encouraged people to eschew this method of alchemy in favor of external sources (the aforementioned Tai Chi Chuan, mastering of the qi, etc.).

7.3.8 Alchemy as a subject of historical research

The history of alchemy has become a significant and recognized subject of academic study.[101] As the language of the alchemists is analyzed, historians are becoming more aware of the intellectual connections between that discipline and other facets of Western cultural history, such as the evolution of science and philosophy, the sociology and psychology of the intellectual communities, kabbalism, spiritualism, Rosicrucianism, and other mystic movements.[102] Institutions involved in this research include The Chymistry of Isaac Newton project at Indiana University, the University of Exeter Centre for the Study of Esotericism (EXESESO), the European Society for the Study of Western Esotericism (ESSWE), and the University of Amsterdam's Sub-department for the History of Hermetic Philosophy and Related Currents. A large collection of books on alchemy is kept in the Bibliotheca Philosophica Hermetica in Amsterdam.

Journals which publish regularly on the topic of Alchemy include 'Ambix', published by the Society for the History of alchemy and Chemistry, and 'Isis', published by The History of Science Society.

7.4 Modern alchemy

Due to the complexity and obscurity of alchemical literature, and the 18th-century disappearance of remaining alchemical practitioners into the area of chemistry; the general understanding of alchemy has been strongly influenced by several distinct and radically different interpretations.[103] Those focusing on the exoteric, such as historians of science Lawrence M. Principe and William R. Newman, have interpreted the 'decknamen' (or code words) of alchemy as physical substances. These practitioners have reconstructed physicochemical experiments that they say are described in medieval and early modern texts.[104]

At the opposite end of the spectrum, esoteric alchemists interpret these same decknamen as spiritual, religious, or psychological concepts. Today new interpretations of alchemy are still perpetuated, sometimes merging in concepts from New Age or radical environmentalism movements.[105] Groups like the rosicrucians and freemasons have a continued interest in alchemy and its symbolism. Since the Victorian revival of alchemy, "occultists reinterpreted alchemy as a spiritual practice, involving the self-transformation of the practitioner and only incidentally or not at all the transformation of laboratory substances.",[84] which has contributed to a merger of magic and alchemy in popular thought.

7.4.1 Alchemy in traditional medicine

Traditional medicine sometimes involves the transmutation of natural substances, using pharmacological or a combination of pharmacological and spiritual techniques. In Ayurveda the samskaras are claimed to transform heavy metals and toxic herbs in a way that removes their toxicity. These processes are actively used to the present day.[106]

Spagyrists of the 20th century, Albert Richard Riedel and Jean Dubuis, merged Paracelsian alchemy with occultism, teaching laboratory pharmaceutical methods. The schools they founded, *Les Philosophes de la Nature* and *The Paracelsus Research Society*, popularized modern spagyrics including the manufacture of herbal tinctures and products.[107] The courses, books, organizations, and conferences generated by their students continue to influence popular applications of alchemy as a new age medicinal practice.

7.4.2 Psychology

Alchemical symbolism has been used by psychologists such as Carl Jung, who reexamined alchemical symbolism and theory and presented the inner meaning of alchemical work as a spiritual path.[108][109] Jung was deeply interested in the occult since his youth, participating in seances, which he used as the basis for his doctoral dissertation "On the Psychology and Pathology of So-Called Occult Phenomena."[110] In 1913, Jung had already adopted a "spiritualist and redemptive interpretation of alchemy", likely reflecting his interest in the occult literature of the 19th century.[111] Jung began writing his views on alchemy from the 1920s and continued until the end of his life. His interpretation of Chinese alchemical texts in terms of his analytical psychology also served the function of comparing Eastern and Western alchemical imagery and core concepts and hence its possible inner sources (archetypes).[112][113][114]

Jung saw alchemy as a Western proto-psychology dedicated to the achievement of individuation.[108][114] In his interpretation, alchemy was the vessel by which Gnosticism survived its various purges into the Renaissance,[114][115] a concept also followed by others such as Stephan A. Hoeller. In this sense, Jung viewed alchemy as comparable to a Yoga of the East, and more adequate to the Western mind than Eastern religions and philosophies. The practice of

Alchemy seemed to change the mind and spirit of the Alchemist. Conversely, spontaneous changes on the mind of Western people undergoing any important stage in individuation seems to produce, on occasion, imagery known to Alchemy and relevant to the person's situation.[116] Jung did not completely reject the material experiments of the alchemists, but he massively downplayed it, writing that the transmutation was performed in the mind of the alchemist. He claimed the material substances and procedures were only a projection of the alchemists' internal state, while the real substance to be transformed was the mind itself.[117]

Marie-Louise von Franz, a disciple of Jung, continued Jung's studies on alchemy and its psychological meaning. Jung's work exercised a great influence on the mainstream perception of alchemy, his approach becoming a stock element in many popular texts on the subject to this day.[118] Modern scholars are sometimes critical of the Jungian approach to alchemy as overly reflective of 19th-century occultism.[21][89][119]

7.5 Magnum opus

Main article: Magnum opus (alchemy)

The Great Work of Alchemy is often described as a series of four stages represented by colors.

- *nigredo*, a blackening or melanosis
- *albedo*, a whitening or leucosis
- *citrinitas*, a yellowing or xanthosis
- *rubedo*, a reddening, purpling, or iosis[120]

7.6 Alchemy in art and entertainment

Main article: Alchemy in art and entertainment

Alchemy has had a long-standing relationship with art, seen both in alchemical texts and in mainstream entertainment. *Literary alchemy* appears throughout the history of English literature from Shakespeare to J. K. Rowling. Here, characters or plot structure follow an alchemical magnum opus. In the 14th century, Chaucer began a trend of alchemical satire that can still be seen in recent fantasy works like those of Terry Pratchett.

Visual artists had a similar relationship with alchemy. While some of them used alchemy as a source of satire, others worked with the alchemists themselves or integrated alchemical thought or symbols in their work. Music was also present in the works of alchemists and continues to influence popular performers. In the last hundred years, alchemists have been portrayed in a magical and spagyric role in fantasy fiction, film, television, novels, comics and video games.

7.7   See also

- Alchemy in art and entertainment
- Biological transmutation
- Chemistry
- Chinese alchemy
- Cupellation
- Hermes Trismegistus
- Historicism
- List of alchemists
- List of topics characterized as pseudoscience
- Magnum opus (alchemy)
- Mary the Jewess
- Nuclear transmutation
- Outline of alchemy
- Philosopher's Stone
- Physics
- Porta Alchemica
- Scientific method
- Superseded scientific theories
- Synthesis of precious metals

7.8   References

[1] Malouin, Paul-Jacques. "Alchemy." The Encyclopedia of Diderot & d'Alembert Collaborative Translation Project. Translated by Lauren Yoder. Ann Arbor: Michigan Publishing, University of Michigan Library, 2003. Web. [fill in today's date in the form 18 Apr. 2009 and remove square brackets]. <http://hdl.handle.net/2027/spo.did2222.0000. 057>. Trans. of "Alchimie," Encyclopédie ou Dictionnaire raisonné des sciences, des arts et des métiers, vol. 1. Paris, 1751.

[2] Matthew Daniel Eddy, Seymour Mauskopf and William R. Newman (Eds.) (2014). *Chemical Knowledge in the Early Modern World*. Chicago: University of Chicago Press.

[3] Alchemy at Dictionary.com.

[4] Linden 1996, pp. 7,11

[5] Linden 1996, pp. 11

[6] For a detailed look into the problems of defining alchemy, see Linden 1996, pp. 6–36

[7] Holmyard 1957, p. 16

[8] Antoine Faivre, Wouter J. Hanegraaff. *Western esotericism and the science of religion*. 1995. p.96

[9] von Franz 1997, p.

[10] Arthur Greenburg. *From alchemy to chemistry in picture and story*.

[11] H. Stanley Redgrove. *Alchemy Ancient and Modern* p.60

[12] Mitch Stokes. *Isaac Newton* p. 57

[13] Principe & Newman 2001, pp. 397–8,400

[14] William R Newman & Lawrence M Principe (1998) "The Etymological Origins of an Historiographic Mistake" in *Early Science and Medicine*, Vol. 3, No. 1 pp. 32–65

[15] Deem, Rich (2005). "The Religious Affiliation of Robert Boyle the father of modern chemistry. From: Famous Scientists Who Believed in God". adherents.com. Archived from the original on 26 March 2009. Retrieved 17 April 2009.

[16] More, Louis Trenchard (January 1941). "Boyle as Alchemist". *Journal of the History of Ideas* (University of Pennsylvania Press) **2** (1): 61–76. doi:10.2307/2707281. JSTOR 2707281.

[17] Allen G. Debus. *Alchemy and early modern chemistry*. The Society for the History of Alchemy and Chemistry. p.34.

[18] Raphael Patai. *The Jewish Alchemists: A History and Source Book*. Princeton University Press. p.4

[19] Raphael Patai. *The Jewish Alchemists: A History and Source Book*. Princeton University Press. p.3

[20] Daniel Merkur. *Gnosis: an esoteric tradition of mystical visions and unions*. State University of New York Press. p.75

[21] Newman & Principe 2002, p. 37

[22] *Newton and Newtonianism* by James E. Force, Sarah Hutton, **p211**

[23] Principe & Newman 2001, pp. 395–6

[24] Calian 2010, p.

[25] alchemy, Oxford Dictionaries

[26] "alchemy". *Oxford English Dictionary* (3rd ed.). Oxford University Press. September 2005. (Subscription or UK public library membership required.) Or see Harper, Douglas. "alchemy". *Online Etymology Dictionary*. Retrieved April 7, 2010..

[27] See, for example, the etymology for χημεία in Liddell, Henry George; Robert Scott (1901). *A Greek-English Lexicon* (Eighth edition, revised throughout ed.). Oxford: Clarendon Press. ISBN 0-19-910205-8.

[28] See, for example, both the etymology given in the Oxford English Dictionary and also that for χυμεία in Liddell, Henry George; Robert Scott; Henry Stuart Jones (1940). *A Greek-English Lexicon* (A new edition, revised and augmented throughout ed.). Oxford: Clarendon Press. ISBN 0-19-910205-8.

[29] *New Scientist*, 24–31 December 1987

[30] Garfinkel, Harold (1986). *Ethnomethodological Studies of Work*. Routledge &Kegan Paul. p. 127. ISBN 0-415-11965-0.

[31] Yves Bonnefoy. 'Roman and European Mythologies'. University of Chicago Press, 1992. pp. 211–213

[32] Clement, *Stromata*, vi. 4.

[33] Linden 1996, p. 12

[34] Partington, James Riddick (1989). *A Short History of Chemistry*. New York: Dover Publications. p. 20. ISBN 0-486-65977-1.

[35] Linden 2003, p. 46

[36] *A History of Chemistry*, Bensaude-Vincent, Isabelle Stengers, *Harvard University Press*, 1996, **p13**

[37] Linden 1996, p. 14

[38] Lindsay, Jack (1970). *The Origins of Alchemy in Graeco-Roman Egypt*. London: Muller. p. 16. ISBN 0-389-01006-5.

[39] Burckhardt, Titus (1967). *Alchemy: Science of the Cosmos, Science of the Soul*. Trans. William Stoddart. Baltimore: Penguin. p. 66. ISBN 0-906540-96-8.

[40] http://www.daviddarling.info/encyclopedia/A/alchemy.html "Alchemy"], in the Encyclopedia of Science by David J. Darling

[41] Fanning, Philip Ashley. *Isaac Newton and the Transmutation of Alchemy: An Alternative View of the Scientific Revolution*. 2009. p.6

[42] F. Sherwood Taylor. *Alchemists, Founders of Modern Chemistry*. p.26.

[43] Allen G. Debus. *Alchemy and early modern chemistry: papers from Ambix*. p. 36

[44] Glen Warren Bowersock, Peter Robert Lamont Brown, Oleg Grabar. *Late antiquity: a guide to the postclassical world*. p. 284–285

[45] Burckhardt, Titus (1967). *Alchemy: Science of the Cosmos, Science of the Soul*. Trans. William Stoddart. Baltimore: Penguin. p. 46. ISBN 0-906540-96-8.

[46] Kraus, Paul, Jâbir ibn Hayyân, *Contribution à l'histoire des idées scientifiques dans l'Islam. I. Le corpus des écrits jâbiriens. II. Jâbir et la science grecque,*. Cairo (1942–1943). Repr. By Fuat Sezgin, (Natural Sciences in Islam. 67–68), Frankfurt. 2002: (cf. Ahmad Y Hassan. "A Critical Reassessment of the Geber Problem: Part Three". Retrieved 16 September 2014.)

[47] Derewenda, Zygmunt S. (2007). "On wine, chirality and crystallography". *Acta Crystallographica Section A: Foundations of Crystallography* **64**: 246–258 [247]. Bibcode:2008AcCrA..64..246D. doi:10.1107/S0108767307054293. PMID 18156689.

[48] Holmyard 1931, p. 60

[49] Burckhardt, Titus (1967). *Alchemy: Science of the Cosmos, Science of the Soul*. Trans. William Stoddart. Baltimore: Penguin. p. 29. ISBN 0-906540-96-8.

[50] Strathern, Paul. (2000), *Mendeleyev's Dream – the Quest for the Elements*, New York: Berkley Books

[51] Moran, Bruce T. (2005). *Distilling knowledge: alchemy, chemistry, and the scientific revolution*. Harvard University Press. p. 146. ISBN 0-674-01495-2. a corpuscularian tradition in alchemy stemming from the speculations of the medieval author Geber (Jabir ibn Hayyan)

[52] Felix Klein-Frank (2001), "Al-Kindi", in Oliver Leaman & Hossein Nasr, *History of Islamic Philosophy*, p. 174. London: Routledge.

[53] Marmura ME (1965). "*An Introduction to Islamic Cosmological Doctrines: Conceptions of Nature and Methods Used for Its Study by the Ikhwan Al-Safa'an, Al-Biruni, and Ibn Sina* by Seyyed Hossein Nasr". *Speculum* **40** (4): 744–6. doi:10.2307/2851429.

[54] Robert Briffault (1938). *The Making of Humanity*, p. 196–197.

[55] Holmyard 1957, pp. 105–108

[56] Holmyard 1957, p. 110

[57] Hollister, C. Warren (1990). *Medieval Europe: A Short History* (6th ed.). Blacklick, Ohio: McGraw–Hill College. pp. 294f. ISBN 0-07-557141-2.

[58] John Read. *From Alchemy to Chemistry*. 1995 p.90

[59] James A. Weisheipl. *Albertus Magnus and the Sciences: Commemorative Essays*. PIMS. 1980. p.187-202

[60] Edmund Brehm. "Roger Bacon's Place in the History of Alchemy." *Ambix*. Vol. 23, Part I, March 1976.

[61] Holmyard 1957, pp. 120–121

[62] Holmyard 1957, pp. 134–141.

[63] Burckhardt, Titus (1967). *Alchemy: Science of the Cosmos, Science of the Soul*. Trans. William Stoddart. Baltimore: Penguin. p. 149. ISBN 0-906540-96-8.

[64] Tara E. Nummedal. *Alchemy and Authority in the Holy Roman Empire*. University of Chicago Press, 2007. p. 49

[65] John Hines, II, R. F. Yeager. *John Gower, Trilingual Poet: Language, Translation, and Tradition*. Boydell & Brewer. 2010. p.170

[66] D. Geoghegan, "A licence of Henry VI to practise Alchemy" Ambix, volume 6, 1957, pages 10-17

[67] Leah DeVun. *From Prophecy, Alchemy, and the End of Time: John of Rupescissa in the late Middle Ages*. Columbia University Press, 2009. p. 104

[68] Linden 2003, p. 123

[69] "Nicolas Flamel. Des Livres et de l'or" by Nigel Wilkins

[70] Burckhardt, Titus (1967). *Alchemy: Science of the Cosmos, Science of the Soul*. Trans. William Stoddart. Baltimore: Penguin. pp. 170–181. ISBN 0-906540-96-8.

[71] Peter J. Forshaw. '"Chemistry, That Starry Science" - Early Modern Conjunctions of Astrology and Alchemy' (2013)

[72] Peter J. Forshaw, 'Cabala Chymica or Chemia Cabalistica - Early Modern Alchemists and Cabala' (2013)

[73] Glenn Alexander Magee. *Hegel and the Hermetic Tradition*. Cornell University Press. 2008. p.30

[74] Nicholas Goodrick-Clarke. *The Western Esoteric Traditions: A Historical Introduction*. Oxford University Press. 2008 p.60

[75] Edwardes, Michael (1977). *The Dark Side of History*. New York: Stein and Day. p. 47. ISBN 0-552-11463-4.

[76] Debus, Allen G. and Multhauf, Robert P. (1966). *Alchemy and Chemistry in the Seventeenth Century*. Los Angeles: William Andrews Clark Memorial Library, University of California. pp. 6–12.

[77] Joseph Needham. *Science and Civilisation in China: Volume 5, Chemistry and Chemical Technology, Part 5, Spagyrical Discovery and Invention: Physiological Alchemy*. Cambridge University Press. P.9

[78] William Royall Newman, Anthony Grafton. *Secrets of Nature: Astrology and Alchemy in Early Modern Europe*. MIT Press, 2001. P.173.

[79]   • *Journal of the History of Ideas, 41*, 1980, **p. 293-318**

     • Principe & Newman 2001, pp. 399

     • *The Aspiring Adept: Robert Boyle and His Alchemical Quest*, by Lawrence M. Principe, 'Princeton University Press', 1998, **pp. 188 90**

[80] Tara E. Nummedal. *Alchemy and authority in the Holy Roman Empire*. p.4

[81] *Accounts of the Lord High Treasurer of Scotland*, vol. iii, (1901), 99, 202, 206, 209, 330, 340, 341, 353, 355, 365, 379, 382, 389, 409.

[82] Tara E. Nummedal. *Alchemy and authority in the Holy Roman Empire*. p.85-98

[83] Tara E. Nummedal. *Alchemy and authority in the Holy Roman Empire*. p.171

[84] Principe, Lawrence M. "Alchemy Restored." Isis 102.2 (2011): 305-12. Web.

[85] Pilkington, Roger (1959). *Robert Boyle: Father of Chemistry*. London: John Murray. p. 11.

[86] Principe & Newman 2001, p. 386

[87] Principe & Newman 2001, pp. 386–7

[88] Principe & Newman 2001, p. 387

[89] Kripal & Shuck 2005, p. 27

[90] Eliade 1994, p. 49

[91] Principe & Newman 2001, p. 388

[92] Principe & Newman 2001, p. 391

[93] Rutkin 2001, p. 143

[94] Daniel Merkur. *Gnosis: An Esoteric Tradition of Mystical Visions and Unions*. SUNY Press. 1993 p.55

[95] Multhauf, Robert P. & Gilbert, Robert Andrew (2008). *Alchemy*. Encyclopædia Britannica (2008).

[96] Meulenbeld, G. Jan (1999–2002). *History of Indian Medical Literature*. Groningen: Egbert Forsten. pp. IIA, 151–155.

[97] See Dominik Wujastyk, "An Alchemical Ghost: The Rasaratnākara of Nāgarjuna" in *Ambix* 31.2 (1984): 70-83. Online at http://univie.academia.edu/DominikWujastyk/Papers/152766/

[98] See bibliographical details and links at https://openlibrary.org/works/OL3266066W/The_Alchemical_Body

[99] DOI: 10.1093/OBO/9780195399318-0046

[100] Meulenbeld, G. Jan (1999–2002). *History of Indian Medical Literature*. Groningen: Egbert Forsten. pp. IIA, 581–738.

[101] Antoine Faivre, Wouter J. Hanegraaff. *Western esotericism and the science of religion*. 1995. p.viii–xvi

[102] See Exeter Centre for the Study of Esotericism website

[103] Principe & Newman 2001, p. 385

[104] Richard Conniff. "Alchemy May Not Have Been the Pseudoscience We All Thought It Was." Smithsonian Magazine. February 2014.

[105] Principe & Newman 2001, p. 396

[106] Junius, Manfred M; *The Practical Handbook of Plant Alchemy: An Herbalist's Guide to Preparing Medicinal Essences, Tinctures, and Elixirs*; Healing Arts Press 1985

[107] Joscelyn Godwin. *The Golden Thread: The Ageless Wisdom of the Western Mystery Traditions.* Quest Books, 2007. p.120

[108] Jung, C. G. (1944). Psychology and Alchemy (2nd ed. 1968 Collected Works Vol. 12 ISBN 0-691-01831-6). London: Routledge.

[109] Jung, C. G., & Hinkle, B. M. (1912). Psychology of the Unconscious : a study of the transformations and symbolisms of the libido, a contribution to the history of the evolution of thought. London: Kegan Paul Trench Trubner. (revised in 1952 as Symbols of Transformation, Collected Works Vol.5 ISBN 0-691-01815-4).

[110] *The Jung Cult*, by Ricard Noll, *Princeton University Press*, 1994, **p144**

[111] Noll. *Aryan Christ.* **p171**

[112] C.-G. Jung Preface to Richard Wilhelm's translation of the I Ching.

[113] C.-G. Jung Preface to the translation of The Secret of The Golden Flower.

[114] Polly Young-Eisendrath, Terence Dawson. *The Cambridge companion to Jung.* Cambridge University Press. 1997. p.33

[115] Jung, C. G., & Jaffe A. (1962). Memories, Dreams, Reflections. London: Collins. This is Jung's autobiography, recorded and edited by Aniela Jaffe, ISBN 0-679-72395-1.

[116] Jung, C. G.—Psychology and Alchemy; Symbols of Transformation.

[117] *Redemption in Alchemy*, by Carl Jung, **p210**

[118] Principe & Newman 2001, p. 401

[119] Principe & Newman 2001, p. 418

[120] Joseph Needham. *Science & Civilisation in China: Chemistry and chemical technology. Spagyrical discovery and invention: magisteries of gold and immortality.* Cambridge. 1974. p.23

7.9   Bibliography

- Calian, George (2010). *Alkimia Operativa and Alkimia Speculativa. Some Modern Controversies on the Historiography of Alchemy.* Annual of Medieval Studies at CEU.

- Eliade, Mircea (1994). *The Forge and the Crucible.* State University of New York Press.

- Forshaw, Peter J. "Chemistry, That Starry Science - Early Modern Conjunctions of Astrology and Alchemy". (2013) *Sky and Symbol* Check |url= scheme (help).

- Forshaw, Peter J. "Cabala Chymica or Chemica Cabalistica - Early Modern Alchemists and Cabala". (2013) *Ambix, Vol. 60:4* Check |url= scheme (help).

- Holmyard, Eric John (1931). *Makers of Chemistry.* Oxford: Clarendon Press.

- Holmyard, Eric John (1957). *Alchemy.* Courier Dover Publications.

- Linden, Stanton J. (1996). *Darke Hierogliphicks: Alchemy in English literature from Chaucer to the Restoration.* University Press of Kentucky.

- Linden, Stanton J. (2003). *The Alchemy Reader: from Hermes Trismegistus to Isaac Newton.* Cambridge University Press.

- Newman, William R.; Principe, Lawrence M. (2002). *Alchemy Tried in the Fire.* University of Chicago Press.

- von Franz, Marie Louise (1997). *Alchemical Active Imagination.* Boston: Shambhala Publications. ISBN 0-87773-589-1.

- Kripal, Jeffrey John; Shuck, Glenn W. (July 2005). *On the Edge of the Future.* Indiana University Press. ISBN 978-0-253-34556-1. Retrieved 17 December 2011.

- Principe, Lawrence M. (2013). *The secrets of alchemy.* Chicago &London: University of Chicago Press. ISBN 9780226682952.

- Principe, Lawrence M.; Newman, William R. (2001). "Some Problems with the Historiography of Alchemy". In Newman, William R.; Grafton, Anthony. *Secrets of Nature, Astrology and Alchemy in Modern Europe.* MIT Press. pp. 385–432. ISBN 978-0-262-14075-1. Retrieved 17 December 2011.

- Rutkin, H. Darrel (2001). "Celestial Offerings: Astrological Motifs in the Dedicatory Letters of Kepler's *Astronomia Nova* and Galileo's *Sidereus Nuncius*". In Newman, William R.; Grafton, Anthony. *Secrets of Nature, Astrology and Alchemy in Modern Europe.* MIT Press. pp. 133–172. ISBN 978-0-262-14075-1. Retrieved 17 December 2011.

- Gallina, Furio (2015). *Miti e storie di alchimisti tra il medioevo e l'età contemporanea.* Resana: mp/edizioni.

7.10   External links

- SHAC: Society for the History of Alchemy and Chemistry

- ESSWE: European Society for the Study of Western Esotericism

- Association for the Study of Esotericism

- The Alchemy Website. – Adam McLean's online collections and academic discussion.

- Inner Garden Alchemy Research Group: a non-profit foundation that aims to transmit the alchemical tradition.

-

- Alchemy on *In Our Time* at the BBC. ((Peter Forshaw, Lauren Kassell and Stephen Pumfrey) listen now)

- *Dictionary of the History of Ideas*: Alchemy

- Book of Secrets: Alchemy and the European Imagination, 1500-2000 – A digital exhibition from the Beinecke Rare Book and Manuscript Library at Yale University

Chapter 8

Alchemy and chemistry in medieval Islam

Alchemy and chemistry in Islam refers to the study of both traditional alchemy and early practical chemistry (the early chemical investigation of nature in general) by scholars in the medieval Islamic world. The word *alchemy* was derived from the Arabic word كيمياء or *kīmiyā* '.[1][2] and may ultimately derive from the ancient Egyptian word *kemi*, meaning black.[2]

After the fall of the Western Roman Empire, the focus of alchemical development moved to the Caliphate and the Islamic civilization. Much more is known about Islamic alchemy as it was better documented; most of the earlier writings that have come down through the years were preserved as Arabic translations.[3]

8.1 The definition of Islamic Alchemy and its relationship with medieval western sciences

In considering Islamic sciences as a distinct, local practice, it is important to define words such as "Arabic," "Islamic," "alchemy," and "chemistry." In order to gain a better grasp on the concepts discussed in this article, it is important to come to an understanding of what these terms mean historically. This may also help to clear up any misconceptions regarding the possible differences between alchemy and early chemistry in the context of medieval times. As A.I. Sabra writes in his article entitled, "Situating Arabic Science: Location versus Essence," "the term Arabic (or Islamic) science denotes the scientific activities of individuals who lived in a region that roughly extended chronologically from the eighth century A.D. to the beginning of the modern era, and geographically from the Iberian Peninsula and North Africa to the Indus valley and from southern Arabia to the Caspian Sea-that is, the region covered for most of that period by what we call Islamic civilization, and in which the results of the activities referred to were for the most part expressed in the Arabic language."[4] This definition of Arabic science provides a sense that there are many dis-

tinguishing factors to contrast with science of the Western hemisphere regarding physical location, culture, and language, though there are also several similarities in the goals pursued by scientists of the Middle Ages, and in the origins of thinking from which both were derived.

Lawrence Principe describes the relationship between alchemy and chemistry in his article entitled, "Alchemy Restored," in which he states, "The search for metallic transmutation—what we call "alchemy" but that is more accurately termed "chrysopoeia"—was ordinarily viewed in the late seventeenth century as synonymous with or as a subset of chemistry." [5] He therefore proposes that the early spelling of chemistry as "chymistry" refers to a unified science including both alchemy and early chemistry. Principe goes on to argue that, "[a]ll their chymical activities were unified by a common focus on the analysis, synthesis, transformation, and production of material substances."[6] Therefore, there is not a defined contrast between the two fields until the early 18th century.[7] Though Principe's discussion is centered on the Western practice of alchemy and chemistry, this argument is supported in the context of Islamic science as well when considering the similarity in methodology and Aristotelian inspirations, as noted in other sections of this article. This distinction between alchemy and early chemistry is one that lies predominately in semantics, though with an understanding of previous uses of the words, we can better understand the historical lack of distinct connotations regarding the terms despite their altered connotations in modern contexts.

The transmission of these sciences throughout the Eastern and Western hemispheres is also important to understand when distinguishing the sciences of both regions. The beginnings of cultural, religious, and scientific diffusion of information between the Western and Eastern societies began with the successful conquests of Alexander the Great (334-323 B.C). By establishing territory throughout the East, Alexander the Great allowed greater communication between the two hemispheres that would continue throughout history. A thousand years later, those Asian territories conquered by Alexander the Great, such as Iraq and

Iran, became a center of religious movements with a focus on Christianity, Manicheism, and Zoroastrianism, which all involve sacred texts as a basis, thus encouraging literacy, scholarship, and the spread of ideas.[8] Aristotelian logic was soon included in the curriculum a center for higher education in Nisibis, located east of the Persian border, and was used to enhance the philosophical discussion of theology taking place at the time.[9] When the Islamic prophet Muhammad and his followers formed the Qur'an, the holy book of Islam, it later became an important source of "theology, morality, law, and cosmology," in what Lindberg describes as "the centerpiece of Islamic education." After the death of Muhammed in 632, Islam was extended throughout the Arabian peninsula, Byzantium, Persia, Syria, Egypt, and Palestine by means of military conquest, solidifying the region as a predominately Muslim one.[10] While the expansion of the Islamic empire was an important factor in diminishing political barriers between such areas, there was still a wide range of religions, beliefs, and philosophies that could move freely and be translated throughout the regions. This development made way for contributions to be made on behalf of the East towards the Western conception of sciences such as alchemy.

While this transmission of information and practices allowed for the further development of the field, and though both were inspired by Aristotelian logic and Hellenic philosophies, it is also important to note that cultural and religious boundaries remained. The mystical and religious elements discussed previously in the article distinguished Islamic alchemy from that of its Western counterpart, given that the West had predominately Christian ideals on which to base their beliefs and results, while the Islamic tradition differed greatly. While the motives differed in some ways, as did the calculations, the practice and development of alchemy and chemistry was similar given the contemporaneous nature of the fields and the ability with which scientists could transmit their beliefs.

8.2   Alchemists and works

8.2.1   Khālid ibn Yazīd

According to the bibliographer Ibn al-Nadīm, the first Muslim alchemist was Khālid ibn Yazīd, who is said to have studied alchemy under the Christian Marianos of Alexandria. The historicity of this story is not clear; according to M. Ullmann, it is a legend.[11][12] According to Ibn al-Nadīm and Ḥajjī Khalīfa, he is the author of the alchemical works *Kitāb al-kharazāt* (*The Book of Pearls*), *Kitāb al-ṣaḥīfa al-kabīr* (*The Big Book of the Roll*), *Kitāb al-ṣaḥīfa al-saghīr* (*The Small Book of the Roll*), *Kitāb Waṣīyatihi ilā bnihi fī-ṣ-ṣanʿa* (*The Book of his Testa-*

ment to his Son about the Craft), and *Firdaws al-ḥikma* (*The Paradise of Wisdom*), but again, these works may be pseudepigraphical.[13][12][11]

8.2.2   Jābir ibn Ḥayyān

15th century European impression of "Geber"

Jābir ibn Ḥayyān (Persian: جابرحیان, Arabic: جابر بن حیان, Latin Geberus; usually rendered in English as Geber) may have been born in 721 or 722, in Persian city of Tus, Iran, and have been the son of Ḥayyān, a druggist from the tribe of al-Azd who originally lived in Kufa. When young Jābir studied in Arabia under Ḥarbī al-Himyarī. Later, he lived in Kufa, and eventually became a court alchemist for Hārūn al-Rashīd, in Baghdad. Jābir was friendly with the Barmecides and became caught up in their disgrace in 803. As a result, he returned to Kufa. According to some sources, he died in Tus in 815.

A large corpus of works is ascribed to Jābir, so large that it's difficult to believe he wrote them all himself. According to the theory of Paul Kraus, many of these works should be ascribed to later Ismaili authors. It includes the following groups of works: *The One Hundred and Twelve Books*; *The Seventy Books*; *The Ten Books of Rectifications*; and *The Books of the Balances*. This article will not distinguish between Jābir and the authors of works attributed to him.[14]

8.2.3 Abū Bakr al-Rāzī

Abū Bakr ibn Zakariyā' al-Rāzī (Latin: Rhazes), born around 864 in Rayy, was mainly known as a Persian physician. He wrote a number of alchemical works, including the *Sirr al-asrār* (Latin: *Secretum secretorum;* English: *Secret of Secrets.*)[15][16]

8.2.4 Ibn Umayl

Muḥammad ibn Umayl al-Tamīmī was an 11th-century alchemist. One of his surviving works is *Kitāb al-mā' al-waraqī wa-l-arḍ al-najmiyya* (*The Book on Silvery Water and Starry Earth.*) This work is a commentary on his poem, the *Risālat al-shams wa-l-hilāl* (*The Epistle of the Sun and the Crescent*) and contains numerous quotations from ancient authors.[17]

8.2.5 Al-Tughrai

Al-Tughrai was an 11th–12th century Persian physician.[18] whose work the *Masabih al-hikma wa-mafatih al-rahma* (The Lanterns of Wisdom and the Keys of Mercy) is one of the earliest works of material sciences.

8.2.6 Al-Jildaki

Al-Jildaki who was a Persian Alchemist urged in his book the need for experimental chemistry and mentioned many experiments *Kanz al-ikhtisas fi ma'rifat al-khawas by Abu 'l-Qasim Aydamir al-Jildaki.*

8.3   Alchemical and chemical theory

Jābir analyzed each Aristotelian element in terms of four basic qualities of *hotness*, *coldness*, *dryness*, and *moistness*. For example, fire is a substance that is hot and dry, as shown in the table.[19] (This scheme was also used by Aristotle.)[20][21] According to Jābir, in each metal two of these qualities were interior and two were exterior. For example, lead was externally cold and dry but internally hot and moist; gold, on the other hand, was externally hot and moist but internally cold and dry. He believed that metals were formed in the Earth by fusion of sulfur (giving the hot and dry qualities) with mercury (giving the cold and moist.) These elements, mercury and sulfur, should be thought of as not the ordinary elements but ideal, hypothetical substances. Which metal is formed depends on the purity of the mercury and sulfur and the proportion in which they come together.[19] The later alchemist al-Rāzī followed Jābir's mercury-sulfur theory, but added a third, salty, component.[22]

Thus, Jābir theorized, by rearranging the qualities of one metal, a different metal would result.[23] By this reasoning, the search for the philosopher's stone was introduced to Western alchemy.[24][25] Jābir developed an elaborate numerology whereby the root letters of a substance's name in Arabic, when treated with various transformations, held correspondences to the element's physical properties.[19]

8.4   Processes and equipment

Al-Rāzī mentions the following chemical processes:

- distillation,

- calcination,

- solution,

- evaporation,

- crystallization,

- sublimation,

- filtration,

- amalgamation,

- and ceration (a process for making solids pasty or fusible.)[26]

Some of these operations (calcination, solution, filtration, crystallization, sublimation and distillation) are also known to have been practiced by pre-Islamic Alexandrian alchemists.[27]

In his *Secretum secretorum*, Al-Rāzī mentions the following equipment:[28]

- Tools for melting substances (*li-tadhwīb*): hearth (*kūr*), bellows (*minfākh* or *ziqq*), crucible (*bawtaqa*), the *būt bar būt* (in Arabic, from Persian) or *botus barbatus* (in Latin), ladle (*mighrafa* or *mil'aqa*), tongs (*māsik* or *kalbatān*), scissors (*miqta'*), hammer (*mukassir*), file (*mibrad*).

- Tools for the preparation of drugs (*li-tadbīr al-'aqāqīr*): cucurbit and still with evacuation tube (*qar'* or *anbīq dhū khatm*), receiving matras (*qābila*), blind still (without evacuation tube) (*al-anbīq al-a'mā*), aludel (*al-uthāl*), goblets (*qadaḥ*), flasks (*qārūra*, plural *quwārīr*), rosewater flasks (*mā' wardiyya*), cauldron (*marjal* or *tanjīr*), earthenware pots varnished on the inside with their lids (*qudūr* and *makabbāt*), water bath or sand bath (*qidr*), oven (*al-tannūr* in Arabic, *athanor* in Latin), small cylindirical oven for heating aludel (*mustawqid*), funnels, sieves, filters, etc.

8.5   See also

- Chinese alchemy
- Islamic science

8.6   References

[1] "alchemy", entry in *The Oxford English Dictionary*, J. A. Simpson and E. S. C. Weiner, vol. 1, 2nd ed., 1989, ISBN 0-19-861213-3.

[2] p. 854, "Arabic alchemy", Georges C. Anawati, pp. 853-885 in *Encyclopedia of the history of Arabic science*, eds. Roshdi Rashed and Régis Morelon, London: Routledge, 1996, vol. 3, ISBN 0-415-12412-3.

[3] Burckhardt, Titus (1967). "Alchemy: science of the cosmos, science of the soul". Stuart & Watkins. p. 46.

[4] Sabra 1996, P. 655

[5] Principe 2011, P. 306

[6] Principe 2011, P. 306

[7] Principe 2011, P. 306

[8] Lindberg 2007, P. 163

[9] Lindberg 2007, P. 164

[10] Lindberg 2007, P. 166

[11] pp. 63-66, *Alchemy*, E. J. Holmyard, New York: Dover Publications, Inc., 1990 (reprint of 1957 Penguin Books edition), ISBN 0-486-26298-7.

[12] M. Ullmann, "Ḵẖālid b. Yazīd b. Muʿāwiya, abū hāṣẖim.", in *Encyclopaedia of Islam*, second edition, edited by P. Bearman, Th. Bianquis, C. E. Bosworth, E. van Donzel, and W.P. Heinrichs, Brill, 2011. Brill Online. Accessed 20 January 2011. <http://www.brillonline.nl/subscriber/entry?entry=islam_SIM-4151>

[13] Anawati 1996, p. 864.

[14] pp. 68-82, Holmyard 1990.

[15] pp. 867-879, Anawati 1996.

[16] pp. 86-92, Holmyard 1990.

[17] pp. 870-872, Anawati 1996.

[18] El Khadem, H. S. (1995). "A Lost Text By Zosimos Reproduced in an Old Alchemy Book". *Journal of Chemical Education* **72** (9): 774. doi:10.1021/ed072p774.

[19] pp. 74-82, Holmyard 1990.

[20] Holmyard 1990, pp. 21-22.

[21] Aristotle, *On Generation and Corruption*, II.3, 330a-330b.

[22] Holmyard 1990, p. 88.

[23] Burckhardt, Titus (1967). "Alchemy: science of the cosmos, science of the soul". Stuart & Watkins. p. 29.

[24] Ragai, Jehane (1992). "The Philosopher's Stone: Alchemy and Chemistry". *Journal of Comparative Poetics* **12** (Metaphor and Allegory in the Middle Ages): 58–77. doi:10.2307/521636.

[25] Holmyard, E. J. (1924). "Maslama al-Majriti and the Rutbatu'l-Hakim". *Isis* **6** (3): 293–305. doi:10.1086/358238.

[26] p. 89, Holmyard 1990.

[27] p. 23, *A short history of chemistry*, James Riddick Partington, 3rd ed., Courier Dover Publications, 1989, ISBN 0-486-65977-1.

[28] Anawati 1996, p. 868

8.7   Further reading

- Lindberg, David C. (2007). "Islamic Science". *The Beginnings of Western Science: The European Scientific Tradition in Philosophical, Religious, and Institutional Context, Prehistory to A.D. 1450*. Chicago: U of Chicago. pp. 163–92. ISBN 978-0-226-48205-7.

- Principe, Lawrence M. (2011). "Alchemy Restored". *Isis* **102** (2): 305–12. doi:10.1086/660139.

- Sabra, A. I. (1996). "Situating Arabic Science: Locality versus Essence". *Isis* **87** (4): 654–70. doi:10.1086/357651. JSTOR 235197.

8.8   External links

- "How Greek Science Passed to the Arabs" by De Lacy O'Leary

Chapter 9

Chemical revolution

*Geoffroy's 1718 **Affinity Table**: at the head of each column is a chemical species with which all the species below can combine. Some historians have defined this table as being the start of the chemical revolution.*[1]

The **chemical revolution**, also called the *first chemical revolution*, was the early modern reformulation of chemistry that culminated in the law of conservation of mass and the oxygen theory of combustion. During the 19th and 20th century, this transformation was credited to the work of the French chemist Antoine Lavoisier (the "father of modern chemistry").[2] However, recent work on the history of early modern chemistry considers the chemical revolution to consist of gradual changes in chemical theory and practice that emerged over a period of two centuries.[3] The so-called scientific revolution took place during the sixteenth and seventeenth centuries whereas the chemical revolution took place during the seventeenth and eighteenth centuries.[4]

Several factors led to this revolution. First, there were the forms of gravimetric analysis that emerged from alchemy and new kinds of instruments that were developed in medical and industrial contexts. In these settings, chemists increasingly challenged hypotheses that had already been presented by the ancient Greeks. For example, chemists began to assert that all structures were composed of more than the four elements of the Greeks or the eight elements of the medieval alchemists. The Irish alchemist, Robert Boyle, laid the foundations for the Chemical Revolution, with his mechanical corpuscular philosophy, which in turn relied heavily on the alchemical corpuscular theory and experimental method dating back to pseudo-Geber.[5]

Other factors included new experimental techniques and the discovery of 'fixed air' (carbon dioxide) by Joseph Black in the middle of the 18th century. This discovery was particularly important because it empirically proved that 'air' did not consist of only one substance and because it established 'gas' as an important experimental substance. Nearer the end of the 18th century, the experiments by Henry Cavendish and Joseph Priestley further proved that air is not an element and is instead composed of several different gases. Lavoisier also translated the names of chemical substance into a new nomenclatural language more appealing to scientists of the nineteenth century. Such changes took place in an atmosphere in which the industrial revolution increased public interest in learning and practicing chemistry. When describing the task of reinventing chemical nomenclature, Lavoisier attempted to harness the new centrality of chemistry by making the rather hyperbolic claim that:[6]

The latter stages of the revolution was fuelled by the 1789 publication of Lavoisier's *Traité Élémentaire de Chimie* (Elements of Chemistry). Beginning with this publication and others to follow, Lavoisier synthesised the work of others and coined the term "oxygen". He also explained the theory of combustion, and challenged the phlogiston theory with his views on caloric. The *Traité* incorporates notions of a "new chemistry" and describes the experiments and reasoning that led to his conclusions. Like Newton's *Principia*, which was the high point of the Scientific Revolution, Lavoisier's *Traité* can be seen as the culmination of the Chemical Revolution.

Lavoisier's work was not immediately accepted and it took several decades for it gain momentum.[7] This transition was aided by the work of Jöns Jakob Berzelius, who came up with a simplified shorthand to describe chemical compounds based on John Dalton's theory of atomic weights.

9.1   References

[1] Kim, Mi Gyung (2003). *Affinity, That Elusive Dream: A Genealogy of the Chemical Revolution*. MIT Press. ISBN 978-0-262-11273-4.

[2] The First Chemical Revolution – the Instrument Project, The College of Wooster

[3] Matthew Daniel Eddy, Seymour Mauskopf, and William R. Newman (2014). "An Introduction to Chemical Knowledge in the Early Modern World". *Osiris* **29**: 1–15. doi:10.1086/678110.

[4] Matthew Daniel Eddy, Seymour Mauskopf and William R. Newman (Eds.) (2014). *Chemical Knowledge in the Early Modern World*. Chicago: University of Chicago Press.

[5] Ursula Klein (July 2007). "Styles of Experimentation and Alchemical Matter Theory in the Scientific Revolution". *Metascience* (Springer) **16** (2): 247–256 [247]. doi:10.1007/s11016-007-9095-8. ISSN 1467-9981.

[6] Jaffe, B. (1976). *Crucibles: The Story of Chemistry from Alchemy to Nuclear Fission* (4th ed.). New York: Dover Publications. ISBN 978-0-486-23342-0.

[7] Eddy, Matthew Daniel (2008). *The Language of Mineralogy: John Walker, Chemistry and the Edinburgh Medical School 1750-1800*. Ashgate.

9.2   Further reading

- William B. Jensen, "Logic, History, and the Chemistry Textbook: III. One Chemical Revolution or Three?", *Journal of Chemical Education*, Vol. 75, No. 8, August 1998

- John G. McEvoy (2010). *Historiography of the Chemical Revolution: Patterns of Interpretation in the History of Science*. Pickering & Chatto. ISBN 978-1-84893-030-8. See also book review by Seymour Mauskopf in *HYLE--International Journal for Philosophy of Chemistry*, Vol. 17, No.1 (2011), pp. 41–46.

9.3   External links

- Chemistry :: The chemical revolution – Encyclopædia Britannica

- A bibliography on the chemical revolution – University of Valencia

Chapter 10

Atomic theory

"Atomic model" redirects here. For the unrelated term in mathematical logic, see Atomic model (mathematical logic).

This article is about the historical models of the atom. For a history of the study of how atoms combine to form molecules, see History of molecular theory.

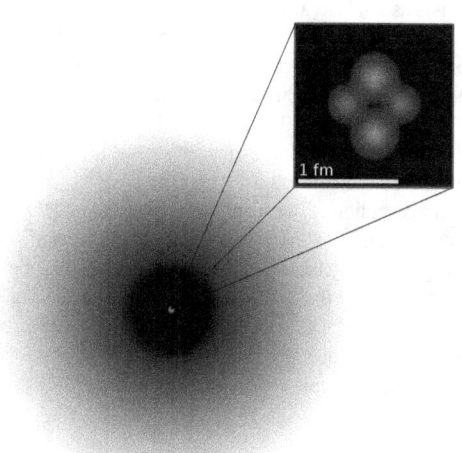

1 Å = 100,000 fm

The current theoretical model of the atom involves a dense nucleus surrounded by a probabilistic "cloud" of electrons

In chemistry and physics, **atomic theory** is a scientific theory of the nature of matter, which states that matter is composed of discrete units called atoms. It began as a philosophical concept in ancient Greece and entered the scientific mainstream in the early 19th century when discoveries in the field of chemistry showed that matter did indeed behave as if it were made up of atoms.

The word *atom* comes from the Ancient Greek adjective *atomos*, meaning "uncuttable".[1] 19th century chemists began using the term in connection with the growing number of irreducible chemical elements. While seemingly apropos, around the turn of the 20th century, through various experiments with electromagnetism and radioactivity, physicists discovered that the so-called "uncuttable atom" was actually a conglomerate of various subatomic particles (chiefly, electrons, protons and neutrons) which can exist separately from each other. In fact, in certain extreme environments, such as neutron stars, extreme temperature and pressure prevents atoms from existing at all. Since atoms were found to be divisible, physicists later invented the term "elementary particles" to describe the "uncuttable", though not indestructible, parts of an atom. The field of science which studies subatomic particles is particle physics, and it is in this field that physicists hope to discover the true fundamental nature of matter.

10.1 History

10.1.1 Philosophical atomism

Main article: Atomism

The idea that matter is made up of discrete units is a very old one, appearing in many ancient cultures such as Greece and India. However, these ideas were founded in philosophical and theological reasoning rather than evidence and experimentation. Because of this, they could not convince everybody, so atomism was but one of a number of competing theories on the nature of matter. It was not until the 19th century that the idea was embraced and refined by scientists, as the blossoming science of chemistry produced discoveries that could easily be explained using the concept of atoms.

10.1.2 Dalton

Near the end of the 18th century, two laws about chemical reactions emerged without referring to the notion of an atomic theory. The first was the law of conservation of mass, formulated by Antoine Lavoisier in 1789, which states that the total mass in a chemical reaction remains constant (that is, the reactants have the same mass as the products).[2] The second was the law of definite proportions. First proven by the French chemist Joseph Louis Proust in 1799,[3] this law states that if a compound is broken down into its constituent elements, then the masses of the constituents will always have the same proportions, regardless of the quantity or source of the original substance.

John Dalton studied and expanded upon this previous work and developed the law of multiple proportions: if two elements can be combined to form a number of possible compounds, then the ratios of the masses of the second element which combine with a fixed mass of the first element will be ratios of small whole numbers. For example: Proust had studied tin oxides and found that their masses were either 88.1% tin and 11.9% oxygen or 78.7% tin and 21.3% oxygen (these were tin(II) oxide and tin dioxide respectively). Dalton noted from these percentages that 100g of tin will combine either with 13.5g or 27g of oxygen; 13.5 and 27 form a ratio of 1:2. Dalton found that an atomic theory of matter could elegantly explain this common pattern in chemistry. In the case of Proust's tin oxides, one tin atom will combine with either one or two oxygen atoms.[4]

Dalton also believed atomic theory could explain why water absorbed different gases in different proportions - for example, he found that water absorbed carbon dioxide far better than it absorbed nitrogen.[5] Dalton hypothesized this was due to the differences in mass and complexity of the gases' respective particles. Indeed, carbon dioxide molecules (CO_2) are heavier and larger than nitrogen molecules (N_2).

Dalton proposed that each chemical element is composed of atoms of a single, unique type, and though they cannot be altered or destroyed by chemical means, they can combine to form more complex structures (chemical compounds). This marked the first truly scientific theory of the atom, since Dalton reached his conclusions by experimentation and examination of the results in an empirical fashion.

In 1803 Dalton orally presented his first list of relative atomic weights for a number of substances. This paper was published in 1805, but he did not discuss there exactly how he obtained these figures.[5] The method was first revealed in 1807 by his acquaintance Thomas Thomson, in the third edition of Thomson's textbook, *A System of Chemistry*. Finally, Dalton published a full account in his own textbook, *A New System of Chemical Philosophy*, 1808 and 1810.

Dalton estimated the atomic weights according to the mass ratios in which they combined, with the hydrogen atom taken as unity. However, Dalton did not conceive that with some elements atoms exist in molecules — e.g. pure oxygen

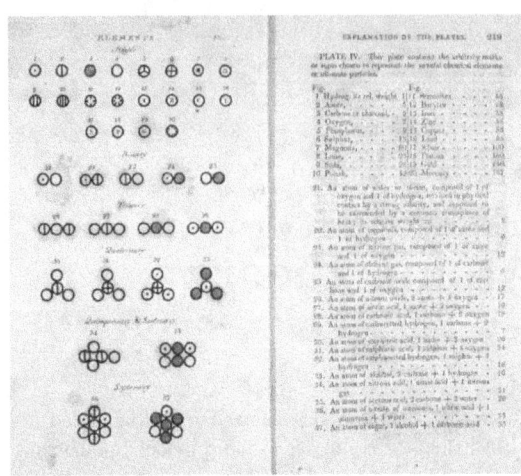

Various atoms and molecules as depicted in John Dalton's A New System of Chemical Philosophy *(1808).*

exists as O_2. He also mistakenly believed that the simplest compound between any two elements is always one atom of each (so he thought water was HO, not H_2O).[6] This, in addition to the crudity of his equipment, flawed his results. For instance, in 1803 he believed that oxygen atoms were 5.5 times heavier than hydrogen atoms, because in water he measured 5.5 grams of oxygen for every 1 gram of hydrogen and believed the formula for water was HO. Adopting better data, in 1806 he concluded that the atomic weight of oxygen must actually be 7 rather than 5.5, and he retained this weight for the rest of his life. Others at this time had already concluded that the oxygen atom must weigh 8 relative to hydrogen equals 1, if one assumes Dalton's formula for the water molecule (HO), or 16 if one assumes the modern water formula (H_2O).[7]

10.1.3   Avogadro

The flaw in Dalton's theory was corrected in principle in 1811 by Amedeo Avogadro. Avogadro had proposed that equal volumes of any two gases, at equal temperature and pressure, contain equal numbers of molecules (in other words, the mass of a gas's particles does not affect the volume that it occupies).[8] Avogadro's law allowed him to deduce the diatomic nature of numerous gases by studying the volumes at which they reacted. For instance: since two liters of hydrogen will react with just one liter of oxygen to produce two liters of water vapor (at constant pressure and temperature), it meant a single oxygen molecule splits in two in order to form two particles of water. Thus, Avogadro was able to offer more accurate estimates of the atomic mass of oxygen and various other elements, and made a clear distinction between molecules and atoms.

10.1.4 Brownian Motion

In 1827, the British botanist Robert Brown observed that dust particles inside pollen grains floating in water constantly jiggled about for no apparent reason. In 1905, Albert Einstein theorized that this Brownian motion was caused by the water molecules continuously knocking the grains about, and developed a hypothetical mathematical model to describe it.[9] This model was validated experimentally in 1908 by French physicist Jean Perrin, thus providing additional validation for particle theory (and by extension atomic theory).

10.1.5 Discovery of subatomic particles

Main articles: Electron and Plum pudding model

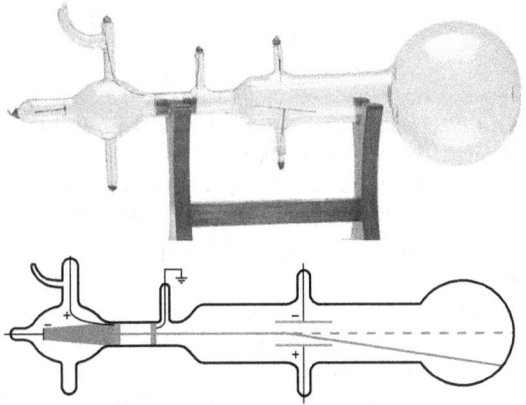

The cathode rays (blue) were emitted from the cathode, sharpened to a beam by the slits, then deflected as they passed between the two electrified plates.

Atoms were thought to be the smallest possible division of matter until 1897 when J.J. Thomson discovered the electron through his work on cathode rays.[10]

A Crookes tube is a sealed glass container in which two electrodes are separated by a vacuum. When a voltage is applied across the electrodes, cathode rays are generated, creating a glowing patch where they strike the glass at the opposite end of the tube. Through experimentation, Thomson discovered that the rays could be deflected by an electric field (in addition to magnetic fields, which was already known). He concluded that these rays, rather than being a form of light, were composed of very light negatively charged particles he called "corpuscles" (they would later be renamed electrons by other scientists). He measured the mass-to-charge ratio and discovered it was 1800 times smaller than that of hydrogen, the smallest atom. These corpuscles were a particle unlike any other previously known.

Thomson suggested that atoms were divisible, and that the corpuscles were their building blocks.[11] To explain the overall neutral charge of the atom, he proposed that the corpuscles were distributed in a uniform sea of positive charge; this was the plum pudding model[12] as the electrons were embedded in the positive charge like plums in a plum pudding (although in Thomson's model they were not stationary).

10.1.6 Discovery of the nucleus

Main article: Rutherford model
Thomson's plum pudding model was disproved in 1909 by

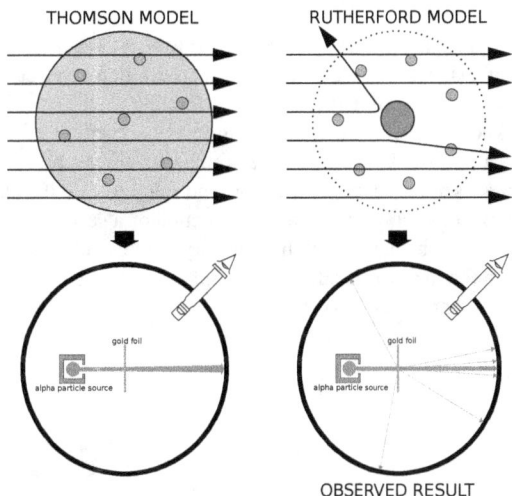

The Geiger-Marsden experiment
Left: *Expected results: alpha particles passing through the plum pudding model of the atom with negligible deflection.*
Right: *Observed results: a small portion of the particles were deflected by the concentrated positive charge of the nucleus.*

one of his former students, Ernest Rutherford, who discovered that most of the mass and positive charge of an atom is concentrated in a very small fraction of its volume, which he assumed to be at the very center.

In the Geiger–Marsden experiment, Hans Geiger and Ernest Marsden (colleagues of Rutherford working at his behest) shot alpha particles at thin sheets of metal and measured their deflection through the use of a fluorescent screen.[13] Given the very small mass of the electrons, the high momentum of the alpha particles, and the low concentration of the positive charge of the plum pudding model, the experimenters expected all the alpha particles to pass through the metal foil without significant deflection. To their astonishment, a small fraction of the alpha particles experienced heavy deflection. Rutherford concluded that the positive charge of the atom must be concentrated in a very tiny volume to produce an electric field sufficiently intense to deflect the alpha particles so strongly.

This led Rutherford to propose a planetary model in which a cloud of electrons surrounded a small, compact nucleus of positive charge. Only such a concentration of charge could produce the electric field strong enough to cause the heavy deflection.[14]

10.1.7   First steps toward a quantum physical model of the atom

Main article: Bohr model

The planetary model of the atom had two significant shortcomings. The first is that, unlike planets orbiting a sun, electrons are charged particles. An accelerating electric charge is known to emit electromagnetic waves according to the Larmor formula in classical electromagnetism. An orbiting charge should steadily lose energy and spiral toward the nucleus, colliding with it in a small fraction of a second. The second problem was that the planetary model could not explain the highly peaked emission and absorption spectra of atoms that were observed.

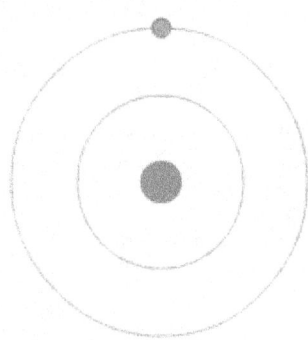

The Bohr model of the atom

Quantum theory revolutionized physics at the beginning of the 20th century, when Max Planck and Albert Einstein postulated that light energy is emitted or absorbed in discrete amounts known as quanta (singular, *quantum*). In 1913, Niels Bohr incorporated this idea into his Bohr model of the atom, in which an electron could only orbit the nucleus in particular circular orbits with fixed angular momentum and energy, its distance from the nucleus (i.e., their radii) being proportional to its energy.[15] Under this model an electron could not spiral into the nucleus because it could not lose energy in a continuous manner; instead, it could

only make instantaneous "quantum leaps" between the fixed energy levels.[15] When this occurred, light was emitted or absorbed at a frequency proportional to the change in energy (hence the absorption and emission of light in discrete spectra).[15]

Bohr's model was not perfect. It could only predict the spectral lines of hydrogen; it couldn't predict those of multielectron atoms. Worse still, as spectrographic technology improved, additional spectral lines in hydrogen were observed which Bohr's model couldn't explain. In 1916, Arnold Sommerfeld added elliptical orbits to the Bohr model to explain the extra emission lines, but this made the model very difficult to use, and it still couldn't explain more complex atoms.

10.1.8   Discovery of isotopes

Main article: Isotope

While experimenting with the products of radioactive decay, in 1913 radiochemist Frederick Soddy discovered that there appeared to be more than one element at each position on the periodic table.[16] The term isotope was coined by Margaret Todd as a suitable name for these elements.

That same year, J.J. Thomson conducted an experiment in which he channeled a stream of neon ions through magnetic and electric fields, striking a photographic plate at the other end. He observed two glowing patches on the plate, which suggested two different deflection trajectories. Thomson concluded this was because some of the neon ions had a different mass.[17] The nature of this differing mass would later be explained by the discovery of neutrons in 1932.

10.1.9   Discovery of nuclear particles

Main article: Atomic nucleus

In 1917 Rutherford bombarded nitrogen gas with alpha particles and observed hydrogen nuclei being emitted from the gas (Rutherford recognized these, because he had previously obtained them bombarding hydrogen with alpha particles, and observing hydrogen nuclei in the products). Rutherford concluded that the hydrogen nuclei emerged from the nuclei of the nitrogen atoms themselves (in effect, he had split a nitrogen).[18]

From his own work and the work of his students Bohr and Henry Moseley, Rutherford knew that the positive charge of any atom could always be equated to that of an integer number of hydrogen nuclei. This, coupled with the atomic mass of many elements being roughly equivalent to an integer number of hydrogen atoms - then assumed to be the lightest particles - led him to conclude that hydrogen nuclei were singular particles and a basic constituent of all atomic nuclei. He named such particles protons. Further experimentation by Rutherford found that the nuclear mass of most atoms exceeded that of the protons it possessed; he speculated that this surplus mass was composed of hitherto unknown neutrally charged particles, which were tentatively dubbed "neutrons".

In 1928, Walter Bothe observed that beryllium emitted a highly penetrating, electrically neutral radiation when bombarded with alpha particles. It was later discovered that this radiation could knock hydrogen atoms out of paraffin wax. Initially it was thought to be high-energy gamma radiation, since gamma radiation had a similar effect on electrons in metals, but James Chadwick found that the ionization effect was too strong for it to be due to electromagnetic radiation, so long as energy and momentum were conserved in the interaction. In 1932, Chadwick exposed various elements, such as hydrogen and nitrogen, to the mysterious "beryllium radiation", and by measuring the energies of the recoiling charged particles, he deduced that the radiation was actually composed of electrically neutral particles which could not be massless like the gamma ray, but instead were required to have a mass similar to that of a proton. Chadwick now claimed these particles as Rutherford's neutrons.[19] For his discovery of the neutron, Chadwick received the Nobel Prize in 1935.

10.1.10 Quantum physical models of the atom

Main article: Atomic orbital

In 1924, Louis de Broglie proposed that all moving parti-

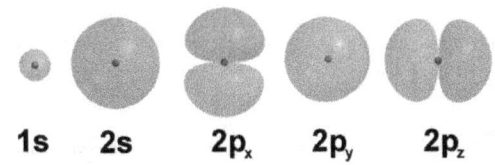

The five filled atomic orbitals of a neon atom separated and arranged in order of increasing energy from left to right, with the last three orbitals being equal in energy. Each orbital holds up to two electrons, which most probably exist in the zones represented by the colored bubbles. Each electron is equally present in both orbital zones, shown here by color only to highlight the different wave phase.

cles — particularly subatomic particles such as electrons — exhibit a degree of wave-like behavior. Erwin Schrödinger, fascinated by this idea, explored whether or not the movement of an electron in an atom could be better explained as a

wave rather than as a particle. Schrödinger's equation, published in 1926,[20] describes an electron as a wavefunction instead of as a point particle. This approach elegantly predicted many of the spectral phenomena that Bohr's model failed to explain. Although this concept was mathematically convenient, it was difficult to visualize, and faced opposition.[21] One of its critics, Max Born, proposed instead that Schrödinger's wavefunction described not the electron but rather all its possible states, and thus could be used to calculate the probability of finding an electron at any given location around the nucleus.[22] This reconciled the two opposing theories of particle versus wave electrons and the idea of wave–particle duality was introduced. This theory stated that the electron may exhibit the properties of both a wave and a particle. For example, it can be refracted like a wave, and has mass like a particle.[23]

A consequence of describing electrons as waveforms is that it is mathematically impossible to simultaneously derive the position and momentum of an electron. This became known as the Heisenberg uncertainty principle after the theoretical physicist Werner Heisenberg, who first described it and published it in 1927.[24] This invalidated Bohr's model, with its neat, clearly defined circular orbits. The modern model of the atom describes the positions of electrons in an atom in terms of probabilities. An electron can potentially be found at any distance from the nucleus, but, depending on its energy level, exists more frequently in certain regions around the nucleus than others; this pattern is referred to as its atomic orbital. The orbitals come in a variety of shapes-sphere, dumbbell, torus, etc.-with the nucleus in the middle.[25]

10.2 See also

- History of the molecule
- Discoveries of the chemical elements
- Introduction to quantum mechanics
- Kinetic theory
- Atomism
- *The Physical Principles of the Quantum Theory*

10.3 Notes

[1] Berryman, Sylvia, "Ancient Atomism", *The Stanford Encyclopedia of Philosophy* (Fall 2008 Edition), Edward N. Zalta (ed.), http://plato.stanford.edu/archives/fall2008/entries/atomism-ancient/

[2] Weisstein, Eric W. "Lavoisier, Antoine (1743-1794)". scienceworld.wolfram.com. Retrieved 2009-08-01.

[3] Proust, Joseph Louis. "Researches on Copper", excerpted from *Ann. chim.* 32, 26-54 (1799) [as translated and reproduced in Henry M. Leicester and Herbert S. Klickstein, *A Source Book in Chemistry, 1400–1900* (Cambridge, Massachusetts: Harvard, 1952)]. Retrieved on August 29, 2007.

[4] Andrew G. van Melsen (1952). *From Atomos to Atom.* Mineola, N.Y.: Dover Publications. ISBN 0-486-49584-1.

[5] Dalton, John. "On the Absorption of Gases by Water and Other Liquids", in *Memoirs of the Literary and Philosophical Society of Manchester*. 1803. Retrieved on August 29, 2007.

[6] Johnson, Chris. "Avogadro - his contribution to chemistry". Archived from the original on 27 June 2009. Retrieved 2009-08-01.

[7] Alan J. Rocke (1984). *Chemical Atomism in the Nineteenth Century*. Columbus: Ohio State University Press.

[8] Avogadro, Amedeo (1811). "Essay on a Manner of Determining the Relative Masses of the Elementary Molecules of Bodies, and the Proportions in Which They Enter into These Compounds". *Journal de Physique* 73: 58–76.

[9] Einstein, A. (1905). "Über die von der molekularkinetischen Theorie der Wärme geforderte Bewegung von in ruhenden Flüssigkeiten suspendierten Teilchen". *Annalen der Physik* 322 (8): 549. Bibcode:1905AnP...322..549E. doi:10.1002/andp.19053220806.

[10] Thomson, J.J. (1897). "Cathode rays" ([facsimile from Stephen Wright, Classical Scientific Papers, Physics (Mills and Boon, 1964)]). *Philosophical Magazine* 44 (269): 293. doi:10.1080/14786449708621070.

[11] Whittaker, E. T. (1951), *A history of the theories of aether and electricity. Vol 1*, Nelson, London

[12] Thomson, J.J. (1904). "On the Structure of the Atom: an Investigation of the Stability and Periods of Oscillation of a number of Corpuscles arranged at equal intervals around the Circumference of a Circle; with Application of the Results to the Theory of Atomic Structure". *Philosophical Magazine* 7 (39): 237. doi:10.1080/14786440409463107.

[13] Geiger, H (1910). "The Scattering of the α-Particles by Matter". *Proceedings of the Royal Society* A 83: 492–504.

[14] Rutherford, Ernest (1911). "The Scattering of α and β Particles by Matter and the Structure of the Atom" (PDF). *Philosophical Magazine* 21 (4): 669. Bibcode:2012PMag...92..379R. doi:10.1080/14786435.2011.617037.

[15] Bohr, Niels (1913). "On the constitution of atoms and molecules" (PDF). *Philosophical Magazine* 26 (153): 476–502. doi:10.1080/14786441308634993.

[16] "Frederick Soddy, The Nobel Prize in Chemistry 1921". Nobel Foundation. Retrieved 2008-01-18.

[17] Thomson, J.J. (1913). "Rays of positive electricity". *Proceedings of the Royal Society* A 89 (607): 1–20. Bibcode:1913RSPSA..89....1T. doi:10.1098/rspa.1913.0057. [as excerpted in Henry A. Boorse & Lloyd Motz, *The World of the Atom*, Vol. 1 (New York: Basic Books, 1966)]. Retrieved on August 29, 2007.

[18] Rutherford, Ernest (1919). "Collisions of alpha Particles with Light Atoms. IV. An Anomalous Effect in Nitrogen". *Philosophical Magazine* 37 (222): 581. doi:10.1080/14786440608635919.

[19] Chadwick, James (1932). "Possible Existence of a Neutron" (PDF). *Nature* 129 (3252): 312. Bibcode:1932Natur.129Q.312C. doi:10.1038/129312a0.

[20] Schrödinger, Erwin (1926). "Quantisation as an Eigenvalue Problem". *Annalen der Physik* 81 (18): 109–139. Bibcode:1926AnP...386..109S. doi:10.1002/andp.19263861802.

[21] Mahanti, Subodh. "Erwin Schrödinger: The Founder of Quantum Wave Mechanics". Retrieved 2009-08-01.

[22] Mahanti, Subodh. "Max Born: Founder of Lattice Dynamics". Retrieved 2009-08-01.

[23] Greiner, Walter. "Quantum Mechanics: An Introduction". Retrieved 2010-06-14.

[24] Heisenberg, W. (1927). "Über den anschaulichen Inhalt der quantentheoretischen Kinematik und Mechanik". *Zeitschrift für Physik* (in German) 43 (3–4): 172–198. Bibcode:1927ZPhy...43..172H. doi:10.1007/BF01397280.

[25] Milton Orchin, Roger Macomber, Allan Pinhas, R. Wilson. "The Vocabulary and Concepts of Organic Chemistry, Second Edition," (PDF). Retrieved 2010-06-14.

10.4   Further reading

- Bernard Pullman (1998) *The Atom in the History of Human Thought*, trans. by Axel Reisinger. Oxford Univ. Press.

- Eric Scerri (2007) *The Periodic Table, Its Story and Its Significance*, Oxford University Press, New York.

- Charles Adolphe Wurtz (1881) *The Atomic Theory*, D. Appleton and Company, New York.

10.5   External links

- Atomism by S. Mark Cohen.

- Atomic Theory - detailed information on atomic theory with respect to electrons and electricity.

Chapter 11

Chemical formula

Not to be confused with the 2-D graphical method of showing atomic spacial relationships called a structural formula.

A **chemical formula** is a way of expressing information

*Structural formula for butane. This is **not** a chemical formula. Examples of chemical formulas for butane are the* empirical formula C_2H_5, *the **molecular formula** C_4H_{10} and the* condensed (or semi-structural) formula $CH_3CH_2CH_2CH_3$.

about the proportions of atoms that constitute a particular chemical compound, using a single line of chemical element symbols, numbers, and sometimes also other symbols, such as parentheses, dashes, brackets, commas and *plus* (+) and *minus* (−) signs. These are limited to a single typographic line of symbols, which may include subscripts and superscripts. A chemical formula is not a chemical name, and it contains no words. Although a chemical formula may imply certain simple chemical structures, it is not the same as a full chemical structural formula. Chemical formulas can fully specify the structure of only the simplest of molecules and chemical substances, and are generally more limited in power than are chemical names and structural formulas.

The simplest types of chemical formulas are called *empirical formulas*, which use letters and numbers indicating the numerical *proportions* of atoms of each type. *Molecular formulas* indicate the simple *numbers* of each type of atom in a molecule, with no information on structure. For example, the empirical formula for glucose is CH_2O (twice as many hydrogen atoms as carbon and oxygen), while its molecular formula is $C_6H_{12}O_6$ (12 hydrogen atoms, six carbon and oxygen atoms).

Sometimes a chemical formula is complicated by being written as a condensed formula (or condensed molecular formula, occasionally called a "semi-structural formula"), which conveys additional information about the particular ways in which the atoms are chemically bonded together, either in covalent bonds, ionic bonds, or various combinations of these types. This is possible if the relevant bonding is easy to show in one dimension. An example is the condensed molecular/chemical formula for ethanol, which is CH_3-CH_2-OH or CH_3CH_2OH. However, even a condensed chemical formula is necessarily limited in its ability to show complex bonding relationships between atoms, especially atoms that have bonds to four or more different substituents.

Since a chemical formula must be expressed as a single line of chemical element symbols, it often cannot be as informative as a true structural formula, which is a graphical representation of the spacial relationship between atoms in chemical compounds (see for example the figure for butane structural and chemical formulas, at right). For reasons of structural complexity, there is no condensed chemical formula (or semi-structural formula) that specifies glucose (and there exist many different molecules, for example fructose and mannose, have the same molecular formula $C_6H_{12}O_6$ as glucose). Linear equivalent chemical *names* exist that can and do specify any complex structural formula (see chemical nomenclature), but such names must use many terms (words), rather than the simple element symbols, numbers, and simple typographical symbols that define a chemical formula.

Chemical formulas may be used in chemical equations to describe chemical reactions and other chemical transformations, such as the dissolving of ionic compounds into solution. While, as noted, chemical formulas do not have the full power of structural formulas to show chemical relationships between atoms, they are sufficient to keep track of numbers of atoms and numbers of electrical charges in chemical reactions, thus balancing chemical equations so that these equations can be used in chemical problems involving conservation of atoms, and conservation of electric charge.

11.1   Overview

A chemical formula identifies each constituent element by its chemical symbol and indicates the proportionate number of atoms of each element. In empirical formulas, these proportions begin with a key element and then assign numbers of atoms of the other elements in the compound, as ratios to the key element. For molecular compounds, these ratio numbers can all be expressed as whole numbers. For example, the empirical formula of ethanol may be written C_2H_6O because the molecules of ethanol all contain two carbon atoms, six hydrogen atoms, and one oxygen atom. Some types of ionic compounds, however, cannot be written with entirely whole-number empirical formulas. An example is boron carbide, whose formula of CB_n is a variable non-whole number ratio with n ranging from over 4 to more than 6.5.

When the chemical compound of the formula consists of simple molecules, chemical formulas often employ ways to suggest the structure of the molecule. These types of formulas are variously known as **molecular formulas** and **condensed formulas**. A molecular formula enumerates the number of atoms to reflect those in the molecule, so that the molecular formula for glucose is $C_6H_{12}O_6$ rather than the glucose empirical formula, which is CH_2O. However, except for very simple substances, molecular chemical formulas lack needed structural information, and are ambiguous.

For simple molecules, a condensed (or semi-structural) formula is a type of chemical formula that may fully imply a correct structural formula. For example, ethanol may be represented by the condensed chemical formula CH_3CH_2OH, and dimethyl ether by the condensed formula CH_3OCH_3. These two molecules have the same empirical and molecular formulas (C_2H_6O), but may be differentiated by the condensed formulas shown, which are sufficient to represent the full structure of these simple organic compounds.

Condensed chemical formulas may also be used to represent ionic compounds that do not exist as discrete molecules, but nonetheless do contain covalently bound clusters within them. These polyatomic ions are groups of atoms that are covalently bound together and have an overall ionic charge, such as the sulfate [SO
4]2−
ion. Each polyatomic ion in a compound is written individually in order to illustrate the separate groupings. For example, the compound dichlorine hexoxide has an empirical formula ClO

3, and molecular formula Cl
2O
6, but in liquid or solid forms, this compound is more correctly shown by an ionic condensed formula [ClO
2]+
[ClO
4]−
, which illustrates that this compound consists of [ClO
2]+
ions and [ClO
4]−
ions. In such cases, the condensed formula only need be complex enough to show at least one of each ionic species.

Chemical formulas must be differentiated from the far more complex chemical systematic names that are used in various systems of chemical nomenclature. For example, one systematic name for glucose is (2R,3S,4R,5R)−2,3,4,5,6-pentahydroxyhexanal. This name, and the rules behind it, fully specify glucose's structural formula, but the name is not a chemical formula, as it uses many extra terms and words that chemical formulas do not permit. Such chemical names may be able to represent full structural formulas without graphs, but in order to do so, they require word terms that are not part of chemical formulas.

11.2   Simple empirical formulas

Main article: Empirical formula

In chemistry, the empirical formula of a chemical is a simple expression of the relative number of each type of atom or ratio of the elements in the compound. Empirical formulas are the standard for ionic compounds, such as CaCl
2, and for macromolecules, such as SiO
2. An empirical formula makes no reference to isomerism, structure, or absolute number of atoms. The term **empirical** refers to the process of elemental analysis, a technique of analytical chemistry used to determine the relative percent composition of a pure chemical substance by element.

For example, hexane has a molecular formula of C
6H
14, or structurally CH
3CH
2CH
2CH
2CH
2CH
3, implying that it has a chain structure of 6 carbon atoms, and 14 hydrogen atoms. However, the empirical formula for hexane is C
3H

7. Likewise the empirical formula for hydrogen peroxide, H
2O
2, is simply HO expressing the 1:1 ratio of component elements. Formaldehyde and acetic acid have the same empirical formula, CH
2O. This is the actual chemical formula for formaldehyde, but acetic acid has double the number of atoms.

11.3   Molecular formulas

Molecular formulas indicate the simple numbers of each type of atom in a molecule of a molecular substance. They are the same as empirical formulas for molecules that only have one atom of a particular type, but otherwise have larger numbers. An example of the difference is the empirical formula for glucose, which is CH_2O (*ratio* 1:2:1), while its molecular formula is $C_6H_{12}O_6$ (*number of atoms* 6:12:6). For water, both formulas are H_2O. A molecular formula provides more information about a molecule than its empirical formula, but is more difficult to establish.

11.4   Condensed formulas in organic chemistry implying molecular geometry and structural formulas

Isobutane structural formula
Molecular formula: C_4H_{10}
Condensed or semi-structural chemical formula: $(CH_3)_3CH$

Butane structural formula
Molecular formula: C_4H_{10}
Condensed or semi-structural formula: $CH_3CH_2CH_2CH_3$

The connectivity of a molecule often has a strong influence on its physical and chemical properties and behavior. Two molecules composed of the same numbers of the same types of atoms (i.e. a pair of isomers) might have completely different chemical and/or physical properties if the atoms are connected differently or in different positions. In such cases, a structural formula is useful, as it illustrates which atoms are bonded to which other ones. From the connectivity, it is often possible to deduce the approximate shape of the molecule.

A condensed chemical formula may represent the types and spatial arrangement of bonds in a simple chemical substance, though it does not necessarily specify isomers or complex structures. For example, ethane consists of two carbon atoms single-bonded to each other, with each carbon atom having three hydrogen atoms bonded to it. Its chemical formula can be rendered as CH_3CH_3. In ethylene there is a double bond between the carbon atoms (and thus each carbon only has two hydrogens), therefore the chemical formula may be written: CH_2CH_2, and the fact that there is a double bond between the carbons is implicit because carbon has a valence of four. However, a more explicit method is to write $H_2C=CH_2$ or less commonly $H_2C::CH_2$. The two lines (or two pairs of dots) indicate that a double bond connects the atoms on either side of them.

A triple bond may be expressed with three lines or pairs of dots, and if there may be ambiguity, a single line or pair of dots may be used to indicate a single bond.

Molecules with multiple functional groups that are the same may be expressed by enclosing the repeated group in round brackets. For example, isobutane may be written $(CH_3)_3CH$. This condensed structural formula implies a different connectivity from other molecules that can be formed using the same atoms in the same proportions (isomers). The formula $(CH_3)_3CH$ implies a central carbon atom attached to one hydrogen atom and three CH_3 groups. The same number of atoms of each element (10 hydrogens and 4 carbons, or C_4H_{10}) may be used to make a straight chain molecule, butane: $CH_3CH_2CH_2CH_3$.

11.5 Chemical names in answer to limitations of chemical formulas

Main article: chemical nomenclature

The alkene called **but-2-ene** has two isomers, which the chemical formula $CH_3CH=CHCH_3$ does not identify. The relative position of the two methyl groups must be indicated by additional notation denoting whether the methyl groups are on the same side of the double bond (*cis* or *Z*) or on the opposite sides from each other (*trans* or *E*). Such extra symbols violate the rules for chemical formulas, and begin to enter the territory of more complex naming systems.

As noted above, in order to represent the full structural formulas of many complex organic and inorganic compounds, chemical nomenclature may be needed which goes well beyond the available resources used above in simple condensed formulas. See IUPAC nomenclature of organic chemistry and IUPAC nomenclature of inorganic chemistry 2005 for examples. In addition, linear naming systems such as International Chemical Identifier (InChI) allow a computer to construct a structural formula, and simplified molecular-input line-entry system (SMILES) allows a more human-readable ASCII input. However, all these nomenclature systems go beyond the standards of chemical formulas, and technically are chemical naming systems, not formula systems.

11.6 Polymers in condensed formulas

For polymers in condensed chemical formulas, parentheses are placed around the repeating unit. For example, a hydrocarbon molecule that is described as $CH_3(CH_2)_{50}CH_3$, is a molecule with fifty repeating units. If the number of repeating units is unknown or variable, the letter *n* may be used to indicate this formula: $CH_3(CH_2)nCH_3$.

11.7 Ions in condensed formulas

For ions, the charge on a particular atom may be denoted with a right-hand superscript. For example, Na^+, or Cu^{2+}. The total charge on a charged molecule or a polyatomic ion may also be shown in this way. For example: H_3O^+ or $SO_4{}^{2-}$.

For more complex ions, brackets [ ] are often used to enclose the ionic formula, as in $[B_{12}H_{12}]^{2-}$, which is found in compounds such as $Cs_2[B_{12}H_{12}]$. Parentheses ( ) can be nested inside brackets to indicate a repeating unit, as in $[Co(NH_3)_6]^{3+}$. Here $(NH_3)_6$ indicates that the ion contains six NH_3 groups, and [ ] encloses the entire formula of the ion with charge +3. ‹The template *Elucidate* is being considered for deletion.›

11.8 Isotopes

Although isotopes are more relevant to nuclear chemistry or stable isotope chemistry than to conventional chemistry, different isotopes may be indicated with a prefixed superscript in a chemical formula. For example, the phosphate ion containing radioactive phosphorus-32 is $^{32}PO_4{}^{3-}$. Also a study involving stable isotope ratios might include the molecule $^{18}O^{16}O$.

A left-hand subscript is sometimes used redundantly to indicate the atomic number. For example, $_8O_2$ for dioxygen, and $^{16}_8O_2$ for the most abundant isotopic species of dioxygen. This is convenient when writing equations for nuclear reactions, in order to show the balance of charge more clearly.

11.9 Trapped atoms

Main article: Endohedral fullerene

The @ symbol (at sign) indicates an atom or molecule trapped inside a cage but not chemically bound to it. For example, a buckminsterfullerene (C_{60}) with an atom (M) would simply be represented as MC_{60} regardless of whether M was inside the fullerene without chemical bonding or outside, bound to one of the carbon atoms. Using the @ symbol, this would be denoted $M@C_{60}$ if M was inside the carbon network. A non-fullerene example is $[As@Ni_{12}As_{20}]^{3-}$, an ion in which one As atom is trapped in a cage formed by the other 32 atoms.

This notation was proposed in 1991[1] with the discovery of fullerene cages (endohedral fullerenes), which can trap atoms such as La to form, for example, $La@C_{60}$ or $La@C_{82}$. The choice of the symbol has been explained by the authors as being concise, readily printed and transmitted electronically (the at sign is included in ASCII, which most modern character encoding schemes are based on), and the visual aspects suggesting the structure of an endo-

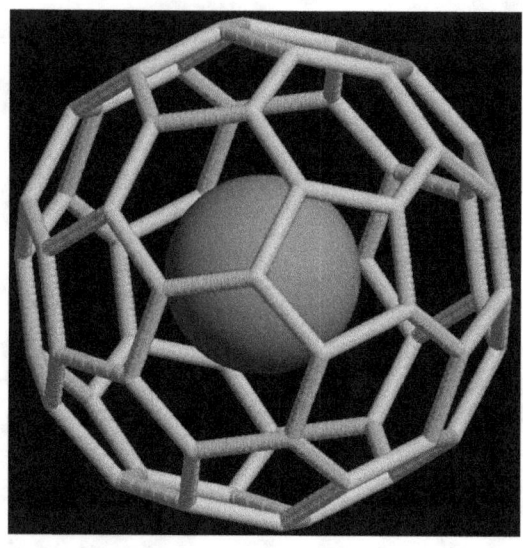

Traditional formula: MC_{60}
The "@" notation: $M@C_{60}$

hedral fullerene.

11.10 Non-stoichiometric chemical formulas

Main article: Non-stoichiometric compound

Chemical formulas most often use integers for each element. However, there is a class of compounds, called non-stoichiometric compounds, that cannot be represented by small integers. Such a formula might be written using decimal fractions, as in $Fe_{0.95}O$, or it might include a variable part represented by a letter, as in $Fe_{1-x}O$, where x is normally much less than 1.

11.11 General forms for organic compounds

A chemical formula used for a series of compounds that differ from each other by a constant unit is called **general formula**. Such a series is called the homologous series, while its members are called homologs.

For example, alcohols may be represented by: $CnH_{(2n+1)}OH$ $(n \geq 1)$

11.12 Hill System

Main article: Hill system

The **Hill system** is a system of writing chemical formulas such that the number of carbon atoms in a molecule is indicated first, the number of hydrogen atoms next, and then the number of all other chemical elements subsequently, in alphabetical order. When the formula contains no carbon, all the elements, including hydrogen, are listed alphabetically. This deterministic system enables straightforward sorting and searching of compounds.

11.13 See also

- Dictionary of chemical formulas

- Element symbol

- Nuclear notation

- Periodic table

- IUPAC nomenclature of inorganic chemistry

11.14 References

[1] Chai, Yan; Guo, Ting; Jin, Changming; Haufler, Robert E.; Chibante, L. P. Felipe; Fure, Jan; Wang, Lihong; Alford, J. Michael; Smalley, Richard E. (1991). "Fullerenes wlth Metals Inside". *Journal of Physical Chemistry* **95** (20): 7564–7568. doi:10.1021/j100173a002.

- Ralph S. Petrucci, William S. Harwood, F. Geoffrey Herring (2002). "3". *General Chemistry: Principles and Modern Applications* (8th ed.). Prentice-Hall. ISBN 0-13-198825-5. OCLC 46872308.

Chapter 12

Electrolysis

This article is about the chemical process. For the cosmetic hair removal procedure, see Electrology.

In chemistry and manufacturing, **electrolysis** is a tech-

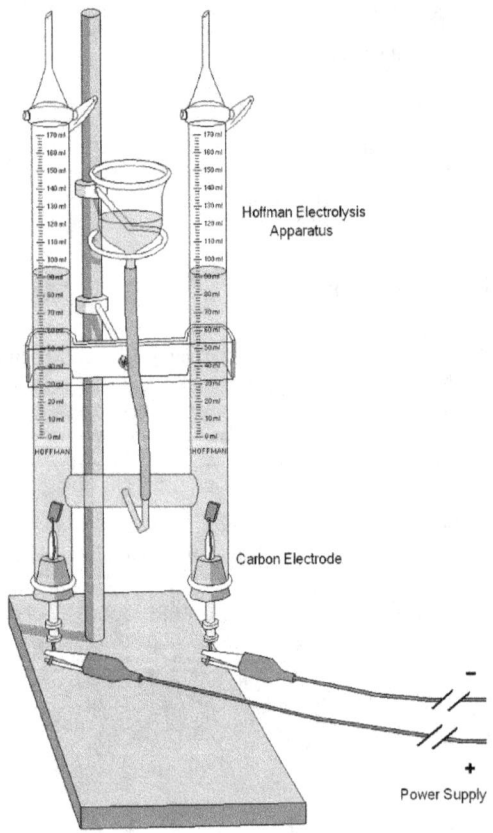

Illustration of an electrolysis apparatus used in a school laboratory.

nique that uses a direct electric current (DC) to drive an otherwise non-spontaneous chemical reaction. Electrolysis is commercially highly important as a stage in the separation of elements from naturally occurring sources such as ores using an electrolytic cell. The voltage that is needed for

electrolysis to occur is called the decomposition potential.

12.1 History

The word electrolysis comes from the Greek ἤλεκτρον [ɛ̆:lektron] "amber" and λύσις [lýsis] "dissolution".

- 1785 – Martinus van Marum's electrostatic generator was used to reduce tin, zinc, and antimony from their salts using electrolysis.[1]

- 1800 – William Nicholson and Anthony Carlisle (view also Johann Ritter), decomposed water into hydrogen and oxygen.

- 1807 – Potassium, sodium, barium, calcium and magnesium were discovered by Sir Humphry Davy using electrolysis.

- 1833 – Michael Faraday develops his two laws of electrolysis, and provides a mathematical explanation of his laws.

- 1875 – Paul Émile Lecoq de Boisbaudran discovered gallium using electrolysis.[2]

- 1886 – Fluorine was discovered by Henri Moissan using electrolysis.

- 1886 – Hall-Héroult process developed for making aluminium

- 1890 – Castner-Kellner process developed for making sodium hydroxide

12.2 Overview

Electrolysis is the passing of a direct electric current through an ionic substance that is either molten or dissolved in a suitable solvent, producing chemical reactions at the electrodes and separation of materials.

The main components required to achieve electrolysis are:

- An electrolyte: a substance, frequently an ion-conducting polymer that contains free ions, which carry electric current in the electrolyte. If the ions are not mobile, as in a solid salt then electrolysis cannot occur.

- A direct current (DC) electrical supply: provides the energy necessary to create or discharge the ions in the electrolyte. Electric current is carried by electrons in the external circuit.

- Two electrodes: electrical conductors that provide the physical interface between the electrolyte and the electrical circuit that provides the energy.

Electrodes of metal, graphite and semiconductor material are widely used. Choice of suitable electrode depends on chemical reactivity between the electrode and electrolyte and manufacturing cost.

12.2.1   Process of electrolysis

The key process of electrolysis is the interchange of atoms and ions by the removal or addition of electrons from the external circuit. The desired products of electrolysis are often in a different physical state from the electrolyte and can be removed by some physical processes. For example, in the electrolysis of brine to produce hydrogen and chlorine, the products are gaseous. These gaseous products bubble from the electrolyte and are collected.[3]

$$2\,NaCl + 2\,H_2O \rightarrow 2\,NaOH + H_2 + Cl_2$$

A liquid containing mobile ions (electrolyte) is produced by:

- Solvation or reaction of an ionic compound with a solvent (such as water) to produce mobile ions

- An ionic compound is fused by heating

An electrical potential is applied across a pair of electrodes immersed in the electrolyte.

Each electrode attracts ions that are of the opposite charge. Positively charged ions (cations) move towards the electron-providing (negative) cathode. Negatively charged ions (anions) move towards the electron-extracting (positive) anode.

In this process electrons are either absorbed or released. Neutral atoms gain or lose electrons and become charged ions that then pass into the electrolyte. The formation of

uncharged atoms from ions is called discharging. When an ion gains or loses enough electrons to become uncharged (neutral) atoms, the newly formed atoms separate from the electrolyte. Positive metal ions like Cu^{++} deposit onto the cathode in a layer. The terms for this are electroplating electrowinning and electrorefining. When an ion gains or loses electrons without becoming neutral, its electronic charge is altered in the process. In chemistry the loss of electrons is called oxidation while electron gain is called reduction.

12.2.2   Oxidation and reduction at the electrodes

Oxidation of ions or neutral molecules occurs at the anode. For example, it is possible to oxidize ferrous ions to ferric ions at the anode:

$$Fe2+ \\ aq \rightarrow Fe3+ \\ aq + e^-$$

Reduction of ions or neutral molecules occurs at the cathode.

It is possible to reduce ferricyanide ions to ferrocyanide ions at the cathode:

$$Fe(CN)3- \\ 6 + e^- \rightarrow Fe(CN)4- \\ 6$$

Neutral molecules can also react at either of the electrodes. For example: p-Benzoquinone can be reduced to hydroquinone at the cathode:

In the last example, H^+ ions (hydrogen ions) also take part in the reaction, and are provided by an acid in the solution, or by the solvent itself (water, methanol etc.). Electrolysis reactions involving H^+ ions are fairly common in acidic solutions. In aqueous alkaline solutions, reactions involving OH^- (hydroxide ions) are common.

Sometimes the solvents themselves (usually water) are oxidized or reduced at the electrodes. It is even possible to have electrolysis involving gases. (Such as when using a Gas diffusion electrode)

12.2.3   Energy changes during electrolysis

The amount of electrical energy that must be added equals the change in Gibbs free energy of the reaction plus the losses in the system. The losses can (in theory) be arbitrarily close to zero, so the maximum thermodynamic efficiency equals the enthalpy change divided by the free energy change of the reaction. In most cases, the electric input is larger than the enthalpy change of the reaction, so some energy is released in the form of heat. In some cases, for instance, in the electrolysis of steam into hydrogen and oxygen at high temperature, the opposite is true and heat energy is absorbed. This heat is absorbed from the surroundings, and the heating value of the produced hydrogen is higher than the electric input.

12.2.4   Related techniques

The following techniques are related to electrolysis:

- Electrochemical cells, including the hydrogen fuel cell, use differences in Standard electrode potential to generate an electrical potential that provides useful power. Though related via the interaction of ions and electrodes, electrolysis and the operation of electrochemical cells are quite distinct. A chemical cell should *not* be thought of as performing *electrolysis in reverse.*

12.3   Faraday's laws of electrolysis

Main article: Faraday's laws of electrolysis

12.3.1   First law of electrolysis

In 1832, Michael Faraday reported that the quantity of elements separated by passing an electric current through a molten or dissolved salt is proportional to the quantity of electric charge passed through the circuit. This became the basis of the first law of electrolysis:

$$m = k.q$$

or

$$m = eQ$$

where; e is known as electrochemical equivalent of the metal deposited or of the gas liberated at the electrode.[4]

12.3.2   Second law of electrolysis

Faraday discovered that when the same amount of electricity is passed through different electrolytes/elements connected in series, the mass of substance liberated/deposited at the electrodes is directly proportional to their equivalent weights.

12.4   Industrial uses

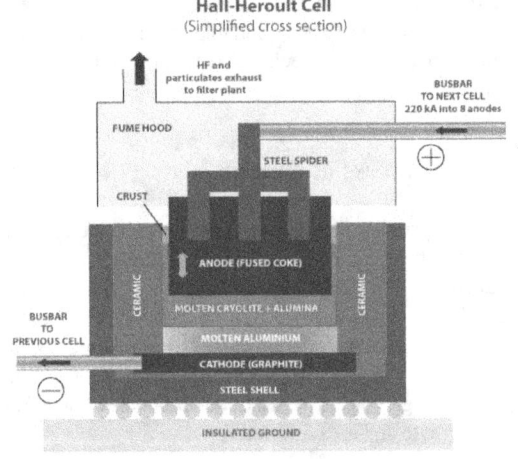

Hall-Heroult process for producing aluminium

- Electrometallurgy is the process of reduction of metals from metallic compounds to obtain the pure form of metal using electrolysis. aluminium, lithium, sodium, potassium, magnesium, calcium, and in some cases copper, are produced in this way.

- Production of chlorine and sodium hydroxide

- Production of sodium chlorate and potassium chlorate

- Production of perfluorinated organic compounds such as trifluoroacetic acid by the process of electrofluorination

- Production of electrolytic copper as a cathode, from refined copper of lower purity as an anode.

Electrolysis has many other uses:

- Production of oxygen for spacecraft and nuclear submarines.

- Electroplating is used in layering metals to fortify them. Electroplating is used in many industries for functional or decorative purposes, as in vehicle bodies and nickel coins.

- Production of hydrogen for fuel, using a cheap source of electrical energy.

- Electrolytic etching of metal surfaces like tools or knives with a permanent mark or logo.

Electrolysis is also used in the cleaning and preservation of old artifacts. Because the process separates the non-metallic particles from the metallic ones, it is very useful for cleaning a wide variety of metallic objects, from old coins to even larger objects including rusted cast iron cylinder blocks and heads when rebuilding automobile engines. Rust removal from small iron or steel objects by electrolysis can be done in a home workshop using simple materials such as a plastic bucket, tap water, lengths of rebar, washing soda, baling wire, and a battery charger.[5]

12.5 Competing half-reactions in solution electrolysis

Using a cell containing inert platinum electrodes, electrolysis of aqueous solutions of some salts leads to reduction of the cations (e.g., metal deposition with, e.g., zinc salts) and oxidation of the anions (e.g. evolution of bromine with bromides). However, with salts of some metals (e.g. sodium) hydrogen is evolved at the cathode, and for salts containing some anions (e.g. sulfate SO_4^{2-}) oxygen is evolved at the anode. In both cases this is due to water being reduced to form hydrogen or oxidized to form oxygen. In principle the voltage required to electrolyze a salt solution can be derived from the standard electrode potential for the reactions at the anode and cathode. The standard electrode potential is directly related to the Gibbs free energy, ΔG, for the reactions at each electrode and refers to an electrode with no current flowing. An extract from the table of standard electrode potentials is shown below.

In terms of electrolysis, this table should be interpreted as follows:

- Oxidized species (often a cation) with a more negative cell potential are more difficult to reduce than oxidized species with a more positive cell potential. For example, it is more difficult to reduce a sodium ion to a sodium metal than it is to reduce a zinc ion to a zinc metal.

- Reduced species (often an anion) with a more positive cell potential are more difficult to oxidize than reduced species with a more negative cell potential. For example, it is more difficult to oxidize sulfate anions than it is to oxidize bromide anions.

Using the Nernst equation the electrode potential can be calculated for a specific concentration of ions, temperature and the number of electrons involved. For pure water (pH 7):

- the electrode potential for the reduction producing hydrogen is −0.41 V

- the electrode potential for the oxidation producing oxygen is +0.82 V.

Comparable figures calculated in a similar way, for 1M zinc bromide, $ZnBr_2$, are −0.76 V for the reduction to Zn metal and +1.10 V for the oxidation producing bromine. The conclusion from these figures is that hydrogen should be produced at the cathode and oxygen at the anode from the electrolysis of water—which is at variance with the experimental observation that zinc metal is deposited and bromine is produced.[8] The explanation is that these calculated potentials only indicate the thermodynamically preferred reaction. In practice many other factors have to be taken into account such as the kinetics of some of the reaction steps involved. These factors together mean that a higher potential is required for the reduction and oxidation of water than predicted, and these are termed overpotentials. Experimentally it is known that overpotentials depend on the design of the cell and the nature of the electrodes.

For the electrolysis of a neutral (pH 7) sodium chloride solution, the reduction of sodium ion is thermodynamically very difficult and water is reduced evolving hydrogen leaving hydroxide ions in solution. At the anode the oxidation of chlorine is observed rather than the oxidation of water since the overpotential for the oxidation of chloride to chlorine is lower than the overpotential for the oxidation of water to oxygen. The hydroxide ions and dissolved chlorine gas react further to form hypochlorous acid. The aqueous solutions resulting from this process is called electrolyzed water and is used as a disinfectant and cleaning agent.

12.6 Research trends

12.6.1 Electrolysis of carbon dioxide

Main article: Electrochemical reduction of carbon dioxide

The electrolysis of carbon dioxide gives formate or carbon monoxide, but sometimes more elaborate organic compounds such as ethylene.[9] This technology is under research as a carbon-neutral route to organic compounds.[10][11]

12.6.2   Electrolysis of water

Main article: Electrolysis of water

Electrolysis of water produces hydrogen.

$$2 \ H_2O(l) \rightarrow 2 \ H_2(g) + O_2(g); E_0 = -1.229 \ V$$

The energy efficiency of water electrolysis varies widely. The efficiency of an electrolyser is a measure of the enthalpy contained in the hydrogen (to undergo combustion with oxygen, or some other later reaction), compared with the input electrical energy. Heat/enthalpy values for hydrogen are well published in science and engineering texts, as 144 MJ/kg. Note that fuel cells (not electrolysers) cannot utilise this full amount of heat/enthalpy, which has led to some confusion when calculating efficiency values for both types of technology. In the reaction, some energy is lost as heat. Some reports quote efficiencies between 50% and 70% for alkaline electrolysers; however, much higher practical efficiencies are available with the use of PEM (Polymer Electrolyte Membrane electrolysis) and catalytic technology, such as 95% efficiency.[12][13]

NREL estimated that 1 kg of hydrogen (roughly equivalent to 3 kg, or 4 L, of petroleum in energy terms) could be produced by wind powered electrolysis for between $5.55 in the near term and $2.27 in the long term.[14]

About 4% of hydrogen gas produced worldwide is generated by electrolysis, and normally used onsite. Hydrogen is used for the creation of ammonia for fertilizer via the Haber process, and converting heavy petroleum sources to lighter fractions via hydrocracking.

12.6.3   Electrocrystallization

A specialized application of electrolysis involves the growth of conductive crystals on one of the electrodes from oxidized or reduced species that are generated in situ. The technique has been used to obtain single crystals of low-dimensional electrical conductors, such as charge-transfer salts.[15][16]

12.7   History

Scientific pioneers of electrolysis include:

- Antoine Lavoisier
- Robert Bunsen
- Humphry Davy
- Michael Faraday
- Paul Héroult
- Svante Arrhenius
- Adolph Wilhelm Hermann Kolbe
- William Nicholson
- Joseph Louis Gay-Lussac
- Alexander von Humboldt
- Johann Wilhelm Hittorf
- Kai Grjotheim-->

Pioneers of batteries:

- Alessandro Volta
- Gaston Planté

More recently, electrolysis of heavy water was performed by Fleischmann and Pons in their famous experiment, resulting in anomalous heat generation and the discredited claim of cold fusion.

12.8   See also

- Alkaline water electrolysis
- Castner-Kellner process
- Electrolytic cell
- Faraday's law of electrolysis
- Faraday constant
- Faraday efficiency
- Galvanic corrosion
- Galvanoluminescence
- Gas cracker

- Hall-Héroult process

- High-pressure electrolysis

- Overpotential

- Patterson Power Cell

- Thermochemical cycle

- Timeline of hydrogen technologies

- PEM electrolysis

12.9 References

[1] The Supplement (1803 edition) to Encyclopedia Britannica 3rd edition (1797), volume 1, page 225, "Mister Van Marum, by means of his great electrical machine, decomposed the calces of tin, zinc, and antimony, and resolved them into their respective metals and oxygen" and gives as a reference Journal de Physiques, 1785.

[2] Sir William Crookes (1875). *The Chemical news and journal of industrial science; with which is incorporated the "Chemical gazette.": A journal of practical chemistry in all its applications to pharmacy, arts and manufactures.* Chemical news office. pp. 294–. Retrieved 27 February 2011.

[3] R. J. D. Tilley (2004). *Understanding solids: the science of materials.* John Wiley and Sons. pp. 281–. ISBN 978-0-470-85276-7. Retrieved 22 October 2011.

[4] Physical Chemistry for JEE and for other Engineering Examinations

[5] "Rust Removal using Electrolysis". *antique-engines.com.* Retrieved April 1, 2015.

[6] Peter Atkins (1997). *Physical Chemistry*, 6th edition (W.H. Freeman and Company, New York).

[7] Vanýsek, Petr (2007). "Electrochemical Series", in *Handbook of Chemistry and Physics: 88th Edition* (Chemical Rubber Company).

[8] A.E. Vogel, 1951, A textbook of Quantitative Inorganic Analysis, Longmans, Green and Co

[9] Y. Hori, in Modern Aspects of Electrochemisty , ed. C. G. Vayeanas, R. White and M. E. Gamboa-Aldeco, Springer, New York, 2008, no. 42. pp. 141–153.

[10] Appel, A. M. et al. "Frontiers, Opportunities, and Challenges in Biochemical and Chemical Catalysis of CO2 Fixation", Chem. Rev. 2013, vol. 113, 6621-6658. doi:10.1021/cr300463y

[11] J. Qiao, et al., A review of catalysts for the electroreduction of carbon dioxide to produce low-carbon fuels, Chem.Soc.Rev., 2014, 43 , 631-675.

[12] Carmo, M; Fritz D; Mergel J; Stolten D (2013). "A comprehensive review on PEM water electrolysis". *Journal of Hydrogen Energy.* doi:10.1016/j.ijhydene.2013.01.151.

[13] Werner Zittel; Reinhold Wurster (8 July 1996). "Chapter 3: Production of Hydrogen. Part 4: Production from electricity by means of electrolysis". *HyWeb: Knowledge – Hydrogen in the Energy Sector.* Ludwig-Bölkow-Systemtechnik GmbH.

[14] J. Levene; B. Kroposki, and G. Sverdrup (March 2006). "Wind Energy and Production of Hydrogen and Electricity – Opportunities for Renewable Hydrogen – Preprint" (PDF). *National Renewable Energy Laboratory.* Retrieved 20 October 2008.

[15] K. Bechgaard, K. Carneiro, F. B. Rasmusen, H. Olsen, G. Rindorf, C. S. Jacobsen, H. Pedersen, J. E. Scott (1981). "Superconductivity in an organic solid. Synthesis, structure, and conductivity of bis(tetramethyltetraselenafulvalenium) perchlorate, (TMTSF)2ClO4". *Journal of the American Chemical Society* **103** (9): 2440. doi:10.1021/ja00399a065.

[16] Williams, Jack M (2007). "Highly Conducting and Superconducting Synthetic Metals". *Inorganic Syntheses* **26**: 386–394. doi:10.1002/9780470132579.ch70.

Chapter 13

Avogadro's law

Avogadro's law (sometimes referred to as **Avogadro's hypothesis** or **Avogadro's principle**) is an experimental gas law relating volume of a gas to the amount of substance of gas present. A modern statement of Avogadro's law is:

> Avogadro's law states that, "equal volumes of all gases, at the same temperature and pressure, have the same number of molecules".
>
> For a given mass of an ideal gas, the volume and amount (moles) of the gas are directly proportional if the temperature and pressure are constant.

which can be written as:

$$V \propto n$$

or

$$\frac{V}{n} = k$$

where:

> V is the volume of the gas
>
> n is the amount of substance of the gas (measured in moles).
>
> k is a constant equal to RT/P, where R is the universal gas constant, T is the Kelvin temperature, and P is the pressure. As temperature and pressure are constant, RT/P is also constant and represented as k. This is derived from the ideal gas law.

This law describes how, under the same condition of temperature and pressure, equal volumes of all gases contain the same number of molecules. For comparing the same substance under two different sets of conditions, the law can be usefully expressed as follows:

$$\frac{V_1}{n_1} = \frac{V_2}{n_2}$$

The equation shows that, as the number of moles of gas increases, the volume of the gas also increases in proportion. Similarly, if the number of moles of gas is decreased, then the volume also decreases. Thus, the number of molecules or atoms in a specific volume of ideal gas is independent of their size or the molar mass of the gas.

The law is named after Amedeo Avogadro who, in 1811,[1] hypothesized that two given samples of an ideal gas, of the same volume and at the same temperature and pressure, contain the same number of molecules. As an example, equal volumes of molecular hydrogen and nitrogen contain the same number of molecules when they are at the same temperature and pressure, and observe ideal gas behavior. In practice, real gases show small deviations from the ideal behavior and the law holds only approximately, but is still a useful approximation for scientists.

13.1 Mathematical definition

Avogadro's law is stated mathematically as:

$$\frac{V}{n} = k$$

Where:

> V is the volume of the gas(es).
>
> n is the amount of substance of the gas.
>
> k is a proportionality constant.

The most significant consequence of Avogadro's law is that the ideal gas constant has the same value for all gases. This means that:

$$\frac{p_1 \cdot V_1}{T_1 \cdot n_1} = \frac{p_2 \cdot V_2}{T_2 \cdot n_2} = constant$$

Where:

p is the pressure of the gas in the cell

T is the temperature in kelvin of the gas

13.2 Ideal gas law

A common rearrangement of this equation is by letting R be the proportionality constant, and rearranging as follows:

$$pV = nRT$$

This equation is known as the ideal gas law.

13.3 Molar volume

Taking STP to be 101.325 kPa and 273.15 K, we can find the volume of one mole of a gas:

$$V_\mathrm{m} = \frac{V}{n} = \frac{RT}{p} = \frac{(8.314 \mathrm{Jmol}^{-1}\mathrm{K}^{-1})(273.15\mathrm{K})}{101.325\mathrm{kPa}}$$

$$= 22.41 \mathrm{dm}^3\mathrm{mol}^{-1} \quad = 22.41 \mathrm{liters/mol}$$

For 100.00 kPa and 273.15 K, the molar volume of an ideal gas is 22.712 $\mathrm{dm}^3\mathrm{mol}^{-1}$. Note that the universal gas constant R is given by the product of Avogadro's number and Boltzmann's constant. (See Gas Constant.)

13.4 Application

1.To Deduce Relationship Between Molecular Mass and Vapour Density

Let us consider the terms as follows: V.D.= vapour density

M.M.=molecular mass

STP=standard temperature and pressure

Now, the definition of molecular mass is,"the ratio of mass of 1 molecule of a substance to the ratio of mass of 1 molecule of hydrogen at STP"

So,M.M.=mass of 1 mole of gas at stp/mass of 1 mole of hygrogen at STP

Since hydrogen is diatomic,

M.M.=mass Of 1 mole of a gas at STP/2×mass of 1 atom of hydrogen at STP

We know from the definition of vapour density that it is the ratio of mass of 1 mole of a gas to the ratio of mass of 1 atom of gas at STP

SO, 2×M.M.=V.D.

Hence molecular mass is twice the vapour density.

2.Relation between molar volume of gas at NTP

Also according to avogadro's hypothesis we have come to know that the molar volume of a gas at NTP is 22.4 litres

13.5 See also

- Boyle's law
- Charles's law
- Combined gas law
- Gay-Lussac's law
- Ideal gas

13.6 References

[1] Avogadro, Amedeo (1810). "Essai d'une maniere de determiner les masses relatives des molecules elementaires des corps, et les proportions selon lesquelles elles entrent dans ces combinaisons". *Journal de Physique* **73**: 58–76. English translation

13.7 External links

- Avogadro's law at the University of Fribourg
- Avogadro's law at the Royal Society of Chemistry

Chapter 14

Vitalism

This article is about the non-mechanist philosophy. For other uses, see Vital (disambiguation).

Vitalism is an obsolete scientific doctrine that "living or-

The synthesis of urea (and other organic substances) from inorganic compounds was counterevidence for the vitalist hypothesis that only organisms could make such compounds.

ganisms are fundamentally different from non-living entities because they contain some non-physical element or are governed by different principles than are inanimate things".[1] a Where vitalism explicitly invokes a vital principle, that element is often referred to as the "vital spark", "energy" or "élan vital", which some equate with the soul.

Although rejected by modern science,[2] vitalism has a long history in medical philosophies: most traditional healing practices posited that disease results from some imbalance in vital forces. In the Western tradition founded by Hippocrates, these vital forces were associated with the four

temperaments and humours; Eastern traditions posited an imbalance or blocking of qi or prana.

14.1 Philosophy

Louis Pasteur argued that only life could catalyse fermentation. (Painting by A. Edelfeldt in 1885.)

The notion that bodily functions are due to a vitalistic principle existing in all living creatures has roots going back at least to ancient Egypt.[3] In Greek philosophy, the Milesian school proposed natural explanations deduced from materialism and mechanism. However, by the time of Lucretius, this account was supplemented, (for example, by the clinamen of Epicurus), and in stoic physics, the pneuma assumed the role of logos. Galen believed the lungs draw pneuma from the air, which the blood communicates throughout the body.[4]

> Plato's world of eternal and unchanging Forms, imperfectly represented in matter by a divine Artisan, contrasts sharply with the various mechanistic Weltanschauungen, of which atomism was, by the fourth century at least, the most prominent... This debate was to persist throughout the ancient world. Atomistic mechanism got a shot in the arm from Epicurus... while the Stoics adopted a divine teleology... The choice seems simple: either show how a structured, regular world could arise out of undirected processes, or inject intelligence into the system.[5]
>
> — R. J. Hankinson, *Cause and Explanation in Ancient Greek Thought* (1997)

14.2 Science

In Europe, medieval physics was influenced by the idea of pneuma, helping to shape later aether theories. In the 17th century, modern science responded to Newton's action at a distance and the mechanism of Cartesian dualism with vitalist theories: that whereas the chemical transformations undergone by non-living substances are reversible, so-called "organic" matter is permanently altered by chemical transformations (such as cooking). Jöns Jakob Berzelius, one of the early 19th century fathers of modern chemistry, argued that a regulative force must exist within living matter to maintain its functions.[6]

Vitalist chemists predicted that organic materials could not be synthesized from inorganic components, but Friedrich Wöhler synthesised urea from inorganic components in 1828.[7] However, contemporary accounts do not support the common belief that vitalism died when Wöhler made urea. This *Wöhler Myth*, as historian Peter Ramberg called it, originated from a popular history of chemistry published in 1931, which, "ignoring all pretense of historical accuracy, turned Wöhler into a crusader who made attempt after attempt to synthesize a natural product that would refute vitalism and lift the veil of ignorance, until 'one afternoon the miracle happened'".[8][9][10] Further discoveries continued to obviate the need for a special "vital force".

Vitalism has long been regarded in the scientific community as a corrupting pseudoscientific influence.[11] Vitalism today is no longer philosophically and scientifically viable, and is sometimes used as a pejorative epithet.[12] Ernst Mayr, co-founder of the modern evolutionary synthesis and a critic of vitalism, wrote:

> It would be ahistorical to ridicule vitalists. When one reads the writings of one of the leading vitalists like Driesch one is forced to agree with him that many of the basic problems of biology simply cannot be solved by a philosophy as that of Descartes, in which the organism is simply considered a machine... The logic of the critique of the vitalists was impeccable.[13]

> Vitalism has become so disreputable a belief in the last fifty years that no biologist alive today would want to be classified as a vitalist. Still, the remnants of vitalist thinking can be found in the work of Alistair Hardy, Sewall Wright, and Charles Birch, who seem to believe in some sort of nonmaterial principle in organisms.[14]

Louis Pasteur, shortly after his famous rebuttal of spontaneous generation, performed several experiments that he felt supported vitalism. According to Bechtel, Pasteur "fitted fermentation into a more general programme describing special reactions that only occur in living organisms. These are irreducibly vital phenomena." In 1858, Pasteur showed that fermentation only occurs when living cells are present and, that fermentation only occurs in the absence of oxygen; he was thus led to describe fermentation as "life without air". Rejecting the claims of Berzelius, Liebig, Traube and others that fermentation resulted from chemical agents or catalysts within cells, he concluded that fermentation was a "vital action".[15]

Other vitalists included English anatomist Francis Glisson (1597–1677) and the Italian doctor Marcello Malpighi (1628–1694).[16] Caspar Friedrich Wolff (1733–1794) is considered to be the father of epigenetic descriptive embryology, that is, he marks the point when embryonic development began to be described in terms of the proliferation of cells rather than the incarnation of a preformed soul. In his *Theoria Generationis* (1759), he endeavored to explain the emergence of the organism by the actions of a "vis essentialis", an organizing, formative force, and declared "All believers in epigenesis are Vitalists." Carl Reichenbach later developed the theory of Odic force, a form of life-energy that permeates living things.

Johann Friedrich Blumenbach established epigenesis as the model of thought in the life sciences in 1781 with his publication of *Über den Bildungstrieb und das Zeugungsgeschäfte*. Blumenbach cut up freshwater polyps and established that the removed parts would regenerate. He inferred the presence of a "formative drive" (*Bildungstrieb*) in living matter. But he pointed out that this name, "like names applied to every other kind of vital power, of itself, explains nothing: it serves merely to designate a peculiar power formed by the combination of the mechanical principle with that which is susceptible of modification". In the early 18th century, the physicians Marie François Xavier Bichat and John Hunter recognized a "living principle" in addition to mechanics.[16]

Between 1833 and 1844, Johannes Peter Müller wrote a book on physiology called *Handbuch der Physiologie*, which became the leading textbook in the field for much of the nineteenth century. The book showed Müller's commitments to vitalism; he questioned why organic matter differs from inorganic, then proceeded to chemical analyses of the blood and lymph. He describes in detail the circulatory, lymphatic, respiratory, digestive, endocrine, nervous, and sensory systems in a wide variety of animals but explains that the presence of a soul makes each organism an indivisible whole. He also claimed the behavior of light and sound waves showed that living organisms possessed a life-energy for which physical laws could never fully account.[17]

Hans Driesch (1867–1941) interpreted his experiments as showing that life is not run by physicochemical laws.[18] His main argument was that when one cuts up an embryo after its first division or two, each part grows into a complete adult. Driesch's reputation as an experimental biologist deteriorated as a result of his vitalistic theories.[18]

Other vitalists included Johannes Reinke and Oscar Hertwig. Reinke used the word *neovitalism* to describe his work, he claimed that it would be eventually verified through experimentation and wanted an improvement over the other vitalistic theories. The work of Reinke was an influence for Carl Jung.[19]

John Scott Haldane adopted an anti-mechanist approach to biology and an idealist philosophy early on in his career. Haldane saw his work as a vindication of his belief that teleology was an essential concept in biology. His views became widely known with his first book *Mechanism, life and personality* in 1913.[20] Haldane borrowed arguments from the vitalists to use against mechanism; however, he was not a vitalist. Haldane treated the organism as fundamental to biology: "we perceive the organism as a self-regulating entity", "every effort to analyze it into components that can be reduced to a mechanical explanation violates this central experience".[20] The work of Haldane was an influence on organicism.

Haldane also stated that a purely mechanist interpretation can not account for the characteristics of life. Haldane wrote a number of books in which he attempted to show the invalidity of both vitalism and mechanist approaches to science. Haldane explained:

> We must find a different theoretical basis of biology, based on the observation that all the phenomena concerned tend towards being so coordinated that they express what is normal for

an adult organism.
— [21]

By 1931, "Biologists have almost unanimously abandoned vitalism as an acknowledged belief."[21]

14.2.1 Relationship to emergentism

Some aspects of contemporary science make reference to emergent processes; those in which the properties of a system cannot be fully described in terms of the properties of the constituents.[22][23] This may be because the properties of the constituents are not fully understood, or because the interactions between the individual constituents are also important for the behavior of the system.

Whether emergent system properties should be grouped with traditional vitalist concepts is a matter of semantic controversy.[24] According to Emmeche *et al.* (1997):

> On the one hand, many scientists and philosophers regard emergence as having only a pseudo-scientific status. On the other hand, new developments in physics, biology, psychology, and cross-disciplinary fields such as cognitive science, artificial life, and the study of non-linear dynamical systems have focused strongly on the high level 'collective behaviour' of complex systems, which is often said to be truly emergent, and the term is increasingly used to characterize such systems.
> — [25]

Emmeche *et al.* (1998) state that "there is a very important difference between the vitalists and the emergentists: the vitalist's creative forces were relevant only in organic substances, not in inorganic matter. Emergence hence is creation of new properties regardless of the substance involved." "The assumption of an extra-physical vitalis (vital force, entelechy, élan vital, etc.), as formulated in most forms (old or new) of vitalism, is usually without any genuine explanatory power. It has served altogether too often as an intellectual tranquilizer or verbal sedative—stifling scientific inquiry rather than encouraging it to proceed in new directions."[26]

14.3 Mesmerism

A popular vitalist theory of the 18th century was "animal magnetism," in the theories of Franz Anton Mesmer (1734–1815). However, the use of the (conventional) English term *animal magnetism* to translate Mesmer's **magnétisme animal** can be misleading for three reasons:

- Mesmer chose his term to clearly distinguish his variant of *magnetic* force from those referred to, at that time, as *mineral magnetism*, *cosmic magnetism* and *planetary magnetism*.

- Mesmer felt that this particular force/power only resided in the bodies of humans and animals.

- Mesmer chose the word "*animal*," for its root meaning (from Latin *animus* = "breath") specifically to identify his force/power as a quality that belonged to all creatures with breath; viz., the animate beings: humans **and** animals.

Mesmer's ideas became so influential that King Louis XVI of France appointed two commissions to investigate mesmerism; one was led by Joseph-Ignace Guillotin, the other, led by Benjamin Franklin, included Bailly and Lavoisier. The commissioners learned about Mesmeric theory, and saw its patients fall into fits and trances. In Franklin's garden, a patient was led to each of five trees, one of which had been "mesmerized"; he hugged each in turn to receive the "vital fluid," but fainted at the foot of a 'wrong' one. At Lavoisier's house, four normal cups of water were held before a "sensitive" woman; the fourth produced convulsions, but she calmly swallowed the mesmerized contents of a fifth, believing it to be plain water. The commissioners concluded that "the fluid without imagination is powerless, whereas imagination without the fluid can produce the effects of the fluid."[27]

14.4 Alternative medicine

The National Center for Complementary and Alternative Medicine (NCCAM) classifies CAM therapies into five categories or domains:[28]

- alternative medical systems, or complete systems of therapy and practice;

- mind-body interventions, or techniques designed to facilitate the mind's effect on bodily functions and symptoms;

- biologically based systems, including herbalism;

- manipulative and body-based methods, such as chiropractic and massage therapy; and

- energy therapy.

The therapies that continue to be most intimately associated with vitalism are bioenergetic medicines, in the category of energy therapies. This field may be further divided into bioelectromagnetic medicines (BEM) and biofield therapies (BT). Compared with bioenergetic medicines, biofield therapies have a stronger identity with vitalism. Examples of biofield therapies include therapeutic touch, Reiki, external qi, chakra healing and SHEN therapy.[29] Biofield therapies are medical treatments in which the "subtle energy" field of a patient is manipulated by a biofield practitioner. The subtle energy is held to exist beyond the electromagnetic (EM) energy that is produced by the heart and brain. Beverly Rubik describes the biofield as a "complex, dynamic, extremely weak EM field within and around the human body...."[29]

The founder of homeopathy, Samuel Hahnemann, promoted an immaterial, vitalistic view of disease: "...they are solely spirit-like (dynamic) derangements of the spirit-like power (the vital principle) that animates the human body." As practised by some homeopaths today, homeopathy simply rests on the premise of treating sick persons with extremely diluted agents that – in undiluted doses – are deemed to produce similar symptoms in a healthy individual. Nevertheless, it remains equally true that the view of disease as a dynamic disturbance of the immaterial and dynamic vital force is taught in many homeopathic colleges and constitutes a fundamental principle for many contemporary practising homeopaths.

14.5 Criticism

Vitalism has sometimes been criticized as begging the question by inventing a name. Molière had famously parodied this fallacy in *Le Malade imaginaire*, where a quack "answers" the question of "Why does opium cause sleep?" with "Because of its soporific power."[30] Thomas Henry Huxley compared vitalism to stating that water is the way it is because of its "aquosity".[31] His grandson Julian Huxley in 1926 compared "vital force" or *élan vital* to explaining a railroad locomotive's operation by its *élan locomotif* ("locomotive force").

Another criticism is that vitalists have failed to rule out mechanistic explanations. This is rather obvious in retrospect for organic chemistry and developmental biology, but this criticism goes back at least a century. In 1912, Jacques Loeb published a landmark work, *The Mechanistic Conception of Life*. He described experiments on how a sea urchin could have a pin for its father, as Bertrand Russell put it (*Religion and Science*). He also offered this challenge:

> "... we must either succeed in producing living matter artificially, or we must find the reasons

why this is impossible." (pp. 5–6)

He also addressed vitalism more explicitly:

> "It is, therefore, unwarranted to continue the statement that in addition to the acceleration of oxidations the beginning of individual life is determined by the entrance of a metaphysical "life principle" into the egg; and that death is determined, aside from the cessation of oxidations, by the departure of this "principle" from the body. In the case of the evaporation of water we are satisfied with the explanation given by the kinetic theory of gases and do not demand that to repeat a well-known jest of Huxley the disappearance of the "aquosity" be also taken into consideration." (pp. 14–15)

Bechtel and Richardson[15] state that today vitalism "is often viewed as unfalsifiable, and therefore a pernicious metaphysical doctrine." For many scientists, "vitalist" theories were unsatisfactory "holding positions" on the pathway to mechanistic understanding. In 1967, Francis Crick, the co-discoverer of the structure of DNA, stated "And so to those of you who may be vitalists I would make this prophecy: what everyone believed yesterday, and you believe today, only cranks will believe tomorrow."[32]

While many vitalistic theories have in fact been falsified, notably Mesmerism, the pseudoscientific retention of untested and untestable theories continues to this day. Alan Sokal published an analysis of the wide acceptance among professional nurses of "scientific theories" of spiritual healing. (Pseudoscience and Postmodernism: Antagonists or Fellow-Travelers?).[33] Use of a technique called therapeutic touch was especially reviewed by Sokal, who concluded, "nearly all the pseudoscientific systems to be examined in this essay are based philosophically on vitalism" and added that "Mainstream science has rejected vitalism since at least the 1930s, for a plethora of good reasons that have only become stronger with time."[33]

Joseph C. Keating, Jr., PhD,[34] discusses vitalism's past and present roles in chiropractic and calls vitalism "a form of bio-theology." He further explains that:

> "Vitalism is that rejected tradition in biology which proposes that life is sustained and explained by an unmeasurable, intelligent force or energy. The supposed effects of vitalism are the manifestations of life itself, which in turn are the basis for inferring the concept in the first place. This circular reasoning offers pseudo-explanation, and may deceive us into believing we have explained some aspect of biology when

in fact we have only labeled our ignorance. 'Explaining an unknown (life) with an unknowable (Innate),' suggests philosopher Joseph Donahue, D.C., 'is absurd'."[35]

Keating views vitalism as incompatible with scientific thinking:

> "Chiropractors are not unique in recognizing a tendency and capacity for self-repair and auto-regulation of human physiology. But we surely stick out like a sore thumb among professions which claim to be scientifically based by our unrelenting commitment to vitalism. So long as we propound the 'One cause, one cure' rhetoric of Innate, we should expect to be met by ridicule from the wider health science community. Chiropractors can't have it both ways. Our theories cannot be both dogmatically held vitalistic constructs and be scientific at the same time. The purposiveness, consciousness and rigidity of the Palmers' Innate should be rejected."[35]

Keating also mentions Skinner's viewpoint:

> "Vitalism has many faces and has sprung up in many areas of scientific inquiry. Psychologist B.F. Skinner, for example, pointed out the irrationality of attributing behavior to mental states and traits. Such 'mental way stations,' he argued, amount to excess theoretical baggage which fails to advance cause-and-effect explanations by substituting an unfathomable psychology of 'mind'."[35]

According to Williams,[36] "today, vitalism is one of the ideas that form the basis for many pseudoscientific health systems that claim that illnesses are caused by a disturbance or imbalance of the body's vital force." "Vitalists claim to be scientific, but in fact they reject the scientific method with its basic postulates of cause and effect and of provability. They often regard subjective experience to be more valid than objective material reality."

Victor Stenger[37] states that the term "bioenergetics" "is applied in biochemistry to refer to the readily measurable exchanges of energy within organisms, and between organisms and the environment, which occur by normal physical and chemical processes. This is not, however, what the new vitalists have in mind. They imagine the bioenergetic field as a holistic living force that goes beyond reductionist physics and chemistry."[38]

Such a field is sometimes explained as electromagnetic(EM), though some advocates also make confused appeals to quantum physics.[29] Joanne Stefanatos states that

"The principles of energy medicine originate in quantum physics."[39] Stenger[38] offers several explanations as to why this line of reasoning may be misplaced. He explains that energy exists in discrete packets called quanta. Energy fields are composed of their component parts and so only exist when quanta are present. Therefore energy fields are not holistic, but are rather a system of discrete parts that must obey the laws of physics. This also means that energy fields are not instantaneous. These facts of quantum physics place limitations on the infinite, continuous field that is used by some theorists to describe so-called "human energy fields".[40] Stenger continues, explaining that the effects of EM forces have been measured by physicists as accurately as one part in a billion and there is yet to be any evidence that living organisms emit a unique field.[38]

Vitalistic thinking has also been identified in the naive biological theories of children: "Recent experimental results show that a majority of preschoolers tend to choose vitalistic explanations as most plausible. Vitalism, together with other forms of intermediate causality, constitute unique causal devices for naive biology as a core domain of thought."[41]

14.6   See also

- Dualism

- Energy (esotericism)

- Etheric body

- Georges Canguilhem

- Hans Adolf Eduard Driesch

- Henri Bergson

- Holism in science

- Homeopathy

- Irreducible complexity

- Odic force

- Orenda

- Philosophy of biology

- Prana

- Ruh

- Qi

- Vis medicatrix naturae

- Vital materialism

14.7   Notes

- ^**a** Stéphane Leduc and D'Arcy Thompson (On Growth and Form) published a series of works that took on the task of uprooting the remaining vestiges of vitalism, essentially by showing that the principles of physics and chemistry were enough, by themselves, to account for the growth and development of biological form.[42]

14.8   References

[1] BECHTEL, WILLIAM and ROBERT C. RICHARDSON (1998). Vitalism. In E. Craig (Ed.), Routledge Encyclopedia of Philosophy. London: Routledge. Vitalism

[2] A Cultural History of Medical Vitalism in Enlightenment Montpellier – Elizabeth Ann Williams – Google Books

[3] Jidenu, Paulin (1996) *African Philosophy, 2nd Ed.* Indiana University Press, ISBN 0-253-21096-8, p.16.

[4] Charles Birch, John B. Cobb, *The Liberation of Life: From the Cell to the Community*, 1985, p. 75

[5] Hankinson, R. J. (1997). *Cause and Explanation in Ancient Greek Thought.* Oxford University Press. p. 125. ISBN 978-0-19-924656-4.

[6] *Andrew Ede (2007), The Rise and Decline of Colloid Science in North America, 1900–1935: The Neglected Dimension*, p. 23

[7] Vitalism and Synthesis of Urea

[8] *The Real Death of Vitalism: Implications of the Wöhler Myth*

[9] cited by Schummer J, op cit

[10] n 1845, Adolph Kolbe succeeded in making acetic acid from inorganic compounds, and in the 1850s, Marcellin Berthelot repeated this feat for numerous organic compounds. In retrospect, Wöhler's work was the beginning of the end of Berzelius's vitalist hypothesis, but only in retrospect, as Ramberg had shown.

[11] Sebastian Normandin; Charles T. Wolfe (2013). *Introduction. Vitalism and the Scientific Image in Post-Enlightenment Life Science, 1800–2010* (Springer). p. 104. ISBN 978-94-007-2445-7. In medicine and biology, vitalism has been seen as a philosophically-charged term, a pseudoscientific gloss that corrupted scientific practice …

[12] "Other writers (eg, Peterfreund, 1971) simply use the term vitalism as a pejorative label." in Galatzer-Levy, RM (1976) Psychic Energy, A Historical Perspective *Ann Psychoanal* 4:41–61

[13] Mayr E (2002) *The Walter Arndt Lecture: The Autonomy of Biology*, adapted for the internet, on

[14] Ernst Mayr *Toward a new philosophy of biology: observations of an evolutionist* 1988, p. 13

[15] Vitalism. Bechtel W, Richardson RC (1998). *Routledge Encyclopedia of Philosophy*. E. Craig (Ed.), London: Routledge.

[16] Charles Birch, John B. Cobb, *The Liberation of Life: From the Cell to the Community*, 1985, pp. 76–78

[17] http://vlp.mpiwg-berlin.mpg.de/pdfgen/essays/enc22.pdf

[18] Developmental Biology 8e Online: A Selective History of Induction

[19] Jung's Concept of Die Dominanten (The Dominants) Online

[20] Peter J. Bowler, Reconciling science and religion: the debate in early-twentieth-century Britain, 2001, pp. 168–169

[21] Mark A. Bedau, Carol E. Cleland, *The Nature of Life: Classical and Contemporary Perspectives from Philosophy and Science*, 2010, p. 95

[22] Schultz, SG (1998). "A century of (epithelial) transport physiology: from vitalism to molecular cloning". *The American journal of physiology* **274** (1 Pt 1): C13–23. PMID 9458708.

[23] Gilbert, SF; Sarkar, S (2000). "Embracing complexity: organicism for the 21st century". *Developmental Dynamics* **219** (1): 1–9. doi:10.1002/1097-0177(2000)9999:9999<::AID-DVDY1036>3.0.CO;2-A. PMID 10974666.

[24] see "Emergent Properties" in the *Stanford Encyclopedia of Philosophy*. online at Stanford University for explicit discussion; briefly, some philosophers see emergentism as midway between traditional spiritual vitalism and mechanistic reductionism; others argue that, structurally, emergentism is equivalent to vitalism. See also Emmeche C (2001) Does a robot have an Umwelt? *Semiotica* 134: 653–693

[25] Emmeche C (1997) Explaining Emergence: towards an ontology of levels. *Journal for General Philosophy of Science* available online

[26] Dictionary of the History of Ideas

[27] Best, M; Neuhauser, D; Slavin, L (2003). "Evaluating Mesmerism, Paris, 1784: the controversy over the blinded placebo controlled trials has not stopped". *Quality & safety in health care* **12** (3): 232–3. doi:10.1136/qhc.12.3.232. PMC 1743715. PMID 12792017.

[28] "Complementary and Alternative Medicine – U.S. National Library of Medicine Collection Development Manual". Retrieved 2008-03-31.

[29] Rubik, *Bioenergetic Medicines*, American Medical Student Association Foundation, viewed 28 November 2006,

[30] *Mihi a docto doctore / Demandatur causam et rationem quare / Opium facit dormire. / A quoi respondeo, / Quia est in eo / Vertus dormitiva, / Cujus est natura / Sensus assoupire.* Le Malade imaginaire, (French Wikisource)

[31] The Physical Basis of Life, *Pall Mall Gazette*, 1869

[32] Crick F (1967) *Of Molecules and Men*; Great Minds Series Prometheus Books 2004, reviewed here. Crick's remark is cited and discussed in: Hein H (2004) Molecular biology vs. organicism: The enduring dispute between mechanism and vitalism. *Synthese* 20:238–253, who describes Crick's remark as "raising spectral red herrings".

[33] Pseudoscience and Postmodernism: Antagonists or Fellow-Travelers?

[34] Joseph C. Keating, Jr., PhD: Biographical sketch

[35] "The Meanings of Innate" Joseph C. Keating, Jr., PhD, J Can Chiropr Assoc 2002; 46(1)

[36] Williams.W. (2000) *The Encyclopedia of Pseudoscience. From Alien Abductions to Zone Therapy*. Facts on File inc. Contributors: Drs D.Conway, L.Dalton, R.Dolby, R.Duval, H.Farrell, J.Frazier, J.McMillan, J.Melton, T.O'Niell, R.Shepherd, S.Utley, W.Williams. ISBN 0-8160-3351-X

[37] Victor J. Stenger's site

[38] Stenger, Victor J. (Spring–Summer 1999). "The Physics of 'Alternative Medicine': Bioenergetic Fields.". *The Scientific Review of Alternative Medicine* **3** (1).

[39] Stefanatos, J. 1997, 'Introduction to Bioenergetic Medicine', Shoen, A.M and S.G. Wynn, *Complementary and Alternative Veterinary Medicine: Principles and Practices*, Mosby-Yearbook, Chicago.

[40] Biley, Francis, C. 2005, *Unitary Health Care: Martha Rogers' Science of Unitary Human Beings*, University of Wales College of Medicine, viewed 30 November 2006,

[41] Inagaki, K; Hatano, G (2004). "'Vitalistic causality in young children's naive biology.'". *Trends Cogn Sci* **8** (8): 356–62. doi:10.1016/j.tics.2004.06.004. PMID 15335462.

[42] EVELYN FOX KELLER, MAKING SENSE OF LIFE Explaining Biological Development with Models, Metaphors, and Machines. H A R V A R D U N I V E R S I T Y P R E S S, 2002.

14.9 External links

-

- Vitalism on *In Our Time* at the BBC. (listen now)

- Vitalism at the Skeptic's Dictionary

Chapter 15

Dmitri Mendeleev

For the Russian Prime Minister with a similar name, see Dmitry Medvedev.

This name uses Eastern Slavic naming customs; the patronymic is *Ivanovich* and the family name is *Mendeleev*.

Dmitri Ivanovich Mendeleev[3] (/ˌmɛndəlˈeɪəf/;[4] Russian: Дми́трий Ива́нович Менделе́ев; IPA: [ˈdmʲitrʲɪj ɪˈvanəvʲɪtɕ mʲɪndʲɪˈlʲejɪf]; 8 February 1834 – 2 February 1907 O.S. 27 January 1834 – 20 January 1907) was a Russian chemist and inventor. He formulated the Periodic Law, created his own version of the periodic table of elements, and used it to correct the properties of some already discovered elements and also to predict the properties of eight elements yet to be discovered.

15.1   Early life

Mendeleev was born in the village of Verkhnie Aremzyani, near Tobolsk in Siberia, to Ivan Pavlovich Mendeleev and Maria Dmitrievna Mendeleeva (née Kornilieva). His grandfather was Pavel Maximovich Sokolov, a priest of the Russian Orthodox Church from the Tver region.[5] Ivan, along with his brothers and sisters, obtained new family names while attending the theological seminary.[6] Mendeleev was raised as an Orthodox Christian, his mother encouraging him to "patiently search divine and scientific truth."[7] His son would later inform that he departed from the Church and embraced a form of deism.[8]

Mendeleev is thought to be the youngest of either 11, 13, 14 or 17 siblings;[9] the exact number differs among sources.[10] His father was a teacher of fine arts, politics and philosophy. Unfortunately for the family's financial well being, his father became blind and lost his teaching position. His mother was forced to work and she restarted her family's abandoned glass factory. At the age of 13, after the passing of his father and the destruction of his mother's factory by fire, Mendeleev attended the Gymnasium in Tobolsk.

In 1849, his mother took Mendeleev across the entire state of Russia from Siberia to Moscow with the aim of getting Mendeleev a higher education. The university in Moscow did not accept him. The mother and son continued to St. Petersburg to the father's alma mater. The now poor Mendeleev family relocated to Saint Petersburg, where he entered the Main Pedagogical Institute in 1850. After graduation, he contracted tuberculosis, causing him to move to the Crimean Peninsula on the northern coast of the Black Sea in 1855. While there he became a science master of the Simferopol gymnasium №1. In 1857, he returned to Saint Petersburg with fully restored health.

15.2   Later life

Between 1859 and 1861, he worked on the capillarity of liquids and the workings of the spectroscope in Heidelberg. In late August 1861 he wrote his first book on the spectroscope. On 4 April 1862 he became engaged to Feozva Nikitichna Leshcheva, and they married on 27 April 1862 at Nikolaev Engineering Institute's church in Saint Petersburg (where he taught).[11] Mendeleev became a professor at the Saint Petersburg Technological Institute and Saint Petersburg State University in 1864 and 1865, respectively. In 1865 he became Doctor of Science for his dissertation "On the Combinations of Water with Alcohol". He achieved tenure in 1867, and by 1871 had transformed Saint Petersburg into an internationally recognized center for chemistry research. In 1876, he became obsessed with Anna Ivanova Popova and began courting her; in 1881 he proposed to her and threatened suicide if she refused. His divorce from Leshcheva was finalized one month after he had married Popova (on 2 April[12]) in early 1882. Even after the divorce, Mendeleev was technically a bigamist; the Russian Orthodox Church required at least seven years before lawful remarriage. His divorce and the surrounding controversy contributed to his failure to be admitted to the Russian Academy of Sciences (despite his international fame by that time). His daughter from his second marriage, Lyubov, became the wife of the famous Russian poet Alexander Blok.

Dmitri Mendeleev

His other children were son Vladimir (a sailor, he took part in the notable Eastern journey of Nicholas II) and daughter Olga, from his first marriage to Feozva, and son Ivan and twins from Anna.

Though Mendeleev was widely honored by scientific organizations all over Europe, including (in 1882) the Davy Medal from the Royal Society of London (which later also awarded him the Copley Medal in 1905), he resigned from Saint Petersburg University on 17 August 1890. He was later elected a Foreign Member of the Royal Society (ForMemRS) in 1892.[2]

Mendeleev also investigated the composition of petroleum, and helped to found the first oil refinery in Russia. He recognized the importance of petroleum as a feedstock for petrochemicals. He is credited with a remark that burning petroleum as a fuel "would be akin to firing up a kitchen stove with bank notes."[13]

In 1905, Mendeleev was elected a member of the Royal Swedish Academy of Sciences. The following year the Nobel Committee for Chemistry recommended to the Swedish Academy to award the Nobel Prize in Chemistry for 1906 to Mendeleev for his discovery of the periodic system. The Chemistry Section of the Swedish Academy supported this recommendation. The Academy was then supposed to approve the Committee's choice, as it has done in almost every case. Unexpectedly, at the full meeting of the

Academy, a dissenting member of the Nobel Committee, Peter Klason, proposed the candidacy of Henri Moissan whom he favored. Svante Arrhenius, although not a member of the Nobel Committee for Chemistry, had a great deal of influence in the Academy and also pressed for the rejection of Mendeleev, arguing that the periodic system was too old to acknowledge its discovery in 1906. According to the contemporaries, Arrhenius was motivated by the grudge he held against Mendeleev for his critique of Arrhenius's dissociation theory. After heated arguments, the majority of the Academy voted for Moissan. The attempts to nominate Mendeleev in 1907 were again frustrated by the absolute opposition of Arrhenius.[14]

In 1907, Mendeleev died at the age of 72 in Saint Petersburg from influenza. The crater Mendeleev on the Moon, as well as element number 101, the radioactive mendelevium, are named after him.

15.3   Periodic table

Reihen	Gruppe I. — R²O	Gruppe II. — RO	Gruppe III. — R²O³	Gruppe IV. RH⁴ RO²	Gruppe V. RH³ R²O⁵	Gruppe VI. RH² RO³	Gruppe VII. RH R²O⁷	Gruppe VIII. — RO⁴
1	H=1							
2	Li=7	Be=9,4	B=11	C=12	N=14	O=16	F=19	
3	Na=23	Mg=24	Al=27,3	Si=28	P=31	S=32	Cl=35,5	
4	K=39	Ca=40	—=44	Ti=48	V=51	Cr=52	Mn=55	Fe=56, Co=59, Ni=59, Cu=63.
5	(Cu=63)	Zn=65	—=68	—=72	As=75	Se=78	Br=80	
6	Rb=85	Sr=87	?Yt=88	Zr=90	Nb=94	Mo=96	—=100	Ru=104, Rh=104, Pd=106, Ag=108.
7	(Ag=108)	Cd=112	In=113	Sn=118	Sb=122	Te=125	J=127	
8	Cs=133	Ba=137	?Di=138	?Ce=140	—	—	—	
9	(—)	—	—	—	—	—	—	
10	—	—	?Er=178	?La=180	Ta=182	W=184	—	Os=195, Ir=197, Pt=198, Au=199.
11	(Au=199)	Hg=200	Tl=204	Pb=207	Bi=208	—	—	
12	—	—	—	Th=231	—	U=240	—	

Mendeleev's 1871 periodic table

Sculpture in honor of Mendeleev and the periodic table, located in Bratislava, Slovakia

In 1863 there were 56 known elements with a new element being discovered at a rate of approximately one per year.

Other scientists had previously identified periodicity of elements. John Newlands described a Law of Octaves, noting their periodicity according to relative atomic weight in 1864, publishing it in 1865. His proposal identified the potential for new elements such as germanium. The concept was criticized and his innovation was not recognized by the Society of Chemists until 1887. Another person to propose a periodic table was Lothar Meyer, who published a paper in 1864 describing 28 elements classified by their valence, but with no prediction of new elements.

After becoming a teacher, Mendeleev wrote the definitive textbook of his time: *Principles of Chemistry* (two volumes, 1868–1870). As he attempted to classify the elements according to their chemical properties, he noticed patterns that led him to postulate his periodic table; he claimed to have envisioned the complete arrangement of the elements in a dream:[15][16][17][18][19]

> "I saw in a dream a table where all elements fell into place as required. Awakening, I immediately wrote it down on a piece of paper, only in one place did a correction later seem necessary."
> — Mendeleev, as quoted by Inostrantzev[20][21]

Unaware of the earlier work on periodic tables going on in the 1860s, he made the following table:

By adding additional elements following this pattern, Dmitri developed his extended version of the periodic table.[22][23] On 6 March 1869, Mendeleev made a formal presentation to the Russian Chemical Society, entitled *The Dependence between the Properties of the Atomic Weights of the Elements*, which described elements according to both atomic weight and valence. This presentation stated that

1. The elements, if arranged according to their atomic weight, exhibit an apparent periodicity of properties.

2. Elements which are similar regarding their chemical properties have atomic weights which are either of nearly the same value (e.g., Pt, Ir, Os) or which increase regularly (e.g., K, Rb, Cs).

3. The arrangement of the elements in groups of elements in the order of their atomic weights corresponds to their so-called valencies, as well as, to some extent, to their distinctive chemical properties; as is apparent among other series in that of Li, Be, B, C, N, O, and F.

4. The elements which are the most widely diffused have small atomic weights.

5. The magnitude of the atomic weight determines the character of the element, just as the magnitude of

the molecule determines the character of a compound body.

6. We must expect the discovery of many yet unknown elements–for example, two elements, analogous to aluminium and silicon, whose atomic weights would be between 65 and 75.

7. The atomic weight of an element may sometimes be amended by a knowledge of those of its contiguous elements. Thus the atomic weight of tellurium must lie between 123 and 126, and cannot be 128. *(Tellurium's atomic mass is 127.6, and Mendeleev was incorrect in his assumption that atomic mass must increase with position within a period.)*

8. Certain characteristic properties of elements can be foretold from their atomic weights.

Mendeleev published his periodic table of all known elements and predicted several new elements to complete the table. Only a few months after, Meyer published a virtually identical table. Some consider Meyer and Mendeleev the co-creators of the periodic table. Mendeleev has the distinction of accurately predicting of the qualities of what he called ekasilicon, ekaaluminium and ekaboron (germanium, gallium and scandium, respectively).

For his predicted eight elements, he used the prefixes of eka, dvi, and tri (Sanskrit one, two, three) in their naming. Mendeleev questioned some of the currently accepted atomic weights (they could be measured only with a relatively low accuracy at that time), pointing out that they did not correspond to those suggested by his Periodic Law. He noted that tellurium has a higher atomic weight than iodine, but he placed them in the right order, incorrectly predicting that the accepted atomic weights at the time were at fault. He was puzzled about where to put the known lanthanides, and predicted the existence of another row to the table which were the actinides which were some of the heaviest in atomic mass. Some people dismissed Mendeleev for predicting that there would be more elements, but he was proven to be correct when Ga (gallium) and Ge (germanium) were found in 1875 and 1886 respectively, fitting perfectly into the two missing spaces.[24]

By giving Sanskrit names to his "missing" elements, Mendeleev showed his appreciation and debt to the Sanskrit grammarians of ancient India, who had created sophisticated theories of language based on their discovery of the two-dimensional patterns in basic sounds. Mendeleev was a friend and colleague of the Sanskritist Böhtlingk, who was preparing the second edition of his book on Pāṇini[25] at about this time, and Mendeleev wished to honor Pāṇini with his nomenclature.[26] Noting that there are striking similarities between the periodic table and the introductory Śiva Sūtras in Pāṇini's grammar, Prof. Kiparsky says:

"The analogies between the two systems are striking. Just as Panini found that the phonological patterning of sounds in the language is a function of their articulatory properties, so Mendeleev found that the chemical properties of elements are a function of their atomic weights. Like Panini, Mendeleev arrived at his discovery through a search for the "grammar" of the elements...[27]"

The original draft made by Mendeleev would be found years later and published under the name *Tentative System of Elements*.[28]

15.4   Other achievements

Mendeleev made other important contributions to chemistry. The Russian chemist and science historian Lev Chugaev has characterized him as "a chemist of genius, first-class physicist, a fruitful researcher in the fields of hydrodynamics, meteorology, geology, certain branches of chemical technology (explosives, petroleum, and fuels, for example) and other disciplines adjacent to chemistry and physics, a thorough expert of chemical industry and industry in general, and an original thinker in the field of economy." Mendeleev was one of the founders, in 1869, of the Russian Chemical Society. He worked on the theory and practice of protectionist trade and on agriculture.

In an attempt at a chemical conception of the Aether, he put forward a hypothesis that there existed two inert chemical elements of lesser atomic weight than hydrogen. Of these two proposed elements, he thought the lighter to be an all-penetrating, all-pervasive gas, and the slightly heavier one to be a proposed element, *coronium*.

Mendeleev devoted much study and made important contributions to the determination of the nature of such indefinite compounds as solutions.

In another department of physical chemistry, he investigated the expansion of liquids with heat, and devised a formula similar to Gay-Lussac's law of the uniformity of the expansion of gases, while in 1861 he anticipated Thomas Andrews' conception of the critical temperature of gases by defining the absolute boiling-point of a substance as the temperature at which cohesion and heat of vaporization become equal to zero and the liquid changes to vapor, irrespective of the pressure and volume.

Mendeleev is given credit for the introduction of the metric system to the Russian Empire.

He invented *pyrocollodion*, a kind of smokeless powder based on nitrocellulose. This work had been commissioned

Mendeleev Medal

by the Russian Navy, which however did not adopt its use. In 1892 Mendeleev organized its manufacture.

Mendeleev studied petroleum origin and concluded hydrocarbons are abiogenic and form deep within the earth – see Abiogenic petroleum origin. He wrote: "*The capital fact to note is that petroleum was born in the depths of the earth, and it is only there that we must seek its origin.*" (Dmitri Mendeleev, 1877)[29]

15.5   Vodka myth

A very popular Russian story is that it was Mendeleev who came up with the 40% standard strength of vodka in 1894, after having been appointed Director of the Bureau of Weights and Measures with the assignment to formulate new state standards for the production of vodka. This story has, for instance, been used in marketing claims by the Russian Standard vodka brand that, *"In 1894, Dmitri Mendeleev, the greatest scientist in all Russia, received the decree to set the Imperial quality standard for Russian vodka and the 'Russian Standard' was born"*,[30] or that the vodka is *"compliant with the highest quality of Russian vodka approved by the royal government commission headed by Mendeleev in 1894."*[31]

While it is true that Mendeleev in 1892 became head of the Archive of Weights and Measures in Saint Petersburg, and evolved it into a government bureau the following year, that institution was never involved in setting any production quality standards, but was issued with standardising Russian trade weights and measuring instruments. Furthermore, the

40% standard strength was already introduced by the Russian government in 1843, when Mendeleev was nine years old.[31]

The basis for the whole story is a popular myth that Mendeleev's 1865 doctoral dissertation "A Discourse on the combination of alcohol and water" contained a statement that 38% is the ideal strength of vodka, and that this number was later rounded to 40% to simplify the calculation of alcohol tax. However, Mendeleev's dissertation was about alcohol concentrations over 70% and he never wrote anything about vodka.[31][32]

15.6 Commemoration

A number of places and objects are associated with the name and achievements of the scientist.

In Saint Petersburg his name was given to the National Metrology Institute[33] dealing with establishing and supporting national and worldwide standards for precise measurements. Next to it there is a monument to him pictured above that consists of his sitting statue and a depiction of his periodic table on the wall of the establishment.

In the Twelve Collegia building, now being the centre of Saint Petersburg State University and in Mendeleev's time – Head Pedagogical Institute – there is Dmitry Mendeleev's Memorial Museum Apartment[34] with his archives. The street in front of these is named after him as Mendeleevskaya liniya (Mendeleev Line).

In Moscow, there is the D. Mendeleyev University of Chemical Technology of Russia.[35]

After him was also named mendelevium, which is a synthetic chemical element with the symbol Md (formerly Mv) and the atomic number 101. It is a metallic radioactive transuranic element in the actinide series, usually synthesized by bombarding einsteinium with alpha particles.

A large lunar impact crater Mendeleev that is located on the far side of the Moon, as seen from the Earth, also bears the name of the scientist.

Russian Academy of Sciences yearly awards since 1998 Mendeleev Golden Medal.

15.7 See also

- List of Russian chemists

- Mendeleev's predicted elements

- Periodic Systems of Small Molecules

15.8 References

[1] Physics Tree profile Dmitri Ivanovich Mendeleev

[2] "Fellows of the Royal Society". London: Royal Society. Archived from the original on 2015-03-16.

[3] Also romanized **Mendeleyev** or **Mendeleef**

[4] "Mendeleev". *Random House Webster's Unabridged Dictionary.*

[5] Dmitriy Mendeleev: A Short CV, and A Story of Life, mendcomm.org

[6] Удомельские корни Дмитрия Ивановича Менделеева (1834–1907), starina.library.tver.ru

[7] Hiebert, Ray Eldon; Hiebert, Roselyn (1975). *Atomic Pioneers: From ancient Greece to the 19th century*. U.S. Atomic Energy Commission. Division of Technical Information. p. 25.

[8] Gordin, Michael D. (2004). *A Well-ordered Thing: Dmitrii Mendeleev And The Shadow Of The Periodic Table*. Basic Books. pp. 229–230. ISBN 9780465027750. Mendeleev seemed to have very few theological commitments. This was not for lack of exposure. His upbringing was actually heavily religious, and his mother — by far the dominating force in his youth - was exceptionally devout. One of his sisters even joined a fanatical religious sect for a time. Despite, or perhaps because of, this background, Mendeleev withheld comment on religious affairs for most of his life, reserving his few words for anti-clerical witticisms... Mendeleev's son Ivan later vehemently denied claims that his father was devoutly Orthodox: "I have also heard the view of my father's 'church religiosity' — and I must reject this categorically. From his earliest years Father practically split from the church — and if he tolerated certain simple everyday rites, then only as an innocent national tradition, similar to Easter cakes, which he didn't consider worth fighting against." ...Mendeleev's opposition to traditional Orthodoxy was not due to either atheism or a scientific materialism. Rather, he held to a form of romanticized deism.

[9] Johnson, George (3 January 2006). "The Nitpicking of the Masses vs. the Authority of the Experts". *New York Times*. Retrieved 14 March 2011.

[10] The number of Mendeleev's siblings is a matter of some historical dispute. When the Princeton historian of science Michael Gordin reviewed this article as part of an analysis of the accuracy of Wikipedia for the 14 December 2005 issue of *Nature*, he cited as one of Wikipedia's errors that "They say Mendeleev is the 14th child. He is the 14th surviving child of 17 total. 14 is right out." However in a *New York Times* article from January 2006, it was noted that in Gordin's own 2004 biography of Mendeleev, he also had the Russian chemist listed as the 17th child, and quoted Gordin's response to this as being: "That's curious. I believe that is a typographical error in my book. Mendeleyev was the final child, that is certain, and the number the reliable sources

have is 13." Gordin's book specifically says that Mendeleev's mother bore her husband "seventeen children, of whom eight survived to young adulthood," with Mendeleev being the youngest. See: Johnson, George (3 January 2006). "The Nitpicking of the Masses vs. the Authority of the Experts". *The New York Times.* and Gordin, Michael (22 December 2005). "Supplementary information to accompany *Nature* news article 'Internet encyclopaedias go head to head' (*Nature* 438, 900–901; 2005)" (PDF). *Blogs.Nature.com* – via 2004, p. 178.

[11] "Rustest.spb.ru". Rustest.spb.ru. Retrieved 13 March 2010.

[12] "Gazeta.ua". Gazeta.ua. 9 March 2010. Retrieved 13 March 2010.

[13] John W. Moore, Conrad L. Stanitski, Peter C. Jurs. *Chemistry: The Molecular Science, Volume 1.* Retrieved 6 September 2011.

[14] Friedman, Robert M. (2001). *The politics of excellence: behind the Nobel Prize in science.* New York: Times Books. pp. 32–34. ISBN 0-7167-3103-7.

[15] John B. Arden (1998). "Science, Theology and Consciousness", Praeger Frederick A. p. 59: *The initial expression of the commonly used chemical periodic table was reportedly envisioned in a dream. In 1869, Dmitri Mendeleev claimed to have had a dream in which he envisioned a table in which all the chemical elements were arranged according to their atomic weight.*

[16] John Kotz, Paul Treichel, Gabriela Weaver (2005). "Chemistry and Chemical Reactivity," Cengage Learning. p. 333

[17] Gerard I. Nierenberg (1986). "The art of creative thinking", Simon & Schuster, p. 201: *Dmitri Mendeleev's solution for the arrangement of the elements that came to him in a dream.*

[18] Helen Palmer (1998). "Inner Knowing: Consciousness, Creativity, Insight, and Intuition". J.P. Tarcher/Putnam. p. 113: *The sewing machine, for instance, invented by Elias Howe, was developed from material appearing in a dream, as was Dmitri Mendeleev's periodic table of elements*

[19] Simon S. Godfrey (2003). "Dreams & Reality". Trafford Publishing. Chapter 2.: *'The Russian chemist, Dmitri Mendeleev (1834-1907), described a dream in which he saw the periodic table of elements in its complete form.* ISBN 1412011434

[20] "The Soviet Review Translations" Summer 1967. Vol. VIII, No. 2, M.E. Sharpe, Incorporated, p. 38

[21] Myron E. Sharpe, (1967). "Soviet Psychology". Volume 5, p. 30.

[22] A brief history of the development of the period table, wou.edu

[23] Mendeleev and the Periodic Table, chemsheets.co.uk

[24] Emsley, John (2001). *Nature's Building Blocks* ((Hardcover, First Edition) ed.). Oxford University Press. pp. 521–522. ISBN 0-19-850340-7.

[25] Otto Böhtlingk, Panini's Grammatik: Herausgegeben, Ubersetzt, Erlautert und MIT Verschiedenen Indices Versehe. St. Petersburg, 1839–40.

[26] Kiparsky, Paul. "Economy and the construction of the Sivasutras." In M. M. Deshpande and S. Bhate (eds.), *Paninian Studies.* Ann Arbor, Michigan, 1991.

[27] Kak, Subhash (2004). "Mendeleev and the Periodic Table of Elements". *Sandhan* **4** (2): 115–123. arXiv:physics/0411080

[28] "The Soviet Review Translations" Summer 1967. Vol. VIII, No. 2, M.E. Sharpe, Incorporated, p. 39

[29] Mendeleev, D., 1877. L'Origine du pétrole. Revue Scientifique, 2e Ser., VIII, p. 409-416.

[30] Sainsburys: *Russian Standard Vodka 1L* Linked 2014-06-28

[31] Evseev, Anton (2011-11-21). "Dmitry Mendeleev and 40 degrees of Russian vodka". *Science.* Moscow: English Pravda.Ru. Retrieved 2014-07-06.

[32] "Prominent Russians: Dmitri Mendeleev". *Prominent Russians: Science and technology.* Moscow: RT. 2011. Retrieved 2014-07-06.

[33] ВНИИМ Дизайн Груп (2011-04-13). "D.I.Mendeleyev Institute for Metrology". Vniim.ru. Retrieved 2012-08-20.

[34] Saint-PetersburgState University. "Museum-Archives n. a. Dmitry Mendeleev - Museums - Culture and Sport - University - Saint-Petersburg state university". Eng.spbu.ru. Retrieved 2012-08-19.

[35] University homepage in English

15.9   Further reading

- Gordin, Michael (2004). *A Well-Ordered Thing: Dmitrii Mendeleev and the Shadow of the Periodic Table.* New York: Basic Books. ISBN 0-465-02775-X.

- Mendeleyev, Dmitry Ivanovich; Jensen, William B. (2005). *Mendeleev on the Periodic Law: Selected Writings, 1869–1905.* Mineola, New York: Dover Publications. ISBN 0-486-44571-2.

- Strathern, Paul (2001). *Mendeleyev's Dream: The Quest For the Elements.* New York: St Martins Press. ISBN 0-241-14065-X.

- Mendeleev, Dmitrii Ivanovich (1901). *Principles of Chemistry.* New York: Collier.

15.10 External links

- Babaev, Eugene V. (February 2009). Dmitriy Mendeleev: A Short CV, and A Story of Life - 2009 biography on the occasion of Mendeleev's 175th anniversary

- Babaev, Eugene V., Moscow State University. Dmitriy Mendeleev Online

- Original Periodic Table, annotated.

- Mendeleev's first draft version of the Periodic Table, 17 February 1869.

- "Everything in its Place", essay by Oliver Sacks

- Works by or about Dmitri Mendeleev in libraries (WorldCat catalog)

Chapter 16

Periodic table

This article is about the table used in chemistry. For other uses, see Periodic table (disambiguation).

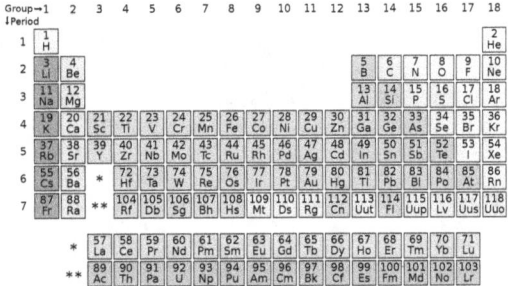

Standard form of the periodic table (color legend below)

The **periodic table** is a tabular arrangement of the chemical elements, ordered by their atomic number (number of protons in the nucleus), electron configurations, and recurring chemical properties. The table also shows four rectangular blocks: s-, p- d- and f-block. In general, within one row (period) the elements are metals on the lefthand side, and non-metals on the righthand side.

The rows of the table are called periods; the columns are called groups. Six groups (columns) have names as well as numbers: for example, group 17 elements are the halogens; and group 18, the noble gases. The periodic table can be used to derive relationships between the properties of the elements, and predict the properties of new elements yet to be discovered or synthesized. The periodic table provides a useful framework for analyzing chemical behavior, and is widely used in chemistry and other sciences.

Although precursors exist, Dmitri Mendeleev is generally credited with the publication, in 1869, of the first widely recognized periodic table. He developed his table to illustrate periodic trends in the properties of the then-known elements. Mendeleev also predicted some properties of then-unknown elements that would be expected to fill gaps in this table. Most of his predictions were proved correct when the elements in question were subsequently discov-

ered. Mendeleev's periodic table has since been expanded and refined with the discovery or synthesis of further new elements and the development of new theoretical models to explain chemical behavior.

All elements from atomic numbers 1 (hydrogen) to 118 (ununoctium) have been discovered or reportedly synthesized, with elements 113, 115, 117, and 118 having yet to be confirmed. The first 94 elements exist naturally, although some are found only in trace amounts and were synthesized in laboratories before being found in nature.[n 1] Elements with atomic numbers from 95 to 118 have only been synthesized in laboratories. It has been shown that elements 95 to 100 once occurred in nature but currently do not.[1] Synthesis of elements having higher atomic numbers is being pursued. Numerous synthetic radionuclides of naturally occurring elements have also been produced in laboratories.

16.1 Overview

For large cell versions, see Periodic table (large cells).

Some presentations include an element zero (i.e. a substance composed purely of neutrons), although this is uncommon. See, for example. Philip Stewart's Chemical Galaxy. Each chemical element has a unique atomic number representing the number of protons in its nucleus. Most elements have differing numbers of neutrons among different atoms, with these variants being referred to as isotopes. For example, carbon has three naturally occurring isotopes: all of its atoms have six protons and most have six neutrons as well, but about one per cent have seven neutrons, and a very small fraction have eight neutrons. Isotopes are never separated in the periodic table; they are always grouped together under a single element. Elements with no stable isotopes have the atomic masses of their most stable isotopes, where such masses are shown, listed in parentheses.[2]

In the standard periodic table, the elements are listed in order of increasing atomic number (the number of protons in the nucleus of an atom). A new row (*period*) is started when a new electron shell has its first electron. Columns (*groups*) are determined by the electron configuration of the atom; elements with the same number of electrons in a particular subshell fall into the same columns (e.g. oxygen and selenium are in the same column because they both have four electrons in the outermost p-subshell). Elements with similar chemical properties generally fall into the same group in the periodic table, although in the f-block, and to some respect in the d-block, the elements in the same period tend to have similar properties, as well. Thus, it is relatively easy to predict the chemical properties of an element if one knows the properties of the elements around it.[3]

As of 2014, the periodic table has 114 confirmed elements, comprising elements 1 (hydrogen) to 112 (copernicium), 114 (flerovium) and 116 (livermorium). Elements 113, 115, 117 and 118 have reportedly been synthesised in laboratories, but none of these claims have been officially confirmed by the International Union of Pure and Applied Chemistry (IUPAC), nor are they named. As such these elements are currently identified by their atomic number (e.g., "element 113"), or by their provisional systematic name ("ununtrium", symbol "Uut").[4]

A total of 94 elements occur naturally; the remaining 20 elements, from americium to copernicium, and flerovium and livermorium, occur only when synthesised in laboratories. Of the 94 elements that occur naturally, 84 are primordial. The other 10 naturally occurring elements occur only in decay chains of primordial elements.[1] No element heavier than einsteinium (element 99) has ever been observed in macroscopic quantities in its pure form, nor has astatine (element 85); francium (element 87) has been only photographed in the form of light emitted from microscopic quantities (300,000 atoms).[5]

16.1.1   Layout variants

In the most common graphic presentation of the periodic table, the main table has 18 columns and the lanthanides and the actinides are shown as two additional rows below the main body of the table,[6] with two placeholders shown in the main table, between barium and hafnium, and radium and rutherfordium, respectively. These placeholders can be asterisk-like markers, or a contracted range description of elements ("57–71"). This convention is entirely a matter of formatting practicality. The same table structure can be shown in a 32-column format, with the lanthanides and actinides in the main table's row 6 and 7.

However, based on the chemical and physical properties of elements, many alternative table *structures* have been constructed.

16.2   Grouping methods

16.2.1   Groups

Main article: Group (periodic table)

A *group* or *family* is a vertical column in the periodic table. Groups usually have more significant periodic trends than periods and blocks, explained below. Modern quantum mechanical theories of atomic structure explain group trends by proposing that elements within the same group generally have the same electron configurations in their valence shell.[7] Consequently, elements in the same group tend to have a shared chemistry and exhibit a clear trend in properties with increasing atomic number.[8] However, in some parts of the periodic table, such as the d-block and the f-block, horizontal similarities can be as important as, or more pronounced than, vertical similarities.[9][10][11]

Under an international naming convention, the groups are numbered numerically from 1 to 18 from the leftmost column (the alkali metals) to the rightmost column (the noble gases).[12] Previously, they were known by roman numerals. In America, the roman numerals were followed by either an "A" if the group was in the s- or p-block, or a "B" if the group was in the d-block. The roman numerals used correspond to the last digit of today's naming convention (e.g. the group 4 elements were group IVB, and the group 14 elements was group IVA). In Europe, the lettering was similar, except that "A" was used if the group was before group 10, and "B" was used for groups including and after group 10. In addition, groups 8, 9 and 10 used to be treated as one triple-sized group, known collectively in both notations as group VIII. In 1988, the new IUPAC naming system was put into use, and the old group names were deprecated.[13]

Some of these groups have been given trivial (unsystematic) names, as seen in the table below, although some are rarely used. Groups 3–10 have no trivial names and are referred to simply by their group numbers or by the name of the first member of their group (such as 'the scandium group' for Group 3), since they display fewer similarities and/or vertical trends.[12]

Elements in the same group tend to show patterns in atomic radius, ionization energy, and electronegativity. From top to bottom in a group, the atomic radii of the elements increase. Since there are more filled energy levels, valence electrons are found farther from the nucleus. From the top, each successive element has a lower ionization energy because it is easier to remove an electron since the atoms are less tightly bound. Similarly, a group has a top to bottom decrease in electronegativity due to an increasing distance between valence electrons and the nucleus.[14] There are exceptions to these trends, however, an example of which occurs in group 11 where electronegativity increases farther down the group.[15]

16.2.2 Periods

Main article: Period (periodic table)

A *period* is a horizontal row in the periodic table. Although groups generally have more significant periodic trends, there are regions where horizontal trends are more significant than vertical group trends, such as the f-block, where the lanthanides and actinides form two substantial horizontal series of elements.[16]

Elements in the same period show trends in atomic radius, ionization energy, electron affinity, and electronegativity. Moving left to right across a period, atomic radius usually decreases. This occurs because each successive element has an added proton and electron, which causes the electron to be drawn closer to the nucleus.[17] This decrease in atomic radius also causes the ionization energy to increase when moving from left to right across a period. The more tightly bound an element is, the more energy is required to remove an electron. Electronegativity increases in the same manner as ionization energy because of the pull exerted on the electrons by the nucleus.[14] Electron affinity also shows a slight trend across a period. Metals (left side of a period) generally have a lower electron affinity than nonmetals (right side of a period), with the exception of the noble gases.[18]

16.2.3 Blocks

Main article: Block (periodic table)
Specific regions of the periodic table can be referred to

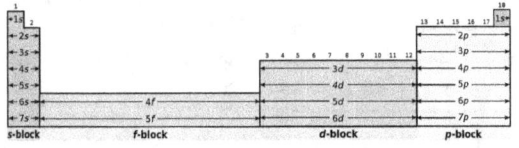

Left to right: s-, f-, d-, p-block in the periodic table

as *blocks* in recognition of the sequence in which the electron shells of the elements are filled. Each block is named according to the subshell in which the "last" electron notionally resides.[19][n 2] The s-block comprises the first two groups (alkali metals and alkaline earth metals) as well as hydrogen and helium. The p-block comprises the last six groups, which are groups 13 to 18 in IUPAC (3A to 8A in American) and contains, among other elements, all of the metalloids. The d-block comprises groups 3 to 12 (or 3B to 2B in American group numbering) and contains all of the

transition metals. The f-block, often offset below the rest of the periodic table, has no group numbers and comprises lanthanides and actinides.[20]

16.2.4 Metals, metalloids and nonmetals

Metals, metalloids, nonmetals, and elements with unknown chemical properties in the periodic table. Sources disagree on the classification of some of these elements.

According to their shared physical and chemical properties, the elements can be classified into the major categories of metals, metalloids and nonmetals. Metals are generally shiny, highly conducting solids that form alloys with one another and salt-like ionic compounds with nonmetals (other than the noble gases). The majority of nonmetals are coloured or colourless insulating gases; nonmetals that form compounds with other nonmetals feature covalent bonding. In between metals and nonmetals are metalloids, which have intermediate or mixed properties.[21]

Metal and nonmetals can be further classified into subcategories that show a gradation from metallic to nonmetallic properties, when going left to right in the rows. The metals are subdivided into the highly reactive alkali metals, through the less reactive alkaline earth metals, lanthanides and actinides, via the archetypal transition metals, and ending in the physically and chemically weak post-transition metals. The nonmetals are simply subdivided into the polyatomic nonmetals, which, being nearest to the metalloids, show some incipient metallic character; the diatomic nonmetals, which are essentially nonmetallic; and the monatomic noble gases, which are nonmetallic and almost completely inert. Specialized groupings such as the refractory metals and the noble metals, which are subsets (in this example) of the transition metals, are also known[22] and occasionally denoted.[23]

Placing the elements into categories and subcategories based on shared properties is imperfect. There is a spectrum of properties within each category and it is not hard to find overlaps at the boundaries, as is the case with most classification schemes.[24] Beryllium, for example, is classified as an alkaline earth metal although its amphoteric chemistry and tendency to mostly form covalent compounds are both attributes of a chemically weak or post transition metal. Radon is classified as a nonmetal and a noble gas yet has some cationic chemistry that is more characteristic of a metal. Other classification schemes are possible such as the division of the elements into mineralogical occurrence categories, or crystalline structures. Categorising the elements in this fashion dates back to at least 1869 when Hinrichs[25] wrote that simple boundary lines could be drawn on the periodic table to show elements having like properties, such as the metals and the nonmetals, or the gaseous elements.

16.3 Periodic trends

Main article: Periodic trends

16.3.1 Electron configuration

Main article: Electronic configuration
The electron configuration or organisation of electrons or-

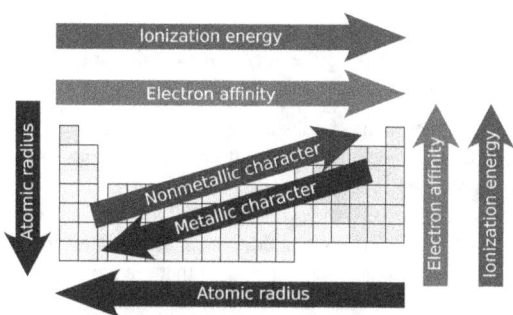

Periodic table trends (arrows direct an increase)

Since the properties of an element are mostly determined by its electron configuration, the properties of the elements likewise show recurring patterns or periodic behaviour, some examples of which are shown in the diagrams below for atomic radii, ionization energy and electron affinity. It is this periodicity of properties, manifestations of which were noticed well before the underlying theory was developed, that led to the establishment of the periodic law (the properties of the elements recur at varying intervals) and the formulation of the first periodic tables.[26][27]

16.3.2 Atomic radii

Main article: Atomic radius
Atomic radii vary in a predictable and explainable manner

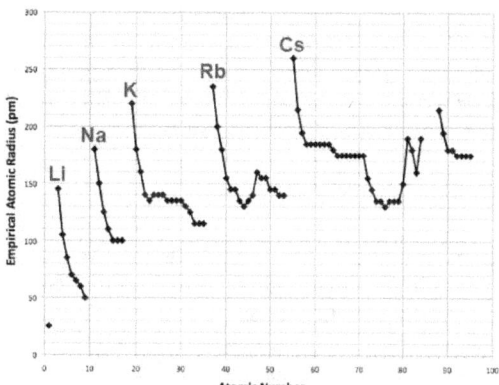

Atomic number plotted against atomic radius[n 3]

biting neutral atoms shows a recurring pattern or periodicity. The electrons occupy a series of electron shells (numbered shell 1, shell 2, and so on). Each shell consists of one or more subshells (named s, p, d, f and g). As atomic number increases, electrons progressively fill these shells and subshells more or less according to the Madelung rule or energy ordering rule, as shown in the diagram. The electron configuration for neon, for example, is $1s^2 2s^2 2p^6$. With an atomic number of ten, neon has two electrons in the first shell, and eight electrons in the second shell—two in the s subshell and six in the p subshell. In periodic table terms, the first time an electron occupies a new shell corresponds to the start of each new period, these positions being occupied by hydrogen and the alkali metals.[26][27]

across the periodic table. For instance, the radii generally decrease along each period of the table, from the alkali metals to the noble gases; and increase down each group. The radius increases sharply between the noble gas at the end of each period and the alkali metal at the beginning of the next period. These trends of the atomic radii (and of various other chemical and physical properties of the elements)

Approximate order in which shells and subshells are arranged by increasing energy according to the Madelung rule can be explained by the electron shell theory of the atom; they provided important evidence for the development and confirmation of quantum theory.[28]

The electrons in the 4f-subshell, which is progressively filled from cerium (element 58) to ytterbium (element 70), are not particularly effective at shielding the increasing nuclear charge from the sub-shells further out. The elements immediately following the lanthanides have atomic radii that are smaller than would be expected and that are almost identical to the atomic radii of the elements immediately above them.[29] Hence hafnium has virtually the same atomic radius (and chemistry) as zirconium, and tantalum has an atomic radius similar to niobium, and so forth. This is known as the lanthanide contraction. The effect of the lanthanide contraction is noticeable up to platinum (element 78), after which it is masked by a relativistic effect known as the inert pair effect.[30] The d-block contraction, which is a similar effect between the d-block and p-block, is less pronounced than the lanthanide contraction but arises from a similar cause.[29]

16.3.3   Ionization energy

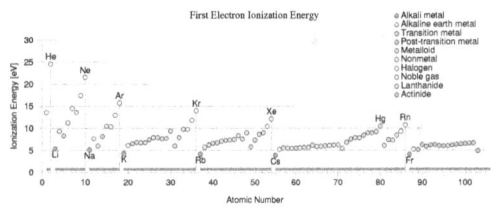

Ionization energy: each period begins at a minimum for the alkali metals, and ends at a maximum for the noble gases

Main article: Ionization energy

The first ionization energy is the energy it takes to remove one electron from an atom, the second ionization energy is the energy it takes to remove a second electron from the atom, and so on. For a given atom, successive ionization energies increase with the degree of ionization. For magnesium as an example, the first ionization energy is 738 kJ/mol and the second is 1450 kJ/mol. Electrons in the closer orbitals experience greater forces of electrostatic attraction; thus, their removal requires increasingly more energy. Ionization energy becomes greater up and to the right of the periodic table.[30]

Large jumps in the successive molar ionization energies occur when removing an electron from a noble gas (complete electron shell) configuration. For magnesium again, the first two molar ionization energies of magnesium given above

correspond to removing the two 3s electrons, and the third ionization energy is a much larger 7730 kJ/mol, for the removal of a 2p electron from the very stable neon-like configuration of Mg^{2+}. Similar jumps occur in the ionization energies of other third-row atoms.[30]

16.3.4   Electronegativity

Main article: Electronegativity

Electronegativity is the tendency of an atom to attract

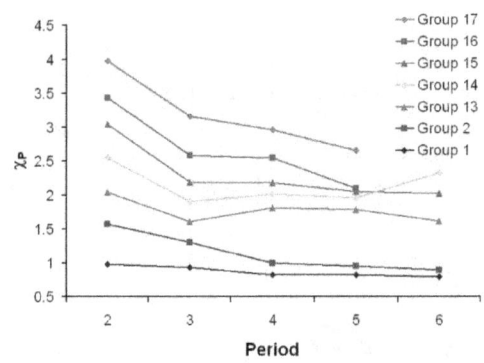

Graph showing increasing electronegativity with growing number of selected groups

electrons.[31] An atom's electronegativity is affected by both its atomic number and the distance between the valence electrons and the nucleus. The higher its electronegativity, the more an element attracts electrons. It was first proposed by Linus Pauling in 1932.[32] In general, electronegativity increases on passing from left to right along a period, and decreases on descending a group. Hence, fluorine is the most electronegative of the elements,[n 4] while caesium is the least, at least of those elements for which substantial data is available.[15]

There are some exceptions to this general rule. Gallium and germanium have higher electronegativities than aluminium and silicon respectively because of the d-block contraction. Elements of the fourth period immediately after the first row of the transition metals have unusually small atomic radii because the 3d-electrons are not effective at shielding the increased nuclear charge, and smaller atomic size correlates with higher electronegativity.[15] The anomalously high electronegativity of lead, particularly when compared to thallium and bismuth, appears to be an artifact of data selection (and data availability)—methods of calculation other than the Pauling method show the normal periodic trends for these elements.[33]

16.3.5 Electron affinity

Main article: Electron affinity

The electron affinity of an atom is the amount of energy re-

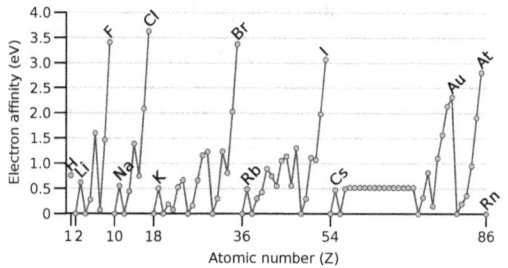

Dependence of electron affinity on atomic number.[34] Values generally increase across each period, culminating with the halogens before decreasing precipitously with the noble gases. Examples of localized peaks seen in hydrogen, the alkali metals and the group 11 elements are caused by a tendency to complete the s-shell (with the 6s shell of gold being further stabilized by relativistic effects and the presence of a filled 4f sub shell). Examples of localized troughs seen in the alkaline earth metals, and nitrogen, phosphorus, manganese and rhenium are caused by filled s-shells, or half-filled p- or d-shells.[35]

leased when an electron is added to a neutral atom to form a negative ion. Although electron affinity varies greatly, some patterns emerge. Generally, nonmetals have more positive electron affinity values than metals. Chlorine most strongly attracts an extra electron. The electron affinities of the noble gases have not been measured conclusively, so they may or may not have slightly negative values.[36]

Electron affinity generally increases across a period. This is caused by the filling of the valence shell of the atom; a group 17 atom releases more energy than group 1 atom on gaining an electron because it obtains a filled valence shell and is therefore more stable.[36]

A trend of decreasing electron affinity going down groups would be expected. The additional electron will be entering an orbital farther away from the nucleus. As such this electron would be less attracted to the nucleus and would release less energy when added. However, in going down a group, around one-third of elements are anomalous, with heavier elements having higher electron affinities than their next lighter congeners. Largely, this is due to the poor shielding by d and f electrons. A uniform decrease in electron affinity only applies to group 1 atoms.[37]

16.3.6 Metallic character

The lower the values of ionization energy, electronegativity and electron affinity, the more metallic character the

element has. Conversely, nonmetallic character increases with higher values of these properties.[38] Given the periodic trends of these three properties, metallic character tends to decrease going across a period (or row) and, with some irregularities (mostly) due to poor screening of the nucleus by d and f electrons, and relativistic effects,[39] tends to increase going down a group (or column or family). Thus, the most metallic elements (such as caesium and francium) are found at the bottom left of traditional periodic tables and the most nonmetallic elements (oxygen, fluorine, chlorine) at the top right. The combination of horizontal and vertical trends in metallic character explains the stair-shaped dividing line between metals and nonmetals found on some periodic tables, and the practice of sometimes categorizing several elements adjacent to that line, or elements adjacent to those elements, as metalloids.[40][41]

16.4 History

Main article: History of the periodic table

16.4.1 First systemization attempts

The discovery of the elements mapped to significant periodic table development dates (pre-, per- and post-)

In 1789, Antoine Lavoisier published a list of 33 chemical elements, grouping them into gases, metals, nonmetals, and earths.[42] Chemists spent the following century searching for a more precise classification scheme. In 1829, Johann Wolfgang Döbereiner observed that many of the elements could be grouped into triads based on their chemical properties. Lithium, sodium, and potassium, for example, were grouped together in a triad as soft, reactive metals. Döbereiner also observed that, when arranged by atomic weight, the second member of each triad was roughly the average of the first and the third;[43] this became known as the Law of

Triads.[44] German chemist Leopold Gmelin worked with this system, and by 1843 he had identified ten triads, three groups of four, and one group of five. Jean-Baptiste Dumas published work in 1857 describing relationships between various groups of metals. Although various chemists were able to identify relationships between small groups of elements, they had yet to build one scheme that encompassed them all.[43]

In 1858, German chemist August Kekulé observed that carbon often has four other atoms bonded to it. Methane, for example, has one carbon atom and four hydrogen atoms. This concept eventually became known as valency; different elements bond with different numbers of atoms.[45]

In 1862, Alexandre-Emile Béguyer de Chancourtois, a French geologist, published an early form of periodic table, which he called the telluric helix or screw. He was the first person to notice the periodicity of the elements. With the elements arranged in a spiral on a cylinder by order of increasing atomic weight, de Chancourtois showed that elements with similar properties seemed to occur at regular intervals. His chart included some ions and compounds in addition to elements. His paper also used geological rather than chemical terms and did not include a diagram; as a result, it received little attention until the work of Dmitri Mendeleev.[46]

In 1864, Julius Lothar Meyer, a German chemist, published a table with 44 elements arranged by valency. The table showed that elements with similar properties often shared the same valency.[47] Concurrently, William Odling (an English chemist) published an arrangement of 57 elements, ordered on the basis of their atomic weights. With some irregularities and gaps, he noticed what appeared to be a periodicity of atomic weights among the elements and that this accorded with 'their usually received groupings.'[48] Odling alluded to the idea of a periodic law but did not pursue it.[49] He subsequently proposed (in 1870) a valence-based classification of the elements.[50]

No.	No.	No.	No.	No.	No.	No.	No.
H 1	F 8	Cl 15	Co & Ni 22	Br 29	Pd 36	I 42	Pt & Ir 50
Li 2	Na 9	K 16	Cu 23	Rb 30	Ag 37	Cs 44	Os 51
G 3	Mg 10	Ca 17	Zn 24	Sr 31	Cd 38	Ba & V 45	Hg 52
Bo 4	Al 11	Cr 19	Y 25	Ce & La 33	U 40	Ta 46	Tl 53
C 5	Si 12	Ti 18	In 26	Zr 32	Sn 39	W 47	Pb 54
N 6	P 13	Mn 20	As 27	Di & Mo 34	Sb 41	Nb 48	Bi 55
O 7	S 14	Fe 21	Se 28	Ro & Ru 35	Te 43	Au 49	Th 56

Newlands's periodic table, as presented to the Chemical Society in 1866, and based on the law of octaves

English chemist John Newlands produced a series of papers from 1863 to 1866 noting that when the elements were listed in order of increasing atomic weight, similar physical and chemical properties recurred at intervals of eight; he likened such periodicity to the octaves of music.[51][52] This so termed Law of Octaves, however, was ridiculed by New-

lands' contemporaries, and the Chemical Society refused to publish his work.[53] Newlands was nonetheless able to draft a table of the elements and used it to predict the existence of missing elements, such as germanium.[54] The Chemical Society only acknowledged the significance of his discoveries five years after they credited Mendeleev.[55]

In 1867, Gustavus Hinrichs, a Danish born academic chemist based in America, published a spiral periodic system based on atomic spectra and weights, and chemical similarities. His work was regarded as idiosyncratic, ostentatious and labyrinthine and this may have militated against its recognition and acceptance.[56][57]

16.4.2 Mendeleev's table

Dmitri Mendeleev

Russian chemistry professor Dmitri Mendeleev and German chemist Julius Lothar Meyer independently published their periodic tables in 1869 and 1870, respectively.[58] Mendeleev's table was his first published version; that of Meyer was an expanded version of his (Meyer's) table of 1864.[59] They both constructed their tables by listing the elements in rows or columns in order of atomic weight and starting a new row or column when the characteristics of the elements began to repeat.[60]

The recognition and acceptance afforded to Mendeleev's table came from two decisions he made. The first was to leave

ОПЫТЪ СИСТЕМЫ ЭЛЕМЕНТОВЪ,
ОСНОВАННОЙ НА ИХЪ АТОМНОМЪ ВѢСЪ И ХИМИЧЕСКОМЪ СХОДСТВѢ.

```
                    Ti=50     Zr=90     ?=180.
                    V=51      Nb=94     Ta=182.
                    Cr=52     Mo=96     W=186.
                    Mn=55     Rh=104,4  Pt=197,1.
                    Fe=56     Ru=104,4  Ir=198.
                    Ni=Co=59  Pd=106,6  Os=199.
     H=1            Cu=63,4   Ag=108    Hg=200.
        Be= 9,4 Mg=24  Zn=65,2  Cd=112
        B=11    Al=27,3  ?=68   Ur=116  Au=197?
        C=12    Si=28   ?=70    Sn=118
        N=14    P=31    As=75   Sb=122  Bi=210?
        O=16    S=32    Se=79,4 Te=128?
        F=19    Cl=35,5 Br=80   I=127
Li=7  Na=23     K=39    Rb=85,4 Cs=133  Tl=204.
                Ca=40   Sr=87,6 Ba=137  Pb=207.
                ?=45    Ce=92
                ?Er=56  La=94
                ?Yt=60  Di=95
                ?In=75,6 Th=118?
```

Д. Менделѣевъ

A version of Mendeleev's 1869 periodic table: An experiment on a system of elements. Based on their atomic weights and chemical similarities. *This early arrangement presents the periods vertically, and the groups horizontally.*

gaps in the table when it seemed that the corresponding element had not yet been discovered.[61] Mendeleev was not the first chemist to do so, but he was the first to be recognized as using the trends in his periodic table to predict the properties of those missing elements, such as gallium and germanium.[62] The second decision was to occasionally ignore the order suggested by the atomic weights and switch adjacent elements, such as tellurium and iodine, to better classify them into chemical families. Later in 1913, Henry Moseley determined experimental values of the nuclear charge or atomic number of each element, and showed that Mendeleev's ordering actually corresponds to the order of increasing atomic number.[63]

The significance of atomic numbers to the organization of the periodic table was not appreciated until the existence and properties of protons and neutrons became understood. Mendeleev's periodic tables used atomic weight instead of atomic number to organize the elements, information determinable to fair precision in his time. Atomic weight worked well enough in most cases to (as noted) give a presentation that was able to predict the properties of missing elements more accurately than any other method then known. Substitution of atomic numbers, once understood, gave a definitive, integer-based sequence for the elements, and Moseley predicted that the only missing elements (in 1913) between aluminum (Z=13) and gold (Z=79) (in 1913) were Z = 43, 61, 72 and 75, which were all later discovered. The sequence of atomic numbers is still used today even as new synthetic elements are being produced and studied.[64]

16.4.3 Second version and further development

Reihen	Gruppe I. — R²O	Gruppe II. — RO	Gruppe III. — R²O³	Gruppe IV. RH⁴ RO²	Gruppe V. RH³ R²O⁵	Gruppe VI. RH² RO³	Gruppe VII. RH R²O⁷	Gruppe VIII. — RO⁴
1	H=1							
2	Li=7	Be=9.4	B=11	C=12	N=14	O=16	F=19	
3	Na=23	Mg=24	Al=27.3	Si=28	P=31	S=32	Cl=35.5	
4	K=39	Ca=40	—=44	Ti=48	V=51	Cr=52	Mn=55	Fe=56, Co=59, Ni=59, Cu=63.
5	(Cu=63)	Zn=65	—=68	—=72	As=75	Se=78	Br=80	
6	Rb=85	Sr=87	?Yt=88	Zr=90	Nb=94	Mo=94	—=100	Ru=104, Rh=104, Pd=106, Ag=108.
7	(Ag=108)	Cd=112	In=113	Sn=118	Sb=122	Te=125	J=127	
8	Cs=133	Ba=137	?Di=138	?Ce=140				— — —
9	(—)							
10			?Er=178	?La=180	Ta=182	W=184		Os=195, Ir=197, Pt=198, Au=199.
11	(Au=199)	Hg=200	Tl=204	Pb=207	Bi=208			— — —
12				Th=231		U=240		

Mendeleev's 1871 periodic table with eight groups of elements. Dashes represented elements unknown in 1871.

Eight-column form of periodic table, updated with all elements discovered to 2015

In 1871, Mendeleev published his periodic table in a new form, with groups of similar elements arranged in columns rather than in rows, and those columns numbered I to VIII corresponding with the element's oxidation state. He also gave detailed predictions for the properties of elements he had earlier noted were missing, but should exist.[65] These gaps were subsequently filled as chemists discovered additional naturally occurring elements.[66] It is often stated that the last naturally occurring element to be discovered was francium (referred to by Mendeleev as *eka-caesium*) in 1939.[67] However, plutonium, produced synthetically in 1940, was identified in trace quantities as a naturally occurring primordial element in 1971.[68][n 5]

The popular[69] periodic table layout, also known as the common or standard form (as shown at various other points in this article), is attributable to Horace Groves Deming. In 1923, Deming, an American chemist, published short (Mendeleev style) and medium (18-column) form periodic tables.[70][n 6] Merck and Company prepared a hand-out form of Deming's 18-column medium table, in 1928, which was widely circulated in American schools. By the 1930s Deming's table was appearing in handbooks and encyclopaedias of chemistry. It was also distributed for many years by the Sargent-Welch Scientific Company.[71][72][73]

With the development of modern quantum mechanical theories of electron configurations within atoms, it became apparent that each period (row) in the table corresponded to the filling of a quantum shell of electrons. Larger atoms have more electron sub-shells, so later tables have required progressively longer periods.[74]

Glenn T. Seaborg who, in 1945, suggested a new periodic table showing the actinides as belonging to a second f-block series

In 1945, Glenn Seaborg, an American scientist, made the suggestion that the actinide elements, like the lanthanides, were filling an f sub-level. Before this time the actinides were thought to be forming a fourth d-block row. Seaborg's colleagues advised him not to publish such a radical suggestion as it would most likely ruin his career. As Seaborg considered he did not then have a career to bring into disrepute, he published anyway. Seaborg's suggestion was found to be correct and he subsequently went on to win the 1951 Nobel Prize in chemistry for his work in synthesizing actinide elements.[75][76][n 7]

Although minute quantities of some transuranic elements occur naturally,[1] they were all first discovered in laboratories. Their production has expanded the periodic table significantly, the first of these being neptunium, synthesized in 1939.[77] Because many of the transuranic elements are highly unstable and decay quickly, they are challenging to detect and characterize when produced. There have been controversies concerning the acceptance of competing discovery claims for some elements, requiring independent review to determine which party has priority, and hence naming rights. The most recently accepted and named elements are flerovium (element 114) and livermorium (element 116), both named on 31 May 2012.[78] In 2010, a joint Russia–US collaboration at Dubna, Moscow Oblast, Russia, claimed to have synthesized six atoms of ununseptium (element 117), making it the most recently claimed discovery.[79]

16.5   Alternative structures

Main article: Alternative periodic tables
There are many periodic tables with structures other than

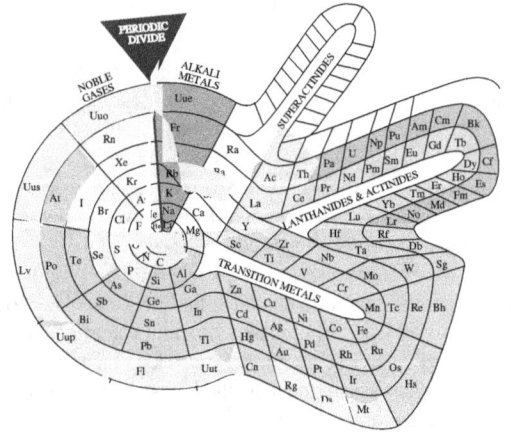

Theodor Benfey's spiral periodic table

that of the standard form. Within 100 years of the appearance of Mendeleev's table in 1869 it has been estimated that around 700 different periodic table versions were published.[80] As well as numerous rectangular variations, other periodic table formats have included, for example,[n 8] circular, cubic, cylindrical, edificial (building-like), helical, lemniscate, octagonal prismatic, pyramidal, separated, spherical, spiral, and triangular forms. Such alternatives are often developed to highlight or emphasize chemical or physical properties of the elements that are not as apparent in traditional periodic tables.[80]

A popular[81] alternative structure is that of Theodor Benfey (1960). The elements are arranged in a continuous spiral, with hydrogen at the center and the transition metals, lanthanides, and actinides occupying peninsulas.[82]

Most periodic tables are two-dimensional;[1] however, three-dimensional tables are known to as far back as at least 1862 (pre-dating Mendeleev's two-dimensional table of 1869). More recent examples include Courtines' Periodic Classification (1925),[83] Wringley's Lamina System (1949),[84] Giguère's Periodic helix (1965)[85][n 9] and Dufour's Periodic Tree (1996).[86] Going one better, Stowe's Physicist's Periodic Table (1989)[87] has been described as being four-dimensional (having three spatial dimensions and one colour dimension).[88]

The various forms of periodic tables can be thought of as lying on a chemistry–physics continuum.[89] Towards the chemistry end of the continuum can be found, as an example, Rayner-Canham's 'unruly'[90] Inorganic Chemist's Periodic Table (2002),[91] which emphasizes trends and patterns, and unusual chemical relationships and properties. Near the physics end of the continuum is Janet's Left-Step Periodic Table (1928). This has a structure that shows a closer connection to the order of electron-shell filling and, by association, quantum mechanics.[92] Somewhere in the middle of the continuum is the ubiquitous common or standard form of periodic table. This is regarded as better expressing empirical trends in physical state, electrical and thermal conductivity, and oxidation numbers, and other properties easily inferred from traditional techniques of the chemical laboratory.[93]

16.6 Open questions and controversies

16.6.1 Elements with unknown chemical properties

Although all elements up to ununoctium have been discovered, of the elements above hassium (element 108), only copernicium (element 112) and flerovium (element 114) have known chemical properties. The other elements may behave differently from what would be predicted by extrapolation, due to relativistic effects; for example, flerovium has been predicted to possibly exhibit some noble-gas-like properties, even though it is currently placed in the carbon group.[94] More recent experiments have suggested, however, that flerovium behaves chemically like lead, as expected from its periodic table position.[95]

16.6.2 Further periodic table extensions

Main article: Extended periodic table

It is unclear whether new elements will continue the pattern of the current periodic table as period 8, or require further adaptations or adjustments. Seaborg expected the eighth period to follow the previously established pattern exactly, so that it would include a two-element s-block for elements 119 and 120, a new g-block for the next 18 elements, and 30 additional elements continuing the current f-, d-, and p-blocks.[97] More recently, physicists such as Pekka Pyykkö have theorized that these additional elements do not follow the Madelung rule, which predicts how electron shells are filled and thus affects the appearance of the present periodic table.[98]

16.6.3 Element with the highest possible atomic number

The number of possible elements is not known. A very early suggestion made by Elliot Adams in 1911, and based on the arrangement of elements in each horizontal periodic table row, was that elements of atomic weight greater than 256± (which would equate to between elements 99 and 100 in modern-day terms) did not exist.[99] A higher—more recent—estimate is that the periodic table may end soon after the island of stability,[100] which is expected to center around element 126, as the extension of the periodic and nuclides tables is restricted by proton and neutron drip lines.[101] Other predictions of an end to the periodic table include at element 128 by John Emsley,[1] at element 137 by Richard Feynman,[102] and at element 155 by Albert Khazan.[1][n 10]

Bohr model

The Bohr model exhibits difficulty for atoms with atomic number greater than 137, as any element with an atomic number greater than 137 would require 1s electrons to be traveling faster than c, the speed of light.[103] Hence the non-relativistic Bohr model is inaccurate when applied to such an element.

Relativistic Dirac equation

The relativistic Dirac equation has problems for elements with more than 137 protons. For such elements, the wave function of the Dirac ground state is oscillatory rather than bound, and there is no gap between the positive and negative energy spectra, as in the Klein paradox.[104] More accurate calculations taking into account the effects of the finite size of the nucleus indicate that the binding energy first exceeds the limit for elements with more than 173 protons. For heavier elements, if the innermost orbital (1s) is not filled, the electric field of the nucleus will pull an electron out of the vacuum, resulting in the spontaneous emission of a positron;[105] however, this does not happen if the innermost orbital is filled, so that element 173 is not necessarily the end of the periodic table.[106]

16.6.4   Placement of hydrogen and helium

Simply following electron configurations, hydrogen (electronic configuration $1s^1$) and helium ($1s^2$) should be placed in groups 1 and 2, above lithium ([He]$2s^1$) and beryllium ([He]$2s^2$).[19] However, such placing is rarely used outside of the context of electron configurations: When the noble gases (then called "inert gases") were first discovered around 1900, they were known as "group 0," reflecting no chemical reactivity of these elements known at that point, and helium was placed on the top that group, as it did share the extreme chemical inertness seen throughout the group. As the group changed its formal number, many authors continued to assign helium directly above neon, in the group 18; one of the examples of such placing is the current IUPAC table.[107]

Hydrogen's chemical properties are not very close to those of the alkali metals, which occupy the group 1, and on that basis hydrogen is sometimes placed elsewhere: one of the most common alternatives is in group 17; one of the factors behind it is the strictly univalent predominantly non-metallic chemistry of hydrogen, and that of fluorine (the element placed on the top of the group 17) is strictly univalent and non-metallic. Sometimes, to show how hydrogen has properties both corresponding to those of the alkali metals and the halogens, it may be shown in two columns simultaneously.[108] Another suggestion is above carbon in group 14: placed that way, it fits well into the trend of increasing trends of ionization potential values and electron affinity values, and is not too stray from the electronegativity trend.[109] Finally, hydrogen is sometimes placed separately from any group; this is based on how general properties of hydrogen differ from that of any group: unlike hydrogen, the other group 1 elements show extremely metallic behavior; the group 17 elements commonly form salts (hence the term "halogen"); elements of any other group show some multivalent chemistry. The other period 1 element, helium, is sometimes placed separately from any group as well.[110] The property that distinguishes helium from the rest of the noble gases (even though the extraordinary inertness of helium is extremely close to that of neon and argon[111]) is that in its closed electron shell, helium has only two electrons in the outermost electron orbital, while the rest of the noble gases have eight.

16.6.5   Groups included in the transition metals

The definition of a transition metal, as given by IUPAC, is an element whose atom has an incomplete d sub-shell, or which can give rise to cations with an incomplete d sub-shell.[112] By this definition all of the elements in groups 3–11 are transition metals. The IUPAC definition therefore excludes group 12, comprising zinc, cadmium and mercury, from the transition metals category.

Some chemists treat the categories "d-block elements" and "transition metals" interchangeably, thereby including groups 3–12 among the transition metals. In this instance the group 12 elements are treated as a special case of transition metal in which the d electrons are not ordinarily involved in chemical bonding. The recent discovery that mercury can use its d electrons in the formation of mercury(IV) fluoride (HgF_4) has prompted some commentators to suggest that mercury can be regarded as a transition metal.[113] Other commentators, such as Jensen,[114] have argued that the formation of a compound like HgF_4 can occur only under highly abnormal conditions. As such, mercury could not be regarded as a transition metal by any reasonable interpretation of the ordinary meaning of the term.[114]

Still other chemists further exclude the group 3 elements from the definition of a transition metal. They do so on the basis that the group 3 elements do not form any ions having a partially occupied d shell and do not therefore exhibit any properties characteristic of transition metal chemistry.[115] In this case, only groups 4–11 are regarded as transition metals.

16.6.6   Period 6 and 7 elements in group 3

Although scandium and yttrium are always the first two elements in group 3 the identity of the next two elements is not settled. They are either lanthanum and actinium; or lutetium and lawrencium. There are strong chemical and physical arguments supporting the latter arrangement[116][117] but not all authors have been convinced.[118] Most working chemists are not aware there is any controversy.[119]

Lanthanum and actinium are traditionally depicted as the remaining group 3 members.[120][121] It has been suggested that this layout originated in the 1940s, with the appearance of periodic tables relying on the electron configurations of the elements and the notion of the differentiating electron. The configurations of caesium, barium and lanthanum are [Xe]$6s^1$, [Xe]$6s^2$ and [Xe]$5d^16s^2$. Lanthanum thus has a $5d$ differentiating electron and this establishes "it in group 3 as the first member of the d-block for period 6."[122] A consistent set of electron configurations is then seen in group 3: scandium [Ar]3d^{1}4s^2, yttrium [Kr]4d^{1}5s^2 and lanthanum [Xe]5d^{1}6s^2. Still in period 6, ytterbium was assigned an electron configuration of [Xe]4f^{13}5d^{1}6s^2 and lutetium [Xe]4f^{14}5d^{1}6s^2, "resulting in a 4f differentiating electron for lutetium and firmly establishing it as the last member of the f-block for period 6."[122]

In other tables, lutetium and lawrencium are the remaining group 3 members.[123] It has been known since the early 20th century that, "yttrium and (to a lesser degree) scandium are closer in their chemical properties to lutetium and the other heavy rare earths [i.e. lanthanides] than they are to lanthanum."[122] Accordingly, lutetium rather than lanthanum was assigned to group 3 by some chemists in the 1920s and 30s. Later spectroscopic work found that the electron configuration of ytterbium was in fact [Xe]4f^{14}6s^2. This meant that ytterbium and lutetium—the latter with [Xe]4f^{14}5d^{1}6s^2—both had 14 f electrons, "resulting in a d rather than an f differentiating electron" for lutetium and making it an "equally valid candidate" with [Xe]5d^{1}6s^2 lanthanum, for the group 3 periodic table position below yttrium.[122] Several physicists in the 1950s and 60s opted for lutetium, in light of a comparison of several of its physical properties with those of lanthanum.[122] This arrangement, in which lanthanum is the first member of the f-block, is disputed by some authors since lanthanum lacks any f electrons. However, it has been argued that this is not valid concern given other periodic table anomalies—thorium, for example, has no f electrons yet is part of the f-block.[124] As for lawrencium, its electron configuration was confirmed in 2015 as [Rn]5f^{14}7s^{2}7p^1. Such a configuration represents another periodic table anomaly, regardless of whether lawrencium is located in the f-block or the d-block, as the only potentially applicable p-block position has been reserved for ununtrium with its predicted electron configuration of [Rn]5f^{14}6d^{10}7s^{2}7p^1.[125]

Some tables, including the table on the IUPAC site,[126][n 11] place footnote markers in the two positions below scandium and yttrium, and show both lanthanum and lutetium, and

actinium and lawrencium as being part of, respectively, the lanthanide series and the actinide series of elements. This arrangement emphasizes similarities in the chemistry of the 15 lanthanide elements (La–Lu) over electron configuration arguments. The actinides are more diverse in their behavior. Most early members show some similarities to transition metals; actinium and the later members are more like lanthanides.[127]

16.6.7   Optimal form

The many different forms of periodic table have prompted the question of whether there is an optimal or definitive form of periodic table. The answer to this question is thought to depend on whether the chemical periodicity seen to occur among the elements has an underlying truth, effectively hard-wired into the universe, or if any such periodicity is instead the product of subjective human interpretation, contingent upon the circumstances, beliefs and predilections of human observers. An objective basis for chemical periodicity would settle the questions about the location of hydrogen and helium, and the composition of group 3. Such an underlying truth, if it exists, is thought to have not yet been discovered. In its absence, the many different forms of periodic table can be regarded as variations on the theme of chemical periodicity, each of which explores and emphasizes different aspects, properties, perspectives and relationships of and among the elements.[n 12] The ubiquity of the standard or medium-long periodic table is thought to be a result of this layout having a good balance of features in terms of ease of construction and size, and its depiction of atomic order and periodic trends.[49][128]

16.7   See also

- Abundance of the chemical elements

- Atomic electron configuration table

- Element collecting

- List of elements

- List of periodic table-related articles

- Table of nuclides

- The Mystery of Matter: Search for the Elements (PBS film)

- Timeline of chemical element discoveries

16.8   Notes

[1] The elements discovered initially by synthesis and later in nature are technetium (Z=43), promethium (61), astatine (85), neptunium (93), and plutonium (94).

[2] There is an inconsistency and some irregularities in this convention. Thus, helium is shown in the p-block but is actually an s-block element, and (for example) the d-subshell in the d-block is actually filled by the time group 11 is reached, rather than group 12.

[3] The noble gases, astatine, francium, and all elements heavier than americium were left out as there is no data for them.

[4] While fluorine is the most electronegative of the elements under the Pauling scale, neon is the most electronegative element under other scales, such as the Allen scale.

[5] John Emsley, in his book, *Nature's Building Blocks,* writes that americium, curium, berkelium and californium (elements 95–98) can occur naturally as trace amounts in uranium ores by neutron capture and beta decay. This assertion appears to lack independent substantiation. See: Emsley J. (2011). *Nature's Building Blocks: An A-Z Guide to the Elements* (New ed.). New York, NY: Oxford University Press, p. 109.

[6] An antecedent of Deming's 18-column table may be seen in Adams' 16-column Periodic Table of 1911. Adams omits the rare earths and the 'radioactive elements' (i.e. the actinides) from the main body of his table and instead shows them as being 'careted in only to save space' (rare earths between Ba and eka-Yt; radioactive elements between eka-Te and eka-I). See: Elliot Q. A. (1911). "A modification of the periodic table". *Journal of the American Chemical Society.* **33**(5): 684–688 (687).

[7] A second extra-long periodic table row, to accommodate known and undiscovered elements with an atomic weight greater than bismuth (thorium, protactinium and uranium, for example), had been postulated as far back as 1892. Most investigators, however, considered that these elements were analogues of the third series transition elements, hafnium, tantalum and tungsten. The existence of a second inner transition series, in the form of the actinides, was not accepted until similarities with the electron structures of the lanthanides had been established. See: van Spronsen, J. W. (1969). *The periodic system of chemical elements.* Amsterdam: Elsevier. p. 315–316, ISBN 0-444-40776-6.

[8] See *The Internet database of periodic tables* for depictions of these kinds of variants.

[9] The animated depiction of Giguère's periodic table that is widely available on the internet (including from here) is erroneous, as it does not include hydrogen and helium. Giguère included hydrogen, above lithium, and helium, above beryllium. See: Giguère P.A. (1966). "The "new look" for the periodic system". *Chemistry in Canada* **18** (12): 36–39 (see p. 37).

[10] Karol (2002, p. 63) contends that gravitational effects would become significant when atomic numbers become astronomically large, thereby overcoming other super-massive nuclei instability phenomena, and that neutron stars (with atomic numbers on the order of 10^{21}) can arguably be regarded as representing the heaviest known elements in the universe. See: Karol P. J. (2002). "The Mendeleev–Seaborg periodic table: Through Z = 1138 and beyond". *Journal of Chemical Education* **79** (1): 60–63.

[11] Although this form of the table is sometimes referred to as the "approved" or "official" IUPAC periodic table, "IUPAC has not approved any specific form of the periodic table..." See: Leigh, G. J. (January–February 2009). "Periodic Tables and IUPAC". *Chemistry International* **31** (1).

[12] Scerri, one of the foremost authorities on the history of the periodic table (Sella 2013), favoured the concept of an optimal form of periodic table but has recently changed his mind and now supports the value of a plurality of periodic tables. See: Sella A. (2013). 'An elementary history lesson'. *New Scientist.* 2929, 13 August: 51, accessed 4 September 2013; and Scerri, E. (2013). 'Is there an optimal periodic table and other bigger questions in the philosophy of science.'. 9 August, accessed 4 September 2013.

16.9   References

[1] Emsley, John (2011). *Nature's Building Blocks: An A-Z Guide to the Elements* (New ed.). New York, NY: Oxford University Press. ISBN 978-0-19-960563-7.

[2] Greenwood, pp. 24–27

[3] Gray, p. 6

[4] Koppenol, W. H. (2002). "Naming of New Elements (IUPAC Recommendations 2002)" (PDF). *Pure and Applied Chemistry* **74** (5): 787–791. doi:10.1351/pac200274050787.

[5] Silva, Robert J. (2006). "Fermium, Mendelevium, Nobelium and Lawrencium". In Morss; Edelstein, Norman M.; Fuger, Jean. *The Chemistry of the Actinide and Transactinide Elements* (3rd ed.). Dordrecht, The Netherlands: Springer Science+Business Media. ISBN 1-4020-3555-1.

[6] Gray, p. 11

[7] Scerri 2007, p. 24

[8] Messler, R. W. (2010). *The essence of materials for engineers.* Sudbury, MA: Jones & Bartlett Publishers. p. 32. ISBN 0-7637-7833-8.

[9] Bagnall, K. W. (1967). "Recent advances in actinide and lanthanide chemistry". In Fields, P.R.; Moeller, T. *Advances in chemistry, Lanthanide/Actinide chemistry.* Advances in Chemistry **71**. American Chemical Society. pp. 1–12. doi:10.1021/ba-1967-0071. ISBN 0-8412-0072-6.

[10] Day, M. C., Jr.; Selbin, J. (1969). *Theoretical inorganic chemistry* (2nd ed.). New York: Nostrand-Rienhold Book Corporation. p. 103. ISBN 0-7637-7833-8.

[11] Holman, J.; Hill, G. C. (2000). *Chemistry in context* (5th ed.). Walton-on-Thames: Nelson Thornes. p. 40. ISBN 0-17-448276-0.

[12] Leigh, G. J. (1990). *Nomenclature of Inorganic Chemistry: Recommendations 1990*. Blackwell Science. ISBN 0-632-02494-1.

[13] Fluck, E. (1988). "New Notations in the Periodic Table" (PDF). *Pure Appl. Chem.* (IUPAC) **60** (3): 431–436. doi:10.1351/pac198860030431. Retrieved 24 March 2012.

[14] Moore, p. 111

[15] Greenwood, p. 30

[16] Stoker, Stephen H. (2007). *General, organic, and biological chemistry*. New York: Houghton Mifflin. p. 68. ISBN 978-0-618-73063-6. OCLC 52445586.

[17] Mascetta, Joseph (2003). *Chemistry The Easy Way* (4th ed.). New York: Hauppauge. p. 50. ISBN 978-0-7641-1978-1. OCLC 52047235.

[18] Kotz, John; Treichel, Paul; Townsend, John (2009). *Chemistry and Chemical Reactivity, Volume 2* (7th ed.). Belmont: Thomson Brooks/Cole. p. 324. ISBN 978-0-495-38712-1. OCLC 220756597.

[19] Gray, p. 12

[20] Jones, Chris (2002). *d- and f-block chemistry*. New York: J. Wiley & Sons. p. 2. ISBN 978-0-471-22476-1. OCLC 300468713.

[21] Silberberg, M. S. (2006). *Chemistry: The molecular nature of matter and change* (4th ed.). New York: McGraw-Hill. p. 536. ISBN 0-07-111658-3.

[22] Manson, S. S.; Halford, G. R. (2006). *Fatigue and durability of structural materials*. Materials Park, Ohio: ASM International. p. 376. ISBN 0-87170-825-6.

[23] Bullinger, Hans-Jörg (2009). *Technology guide: Principles, applications, trends*. Berlin: Springer-Verlag. p. 8. ISBN 978-3-540-88545-0.

[24] Jones, B. W. (2010). *Pluto: Sentinel of the outer solar system*. Cambridge: Cambridge University Press. pp. 169–71. ISBN 978-0-521-19436-5.

[25] Hinrichs, G. D. (1869). "On the classification and the atomic weights of the so-called chemical elements, with particular reference to Stas's determinations". *Proceedings of the American Association for the Advancement of Science* **18** (5): 112–124.

[26] Myers, R. (2003). *The basics of chemistry*. Westport, CT: Greenwood Publishing Group. pp. 61–67. ISBN 0-313-31664-3.

[27] Chang, Raymond (2002). *Chemistry* (7 ed.). New York: McGraw-Hill. pp. 289–310; 340–42. ISBN 0-07-112072-6.

[28] Greenwood, p. 27

[29] Jolly, W. L. (1991). *Modern Inorganic Chemistry* (2nd ed.). McGraw-Hill. p. 22. ISBN 978-0-07-112651-9.

[30] Greenwood, p. 28

[31] IUPAC, *Compendium of Chemical Terminology*, 2nd ed. (the "Gold Book") (1997). Online corrected version: (2006–) "Electronegativity".

[32] Pauling, L. (1932). "The Nature of the Chemical Bond. IV. The Energy of Single Bonds and the Relative Electronegativity of Atoms". *Journal of the American Chemical Society* **54** (9): 3570–3582. doi:10.1021/ja01348a011.

[33] Allred, A. L. (1960). "Electronegativity values from thermochemical data". *Journal of Inorganic and Nuclear Chemistry* (Northwestern University) **17** (3–4): 215–221. doi:10.1016/0022-1902(61)80142-5. Retrieved 11 June 2012.

[34] Huheey, Keiter & Keiter, p. 42

[35] Siekierski, Slawomir; Burgess, John (2002). *Concise chemistry of the elements*. Chichester: Horwood Publishing. pp. 35–36. ISBN 1-898563-71-3.

[36] Chang, pp. 307–309

[37] Huheey, Keiter & Keiter, pp. 42, 880–81

[38] Yoder, C. H.; Suydam, F. H.; Snavely, F. A. (1975). *Chemistry* (2nd ed.). Harcourt Brace Jovanovich. p. 58. ISBN 0-15-506465-7.

[39] Huheey, Keiter & Keiter, pp. 880–85

[40] Sacks, O (2009). *Uncle Tungsten: Memories of a chemical boyhood*. New York: Alfred A. Knopf. pp. 191, 194. ISBN 0-375-70404-3.

[41] Gray, p. 9

[42] Siegfried, Robert (2002). *From elements to atoms a history of chemical composition*. Philadelphia, Pennsylvania: Library of Congress Cataloging-in-Publication Data. p. 92. ISBN 0-87169-924-9.

[43] Ball, p. 100

[44] Horvitz, Leslie (2002). *Eureka!: Scientific Breakthroughs That Changed The World*. New York: John Wiley. p. 43. ISBN 978-0-471-23341-1. OCLC 50766822.

[45] van Spronsen, J. W. (1969). *The periodic system of chemical elements*. Amsterdam: Elsevier. p. 19. ISBN 0-444-40776-6.

[46] "Alexandre-Emile Bélguier de Chancourtois (1820-1886)" (in French). Annales des Mines history page. Retrieved 18 September 2014.

[47] Venable, pp. 85–86; 97

[48] Odling, W. (2002). "On the proportional numbers of the elements". *Quarterly Journal of Science* **1**: 642–648 (643).

[49] Scerri, Eric R. (2011). *The periodic table: A very short introduction.* Oxford: Oxford University Press. ISBN 978-0-19-958249-5.

[50] Kaji, M. (2004). "Discovery of the periodic law: Mendeleev and other researchers on element classification in the 1860s". In Rouvray, D. H.; King, R. Bruce. *The periodic table: Into the 21st Century.* Research Studies Press. pp. 91–122 (95). ISBN 0-86380-292-3.

[51] Newlands, John A. R. (20 August 1864). "On Relations Among the Equivalents". *Chemical News* **10**: 94–95.

[52] Newlands, John A. R. (18 August 1865). "On the Law of Octaves". *Chemical News* **12**: 83.

[53] Bryson, Bill (2004). *A Short History of Nearly Everything.* Black Swan. pp. 141–142. ISBN 978-0-552-15174-0.

[54] Scerri 2007, p. 306

[55] Brock, W. H.; Knight, D. M. (1965). "The Atomic Debates: 'Memorable and Interesting Evenings in the Life of the Chemical Society'". *Isis* (The University of Chicago Press) **56** (1): 5–25. doi:10.1086/349922.

[56] Scerri 2007, pp. 87, 92

[57] Kauffman, George B. (March 1969). "American forerunners of the periodic law". *Journal of Chemical Education* **46** (3): 128–135 (132). Bibcode:1969JChEd..46..128K. doi:10.1021/ed046p128.

[58] Mendelejew, Dimitri (1869). "Über die Beziehungen der Eigenschaften zu den Atomgewichten der Elemente". *Zeitschrift für Chemie* (in German): 405–406.

[59] Venable, pp. 96–97; 100–102

[60] Ball, pp. 100–102

[61] Pullman, Bernard (1998). *The Atom in the History of Human Thought.* Translated by Axel Reisinger. Oxford University Press. p. 227. ISBN 0-19-515040-6.

[62] Ball, p. 105

[63] Atkins, P. W. (1995). *The Periodic Kingdom.* HarperCollins Publishers, Inc. p. 87. ISBN 0-465-07265-8.

[64] Samanta, C.; Chowdhury, P. Roy; Basu, D.N. (2007). "Predictions of alpha decay half lives of heavy and superheavy elements". *Nucl. Phys.* A **789**: 142–154. arXiv:nucl-th/0703086. Bibcode:2007NuPhA.789..142S. doi:10.1016/j.nuclphysa.2007.04.001.

[65] Scerri 2007, p. 112

[66] Kaji, Masanori (2002). "D.I. Mendeleev's Concept of Chemical Elements and the Principle of Chemistry" (PDF). *Bull. Hist. Chem.* (Tokyo Institute of Technology) **27** (1): 4–16. Retrieved 11 June 2012.

[67] Adloff, Jean-Pierre; Kaufman, George B. (25 September 2005). "Francium (Atomic Number 87), the Last Discovered Natural Element". The Chemical Educator. Retrieved 26 March 2007.

[68] Hoffman, D. C.; Lawrence, F. O.; Mewherter, J. L.; Rourke, F. M. (1971). "Detection of Plutonium-244 in Nature". *Nature* **234** (5325): 132–134. Bibcode:1971Natur.234..132H. doi:10.1038/234132a0.

[69] Gray, p. 12

[70] Deming, Horace G (1923). *General chemistry: An elementary survey.* New York: J. Wiley & Sons. pp. 160, 165.

[71] Abraham, M; Coshow, D; Fix, W. *Periodicity:A source book module, version 1.0* (PDF). New York: Chemsource, Inc. p. 3.

[72] Emsley, J (7 March 1985). "Mendeleyev's dream table". *New Scientist*: 32–36(36).

[73] Fluck, E (1988). "New notations in the period table". *Pure & Applied Chemistry* **60** (3): 431–436 (432). doi:10.1351/pac198860030431.

[74] Ball, p. 111

[75] Scerri 2007, pp. 270–71

[76] Masterton, William L.; Hurley, Cecile N.; Neth, Edward J. *Chemistry: Principles and reactions* (7th ed.). Belmont, CA: Brooks/Cole Cengage Learning. p. 173. ISBN 1-111-42710-0.

[77] Ball, p. 123

[78] Barber, Robert C.; Karol, Paul J; Nakahara, Hiromichi; Vardaci, Emanuele; Vogt, Erich W. (2011). "Discovery of the elements with atomic numbers greater than or equal to 113 (IUPAC Technical Report)". *Pure Appl. Chem.* **83** (7): 1485. doi:10.1351/PAC-REP-10-05-01.

[79] Эксперимент по синтезу 117-го элемента получает продолжение[Experiment on sythesis of the 117th element is to be continued] (in Russian). JINR. 2012.

[80] Scerri 2007, p. 20

[81] Emsely, J; Sharp, R (21 June 2010). "The periodic table: Top of the charts". *The Independent.*

[82] Seaborg, Glenn (1964). "Plutonium: The Ornery Element". *Chemistry* **37** (6): 14.

[83] Mark R. Leach. "1925 Courtines' Periodic Classification". Retrieved 16 October 2012.

[84] Mark R. Leach. "1949 Wringley's Lamina System". Retrieved 16 October 2012.

[85] Mazurs, E.G. (1974). *Graphical Representations of the Periodic System During One Hundred Years*. Alabama: University of Alabama Press. p. 111. ISBN 978-0-8173-3200-6.

[86] Mark R. Leach. "1996 Dufour's Periodic Tree". Retrieved 16 October 2012.

[87] Mark R. Leach. "1989 Physicist's Periodic Table by Timothy Stowe". Retrieved 16 October 2012.

[88] Bradley, David (20 July 2011). "At last, a definitive periodic table?". *ChemViews Magazine*. doi:10.1002/chemv.201000107.

[89] Scerri 2007, pp. 285–86

[90] Scerri 2007, p. 285

[91] Mark R. Leach. "2002 Inorganic Chemist's Periodic Table". Retrieved 16 October 2012.

[92] Scerri, Eric (2008). "The role of triads in the evolution of the periodic table: Past and present". *Journal of Chemical Education* **85** (4): 585–89 (see p.589). Bibcode:2008JChEd..85..585S. doi:10.1021/ed085p585.

[93] Bent, H. A.; Weinhold, F (2007). "Supporting information: News from the periodic table: An introduction to "Periodicity symbols, tables, and models for higher-order valency and donor–acceptor kinships"". *Journal of Chemical Education* **84** (7): 3–4. doi:10.1021/ed084p1145.

[94] Schändel, Matthias (2003). *The Chemistry of Superheavy Elements*. Dordrecht: Kluwer Academic Publishers. p. 277. ISBN 1-4020-1250-0.

[95] Scerri 2011, pp. 142–143

[96] Fricke, B.; Greiner, W.; Waber, J. T. (1971). "The continuation of the periodic table up to $Z = 172$. The chemistry of superheavy elements". *Theoretica chimica acta* (Springer-Verlag) **21** (3): 235–260. doi:10.1007/BF01172015. Retrieved 28 November 2012.

[97] Frazier, K. (1978). "Superheavy Elements". *Science News* **113** (15): 236–238. doi:10.2307/3963006. JSTOR 3963006.

[98] Pyykkö, Pekka (2011). "A suggested periodic table up to $Z \leq 172$, based on Dirac–Fock calculations on atoms and ions". *Physical Chemistry Chemical Physics* **13** (1): 161–168. Bibcode:2011PCCP...13..161P. doi:10.1039/c0cp01575j. PMID 20967377.

[99] Elliot, Q. A. (1911). "A modification of the periodic table". *Journal of the American Chemical Society* **33** (5): 684–688 (688). doi:10.1021/ja02218a004.

[100] Glenn Seaborg (c. 2006). "transuranium element (chemical element)". Encyclopædia Britannica. Retrieved 16 March 2010.

[101] Cwiok, S.; Heenen, P.-H.; Nazarewicz, W. (2005). "Shape coexistence and triaxiality in the superheavy nuclei". *Nature* **433** (7027): 705–9. Bibcode:2005Natur.433..705C. doi:10.1038/nature03336. PMID 15716943.

[102] Column: The crucible Ball, Philip in Chemistry World, Royal Society of Chemistry, Nov. 2010

[103] Eisberg, R.; Resnick, R. (1985). *Quantum Physics of Atoms, Molecules, Solids, Nuclei and Particles*. Wiley.

[104] Bjorken, J. D.; Drell, S. D. (1964). *Relativistic Quantum Mechanics*. McGraw-Hill.

[105] Greiner, W.; Schramm, S. (2008). "American Journal of Physics" **76**. p. 509., and references therein.

[106] Ball, Philip (November 2010). "Would Element 137 Really Spell the End of the Periodic Table? Philip Ball Examines the Evidence". Royal Society of Chemistry. Retrieved 30 September 2012.

[107] IUPAC (2013-05-01). "IUPAC Periodic Table of the Elements" (PDF). *iupac.org*. IUPAC. Retrieved 2015-09-20.

[108] Seaborg, Glenn Theodore (1945). "The chemical and radioactive properties of the heavy elements". *Chemical English Newspaper* **23** (23): 2190–2193.

[109] Cronyn, Marshall W. (August 2003). "The Proper Place for Hydrogen in the Periodic Table". *Journal of Chemical Education* **80** (8): 947–951. Bibcode:2003JChEd..80..947C. doi:10.1021/ed080p947.

[110] Greenwood, throughout the book

[111] Lewars, Errol G. (2008-12-05). *Modeling Marvels: Computational Anticipation of Novel Molecules*. Springer Science & Business Media. p. 69–71. ISBN 9781402069734.

[112] IUPAC, *Compendium of Chemical Terminology*, 2nd ed. (the "Gold Book") (1997). Online corrected version: (2006–) "transition element".

[113] Xuefang Wang; Lester Andrews; Sebastian Riedel; Martin Kaupp (2007). "Mercury Is a Transition Metal: The First Experimental Evidence for HgF_4". *Angew. Chem. Int. Ed.* **46** (44): 8371–8375. doi:10.1002/anie.200703710. PMID 17899620.

[114] William B. Jensen (2008). "Is Mercury Now a Transition Element?". *J. Chem. Educ.* **85** (9): 1182–1183. Bibcode:2008JChEd..85.1182J. doi:10.1021/ed085p1182.

[115] Rayner-Canham, G; Overton, T. *Descriptive inorganic chemistry* (4th ed.). New York: W H Freeman. pp. 484–485. ISBN 0-7167-8963-9.

[116] Thyssen, P.; Binnemanns, K. (2011). "1: Accommodation of the rare earths in the periodic table: A historical analysis". In Gschneidner Jr., K. A.; Büzli, J-C. J.; Pecharsky, V. K. *Handbook on the Physics and Chemistry of Rare Earths* **41**. Amsterdam: Elsevier. pp. 80–81. ISBN 978-0-444-53590-0.

[117] Keeler, J.; Wothers, P. (2014). *Chemical Structure and Reactivity: An Integrated Approach*. Oxford: Oxford University. p. 259. ISBN 978-0-19-9604135.

[118] Scerri, E. (2012). "Mendeleev's Periodic Table Is Finally Completed and What To Do about Group 3?". *Chemistry International* **34** (4).

[119] Castelvecchi, Davide (8 April 2015). "Exotic atom struggles to find its place in the periodic table". *Nature News*. Retrieved 20 Sep 2015.

[120] Emsley, J. (2011). *Nature's Building Blocks* (new ed.). Oxford: Oxford University. p. 651. ISBN 978-0-19-960563-7.

[121] See, for example: "Periodic Table". Royal Society of Chemistry. Retrieved 20 Sep 2015.

[122] William B. Jensen (1982). "The Positions of Lanthanum (Actinium) and Lutetium (Lawrencium) in the Periodic Table". *J. Chem. Educ.* **59** (8): 634–636. doi:10.1021/ed059p634.

[123] See, for example: Brown, T. L.; LeMay Jr., H. E.; Bursten, B. E.; Murphy, C. J. (2009). *Chemistry: The Central Science* (11th ed.). Upper Saddle River, New Jersey: Pearson Education. p. endpapers. ISBN 0-13-235848-4.

[124] Scerri, E (2015). "Five ideas in chemical education that must die - part five". *educationinchemistryblog*. Royal Society of Chemistry. Retrieved Sep 19, 2015. It is high time that the idea of group 3 consisting of Sc, Y, La and Ac is abandoned

[125] Jensen, W. B. (2015). "Some Comments on the Position of Lawrencium in the Periodic Table" (PDF). Retrieved 20 Sep 2015.

[126] "Periodic Table of the Elements". International Union of Pure and Applied Chemistry. Retrieved 3 April 2010.

[127] Owen, S. M. (1991). *A Guide to Modern Inorganic Chemistry*. Harlow, Essex: Longman Scientific & Technical. p. 190. ISBN 0-58-206439-2.

[128] Francl, Michelle (May 2009). "Table manners" (PDF). *Nature Chemistry* **1** (2): 97–98. Bibcode:2009NatCh...1...97F. doi:10.1038/nchem.183. PMID 21378810.

16.10   Bibliography

- Ball, Philip (2002). *The Ingredients: A Guided Tour of the Elements*. Oxford: Oxford University Press. ISBN 0-19-284100-9.

- Chang, Raymond (2002). *Chemistry* (7th ed.). New York: McGraw-Hill Higher Education. ISBN 978-0-19-284100-1.

- Gray, Theodore (2009). *The Elements: A Visual Exploration of Every Known Atom in the Universe*. New York: Black Dog & Leventhal Publishers. ISBN 978-1-57912-814-2.

- Greenwood, Norman N.; Earnshaw, Alan (1984). *Chemistry of the Elements*. Oxford: Pergamon Press. ISBN 0-08-022057-6.

- Huheey, JE; Keiter, EA; Keiter, RL. *Principles of structure and reactivity* (4th ed.). New York: Harper Collins College Publishers. ISBN 0-06-042995-X.

- Moore, John (2003). *Chemistry For Dummies*. New York: Wiley Publications. p. 111. ISBN 978-0-7645-5430-8. OCLC 51168057.

- Scerri, Eric (2007). *The periodic table: Its story and its significance*. Oxford: Oxford University Press. ISBN 0-19-530573-6.

- Scerri, Eric R. (2011). *The periodic table: A very short introduction*. Oxford: Oxford University Press. ISBN 978-0-19-958249-5.

- Venable, F P (1896). *The Development of the Periodic Law*. Easton PA: Chemical Publishing Company.

16.11   External links

- M. Dayah. "Dynamic Periodic Table". Retrieved 14 May 2012.

- Brady Haran. "The Periodic Table of Videos". University of Nottingham. Retrieved 14 May 2012.

- Mark Winter. "WebElements: the periodic table on the web". University of Sheffield. Retrieved 14 May 2012.

- Mark R. Leach. "The INTERNET Database of Periodic Tables". Retrieved 14 May 2012.

Chapter 17

History of the periodic table

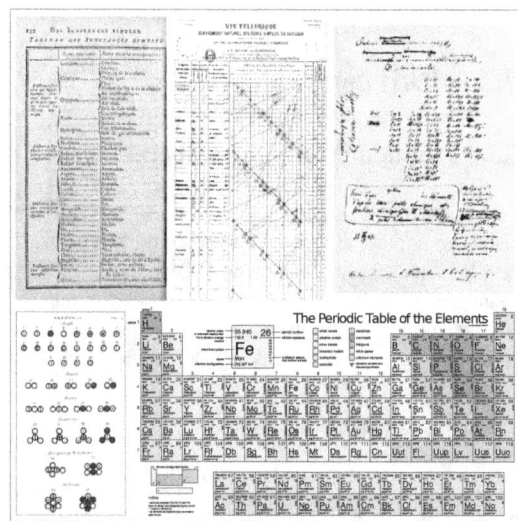

A collection of historic documents that led to the development of the modern periodic table (clockwise from top left) - Lavoisier's 'Table of Simple substances'; de Chancourtois' 'Vis Tellurique'; Mendeleev's hand-written periodic table; a modern periodic table; John Dalton's list of atomic weights & symbols.

The periodic table is an arrangement of the chemical elements, organized on the basis of their atomic numbers, electron configurations and recurring chemical properties. Elements are presented in order of increasing atomic number. The standard form of the table consists of a grid of elements, with rows called periods and columns called groups.

The **history of the periodic table** reflects over a century of growth in the understanding of chemical properties. The most important event in its history occurred in 1869, when the table was published by Dmitri Mendeleev,[1] who built upon earlier discoveries by scientists such as Antoine-Laurent de Lavoisier and John Newlands, but who is nevertheless generally given sole credit for its development.

17.1 Ancient times

A number of physical elements (such as gold, silver and copper) have been known from antiquity, as they are found in their native form and are relatively simple to mine with primitive tools.[2] However, the notion that there were a limited number of elements from which everything was composed originated in around 330 BCE, when the Greek philosopher Aristotle proposed that everything is made up of a mixture of one or more *roots*, an idea that had originally been suggested by the Sicilian philosopher Empedocles. The four roots, which were later renamed as *elements* by Plato, were *earth*, *water*, *air* and *fire*. While Aristotle and Plato introduced the concept of an element, their ideas did nothing to advance the understanding of the nature of matter.

17.2 Age of Enlightenment

17.2.1 Hennig Brand

The history of the periodic table is also a history of the discovery of the chemical elements. The first person in history to discover a new element was Hennig Brand, a bankrupt German merchant. Brand tried to discover the Philosopher's Stone — a mythical object that was supposed to turn inexpensive base metals into gold. In 1649, his experiments with distilled human urine resulted in the production of a glowing white substance, which he named phosphorus.[3] He kept his discovery secret until 1680, when Robert Boyle rediscovered phosphorus and published his findings. The discovery of phosphorus helped to raise the question of what it meant for a substance to be an element.

In 1661, Boyle defined an element as "a substance that cannot be broken down into a simpler substance by a chemical reaction". This simple definition served for three centuries and lasted until the discovery of subatomic particles.

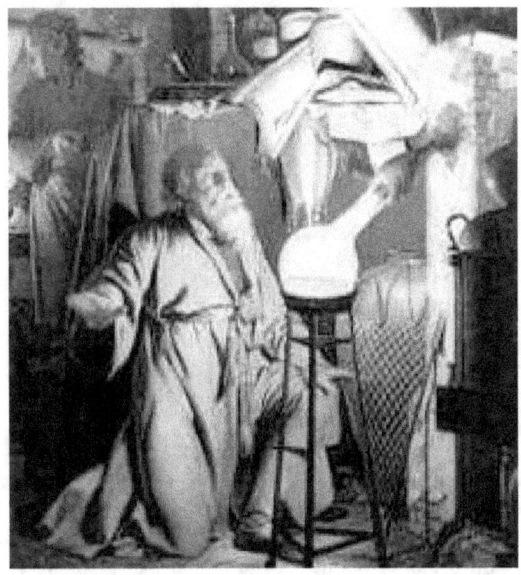

Hennig Brand, as shown in The Alchemist Discovering Phosphorus

17.2.2   Antoine-Laurent de Lavoisier

Antoine Laurent de Lavoisier

Lavoisier's *Traité Élémentaire de Chimie* (*Elementary Treatise of Chemistry*), which was written in 1789 and first translated into English by the writer Robert Kerr, is considered to be the first modern textbook about chemistry. It contained a list of "simple substances" that Lavoisier believed could not be broken down further, which included oxygen, nitrogen, hydrogen, phosphorus, mercury, zinc and sulfur, which formed the basis for the modern list of elements. Lavoisier's list also included 'light' and 'caloric', which at the time were believed to be material substances. He has classified these substances into metals and non metals. While many leading chemists refused to believe Lavoisier's new revelations, the *Elementary Treatise* was written well enough to convince the younger generation. However, Lavoisier's descriptions of his elements lack completeness, as he only classified them as metals and non-metals.

17.3   19th century

17.3.1   Johann Wolfgang Döbereiner

In 1817, Johann Wolfgang Döbereiner began to formulate one of the earliest attempts to classify the elements. In 1829, he found that he could form some of the elements into groups of three, with the members of each group having related properties. He termed these groups *triads*. Some of the triads that were classified by Döbereiner are:

1. chlorine, bromine, and iodine

2. calcium, strontium, and barium

3. sulfur, selenium, and tellurium

4. lithium, sodium, and potassium

In all of the triads, the atomic weight of the middle element was almost exactly the average of the atomic weights of the other two elements.[4]

17.3.2   Alexandre-Emile Béguyer de Chancourtois

Alexandre-Emile Béguyer de Chancourtois, a French geologist, was the first person to notice the periodicity of the elements — similar elements occurring at regular intervals when they are ordered by their atomic weights. In 1862 he devised an early form of periodic table, which he named *Vis tellurique* (the 'telluric helix'), after the element tellurium, which fell near the center of his diagram.[5] With the elements arranged in a spiral on a cylinder by order of increasing atomic weight, de Chancourtois saw that elements with similar properties lined up vertically. His 1863 publication included a chart (which contained ions and compounds, in addition to elements), but his original paper in the *Comptes*

Rendus Academie des Scéances used geological rather than chemical terms and did not include a diagram. As a result, de Chancourtois' ideas received little attention until after the work of Dmitri Mendeleev had been publicised.[6]

17.3.3 John Newlands

No.		No.		No.		No.		No.		No.		No.		No.	
H	1	F	8	Cl	15	Co & Ni	22	Br	29	Pd	36	I	42	Pt & Ir	50
Li	2	Na	9	K	16	Cu	23	Rb	30	Ag	37	Cs	44	Os	51
G	3	Mg	10	Ca	17	Zn	24	Sr	31	Cd	38	Ba & V	45	Hg	52
Bo	4	Al	11	Cr	19	Y	25	Ce & La	33	U	40	Ta	46	Tl	53
C	5	Si	12	Ti	18	In	26	Zr	32	Sn	39	W	47	Pb	54
N	6	P	13	Mn	20	As	27	Di & Mo	34	Sb	41	Nb	48	Bi	55
O	7	S	14	Fe	21	Se	28	Ro & Ru	35	Te	43	Au	49	Th	56

Newlands' law of octaves

In 1864, the English chemist John Newlands classified the sixty-two known elements into seven groups, based on their physical properties.[7][8]

Newlands noted that many pairs of similar elements existed, which differed by some multiple of eight in mass number, and was the first to assign them an atomic number.[9] When his 'law of octaves' was printed in *Chemistry News*, likening this periodicity of eights to the musical scale, it was ridiculed by some of his contemporaries. His lecture to the Chemistry Society on 1 March 1866 was not published, the Society defending their decision by saying that such 'theoretical' topics might be controversial.

The importance of Newlands' analysis was eventually recognised by the Chemistry Society with a Gold Medal five years after they recognised Mendeleev's work. It was not until the following century, with Gilbert N. Lewis' valence bond theory (1916) and Irving Langmuir's octet theory of chemical bonding (1919), that the importance of the periodicity of eight would be accepted.[10][11] The Royal Chemistry Society acknowledged Newlands' contribution to science in 2008, when they put a Blue Plaque on the house where he was born, which described him as the "discoverer of the Periodic Law for the chemical elements".[9] He contributed the word 'periodic' in chemistry.

17.3.4 Dmitri Mendeleev

The Russian chemist Dmitri Mendeleev was the first scientist to make a periodic table similar to the one used today. Mendeleev arranged the elements by atomic mass, corresponding to relative molar mass. It is sometimes said that he played 'chemical solitaire' on long train journeys, using cards with various facts about the known elements.[12] On March 6, 1869, a formal presentation was made to the Russian Chemical Society, entitled *The Dependence Between the Properties of the Atomic Weights of the Elements*. In

Dmitri Ivanovich Mendeleev

Zeitschrift für Chemie *(1869, pages 405-6), in which Mendeleev's periodic table is first published outside Russia.*

1869, the table was published in an obscure Russian journal and then republished in a German journal, *Zeitschrift für Chemie.*[13] In it, Mendeleev stated that:

1. The elements, if arranged according to their atomic mass, exhibit an apparent periodicity of properties.

Mendeleev's 1871 periodic table. Dashes: unknown elements. Group I-VII: modern group 1–2 and 3–7 with transition metals added; some of these extend into a group VIII. Noble gasses unknown (and unpredicted).

2. Elements which are similar as regards to their chemical properties have atomic weights which are either of nearly the same value (e.g., Pt, Ir, Os) or which increase regularly (e.g., K, Rb, Cs).

3. The arrangement of the elements, or of groups of elements in the order of their atomic masses, corresponds to their so-called valencies, as well as, to some extent, to their distinctive chemical properties; as is apparent among other series in that of Li, Be, B, C, N, O, and F.

4. The elements which are the most widely diffused have small atomic weights.

5. The magnitude of the atomic weight determines the character of the element, just as the magnitude of the molecule determines the character of a compound body.

6. We must expect the discovery of many yet unknown elements – for example, elements analogous to aluminium and silicon – whose atomic weight would be between 65 and 75.

7. The atomic weight of an element may sometimes be amended by a knowledge of those of its contiguous elements. Thus the atomic weight of tellurium must lie between 123 and 126, and cannot be 128.

8. Certain characteristic properties of elements can be foretold from their atomic masses.

Scientific benefits of Mendeleev's table

- It enabled Mendeleev to predict the discovery of new elements and left spaces for them, namely eka-silicon (germanium), eka-aluminium (gallium), and eka-boron (scandium). Thus, there was no disturbance in the periodic table.

- It could be used by Mendeleev to point out that some of the atomic weights being used at the time were incorrect.

- It provided for variance from atomic weight order.

Shortcomings of Mendeleev's table

- The table was not able to predict the existence of the noble gases. However, when this entire family of elements was discovered, Sir William Ramsay was able to add them to the table as Group 0, without the basic concept of the periodic table being disturbed.

- A single position could not be assigned to hydrogen, which could be placed either in the alkali metals group, the halogens group or separately above the table between boron and carbon.[14]

17.3.5   Lothar Meyer

Unknown to Mendeleev, the German chemist Lothar Meyer was also working on a periodic table. Although his work was published in 1864, and was done independently of Mendeleev, few historians regard him as an equal co-creator of the periodic table. Meyer's table only included twenty-eight elements, which were not classified by atomic weight, but by valence and he never reached the idea of predicting new elements and correcting atomic weights. A few months after Mendeleev published his periodic table of the known elements, predicted new elements to help complete his table and corrected the atomic weights of some of the elements, Meyer published a virtually identical periodic table.

Meyer and Mendeleev are considered by some historians of science to be the co-creators of the periodic table, but Mendeleev's accurate prediction of the qualities of undiscovered elements enables him to have the larger share of the credit.

17.3.6   William Odling

In 1864, the English chemist William Odling also drew up a table that was remarkably similar to the table produced by Mendeleev. Odling overcame the tellurium-iodine problem and even managed to get thallium, lead, mercury and platinum into the right groups, which is something that Mendeleev failed to do at his first attempt. Odling failed to achieve recognition, however, since it is suspected that he, as Secretary of the Chemical Society of London, was instrumental in discrediting Newlands' earlier work on the periodic table.

17.4   20th century

17.4.1   Henry Moseley

Henry Moseley

In 1914, a year before he was killed in action at Gallipoli, the English physicist Henry Moseley found a relationship between the X-ray wavelength of an element and its atomic number. He was then able to re-sequence the periodic table by nuclear charge, rather than by atomic weight. Before this discovery, atomic numbers were sequential numbers based on an element's atomic weight. Moseley's discovery showed that atomic numbers were in fact based upon experimental measurements.

Using information about their X-ray wavelengths, Moseley placed argon (with an atomic number Z=18) before potassium (Z=19), despite the fact that argon's atomic weight of 39.9 is greater than the atomic weight of potassium (39.1). The new order was in agreement with the chemical properties of these elements, since argon is a noble gas and potassium is an alkali metal. Similarly, Moseley placed cobalt before nickel and was able to explain that tellurium occurs before iodine, without revising the experimental atomic weight of tellurium, as had been proposed by Mendeleev.

Moseley's research showed that there were gaps in the periodic table at atomic numbers 43 and 61, which are now known to be occupied by technetium and promethium respectively.

17.4.2   Glenn T. Seaborg

During his Manhattan Project research in 1943, Glenn T. Seaborg experienced unexpected difficulties in isolating the elements americium and curium. Seaborg wondered if these elements belonged to a different series, which would explain why their chemical properties were different from what was expected. In 1945, against the advice of colleagues, he proposed a significant change to Mendeleev's table: the actinide series.

Seaborg's actinide concept of heavy element electronic structure, predicting that the actinides form a transition series analogous to the rare earth series of lanthanide elements, is now well accepted and included in the periodic table. The actinide series is the second row of the f-block (5f series). In both the actinide and lanthanide series, an inner electron shell is being filled. The actinide series comprises the elements from actinium to lawrencium. Seaborg's subsequent elaborations of the actinide concept theorized a series of superheavy elements in a transactinide series comprising elements from 104 to 121 and a superactinide series of elements from 122 to 153.

17.5   See also

- Alternative periodic tables

- History of chemistry

- Periodic Systems of Small Molecules

- Prout's hypothesis

- The Mystery of Matter: Search for the Elements (PBS film)

- Timeline of chemical element discoveries

17.6   References

[1] IUPAC article on periodic table

[2] Scerri, E. R. (2006). *The Periodic Table: Its Story ad Its Significance*; New York City, New York; Oxford University Press.

[3] "A Brief History of the Development of Periodic Table".

[4] Leicester, Henry M. (1971). *The Historical Background of Chemistry*; New York City, New York; Dover Publications.

[5] Chancourtois, *Comptes rendus Academie des sciences*, volume 55, p. 600.

[6] Annales des Mines history page.

[7] in a letter published in *Chemistry News* in February 1863, according to the Notable Names Data Base

[8] Newlands on classification of elements

[9] John Newlands, Chemistry Review, November 2003, pp15-16

[10] Irving Langmuir, "The Structure of Atoms and the Octet Theory of Valence", Proceedings of the National Academy of Science, Vol. V, 252, Letters (1919) – online at

[11] Irving Langmuir, "The Arrangement of Electrons in Atoms and Molecules", Journal of the American Chemical Society, Vol. 41, No, 6, pg. 868 (June 1919) – beginning and ending of the paper are transcribed online at ; the middle is missing

[12] *Physical Science*, Holt Rinehart & Winston (January 2004), page 302 ISBN 0-03-073168-2

[13] Mendeleev, Dmitri (1869). "Ueber die Beziehungen der Eigenschaften zu den Atomgewichten der Elemente". *Zeitschrift für Chemie* **12**: 405–406. Retrieved 29 November 2013.

[14] http://www.reed.edu/reed_magazine/summer2009/columns/NoAA/from_the_archives.html

17.7   External links

- Development of the periodic table (part of a collection of pages that explores the periodic table and the elements) by the Royal Society of Chemistry

- The path to the periodic table by the Chemical Heritage Foundation

- Dr. Eric Scerri's web page, which has links to interviews, lectures and articles on various aspects of the periodic system, including the history of the periodic table.

- The Internet Database of Periodic Tables - a large collection of periodic tables and periodic system formulations.

- History of Mendeleev periodic table of elements as a data visualization at CrossValidated Stack Exchange

Chapter 18

Statistical mechanics

Statistical mechanics is a branch of theoretical physics that studies, using probability theory, the average behaviour of a mechanical system where the state of the system is uncertain.[1][2][3][note 1]

The classical view of the universe was that its fundamental laws are mechanical in nature, and that all physical systems are therefore governed by mechanical laws at a microscopic level. These laws are precise equations of motion that map any given initial state to a corresponding future state at a later time. There is however a disconnection between these laws and everyday life experiences, as we do not find it necessary (nor even theoretically possible) to know exactly at a microscopic level the simultaneous positions and velocities of each molecule while carrying out processes at the human scale (for example, when performing a chemical reaction). Statistical mechanics is a collection of mathematical tools that are used to fill this disconnection between the laws of mechanics and the practical experience of incomplete knowledge.

A common use of statistical mechanics is in explaining the thermodynamic behaviour of large systems. Microscopic mechanical laws do not contain concepts such as temperature, heat, or entropy, however, statistical mechanics shows how these concepts arise from the natural uncertainty that arises about the state of a system when that system is prepared in practice. The benefit of using statistical mechanics is that it provides exact methods to connect thermodynamic quantities (such as heat capacity) to microscopic behaviour, whereas in classical thermodynamics the only available option would be to just measure and tabulate such quantities for various materials. Statistical mechanics also makes it possible to *extend* the laws of thermodynamics to cases which are not considered in classical thermodynamics, for example microscopic systems and other mechanical systems with few degrees of freedom.[1] This branch of statistical mechanics which treats and extends classical thermodynamics is known as **statistical thermodynamics** or **equilibrium statistical mechanics**.

Statistical mechanics also finds use outside equilibrium. An important subbranch known as **non-equilibrium statisti-cal mechanics** deals with the issue of microscopically modelling the speed of irreversible processes that are driven by imbalances. Examples of such processes include chemical reactions, or flows of particles and heat. Unlike with equilibrium, there is no exact formalism that applies to non-equilibrium statistical mechanics in general and so this branch of statistical mechanics remains an active area of theoretical research.

18.1 Principles: mechanics and ensembles

Main articles: Mechanics and Statistical ensemble

In physics there are two types of mechanics usually examined: classical mechanics and quantum mechanics. For both types of mechanics, the standard mathematical approach is to consider two ingredients:

1. The complete state of the mechanical system at a given time, mathematically encoded as a phase point (classical mechanics) or a pure quantum state vector (quantum mechanics).

2. An equation of motion which carries the state forward in time: Hamilton's equations (classical mechanics) or the time-dependent Schrödinger equation (quantum mechanics)

Using these two ingredients, the state at any other time, past or future, can in principle be calculated.

Whereas ordinary mechanics only considers the behaviour of a single state, statistical mechanics introduces the statistical ensemble, which is a large collection of virtual, independent copies of the system in various states. The statistical ensemble is a probability distribution over all possible states of the system. In classical statistical mechanics, the ensemble is a probability distribution over phase points (as opposed to a single phase point in ordinary mechanics), usually represented as a distribution in a phase space with canonical coordinates. In quantum statistical mechanics, the ensemble is a probability distribution over pure states,[note 2] and can be compactly summarized as a density matrix.

As is usual for probabilities, the ensemble can be interpreted in different ways:[1]

- an ensemble can be taken to represent the various possible states that a *single system* could be in (epistemic probability, a form of knowledge), or

- the members of the ensemble can be understood as the states of the systems in experiments repeated on independent systems which have been prepared in a similar but imperfectly controlled manner (empirical probability), in the limit of an infinite number of trials.

These two meanings are equivalent for many purposes, and will be used interchangeably in this article.

However the probability is interpreted, each state in the ensemble evolves over time according to the equation of motion. Thus, the ensemble itself (the probability distribution over states) also evolves, as the virtual systems in the ensemble continually leave one state and enter another. The ensemble evolution is given by the Liouville equation (classical mechanics) or the von Neumann equation (quantum mechanics). These equations are simply derived by the application of the mechanical equation of motion separately to each virtual system contained in the ensemble, with the probability of the virtual system being conserved over time as it evolves from state to state.

One special class of ensemble is those ensembles that do not evolve over time. These ensembles are known as *equilibrium ensembles* and their condition is known as *statistical equilibrium*. Statistical equilibrium occurs if, for each state in the ensemble, the ensemble also contains all of its future and past states with probabilities equal to the probability of being in that state.[note 3] The study of equilibrium ensembles of isolated systems is the focus of statistical thermodynamics. Non-equilibrium statistical mechanics addresses the more general case of ensembles that change over time, and/or ensembles of non-isolated systems.

18.2   Statistical thermodynamics

The primary goal of statistical thermodynamics (also known as equilibrium statistical mechanics) is to explain the classical thermodynamics of materials in terms of the properties of their constituent particles and the interactions between them. In other words, statistical thermodynamics provides a connection between the macroscopic properties of materials in thermodynamic equilibrium, and the microscopic behaviours and motions occurring inside the material.

As an example, one might ask what is it about a thermodynamic system of NH_3 molecules that determines the free energy characteristic of that compound? Classical thermodynamics does not provide the answer. If, for example, we were given spectroscopic data, of this body of gas molecules, such as bond length, bond angle, bond rotation, and flexibility of the bonds in NH_3 we should see that the free energy could not be other than it is. To prove this true, we need to bridge the gap between the microscopic realm of atoms and molecules and the macroscopic realm of classical thermodynamics. Statistical mechanics demonstrates how the thermodynamic parameters of a system, such as temperature and pressure, are related to microscopic behaviours of such constituent atoms and molecules.[4]

Although we may understand a system generically, in general we lack information about the state of a specific instance of that system. For this reason the notion of statistical ensemble (a probability distribution over possible states) is necessary. Furthermore, in order to reflect that the material is in a thermodynamic equilibrium, it is necessary to introduce a corresponding statistical mechanical definition of equilibrium. The analogue of thermodynamic equilibrium in statistical thermodynamics is the ensemble property of statistical equilibrium, described in the previous section. An additional assumption in statistical thermodynamics is that the system is isolated (no varying external forces are acting on the system), so that its total energy does not vary over time. A sufficient (but not necessary) condition for statistical equilibrium with an isolated system is that the probability distribution is a function only of conserved properties (total energy, total particle numbers, etc.).[1]

18.2.1   Fundamental postulate

There are many different equilibrium ensembles that can be considered, and only some of them correspond to thermodynamics.[1] An additional postulate is necessary to motivate why the ensemble for a given system should have one form or another.

A common approach found in many textbooks is to take the *equal a priori probability postulate*.[2] This postulate states that

> *For an isolated system with an exactly known energy and exactly known composition, the system can be found with* equal *probability in any microstate consistent with that knowledge.*

The equal a priori probability postulate therefore provides a motivation for the microcanonical ensemble described below. There are various arguments in favour of the equal a priori probability postulate:

- Ergodic hypothesis: An ergodic state is one that evolves over time to explore "all accessible" states: all those with the same energy and composition. In an ergodic system, the microcanonical ensemble is the only possible equilibrium ensemble with fixed energy. This approach has limited applicability, since most systems are not ergodic.

- Principle of indifference: In the absence of any further information, we can only assign equal probabilities to each compatible situation.

- Maximum information entropy: A more elaborate version of the principle of indifference states that the correct ensemble is the ensemble that is compatible with the known information and that has the largest Gibbs entropy (information entropy).[5]

Other fundamental postulates for statistical mechanics have also been proposed.[6]

In any case, the reason for establishing the microcanonical ensemble is mainly axiomatic.[6] The microcanonical ensemble itself is mathematically awkward to use for real calculations, and even very simple finite systems can only be solved approximately. However, it is possible to use the microcanonical ensemble to construct a hypothetical infinite thermodynamic reservoir that has an exactly defined notion of temperature and chemical potential. Once this reservoir has been established, it can be used to justify exactly the canonical ensemble or grand canonical ensemble (see below) for any other system by considering the contact of this system with the reservoir.[1] These other ensembles are those actually used in practical statistical mechanics calculations as they are mathematically simpler and also correspond to a much more realistic situation (energy not known exactly).[2]

18.2.2   Three thermodynamic ensembles

Main articles: Microcanonical ensemble, Canonical ensemble and Grand canonical ensemble

There are three equilibrium ensembles with a simple form that can be defined for any isolated system bounded inside a finite volume.[1] These are the most often discussed ensembles in statistical thermodynamics. In the macroscopic limit (defined below) they all correspond to classical thermodynamics.

- The microcanonical ensemble describes a system with a precisely given energy and fixed composition (precise number of particles). The microcanonical ensemble contains with equal probability each possible state that is consistent with that energy and composition.

- The canonical ensemble describes a system of fixed composition that is in thermal equilibrium[note 4] with a heat bath of a precise temperature. The canonical ensemble contains states of varying energy but identical composition; the different states in the ensemble are accorded different probabilities depending on their total energy.

- The grand canonical ensemble describes a system with non-fixed composition (uncertain particle numbers) that is in thermal and chemical equilibrium with a thermodynamic reservoir. The reservoir has a precise temperature, and precise chemical potentials for various types of particle. The grand canonical ensemble contains states of varying energy and varying numbers of particles; the different states in the ensemble are accorded different probabilities depending on their total energy and total particle numbers.

Statistical fluctuations and the macroscopic limit

Main article: Thermodynamic limit

The thermodynamic ensembles' most significant difference is that they either admit uncertainty in the variables of energy or particle number, or that those variables are fixed to particular values. While this difference can be observed in some cases, for macroscopic systems the thermodynamic ensembles are usually observationally equivalent.

The limit of large systems in statistical mechanics is known as the thermodynamic limit. In the thermodynamic limit the microcanonical, canonical, and grand canonical ensembles tend to give identical predictions about thermodynamic characteristics. This means that one can specify either total energy or temperature and arrive at the same result; likewise one can specify either total particle number or chemical potential. Given these considerations, the best ensemble to choose for the calculation of the properties of a macroscopic system is usually just the ensemble which allows the result to be derived most easily.[7]

Important cases where the thermodynamic ensembles *do not* give identical results include:

- Systems at a phase transition.

- Systems with long-range interactions.

- Microscopic systems.

In these cases the correct thermodynamic ensemble must be chosen as there are observable differences between these ensembles not just in the size of fluctuations, but also in average quantities such as the distribution of particles. The correct ensemble is that which corresponds to the way the system has been prepared and characterized—in other words, the ensemble that reflects the knowledge about that system.[2]

18.2.3   Illustrative example (a gas)

The above concepts can be illustrated for the specific case of one liter of ammonia gas at standard conditions. (Note that statistical thermodynamics is not restricted to the study of macroscopic gases, and the example of a gas is given here to illustrate concepts. Statistical mechanics and statistical thermodynamics apply to all mechanical systems (including microscopic systems) and to all phases of matter: liquids, solids, plasmas, gases, nuclear matter, quark matter.)

A simple way to prepare one litre sample of ammonia in a standard condition is to take a very large reservoir of ammonia at those standard conditions, and connect it to a previously evacuated one-litre container. After ammonia gas has entered the container and the container has been given time to reach thermodynamic equilibrium with the reservoir, the container is then sealed and isolated. In thermodynamics, this is a repeatable process resulting in a very well defined sample of gas with a precise description. We now consider the corresponding precise description in statistical thermodynamics.

Although this process is well defined and repeatable in a macroscopic sense, we have no information about the exact locations and velocities of each and every molecule in the container of gas. Moreover, we do not even know exactly how many molecules are in the container; even supposing we knew exactly the average density of the ammonia gas in general, we do not know how many molecules of the gas happened to be inside our container at the moment when we sealed it. The sample is in equilibrium and is in equilibrium with the reservoir: we could reconnect it to the reservoir for some time, and then re-seal it, and our knowledge about the state of the gas would not change. In this case, our knowledge about the state of the gas is precisely described by the grand canonical ensemble. Provided we have an accurate microscopic model of the ammonia gas, we could in principle compute all thermodynamic properties of this sample of gas by using the distribution provided by the grand canonical ensemble.

Hypothetically, we could use an extremely sensitive weight scale to measure exactly the mass of the container before and after introducing the ammonia gas, so that we can exactly know the number of ammonia molecules. After we make this measurement, then our knowledge about the gas would correspond to the canonical ensemble. Finally, suppose by some hypothetical apparatus we can measure exactly the number of molecules and also measure exactly the total energy of the system. Supposing furthermore that this apparatus gives us no further information about the molecules' positions and velocities, our knowledge about the system would correspond to the microcanonical ensemble.

Even after making such measurements, however, our expectations about the behaviour of the gas do not change appreciably. This is because the gas sample is macroscopic and approximates very well the thermodynamic limit, so the different ensembles behave similarly. This can be demonstrated by considering how small the actual fluctuations would be. Suppose that we knew the number density of ammonia gas was exactly 3.04×10^{22} molecules per liter inside the reservoir of ammonia gas used to fill the one-litre container. In describing the container with the grand canonical ensemble, then, the average number of molecules would be $\langle N \rangle = 3.04 \times 10^{22}$ and the uncertainty (standard deviation) in the number of molecules would be $\sigma_N = \sqrt{\langle N \rangle} \approx 2 \times 10^{11}$ (assuming Poisson distribution), which is relatively very small compared to the total number of molecules. Upon measuring the particle number (thus arriving at a canonical ensemble) we should find very nearly 3.04×10^{22} molecules. For example, the probability of finding more than 3.040001×10^{22} or less than 3.039999×10^{22} molecules would be about 1 in $10^{3000000000}$.[note 5]

18.2.4   Calculation methods

Once the characteristic state function for an ensemble has been calculated for a given system, that system is 'solved' (macroscopic observables can be extracted from the characteristic state function). Calculating the characteristic state function of a thermodynamic ensemble is not necessarily a simple task, however, since it involves considering every possible state of the system. While some hypothetical systems have been exactly solved, the most general (and realistic) case is too complex for exact solution. Various approaches exist to approximate the true ensemble and allow calculation of average quantities.

Exact

There are some cases which allow exact solutions.

- For very small microscopic systems, the ensembles can be directly computed by simply enumerating over all possible states of the system (using exact diagonalization in quantum mechanics, or integral over all phase space in classical mechanics).

- Some large systems consist of many separable microscopic systems, and each of the subsystems can be analysed independently. Notably, idealized gases of non-interacting particles have this property, allowing exact derivations of Maxwell–Boltzmann statistics, Fermi–Dirac statistics, and Bose–Einstein statistics.[2]

- A few large systems with interaction have been solved. By the use of subtle mathematical techniques, exact solutions have been found for a few toy models.[8] Some examples include the Bethe ansatz, square-lattice Ising model in zero field, hard hexagon model.

Monte Carlo

Main article: Monte Carlo method

One approximate approach that is particularly well suited to computers is the Monte Carlo method, which examines just a few of the possible states of the system, with the states chosen randomly (with a fair weight). As long as these states form a representative sample of the whole set of states of the system, the approximate characteristic function is obtained. As more and more random samples are included, the errors are reduced to an arbitrarily low level.

- The Metropolis–Hastings algorithm is a classic Monte Carlo method which was initially used to sample the canonical ensemble.

- Path integral Monte Carlo, also used to sample the canonical ensemble.

Other

- For rarefied non-ideal gases, approaches such as the cluster expansion use perturbation theory to include the effect of weak interactions, leading to a virial expansion.[3]

- For dense fluids, another approximate approach is based on reduced distribution functions, in particular the radial distribution function.[3]

- Molecular dynamics computer simulations can be used to calculate microcanonical ensemble averages, in ergodic systems. With the inclusion of a connection to a stochastic heat bath, they can also model canonical and grand canonical conditions.

- Mixed methods involving non-equilibrium statistical mechanical results (see below) may be useful.

18.3 Non-equilibrium statistical mechanics

See also: Non-equilibrium thermodynamics

There are many physical phenomena of interest that involve quasi-thermodynamic processes out of equilibrium, for example:

- heat transport by the internal motions in a material, driven by a temperature imbalance,

- electric currents carried by the motion of charges in a conductor, driven by a voltage imbalance,

- spontaneous chemical reactions driven by a decrease in free energy,

- friction, dissipation, quantum decoherence,

- systems being pumped by external forces (optical pumping, etc.),

- and irreversible processes in general.

All of these processes occur over time with characteristic rates, and these rates are of importance for engineering. The field of non-equilibrium statistical mechanics is concerned with understanding these non-equilibrium processes at the microscopic level. (Statistical thermodynamics can only be used to calculate the final result, after the external imbalances have been removed and the ensemble has settled back down to equilibrium.)

In principle, non-equilibrium statistical mechanics could be mathematically exact: ensembles for an isolated system evolve over time according to deterministic equations such as Liouville's equation or its quantum equivalent, the von Neumann equation. These equations are the result of applying the mechanical equations of motion independently to each state in the ensemble. Unfortunately, these ensemble evolution equations inherit much of the complexity of the underlying mechanical motion, and so exact solutions are very difficult to obtain. Moreover, the ensemble evolution equations are fully reversible and do not destroy information (the ensemble's Gibbs entropy is preserved). In order to make headway in modelling irreversible processes, it is necessary to add additional ingredients besides probability and reversible mechanics.

Non-equilibrium mechanics is therefore an active area of theoretical research as the range of validity of these additional assumptions continues to be explored. A few approaches are described in the following subsections.

18.3.1   Stochastic methods

One approach to non-equilibrium statistical mechanics is to incorporate stochastic (random) behaviour into the system. Stochastic behaviour destroys information contained in the ensemble. While this is technically inaccurate (aside from hypothetical situations involving black holes, a system cannot in itself cause loss of information), the randomness is added to reflect that information of interest becomes converted over time into subtle correlations within the system, or to correlations between the system and environment. These correlations appear as chaotic or pseudorandom influences on the variables of interest. By replacing these correlations with randomness proper, the calculations can be made much easier.

- *Boltzmann transport equation*: An early form of stochastic mechanics appeared even before the term "statistical mechanics" had been coined, in studies of kinetic theory. James Clerk Maxwell had demonstrated that molecular collisions would lead to apparently chaotic motion inside a gas. Ludwig Boltzmann subsequently showed that, by taking this molecular chaos for granted as a complete randomization, the motions of particles in a gas would follow a simple Boltzmann transport equation that would rapidly restore a gas to an equilibrium state (see H-theorem).

  The Boltzmann transport equation and related approaches are important tools in non-equilibrium statistical mechanics due to their extreme simplicity. These approximations work well in systems where the "interesting" information is immediately (after just one collision) scrambled up into subtle correlations, which essentially restricts them to rarefied gases. The Boltzmann transport equation has been found to be very useful in simulations of electron transport in lightly doped semiconductors (in transistors), where the electrons are indeed analogous to a rarefied gas.

  A quantum technique related in theme is the random phase approximation.

- *BBGKY hierarchy*: In liquids and dense gases, it is not valid to immediately discard the correlations between particles after one collision. The BBGKY hierarchy (Bogoliubov–Born–Green–Kirkwood–Yvon hierarchy) gives a method for deriving Boltzmann-type equations but also extending them beyond the dilute gas case, to include correlations after a few collisions.

- *Keldysh formalism* (a.k.a. NEGF—non-equilibrium Green functions): A quantum approach to including stochastic dynamics is found in the Keldysh formalism. This approach often used in electronic quantum transport calculations.

18.3.2   Near-equilibrium methods

Another important class of non-equilibrium statistical mechanical models deals with systems that are only very slightly perturbed from equilibrium. With very small perturbations, the response can be analysed in linear response theory. A remarkable result, as formalized by the fluctuation-dissipation theorem, is that the response of a system when near equilibrium is precisely related to the fluctuations that occur when the system is in total equilibrium. Essentially, a system that is slightly away from equilibrium—whether put there by external forces or by fluctuations—relaxes towards equilibrium in the same way, since the system cannot tell the difference or "know" how it came to be away from equilibrium.[3]:664

This provides an indirect avenue for obtaining numbers such as ohmic conductivity and thermal conductivity by extracting results from equilibrium statistical mechanics. Since equilibrium statistical mechanics is mathematically well defined and (in some cases) more amenable for calculations, the fluctuation-dissipation connection can be a convenient shortcut for calculations in near-equilibrium statistical mechanics.

A few of the theoretical tools used to make this connection include:

- Fluctuation–dissipation theorem

- Onsager reciprocal relations

- Green–Kubo relations

- Landauer–Büttiker formalism

- Mori–Zwanzig formalism

18.3.3   Hybrid methods

An advanced approach uses a combination of stochastic methods and linear response theory. As an example, one approach to compute quantum coherence effects (weak localization, conductance fluctuations) in the conductance of an electronic system is the use of the Green-Kubo relations, with the inclusion of stochastic dephasing by interactions between various electrons by use of the Keldysh method.[9][10]

18.4 Applications outside thermodynamics

The ensemble formalism also can be used to analyze general mechanical systems with uncertainty in knowledge about the state of a system. Ensembles are also used in:

- propagation of uncertainty over time,[1]

- regression analysis of gravitational orbits,

- ensemble forecasting of weather,

- dynamics of neural networks,

- bounded-rational potential games in game theory and economics.

18.5 History

In 1738, Swiss physicist and mathematician Daniel Bernoulli published *Hydrodynamica* which laid the basis for the kinetic theory of gases. In this work, Bernoulli posited the argument, still used to this day, that gases consist of great numbers of molecules moving in all directions, that their impact on a surface causes the gas pressure that we feel, and that what we experience as heat is simply the kinetic energy of their motion.[6]

In 1859, after reading a paper on the diffusion of molecules by Rudolf Clausius, Scottish physicist James Clerk Maxwell formulated the Maxwell distribution of molecular velocities, which gave the proportion of molecules having a certain velocity in a specific range. This was the first-ever statistical law in physics.[11] Five years later, in 1864, Ludwig Boltzmann, a young student in Vienna, came across Maxwell's paper and was so inspired by it that he spent much of his life developing the subject further.

Statistical mechanics proper was initiated in the 1870s with the work of Boltzmann, much of which was collectively published in his 1896 *Lectures on Gas Theory*.[12] Boltzmann's original papers on the statistical interpretation of thermodynamics, the H-theorem, transport theory, thermal equilibrium, the equation of state of gases, and similar subjects, occupy about 2,000 pages in the proceedings of the Vienna Academy and other societies. Boltzmann introduced the concept of an equilibrium statistical ensemble and also investigated for the first time non-equilibrium statistical mechanics, with his *H*-theorem.

The term "statistical mechanics" was coined by the American mathematical physicist J. Willard Gibbs in 1884.[13][note 6] "Probabilistic mechanics" might today seem a more appropriate term, but "statistical mechanics" is

firmly entrenched.[14] Shortly before his death, Gibbs published in 1902 *Elementary Principles in Statistical Mechanics*, a book which formalized statistical mechanics as a fully general approach to address all mechanical systems—macroscopic or microscopic, gaseous or non-gaseous.[1] Gibbs' methods were initially derived in the framework classical mechanics, however they were of such generality that they were found to adapt easily to the later quantum mechanics, and still form the foundation of statistical mechanics to this day.[2]

18.6 See also

- Thermodynamics: non-equilibrium, chemical

- Mechanics: classical, quantum

- Probability, statistical ensemble

- Numerical methods: Monte Carlo method, molecular dynamics

- Statistical physics

- Quantum statistical mechanics

- List of notable textbooks in statistical mechanics

- List of important publications in statistical mechanics

Fundamentals of Statistical Mechanics – Wikipedia book

18.7 Notes

[1] The term *statistical mechanics* is sometimes used to refer to only *statistical thermodynamics*. This article takes the broader view. By some definitions, *statistical physics* is an even broader term which statistically studies any type of physical system, but is often taken to be synonymous with statistical mechanics.

[2] The probabilities in quantum statistical mechanics should not be confused with quantum superposition. While a quantum ensemble can contain states with quantum superpositions, a single quantum state cannot be used to represent an ensemble.

[3] Statistical equilibrium should not be confused with *mechanical equilibrium*. The latter occurs when a mechanical system has completely ceased to evolve even on a microscopic scale, due to being in a state with a perfect balancing of forces. Statistical equilibrium generally involves states that are very far from mechanical equilibrium.

[4] The transitive thermal equilibrium (as in, "X is thermal equilibrium with Y") used here means that the ensemble for the first system is not perturbed when the system is allowed to weakly interact with the second system.

[5] This is so unlikely as to be practically impossible. The statistical physicist Émile Borel noted that, compared to the improbabilities found in statistical mechanics, it would be more likely that monkeys typing randomly on a typewriter would happen to reproduce the books of the world. See infinite monkey theorem.

[6] According to Gibbs, the term "statistical", in the context of mechanics, i.e. statistical mechanics, was first used by the Scottish physicist James Clerk Maxwell in 1871. From: J. Clerk Maxwell, *Theory of Heat* (London, England: Longmans, Green, and Co., 1871), p. 309: "In dealing with masses of matter, while we do not perceive the individual molecules, we are compelled to adopt what I have described as the statistical method of calculation, and to abandon the strict dynamical method, in which we follow every motion by the calculus."

18.8   References

[1] Gibbs, Josiah Willard (1902). *Elementary Principles in Statistical Mechanics*. New York: Charles Scribner's Sons.

[2] Tolman, R. C. (1938). *The Principles of Statistical Mechanics*. Dover Publications. ISBN 9780486638966.

[3] Balescu, Radu (1975). *Equilibrium and Non-Equilibrium Statistical Mechanics*. John Wiley & Sons. ISBN 9780471046004.

[4] Nash, Leonard K. (1974). *Elements of Statistical Thermodynamics, 2nd Ed*. Dover Publications, Inc. ISBN 0-486-44978-5. OCLC 61513215.

[5] Jaynes, E. (1957). "Information Theory and Statistical Mechanics". *Physical Review* **106** (4): 620. doi:10.1103/PhysRev.106.620.

[6] J. Uffink, "Compendium of the foundations of classical statistical physics." (2006)

[7] Reif, F. (1965). *Fundamentals of Statistical and Thermal Physics*. McGraw–Hill. p. 227. ISBN 9780070518001.

[8] Baxter, Rodney J. (1982). *Exactly solved models in statistical mechanics*. Academic Press Inc. ISBN 9780120831807.

[9] Altshuler, B. L.; Aronov, A. G.; Khmelnitsky, D. E. (1982). "Effects of electron-electron collisions with small energy transfers on quantum localisation". *Journal of Physics C: Solid State Physics* **15** (36): 7367. doi:10.1088/0022-3719/15/36/018.

[10] Aleiner, I.; Blanter, Y. (2002). "Inelastic scattering time for conductance fluctuations". *Physical Review B* **65** (11). doi:10.1103/PhysRevB.65.115317.

[11] Mahon, Basil (2003). *The Man Who Changed Everything – the Life of James Clerk Maxwell*. Hoboken, NJ: Wiley. ISBN 0-470-86171-1. OCLC 52358254 62045217.

[12] Ebeling, Werner; Sokolov, Igor M. (2005). *Statistical Thermodynamics and Stochastic Theory of Nonequilibrium Systems*. World Scientific Publishing Co. Pte. Ltd. pp. 3–12. ISBN 978-90-277-1674-3. (section 1.2)

[13] J. W. Gibbs, "On the Fundamental Formula of Statistical Mechanics, with Applications to Astronomy and Thermodynamics." Proceedings of the American Association for the Advancement of Science, **33**, 57-58 (1884). Reproduced in *The Scientific Papers of J. Willard Gibbs, Vol II* (1906), pp. 16.

[14] Mayants, Lazar (1984). *The enigma of probability and physics*. Springer. p. 174. ISBN 978-90-277-1674-3.

18.9   External links

- Philosophy of Statistical Mechanics article by Lawrence Sklar for the Stanford Encyclopedia of Philosophy.

- Sklogwiki - Thermodynamics, statistical mechanics, and the computer simulation of materials. SklogWiki is particularly orientated towards liquids and soft condensed matter.

- Statistical Thermodynamics - Historical Timeline

- Thermodynamics and Statistical Mechanics by Richard Fitzpatrick

- Lecture Notes in Statistical Mechanics and Mesoscopics by Doron Cohen

- Videos of lecture series in statistical mechanics on YouTube taught by Leonard Susskind.

- Vu-Quoc, L., Configuration integral (statistical mechanics), 2008. this wiki site is down; see this article in the web archive on 2012 April 28.

Chapter 19

Thermodynamics

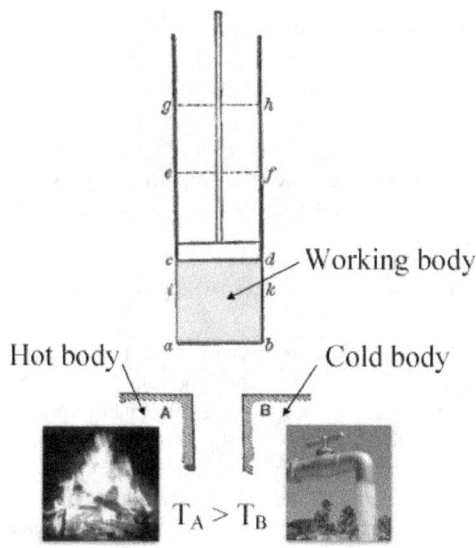

Working body

Hot body

Cold body

$T_A > T_B$

Annotated color version of the original 1824 Carnot heat engine showing the hot body (boiler), working body (system, steam), and cold body (water), the letters labeled according to the stopping points in Carnot cycle

Thermodynamics is a branch of physics concerned with heat and temperature and their relation to energy and work. It defines macroscopic variables, such as internal energy, entropy, and pressure, that partly describe a body of matter or radiation. It states that the behavior of those variables is subject to general constraints, that are common to all materials, not the peculiar properties of particular materials. These general constraints are expressed in the four laws of thermodynamics. Thermodynamics describes the bulk behavior of the body, not the microscopic behaviors of the very large numbers of its microscopic constituents, such as molecules. The basic results of thermodynamics rely on the existence of idealized states of thermodynamic equilibrium. Its laws are explained by statistical mechanics, in terms of the microscopic constituents.

Thermodynamics applies to a wide variety of topics in science and engineering, especially physical chemistry, chemical engineering and mechanical engineering.

Historically, the distinction between heat and temperature was studied in the 1750s by Joseph Black. Characteristically thermodynamic thinking began in the work of Carnot (1824) who believed that the efficiency of heat engines was the key that could help France win the Napoleonic Wars.[1] The Irish-born British physicist Lord Kelvin was the first to formulate a concise definition of thermodynamics in 1854:[2]

> "Thermo-dynamics is the subject of the relation of heat to forces acting between contiguous parts of bodies, and the relation of heat to electrical agency."

Initially, thermodynamics, as applied to heat engines, was concerned with the thermal properties of their 'working materials', such as steam, in an effort to increase the efficiency and power output of engines. Thermodynamics was later expanded to the study of energy transfers in chemical processes, such as the investigation, published in 1840, of the heats of chemical reactions[3] by Germain Hess, which was not originally explicitly concerned with the relation between energy exchanges by heat and work. From this evolved the study of Chemical thermodynamics and the role of entropy in chemical reactions.[4][5][6][7][8][9][10][11][12]

19.1 Introduction

Historically, thermodynamics arose from the study of two distinct kinds of transfer of energy, as heat and as work, and the relation of those to the system's macroscopic variables of volume, pressure and temperature.[13][14] As it developed, thermodynamics began also to study transfers of matter.

The plain term 'thermodynamics' refers to a macroscopic description of bodies and processes.[15] Reference to atomic constitution is foreign to classical thermodynamics.[16] The qualified term 'statistical thermodynamics' refers to descriptions of bodies and processes in terms of the atomic or other microscopic constitution of matter, using statistical and probabilistic reasoning.

Thermodynamic equilibrium is one of the most important concepts for thermodynamics.[17] The temperature of a thermodynamic system is well defined, and is perhaps the most characteristic quantity of thermodynamics. As the systems and processes of interest are taken further from thermodynamic equilibrium, their exact thermodynamical study becomes more difficult. Relatively simple approximate calculations, however, using the variables of equilibrium thermodynamics, are of much practical value. Many important practical engineering cases, as in heat engines or refrigerators, can be approximated as systems consisting of many subsystems at different temperatures and pressures. If a physical process is too fast, the equilibrium thermodynamic variables, for example temperature, may not be well enough defined to provide a useful approximation.

Central to thermodynamic analysis are the definitions of the system, which is of interest, and of its surroundings.[8][18] The surroundings of a thermodynamic system consist of physical devices and of other thermodynamic systems that can interact with it. An example of a thermodynamic surrounding is a heat bath, which is held at a prescribed temperature, regardless of how much heat might be drawn from it.

There are four fundamental kinds of physical entities in thermodynamics, states of a system, walls of a system,[19][20][21][22][23] thermodynamic processes of a system, and thermodynamic operations. This allows two fundamental approaches to thermodynamic reasoning, that in terms of states of a system, and that in terms of cyclic processes of a system.

A thermodynamic system can be defined in terms of its states. In this way, a thermodynamic system is a macroscopic physical object, explicitly specified in terms of macroscopic physical and chemical variables that describe its macroscopic properties. The macroscopic state variables of thermodynamics have been recognized in the course of empirical work in physics and chemistry.[9] Always associated with the material that constitutes a system, its working substance, are the walls that delimit the system, and connect it with its surroundings. The state variables chosen for the system should be appropriate for the natures of the walls and surroundings.[24]

A thermodynamic operation is an artificial physical manipulation that changes the definition of a system or its surroundings. Usually it is a change of the permeability or some other feature of a wall of the system,[25] that allows energy (as heat or work) or matter (mass) to be exchanged

with the environment. For example, the partition between two thermodynamic systems can be removed so as to produce a single system. A thermodynamic operation usually leads to a thermodynamic process of transfer of mass or energy that changes the state of the system, and the transfer occurs in natural accord with the laws of thermodynamics. Besides thermodynamic operations, changes in the surroundings can also initiate thermodynamic processes.

A thermodynamic system can also be defined in terms of the cyclic processes that it can undergo.[26] A cyclic process is a cyclic sequence of thermodynamic operations and processes that can be repeated indefinitely often without changing the final state of the system.

For thermodynamics and statistical thermodynamics to apply to a physical system, it is necessary that its internal atomic mechanisms fall into one of two classes:

- those so rapid that, in the time frame of the process of interest, the atomic states rapidly bring system to its own state of internal thermodynamic equilibrium; and

- those so slow that, in the time frame of the process of interest, they leave the system unchanged.[27][28]

The rapid atomic mechanisms account for the internal energy of the system. They mediate the macroscopic changes that are of interest for thermodynamics and statistical thermodynamics, because they quickly bring the system near enough to thermodynamic equilibrium. "When intermediate rates are present, thermodynamics and statistical mechanics cannot be applied."[27] Such intermediate rate atomic processes do not bring the system near enough to thermodynamic equilibrium in the time frame of the macroscopic process of interest. This separation of time scales of atomic processes is a theme that recurs throughout the subject.

For example, classical thermodynamics is characterized by its study of materials that have equations of state or characteristic equations. They express equilibrium relations between macroscopic mechanical variables and temperature and internal energy. They express the constitutive peculiarities of the material of the system. A classical material can usually be described by a function that makes pressure dependent on volume and temperature, the resulting pressure being established much more rapidly than any imposed change of volume or temperature.[29][30][31][32]

The present article takes a gradual approach to the subject, starting with a focus on cyclic processes and thermodynamic equilibrium, and then gradually beginning to further consider non-equilibrium systems.

Thermodynamic facts can often be explained by viewing macroscopic objects as assemblies of very many microscopic or atomic objects that obey Hamiltonian dynamics.[8][33][34] The microscopic or atomic objects exist in species, the objects of each species being all alike. Because of this likeness, statistical methods can be used to account for the macroscopic properties of the thermodynamic system in terms of the properties of the microscopic species. Such explanation is called statistical thermodynamics; also often it is referred to by the term 'statistical mechanics', though this term can have a wider meaning, referring to 'microscopic objects', such as economic quantities, that do not obey Hamiltonian dynamics.[33]

19.2   History

The thermodynamicists representative of the original eight founding schools of thermodynamics. The schools with the most-lasting effect in founding the modern versions of thermodynamics are the Berlin school, particularly as established in Rudolf Clausius's 1865 textbook The Mechanical Theory of Heat, *the Vienna school, while the statistical mechanics of Ludwig Boltzmann, and the Gibbsian school at Yale University, led by the American engineer Willard Gibbs' 1876* On the Equilibrium of Heterogeneous Substances *launched chemical thermodynamics.*

The history of thermodynamics as a scientific discipline generally begins with Otto von Guericke who, in 1650, built and designed the world's first vacuum pump and demonstrated a vacuum using his Magdeburg hemispheres. Guericke was driven to make a vacuum in order to disprove Aristotle's long-held supposition that 'nature abhors a vacuum'. Shortly after Guericke, the physicist and chemist Robert Boyle had learned of Guericke's designs and, in 1656, in coordination with the scientist Robert Hooke, built an air pump.[35] Using this pump, Boyle and Hooke noticed a correlation between pressure, temperature, and volume. In time, they formulated Boyle's Law, which states that for

a gas at constant temperature, its pressure and volume are inversely proportional. In 1679, based on these concepts, an associate of Boyle's named Denis Papin built a steam digester, which was a closed vessel with a tightly fitting lid that confined steam until a high pressure was generated. Later versions of this design implemented a steam release valve that kept the machine from exploding. By watching the valve rhythmically move up and down, Papin conceived of the idea of a piston and a cylinder engine. He did not, however, follow through with his design. Nevertheless, in 1697, based on Papin's designs, the engineer Thomas Savery built the first engine, followed by Thomas Newcomen in 1712. Although these early engines were crude and inefficient, they attracted the attention of the leading scientists of the time.

The concepts of heat capacity and latent heat, which were necessary for development of thermodynamics, were developed by Professor Joseph Black at the University of Glasgow, where James Watt worked as an instrument maker. Watt consulted with Black on tests of his steam engine, but it was Watt who conceived the idea of the external condenser, greatly raising the steam engine's efficiency.[36] All the previous work led Sadi Carnot, the "father of thermodynamics", to publish *Reflections on the Motive Power of Fire* (1824), a discourse on heat, power, energy, and engine efficiency. The paper outlined the basic energetic relations between the Carnot engine, the Carnot cycle, and motive power. It marked the start of thermodynamics as a modern science.[11]

The first thermodynamic textbook was written in 1859 by William Rankine, originally trained as a physicist and a professor of civil and mechanical engineering at the University of Glasgow.[37] The first and second laws of thermodynamics emerged simultaneously in the 1850s, primarily out of the works of William Rankine, Rudolf Clausius, and William Thomson (Lord Kelvin).

The foundations of statistical thermodynamics were set out by physicists such as James Clerk Maxwell, Ludwig Boltzmann, Max Planck, Rudolf Clausius and J. Willard Gibbs.

From 1873 to '76, the American mathematical physicist Josiah Willard Gibbs published a series of three papers, the most famous being "On the equilibrium of heterogeneous substances".[4] Gibbs showed how thermodynamic processes, including chemical reactions, could be graphically analyzed. By studying the energy, entropy, volume, chemical potential, temperature and pressure of the thermodynamic system, one can determine whether a process would occur spontaneously.[38] Chemical thermodynamics was further developed by Pierre Duhem,[5] Gilbert N. Lewis, Merle Randall,[6] and E. A. Guggenheim,[7][8] who applied the mathematical methods of Gibbs.

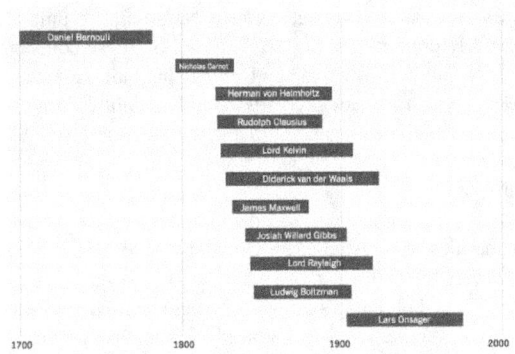

The lifetimes of some of the most important contributors to thermodynamics.

19.2.1   Etymology

The etymology of *thermodynamics* has an intricate history. It was first spelled in a hyphenated form as an adjective (*thermo-dynamic*) in 1849 and from 1854 to 1859 as the hyphenated noun *thermo-dynamics* to represent the science of heat and motive power and thereafter as *thermodynamics*.

The components of the word *thermo-dynamic* are derived from the Greek words θέρμη *therme*, meaning "heat," and δύναμις *dynamis*, meaning "power" (Haynie claims that the word was coined around 1840).[39][40]

The term **thermo-dynamic** was first used in January 1849 by William Thomson, later Lord Kelvin, in the phrase *a perfect thermo-dynamic engine* to describe Sadi Carnot's heat engine.[41]:545 In April 1849, Thomson added an appendix to his paper and used the term **thermodynamic** in the phrase *the object of a thermodynamic engine*.[41]:569

Pierre Perrot claims that the term *thermodynamics* was coined by James Joule in 1858 to designate the science of relations between heat and power.[11] Joule, however, never used that term, but did use the term *perfect thermo-dynamic engine* in reference to Thomson's 1849 phraseology,[41]:545 and Thomson's note on Joules' 1851 paper *On the Air-Engine*.

In 1854, **thermo-dynamics**, as a functional term to denote the general study of the action of heat, was first used by William Thomson in his paper "On the Dynamical Theory of Heat".[2]

In 1859, the closed compound form **thermodynamics** was first used by William Rankine in *A Manual of the Steam Engine* in a chapter on the Principles of Thermodynamics.[42]

19.3   Branches of description

Thermodynamic systems are theoretical constructions used to model physical systems that exchange matter and energy in terms of the laws of thermodynamics. The study of thermodynamical systems has developed into several related branches, each using a different fundamental model as a theoretical or experimental basis, or applying the principles to varying types of systems.

19.3.1   Classical thermodynamics

Classical thermodynamics accounts for the adventures of a thermodynamic system in terms, either of its time-invariant equilibrium states, or else of its continually repeated cyclic processes, but, formally, not both in the same account. It uses only time-invariant, or equilibrium, macroscopic quantities measurable in the laboratory, counting as time-invariant a long-term time-average of a quantity, such as a flow, generated by a continually repetitive process.[43][44] In classical thermodynamics, rates of change are not admitted as variables of interest. An equilibrium state stands endlessly without change over time, while a continually repeated cyclic process runs endlessly without a net change in the system over time.

In the account in terms of equilibrium states of a system, a state of thermodynamic equilibrium in a simple system is spatially homogeneous.

In the classical account solely in terms of a cyclic process, the spatial interior of the 'working body' of that process is not considered; the 'working body' thus does not have a defined internal thermodynamic state of its own because no assumption is made that it should be in thermodynamic equilibrium; only its inputs and outputs of energy as heat and work are considered.[45] It is common to describe a cycle theoretically as composed of a sequence of very many thermodynamic operations and processes. This creates a link to the description in terms of equilibrium states. The cycle is then theoretically described as a continuous progression of equilibrium states.

Classical thermodynamics was originally concerned with the transformation of energy in a cyclic process, and the exchange of energy between closed systems defined only by their equilibrium states. The distinction between transfers of energy as heat and as work was central.

As classical thermodynamics developed, the distinction between heat and work became less central. This was because there was more interest in open systems, for which the distinction between heat and work is not simple, and is beyond the scope of the present article. Alongside the amount of heat transferred as a fundamental quantity, entropy was gradually found to be a more generally applicable concept, especially when considering chemical reactions. Massieu in 1869 considered entropy as the basic dependent thermodynamic variable, with energy potentials and the reciprocal of the thermodynamic temperature as fundamental independent variables. Massieu functions can be useful in present-day non-equilibrium thermodynamics. In 1875, in the work of Josiah Willard Gibbs, entropy was considered a fundamental independent variable, while internal energy was a dependent variable.[46]

All actual physical processes are to some degree irreversible. Classical thermodynamics can consider irreversible processes, but its account in exact terms is restricted to variables that refer only to initial and final states of thermodynamic equilibrium, or to rates of input and output that do not change with time. For example, classical thermodynamics can consider time-average rates of flows generated by continually repeated irreversible cyclic processes. Also it can consider irreversible changes between equilibrium states of systems consisting of several phases (as defined below in this article), or with removable or replaceable partitions. But for systems that are described in terms of equilibrium states, it considers neither flows, nor spatial inhomogeneities in simple systems with no externally imposed force fields such as gravity. In the account in terms of equilibrium states of a system, descriptions of irreversible processes refer only to initial and final static equilibrium states; the time it takes to change thermodynamic state is not considered.[47][48]

19.3.2   Local equilibrium thermodynamics

Local equilibrium thermodynamics is concerned with the time courses and rates of progress of irreversible processes in systems that are smoothly spatially inhomogeneous. It admits time as a fundamental quantity, but only in a restricted way. Rather than considering time-invariant flows as long-term-average rates of cyclic processes, local equilibrium thermodynamics considers time-varying flows in systems that are described by states of local thermodynamic equilibrium, as follows.

For processes that involve only suitably small and smooth spatial inhomogeneities and suitably small changes with time, a good approximation can be found through the assumption of local thermodynamic equilibrium. Within the large or global region of a process, for a suitably small local region, this approximation assumes that a quantity known as the entropy of the small local region can be defined in a particular way. That particular way of definition of entropy is largely beyond the scope of the present article, but here it may be said that it is entirely derived from the concepts of classical thermodynamics; in particular, neither flow rates

nor changes over time are admitted into the definition of the entropy of the small local region. It is assumed without proof that the instantaneous global entropy of a non-equilibrium system can be found by adding up the simultaneous instantaneous entropies of its constituent small local regions. Local equilibrium thermodynamics considers processes that involve the time-dependent production of entropy by dissipative processes, in which kinetic energy of bulk flow and chemical potential energy are converted into internal energy at time-rates that are explicitly accounted for. Time-varying bulk flows and specific diffusional flows are considered, but they are required to be dependent variables, derived only from material properties described only by static macroscopic equilibrium states of small local regions. The independent state variables of a small local region are only those of classical thermodynamics.

19.3.3   Generalized or extended thermodynamics

Like local equilibrium thermodynamics, generalized or extended thermodynamics also is concerned with the time courses and rates of progress of irreversible processes in systems that are smoothly spatially inhomogeneous. It describes time-varying flows in terms of states of suitably small local regions within a global region that is smoothly spatially inhomogeneous, rather than considering flows as time-invariant long-term-average rates of cyclic processes. In its accounts of processes, generalized or extended thermodynamics admits time as a fundamental quantity in a more far-reaching way than does local equilibrium thermodynamics. The states of small local regions are defined by macroscopic quantities that are explicitly allowed to vary with time, including time-varying flows. Generalized thermodynamics might tackle such problems as ultrasound or shock waves, in which there are strong spatial inhomogeneities and changes in time fast enough to outpace a tendency towards local thermodynamic equilibrium. Generalized or extended thermodynamics is a diverse and developing project, rather than a more or less completed subject such as is classical thermodynamics.[49][50]

For generalized or extended thermodynamics, the definition of the quantity known as the entropy of a small local region is in terms beyond those of classical thermodynamics; in particular, flow rates are admitted into the definition of the entropy of a small local region. The independent state variables of a small local region include flow rates, which are not admitted as independent variables for the small local regions of local equilibrium thermodynamics.

Outside the range of classical thermodynamics, the definition of the entropy of a small local region is no simple matter. For a thermodynamic account of a process in terms of the entropies of small local regions, the definition of entropy should be such as to ensure that the second law of thermodynamics applies in each small local region. It is often assumed without proof that the instantaneous global entropy of a non-equilibrium system can be found by adding up the simultaneous instantaneous entropies of its constituent small local regions. For a given physical process, the selection of suitable independent local non-equilibrium macroscopic state variables for the construction of a thermodynamic description calls for qualitative physical understanding, rather than being a simply mathematical problem concerned with a uniquely determined thermodynamic description. A suitable definition of the entropy of a small local region depends on the physically insightful and judicious selection of the independent local non-equilibrium macroscopic state variables, and different selections provide different generalized or extended thermodynamical accounts of one and the same given physical process. This is one of the several good reasons for considering entropy as an epistemic physical variable, rather than as a simply material quantity. According to a respected author: "There is no compelling reason to believe that the classical thermodynamic entropy is a measurable property of nonequilibrium phenomena, ..."[51]

19.3.4   Statistical thermodynamics

Statistical thermodynamics, also called statistical mechanics, emerged with the development of atomic and molecular theories in the second half of the 19th century and early 20th century. It provides an explanation of classical thermodynamics. It considers the microscopic interactions between individual particles and their collective motions, in terms of classical or of quantum mechanics. Its explanation is in terms of statistics that rest on the fact the system is composed of several species of particles or collective motions, the members of each species respectively being in some sense all alike.

19.4   Thermodynamic equilibrium

Equilibrium thermodynamics studies transformations of matter and energy in systems at or near thermodynamic equilibrium. In thermodynamic equilibrium, a system's properties are, by definition, unchanging in time. In thermodynamic equilibrium no macroscopic change is occurring or can be triggered; within the system, every microscopic process is balanced by its opposite; this is called the principle of detailed balance. A central aim in equilibrium thermodynamics is: given a system in a well-defined initial state, subject to specified constraints, to calculate what the equilibrium state of the system is.[52]

In theoretical studies, it is often convenient to consider the simplest kind of thermodynamic system. This is defined variously by different authors.[47][53][54][55][56][57] For the present article, the following definition is convenient, as abstracted from the definitions of various authors. A region of material with all intensive properties continuous in space and time is called a phase. A simple system is for the present article defined as one that consists of a single phase of a pure chemical substance, with no interior partitions.

Within a simple isolated thermodynamic system in thermodynamic equilibrium, in the absence of externally imposed force fields, all properties of the material of the system are spatially homogeneous.[58] Much of the basic theory of thermodynamics is concerned with homogeneous systems in thermodynamic equilibrium.[4][59]

Most systems found in nature or considered in engineering are not in thermodynamic equilibrium, exactly considered. They are changing or can be triggered to change over time, and are continuously and discontinuously subject to flux of matter and energy to and from other systems.[22] For example, according to Callen, "in absolute thermodynamic equilibrium all radioactive materials would have decayed completely and nuclear reactions would have transmuted all nuclei to the most stable isotopes. Such processes, which would take cosmic times to complete, generally can be ignored.".[22] Such processes being ignored, many systems in nature are close enough to thermodynamic equilibrium that for many purposes their behaviour can be well approximated by equilibrium calculations.

19.4.1   Quasi-static transfers between simple systems are nearly in thermodynamic equilibrium and are reversible

It very much eases and simplifies theoretical thermodynamical studies to imagine transfers of energy and matter between two simple systems that proceed so slowly that at all times each simple system considered separately is near enough to thermodynamic equilibrium. Such processes are sometimes called quasi-static and are near enough to being reversible.[60][61]

19.4.2   Natural processes are partly described by tendency towards thermodynamic equilibrium and are irreversible

If not initially in thermodynamic equilibrium, simple isolated thermodynamic systems, as time passes, tend to evolve naturally towards thermodynamic equilibrium. In the absence of externally imposed force fields, they become homogeneous in all their local properties. Such homogeneity is an important characteristic of a system in thermodynamic equilibrium in the absence of externally imposed force fields.

Many thermodynamic processes can be modeled by compound or composite systems, consisting of several or many contiguous component simple systems, initially not in thermodynamic equilibrium, but allowed to transfer mass and energy between them. Natural thermodynamic processes are described in terms of a tendency towards thermodynamic equilibrium within simple systems and in transfers between contiguous simple systems. Such natural processes are irreversible.[62]

19.5 Non-equilibrium thermodynamics

Non-equilibrium thermodynamics[63] is a branch of thermodynamics that deals with systems that are not in thermodynamic equilibrium; it is also called thermodynamics of irreversible processes.

19.6 Laws of thermodynamics

Main article: Laws of thermodynamics

Thermodynamics states a set of four laws that are valid for all systems that fall within the constraints implied by each. In the various theoretical descriptions of thermodynamics these laws may be expressed in seemingly differing forms, but the most prominent formulations are the following:

- Zeroth law of thermodynamics: *If two systems are each in thermal equilibrium with a third, they are also in thermal equilibrium with each other.*

This statement implies that thermal equilibrium is an equivalence relation on the set of thermodynamic systems under consideration. Systems are said to be in thermal equilibrium with each other if spontaneous molecular thermal energy exchanges between them do not lead to a net exchange of energy. This law is tacitly assumed in every measurement of temperature. For two bodies known to be at the same temperature, deciding if they are in thermal equilibrium when put into thermal contact does not require actually bringing them into contact and measuring any changes of their observable properties in time.[64] In traditional statements, the law provides an empirical definition of temperature and justification for the construction of practical ther-

mometers. In contrast to absolute thermodynamic temperatures, empirical temperatures are measured just by the mechanical properties of bodies, such as their volumes, without reliance on the concepts of energy, entropy or the first, second, or third laws of thermodynamics.[55][65] Empirical temperatures lead to calorimetry for heat transfer in terms of the mechanical properties of bodies, without reliance on mechanical concepts of energy.

The physical content of the zeroth law has long been recognized. For example, Rankine in 1853 defined temperature as follows: "Two portions of matter are said to have equal temperatures when neither tends to communicate heat to the other."[66] Maxwell in 1872 stated a "Law of Equal Temperatures".[67] He also stated: "All Heat is of the same kind."[68] Planck explicitly assumed and stated it in its customary present-day wording in his formulation of the first two laws.[69] By the time the desire arose to number it as a law, the other three had already been assigned numbers, and so it was designated the *zeroth law*.

- First law of thermodynamics: *The increase in internal energy of a closed system is equal to the difference of the heat supplied to the system and the work done by the system: $\Delta U = Q - W$* [70][71][72][73][74][75][76][77][78][79] (Note that due to the ambiguity of what constitutes positive work, some sources state that $\Delta U = Q + W$, in which case work done on the system is positive.)

The first law of thermodynamics asserts the existence of a state variable for a system, the internal energy, and tells how it changes in thermodynamic processes. The law allows a given internal energy of a system to be reached by any combination of heat and work. It is important that internal energy is a variable of state of the system (see Thermodynamic state) whereas heat and work are variables that describe processes or changes of the state of systems.

The first law observes that the internal energy of an isolated system obeys the principle of conservation of energy, which states that energy can be transformed (changed from one form to another), but cannot be created or destroyed.[80][81][82][83][84]

- Second law of thermodynamics: *Heat cannot spontaneously flow from a colder location to a hotter location.*

The second law of thermodynamics is an expression of the universal principle of dissipation of kinetic and potential energy observable in nature. The second law is an observation of the fact that over time, differences in temperature, pressure, and chemical potential tend to even out in a physical system that is isolated from the outside world. Entropy is a measure of how much this process has progressed. The entropy of an isolated system that is not in equilibrium tends to increase over time, approaching a maximum value at equilibrium.

In classical thermodynamics, the second law is a basic postulate applicable to any system involving heat energy transfer; in statistical thermodynamics, the second law is a consequence of the assumed randomness of molecular chaos. There are many versions of the second law, but they all have the same effect, which is to explain the phenomenon of irreversibility in nature.

- Third law of thermodynamics: *As a system approaches absolute zero the entropy of the system approaches a minimum value.*

The third law of thermodynamics is a statistical law of nature regarding entropy and the impossibility of reaching absolute zero of temperature. This law provides an absolute reference point for the determination of entropy. The entropy determined relative to this point is the absolute entropy. Alternate definitions of the third law are, "the entropy of all systems and of all states of a system is smallest at absolute zero," or equivalently "it is impossible to reach the absolute zero of temperature by any finite number of processes".

Absolute zero is −273.15 °C (degrees Celsius), −459.67 °F (degrees Fahrenheit), 0 K (kelvin), or 0 R (Rankine).

19.7   System models

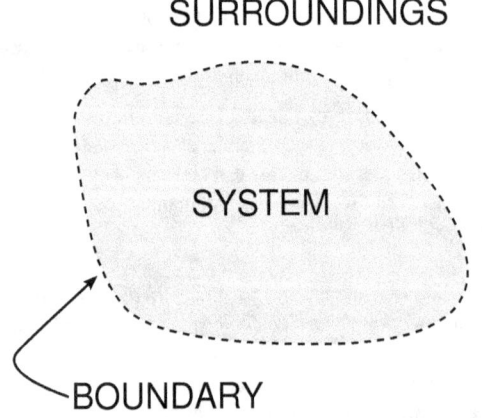

A diagram of a generic thermodynamic system

The thermodynamic system is an important concept of thermodynamics. It is a precisely defined region of the universe under study. Everything in the universe except the system is known as the *surroundings*. A system is separated from the remainder of the universe by a *boundary*, which may be actual, or merely notional and fictive, but by convention delimits a finite volume. Transfers of work, heat, or matter between the system and the surroundings take place across this boundary, which may or may not have properties that restrict what can be transferred across it. A system may have several distinct boundary sectors or partitions separating it from the surroundings, each characterized by how it restricts transfers, and being permeable to its characteristic transferred quantities.

The volume can be the region surrounding a single atom resonating energy, as Max Planck defined in 1900; it can be a body of steam or air in a steam engine, such as Sadi Carnot defined in 1824; it can be the body of a tropical cyclone, as Kerry Emanuel theorized in 1986 in the field of atmospheric thermodynamics; it could also be just one nuclide (i.e. a system of quarks) as hypothesized in quantum thermodynamics.

Anything that passes across the boundary needs to be accounted for in a proper transfer balance equation. Thermodynamics is largely about such transfers.

Boundary sectors are of various characters: rigid, flexible, fixed, moveable, actually restrictive, and fictive or not actually restrictive. For example, in an engine, a fixed boundary sector means the piston is locked at its position; then no pressure-volume work is done across it. In that same engine, a moveable boundary allows the piston to move in and out, permitting pressure-volume work. There is no restrictive boundary sector for the whole earth including its atmosphere, and so roughly speaking, no pressure-volume work is done on or by the whole earth system. Such a system is sometimes said to be diabatically heated or cooled by radiation.[85][86]

Thermodynamics distinguishes classes of systems by their boundary sectors.

- An **open** system has a boundary sector that is permeable to matter; such a sector is usually permeable also to energy, but the energy that passes cannot in general be uniquely sorted into heat and work components. Open system boundaries may be either actually restrictive, or else non-restrictive.

- A **closed** system has no boundary sector that is permeable to matter, but in general its boundary is permeable to energy. For closed systems, boundaries are totally prohibitive of matter transfer.

- An **adiabatically isolated** system has only adiabatic boundary sectors. Energy can be transferred as work, but transfers of matter and of energy as heat are prohibited.

- A **purely diathermically isolated** system has only boundary sectors permeable only to heat; it is sometimes said to be adynamically isolated and closed to matter transfer. A process in which no work is transferred is sometimes called adynamic.[87]

- An **isolated** system has only isolating boundary sectors. Nothing can be transferred into or out of it.

Engineering and natural processes are often described as composites of many different component simple systems, sometimes with unchanging or changing partitions between them. A change of partition is an example of a thermodynamic operation.

19.8 States and processes

There are four fundamental kinds of entity in thermodynamics—states of a system, walls between systems, thermodynamic processes, and thermodynamic operations. This allows three fundamental approaches to thermodynamic reasoning—that in terms of states of thermodynamic equilibrium of a system, and that in terms of time-invariant processes of a system, and that in terms of cyclic processes of a system.

The approach through states of thermodynamic equilibrium of a system requires a full account of the state of the system as well as a notion of process from one state to another of a system, but may require only an idealized or partial account of the state of the surroundings of the system or of other systems.

The method of description in terms of states of thermodynamic equilibrium has limitations. For example, processes in a region of turbulent flow, or in a burning gas mixture, or in a Knudsen gas may be beyond "the province of thermodynamics".[88][89][90] This problem can sometimes be circumvented through the method of description in terms of cyclic or of time-invariant flow processes. This is part of the reason why the founders of thermodynamics often preferred the cyclic process description.

Approaches through processes of time-invariant flow of a system are used for some studies. Some processes, for example Joule-Thomson expansion, are studied through steady-flow experiments, but can be accounted for by distinguishing the steady bulk flow kinetic energy from the internal energy, and thus can be regarded as within the scope of classical thermodynamics defined in terms of equilibrium states or of cyclic processes.[43][91] Other flow processes, for example thermoelectric effects, are essentially defined by the presence of differential flows or diffusion so that they cannot be adequately accounted for in terms of equilibrium states or classical cyclic processes.[92][93]

The notion of a cyclic process does not require a full account of the state of the system, but does require a full account of how the process occasions transfers of matter and energy between the principal system (which is often called the *working body*) and its surroundings, which must include at least two heat reservoirs at different known and fixed temperatures, one hotter than the principal system and the other colder than it, as well as a reservoir that can receive energy from the system as work and can do work on the system. The reservoirs can alternatively be regarded as auxiliary idealized component systems, alongside the principal system. Thus an account in terms of cyclic processes requires at least four contributory component systems. The independent variables of this account are the amounts of energy that enter and leave the idealized auxiliary systems. In this kind of account, the working body is often regarded as a "black box",[94] and its own state is not specified. In this approach, the notion of a properly numerical scale of empirical temperature is a presupposition of thermodynamics, not a notion constructed by or derived from it.

19.8.1 Account in terms of states of thermodynamic equilibrium

When a system is at thermodynamic equilibrium under a given set of conditions of its surroundings, it is said to be in a definite thermodynamic state, which is fully described by its state variables.

If a system is simple as defined above, and is in thermodynamic equilibrium, and is not subject to an externally imposed force field, such as gravity, electricity, or magnetism, then it is homogeneous, that is say, spatially uniform in all respects.[95]

In a sense, a homogeneous system can be regarded as spatially zero-dimensional, because it has no spatial variation.

If a system in thermodynamic equilibrium is homogeneous, then its state can be described by a few physical variables, which are mostly classifiable as intensive variables and extensive variables.[8][33][96][97][98]

An intensive variable is one that is unchanged with the thermodynamic operation of scaling of a system.

An extensive variable is one that simply scales with the scaling of a system, without the further requirement used just below here, of additivity even when there is inhomogeneity of the added systems.

Examples of extensive thermodynamic variables are total mass and total volume. Under the above definition, entropy is also regarded as an extensive variable. Examples of intensive thermodynamic variables are temperature, pressure, and chemical concentration; intensive thermodynamic variables are defined at each spatial point and each instant of time in a system. Physical macroscopic variables can be mechanical, material, or thermal.[33] Temperature is a thermal variable; according to Guggenheim, "the most important conception in thermodynamics is temperature."[8]

Intensive variables have the property that if any number of systems, each in its own separate homogeneous thermodynamic equilibrium state, all with the same respective values of all of their intensive variables, regardless of the values of their extensive variables, are laid contiguously with no partition between them, so as to form a new system, then the values of the intensive variables of the new system are the same as those of the separate constituent systems. Such a composite system is in a homogeneous thermodynamic equilibrium. Examples of intensive variables are temperature, chemical concentration, pressure, density of mass, density of internal energy, and, when it can be properly defined, density of entropy.[99] In other words, intensive variables are not altered by the thermodynamic operation of scaling.

For the immediately present account just below, an alternative definition of extensive variables is considered, that requires that if any number of systems, regardless of their possible separate thermodynamic equilibrium or non-equilibrium states or intensive variables, are laid side by side with no partition between them so as to form a new system, then the values of the extensive variables of the new system are the sums of the values of the respective extensive variables of the individual separate constituent systems. Obviously, there is no reason to expect such a composite system to be in a homogeneous thermodynamic equilibrium. Examples of extensive variables in this alternative definition are mass, volume, and internal energy. They depend on the total quantity of mass in the system.[100] In other words, although extensive variables scale with the system under the thermodynamic operation of scaling, nevertheless the present alternative definition of an extensive variable requires more than this: it requires also its additivity regardless of the inhomogeneity (or equality or inequality of the values of the intensive variables) of the component systems.

Though, when it can be properly defined, density of entropy is an intensive variable, for inhomogeneous systems, entropy itself does not fit into this alternative classification of state variables.[101][102] The reason is that entropy is a property of a system as a whole, and not necessarily related simply to its constituents separately. It is true that for any number of systems each in its own separate homogeneous thermodynamic equilibrium, all with the same values of intensive variables, removal of the partitions between the separate systems results in a composite homogeneous system in thermodynamic equilibrium, with all the values of its intensive variables the same as those of the constituent systems, and it is reservedly or conditionally true that the entropy

of such a restrictively defined composite system is the sum of the entropies of the constituent systems. But if the constituent systems do not satisfy these restrictive conditions, the entropy of a composite system cannot be expected to be the sum of the entropies of the constituent systems, because the entropy is a property of the composite system as a whole. Therefore, though under these restrictive reservations, entropy satisfies some requirements for extensivity defined just above, entropy in general does not fit the immediately present definition of an extensive variable.

Being neither an intensive variable nor an extensive variable according to the immediately present definition, entropy is thus a stand-out variable, because it is a state variable of a system as a whole.[101] A non-equilibrium system can have a very inhomogeneous dynamical structure. This is one reason for distinguishing the study of equilibrium thermodynamics from the study of non-equilibrium thermodynamics.

The physical reason for the existence of extensive variables is the time-invariance of volume in a given inertial reference frame, and the strictly local conservation of mass, momentum, angular momentum, and energy. As noted by Gibbs, entropy is unlike energy and mass, because it is not locally conserved.[101] The stand-out quantity entropy is never conserved in real physical processes; all real physical processes are irreversible.[103] The motion of planets seems reversible on a short time scale (millions of years), but their motion, according to Newton's laws, is mathematically an example of deterministic chaos. Eventually a planet suffers an unpredictable collision with an object from its surroundings, outer space in this case, and consequently its future course is radically unpredictable. Theoretically this can be expressed by saying that every natural process dissipates some information from the predictable part of its activity into the unpredictable part. The predictable part is expressed in the generalized mechanical variables, and the unpredictable part in heat.

Other state variables can be regarded as conditionally 'extensive' subject to reservation as above, but not extensive as defined above. Examples are the Gibbs free energy, the Helmholtz free energy, and the enthalpy. Consequently, just because for some systems under particular conditions of their surroundings such state variables are conditionally conjugate to intensive variables, such conjugacy does not make such state variables extensive as defined above. This is another reason for distinguishing the study of equilibrium thermodynamics from the study of non-equilibrium thermodynamics. In another way of thinking, this explains why heat is to be regarded as a quantity that refers to a process and not to a state of a system.

A system with no internal partitions, and in thermodynamic equilibrium, can be inhomogeneous in the following respect: it can consist of several so-called 'phases', each homogeneous in itself, in immediate contiguity with other phases of the system, but distinguishable by their having various respectively different physical characters, with discontinuity of intensive variables at the boundaries between the phases; a mixture of different chemical species is considered homogeneous for this purpose if it is physically homogeneous.[104] For example, a vessel can contain a system consisting of water vapour overlying liquid water; then there is a vapour phase and a liquid phase, each homogeneous in itself, but still in thermodynamic equilibrium with the other phase. For the immediately present account, systems with multiple phases are not considered, though for many thermodynamic questions, multiphase systems are important.

Equation of state

The macroscopic variables of a thermodynamic system in thermodynamic equilibrium, in which temperature is well defined, can be related to one another through equations of state or characteristic equations.[29][30][31][32] They express the **constitutive** peculiarities of the material of the system. The equation of state must comply with some thermodynamic constraints, but cannot be derived from the general principles of thermodynamics alone.

19.8.2 Thermodynamic processes between states of thermodynamic equilibrium

A thermodynamic process is defined by changes of state internal to the system of interest, combined with transfers of matter and energy to and from the surroundings of the system or to and from other systems. A system is demarcated from its surroundings or from other systems by partitions that more or less separate them, and may move as a piston to change the volume of the system and thus transfer work.

Dependent and independent variables for a process

A process is described by changes in values of state variables of systems or by quantities of exchange of matter and energy between systems and surroundings. The change must be specified in terms of prescribed variables. The choice of which variables are to be used is made in advance of consideration of the course of the process, and cannot be changed. Certain of the variables chosen in advance are called the independent variables.[105] From changes in independent variables may be derived changes in other variables called dependent variables. For example, a process may occur at constant pressure with pressure prescribed as an independent variable, and temperature changed as another independent variable, and then changes in volume are

considered as dependent. Careful attention to this principle is necessary in thermodynamics.[106][107]

Changes of state of a system

In the approach through equilibrium states of the system, a process can be described in two main ways.

In one way, the system is considered to be connected to the surroundings by some kind of more or less separating partition, and allowed to reach equilibrium with the surroundings with that partition in place. Then, while the separative character of the partition is kept unchanged, the conditions of the surroundings are changed, and exert their influence on the system again through the separating partition, or the partition is moved so as to change the volume of the system; and a new equilibrium is reached. For example, a system is allowed to reach equilibrium with a heat bath at one temperature; then the temperature of the heat bath is changed and the system is allowed to reach a new equilibrium; if the partition allows conduction of heat, the new equilibrium is different from the old equilibrium.

In the other way, several systems are connected to one another by various kinds of more or less separating partitions, and to reach equilibrium with each other, with those partitions in place. In this way, one may speak of a 'compound system'. Then one or more partitions is removed or changed in its separative properties or moved, and a new equilibrium is reached. The Joule-Thomson experiment is an example of this; a tube of gas is separated from another tube by a porous partition; the volume available in each of the tubes is determined by respective pistons; equilibrium is established with an initial set of volumes; the volumes are changed and a new equilibrium is established.[108][109][110][111][112] Another example is in separation and mixing of gases, with use of chemically semi-permeable membranes.[113]

Commonly considered thermodynamic processes

It is often convenient to study a thermodynamic process in which a single variable, such as temperature, pressure, or volume, etc., is held fixed. Furthermore, it is useful to group these processes into pairs, in which each variable held constant is one member of a conjugate pair.

Several commonly studied thermodynamic processes are:

- Isobaric process: occurs at constant pressure

- Isochoric process: occurs at constant volume (also called isometric/isovolumetric)

- Isothermal process: occurs at a constant temperature

- Adiabatic process: occurs without loss or gain of energy as heat

- Isentropic process: a reversible adiabatic process occurs at a constant entropy, but is a fictional idealization. Conceptually it is possible to actually physically conduct a process that keeps the entropy of the system constant, allowing systematically controlled removal of heat, by conduction to a cooler body, to compensate for entropy produced within the system by irreversible work done on the system. Such isentropic conduct of a process seems called for when the entropy of the system is considered as an independent variable, as for example when the internal energy is considered as a function of the entropy and volume of the system, the natural variables of the internal energy as studied by Gibbs.

- Isenthalpic process: occurs at a constant enthalpy

- Isolated process: no matter or energy (neither as work nor as heat) is transferred into or out of the system

It is sometimes of interest to study a process in which several variables are controlled, subject to some specified constraint. In a system in which a chemical reaction can occur, for example, in which the pressure and temperature can affect the equilibrium composition, a process might occur in which temperature is held constant but pressure is slowly altered, just so that chemical equilibrium is maintained all the way. There is a corresponding process at constant temperature in which the final pressure is the same but is reached by a rapid jump. Then it can be shown that the volume change resulting from the rapid jump process is smaller than that from the slow equilibrium process.[114] The work transferred differs between the two processes.

19.8.3   Account in terms of cyclic processes

A cyclic process[26] is a process that can be repeated indefinitely often without changing the final state of the system in which the process occurs. The only traces of the effects of a cyclic process are to be found in the surroundings of the system or in other systems. This is the kind of process that concerned early thermodynamicists such as Sadi Carnot, and in terms of which Kelvin defined absolute temperature,[115] before the use of the quantity of entropy by Rankine and its clear identification by Clausius.[116] For some systems, for example with some plastic working substances, cyclic processes are practically nearly unfeasible because the working substance undergoes practically irreversible changes.[117] This is why mechanical devices are lubricated with oil and one of the reasons why electrical devices are often useful.

A cyclic process of a system requires in its surroundings at least two heat reservoirs at different temperatures, one at a higher temperature that supplies heat to the system, the other at a lower temperature that accepts heat from the system. The early work on thermodynamics tended to use the cyclic process approach, because it was interested in machines that converted some of the heat from the surroundings into mechanical power delivered to the surroundings, without too much concern about the internal workings of the machine. Such a machine, while receiving an amount of heat from a higher temperature reservoir, always needs a lower temperature reservoir that accepts some lesser amount of heat. The difference in amounts of heat is equal to the amount of heat converted to work.[82] Later, the internal workings of a system became of interest, and they are described by the states of the system. Nowadays, instead of arguing in terms of cyclic processes, some writers are inclined to derive the concept of absolute temperature from the concept of entropy, a variable of state.

19.9   Instrumentation

There are two types of thermodynamic instruments, the **meter** and the **reservoir**. A thermodynamic meter is any device that measures any parameter of a thermodynamic system. In some cases, the thermodynamic parameter is actually defined in terms of an idealized measuring instrument. For example, the zeroth law states that if two bodies are in thermal equilibrium with a third body, they are also in thermal equilibrium with each other. This principle, as noted by James Maxwell in 1872, asserts that it is possible to measure temperature. An idealized thermometer is a sample of an ideal gas at constant pressure. From the ideal gas law $PV=nRT$, the volume of such a sample can be used as an indicator of temperature; in this manner it defines temperature. Although pressure is defined mechanically, a pressure-measuring device, called a barometer may also be constructed from a sample of an ideal gas held at a constant temperature. A calorimeter is a device that measures and define the internal energy of a system.

A thermodynamic reservoir is a system so large that it does not appreciably alter its state parameters when brought into contact with the test system. It is used to impose a particular value of a state parameter upon the system. For example, a pressure reservoir is a system at a particular pressure, which imposes that pressure upon any test system that it is mechanically connected to. The Earth's atmosphere is often used as a pressure reservoir.

19.10   Conjugate variables

Main article: Conjugate variables

A central concept of thermodynamics is that of energy. By the First Law, the total energy of a system and its surroundings is conserved. Energy may be transferred into a system by heating, compression, or addition of matter, and extracted from a system by cooling, expansion, or extraction of matter. In mechanics, for example, energy transfer equals the product of the force applied to a body and the resulting displacement.

Conjugate variables are pairs of thermodynamic concepts, with the first being akin to a "force" applied to some thermodynamic system, the second being akin to the resulting "displacement," and the product of the two equalling the amount of energy transferred. The common conjugate variables are:

- Pressure-volume (the mechanical parameters);

- Temperature-entropy (thermal parameters);

- Chemical potential-particle number (material parameters).

19.11   Potentials

Thermodynamic potentials are different quantitative measures of the stored energy in a system. Potentials are used to measure energy changes in systems as they evolve from an initial state to a final state. The potential used depends on the constraints of the system, such as constant temperature or pressure. For example, the Helmholtz and Gibbs energies are the energies available in a system to do useful work when the temperature and volume or the pressure and temperature are fixed, respectively.

The five most well known potentials are:

where T is the temperature, S the entropy, p the pressure, V the volume, μ the chemical potential, N the number of particles in the system, and i is the count of particles types in the system.

Thermodynamic potentials can be derived from the energy balance equation applied to a thermodynamic system. Other thermodynamic potentials can also be obtained through Legendre transformation.

19.12   Axiomatics

Most accounts of thermodynamics presuppose the law of conservation of mass, sometimes with,[118] and sometimes without,[119][120] explicit mention. Particular attention is paid to the law in accounts of non-equilibrium thermodynamics.[121][122] One statement of this law is "The total mass of a closed system remains constant."[9] Another statement of it is "In a chemical reaction, matter is neither created nor destroyed."[123] Implied in this is that matter and energy are not considered to be interconverted in such accounts. The full generality of the law of conservation of energy is thus not used in such accounts.

In 1909, Constantin Carathéodory presented[55] a purely mathematical axiomatic formulation, a description often referred to as *geometrical thermodynamics*, and sometimes said to take the "mechanical approach"[78] to thermodynamics. The Carathéodory formulation is restricted to equilibrium thermodynamics and does not attempt to deal with non-equilibrium thermodynamics, forces that act at a distance on the system, or surface tension effects.[124] Moreover, Carathéodory's formulation does not deal with materials like water near 4 °C, which have a density extremum as a function of temperature at constant pressure.[125][126] Carathéodory used the law of conservation of energy as an axiom from which, along with the contents of the zeroth law, and some other assumptions including his own version of the second law, he derived the first law of thermodynamics.[127] Consequently, one might also describe Carathéodory's work as lying in the field of energetics,[128] which is broader than thermodynamics. Carathéodory presupposed the law of conservation of mass without explicit mention of it.

Since the time of Carathéodory, other influential axiomatic formulations of thermodynamics have appeared, which like Carathéodory's, use their own respective axioms, different from the usual statements of the four laws, to derive the four usually stated laws.[129][130][131]

Many axiomatic developments assume the existence of states of thermodynamic equilibrium and of states of thermal equilibrium. States of thermodynamic equilibrium of compound systems allow their component simple systems to exchange heat and matter and to do work on each other on their way to overall joint equilibrium. Thermal equilibrium allows them only to exchange heat. The physical properties of glass depend on its history of being heated and cooled and, strictly speaking, glass is not in thermodynamic equilibrium.[132]

According to Herbert Callen's widely cited 1985 text on thermodynamics: "An essential prerequisite for the measurability of energy is the existence of walls that do not permit transfer of energy in the form of heat.".[133] According to Werner Heisenberg's mature and careful examination of the basic concepts of physics, the theory of heat has a self-standing place.[134]

From the viewpoint of the axiomatist, there are several different ways of thinking about heat, temperature, and the second law of thermodynamics. The Clausius way rests on the empirical fact that heat is conducted always down, never up, a temperature gradient. The Kelvin way is to assert the empirical fact that conversion of heat into work by cyclic processes is never perfectly efficient. A more mathematical way is to assert the existence of a function of state called the entropy that tells whether a hypothesized process occurs spontaneously in nature. A more abstract way is that of Carathéodory that in effect asserts the irreversibility of some adiabatic processes. For these different ways, there are respective corresponding different ways of viewing heat and temperature.

The Clausius–Kelvin–Planck way This way prefers ideas close to the empirical origins of thermodynamics. It presupposes transfer of energy as heat, and empirical temperature as a scalar function of state. According to Gislason and Craig (2005): "Most thermodynamic data come from calorimetry..."[135] According to Kondepudi (2008): "Calorimetry is widely used in present day laboratories."[136] In this approach, what is often currently called the zeroth law of thermodynamics is deduced as a simple consequence of the presupposition of the nature of heat and empirical temperature, but it is not named as a numbered law of thermodynamics. Planck attributed this point of view to Clausius, Kelvin, and Maxwell. Planck wrote (on page 90 of the seventh edition, dated 1922, of his treatise) that he thought that no proof of the second law of thermodynamics could ever work that was not based on the impossibility of a perpetual motion machine of the second kind. In that treatise, Planck makes no mention of the 1909 Carathéodory way, which was well known by 1922. Planck for himself chose a version of what is just above called the Kelvin way.[137] The development by Truesdell and Bharatha (1977) is so constructed that it can deal naturally with cases like that of water near 4 °C.[130]

The way that assumes the existence of entropy as a function of state This way also presupposes transfer of energy as heat, and it presupposes the usually stated form of the zeroth law of thermodynamics, and from these two it deduces the existence of empirical temperature. Then from the existence of entropy it deduces the existence of absolute thermodynamic temperature.[8][129]

The Carathéodory way This way presupposes that the state of a simple one-phase system is fully specifiable by just one more state variable than the known exhaustive list of mechanical variables of state. It does not explicitly name empirical temperature, but speaks of the one-

dimensional "non-deformation coordinate". This satisfies the definition of an empirical temperature, that lies on a one-dimensional manifold. The Carathéodory way needs to assume moreover that the one-dimensional manifold has a definite sense, which determines the direction of irreversible adiabatic process, which is effectively assuming that heat is conducted from hot to cold. This way presupposes the often currently stated version of the zeroth law, but does not actually name it as one of its axioms.[124] According to one author, Carathéodory's principle, which is his version of the second law of thermodynamics, does not imply the increase of entropy when work is done under adiabatic conditions (as was noted by Planck[138]). Thus Carathéodory's way leaves unstated a further empirical fact that is needed for a full expression of the second law of thermodynamics.[139]

19.13   Scope of thermodynamics

Originally thermodynamics concerned material and radiative phenomena that are experimentally reproducible. For example, a state of thermodynamic equilibrium is a steady state reached after a system has aged so that it no longer changes with the passage of time. But more than that, for thermodynamics, a system, defined by its being prepared in a certain way must, consequent on every particular occasion of preparation, upon aging, reach one and the same eventual state of thermodynamic equilibrium, entirely determined by the way of preparation. Such reproducibility is because the systems consist of so many molecules that the molecular variations between particular occasions of preparation have negligible or scarcely discernable effects on the macroscopic variables that are used in thermodynamic descriptions. This led to Boltzmann's discovery that entropy had a statistical or probabilistic nature. Probabilistic and statistical explanations arise from the experimental reproducibility of the phenomena.[140]

Gradually, the laws of thermodynamics came to be used to explain phenomena that occur outside the experimental laboratory. For example, phenomena on the scale of the earth's atmosphere cannot be reproduced in a laboratory experiment. But processes in the atmosphere can be modeled by use of thermodynamic ideas, extended well beyond the scope of laboratory equilibrium thermodynamics.[141][142][143] A parcel of air can, near enough for many studies, be considered as a closed thermodynamic system, one that is allowed to move over significant distances. The pressure exerted by the surrounding air on the lower face of a parcel of air may differ from that on its upper face. If this results in rising of the parcel of air, it can be considered to have gained potential energy as a result of work being done on it by the combined surrounding air below and above it. As it rises, such a parcel usually expands because the pressure is lower at the higher altitudes that it reaches. In that way, the rising parcel also does work on the surrounding atmosphere. For many studies, such a parcel can be considered nearly to neither gain nor lose energy by heat conduction to its surrounding atmosphere, and its rise is rapid enough to leave negligible time for it to gain or lose heat by radiation; consequently the rising of the parcel is near enough adiabatic. Thus the adiabatic gas law accounts for its internal state variables, provided that there is no precipitation into water droplets, no evaporation of water droplets, and no sublimation in the process. More precisely, the rising of the parcel is likely to occasion friction and turbulence, so that some potential and some kinetic energy of bulk converts into internal energy of air considered as effectively stationary. Friction and turbulence thus oppose the rising of the parcel.[144][145]

19.14   Applied fields

- Atmospheric thermodynamics
- Biological thermodynamics
- Black hole thermodynamics
- Chemical thermodynamics
- Equilibrium thermodynamics
- Geology
- Industrial ecology (re: Exergy)
- Maximum entropy thermodynamics
- Non-equilibrium thermodynamics
- Philosophy of thermal and statistical physics
- Psychrometrics
- Quantum thermodynamics
- Statistical thermodynamics
- Thermoeconomics

19.15   See also

Entropy production

19.15.1   Lists and timelines

- List of important publications in thermodynamics
- List of textbooks in statistical mechanics
- List of thermal conductivities
- List of thermodynamic properties
- Table of thermodynamic equations
- Timeline of thermodynamics

19.15.2   Wikibooks

- Engineering Thermodynamics
- Entropy for Beginners

19.16   References

[1] Clausius, Rudolf (1850). *On the Motive Power of Heat, and on the Laws which can be deduced from it for the Theory of Heat.* Poggendorff's *Annalen der Physik*, LXXIX (Dover Reprint). ISBN 0-486-59065-8.

[2] Thomson, W. (1854). "On the Dynamical Theory of Heat. Part V. Thermo-electric Currents". *Transactions of the Royal Society of Edinburgh* **21** (part I): 123. doi:10.1017/s0080456800032014. reprinted in Sir William Thomson, LL.D. D.C.L., F.R.S. (1882). *Mathematical and Physical Papers* **1**. London, Cambridge: C.J. Clay, M.A. & Son, Cambridge University Press. p. 232. Hence Thermodynamics falls naturally into two divisions, of which the subjects are respectively, *the relation of heat to the forces acting between contiguous parts of bodies, and the relation of heat to electrical agency.*

[3] Hess, H. (1840). Thermochemische Untersuchungen, *Annalen der Physik und Chemie* (Poggendorff, Leipzig) **126**(6): 385–404.

[4] Gibbs, Willard, J. (1876). *Transactions of the Connecticut Academy*, III, pp. 108–248, Oct. 1875 – May 1876, and pp. 343–524, May 1877 – July 1878.

[5] Duhem, P.M.M. (1886). *Le Potential Thermodynamique et ses Applications*, Hermann, Paris.

[6] Lewis, Gilbert N.; Randall, Merle (1923). *Thermodynamics and the Free Energy of Chemical Substances.* McGraw-Hill Book Co. Inc.

[7] Guggenheim, E.A. (1933). *Modern Thermodynamics by the Methods of J.W. Gibbs*, Methuen, London.

[8] Guggenheim, E.A. (1949/1967)

[9] Ilya Prigogine, I. & Defay, R., translated by D.H. Everett (1954). *Chemical Thermodynamics*. Longmans, Green & Co., London. Includes classical non-equilibrium thermodynamics.

[10] Enrico Fermi (1956). *Thermodynamics*. Courier Dover Publications. p. ix. ISBN 0-486-60361-X. OCLC 230763036 54033021.

[11] Perrot, Pierre (1998). *A to Z of Thermodynamics*. Oxford University Press. ISBN 0-19-856552-6. OCLC 123283342 38073404.

[12] Clark, John, O.E. (2004). *The Essential Dictionary of Science*. Barnes & Noble Books. ISBN 0-7607-4616-8. OCLC 58732844 63473130.

[13] Bridgman, P.W. (1943). *The Nature of Thermodynamics*, Harvard University Press, Cambridge MA, p. 48.

[14] Partington, J.R. (1949), page 118.

[15] Reif, F. (1965). *Fundamentals of Statistical and Thermal Physics*, McGraw-Hill Book Company, New York, page 122.

[16] Fowler, R., Guggenheim, E.A. (1939), p. 3.

[17] Tisza, L. (1966), p. 18.

[18] Adkins, C.J. (1968/1983), p. 4.

[19] Born, M. (1949), p. 44.

[20] Guggenheim, E.A. (1949/1967), pp. 7–8.

[21] Tisza, L. (1966), pp. 109, 112.

[22] Callen, H.B. (1960/1985), pp. 15, 17.

[23] Bailyn, M. (1994), p. 21.

[24] Callen, H.B. (1960/1985), p. 427.

[25] Tisza, L. (1966), pp. 41, 109, 121, originally published as 'The thermodynamics of phase equilibrium', *Annals of Physics*, **13**: 1–92.

[26] Serrin, J. (1986). Chapter 1, 'An Outline of Thermodynamical Structure', pp. 3–32, especially p. 8, in Serrin, J. (1986).

[27] Fowler, R., Guggenheim, E.A. (1939), p. 13.

[28] Tisza, L. (1966), pp. 79–80.

[29] Planck, M. 1923/1926, page 5.

[30] Partington, p. 121.

[31] Adkins, pp. 19–20.

[32] Haase, R. (1971), pages 11–16.

[33] Balescu, R. (1975). *Equilibrium and Nonequilibrium Statistical Mechanics*, Wiley-Interscience, New York, ISBN 0-471-04600-0.

[34] Schrödinger, E. (1946/1967). *Statistical Thermodynamics. A Course of Seminar Lectures*, Cambridge University Press, Cambridge UK.

[35] Partington, J.R. (1989). *A Short History of Chemistry*. Dover. OCLC 19353301.

[36] The Newcomen engine was improved from 1711 until Watt's work, making the efficiency comparison subject to qualification, but the increase from the Newcomen 1765 version was on the order of 100%.

[37] Cengel, Yunus A.; Boles, Michael A. (2005). *Thermodynamics – an Engineering Approach*. McGraw-Hill. ISBN 0-07-310768-9.

[38] Gibbs, Willard (1993). *The Scientific Papers of J. Willard Gibbs, Volume One: Thermodynamics*. Ox Bow Press. ISBN 0-918024-77-3. OCLC 27974820.

[39] *Oxford English Dictionary*, Oxford University Press, Oxford UK.

[40] Donald T. Haynie (2001). *Biological Thermodynamics* (2 ed.). Cambridge University Press. p. 22.

[41] Thomson, W. (1849). "An Account of Carnot's Theory of the Motive Power of Heat; with Numerical Results deduced from Regnault's Experiments on Steam". *Transactions of the Royal Society of Edinburgh* **16** (part V): 541–574. doi:10.1017/s0080456800022481.

[42] Rankine, William (1859). "3: Principles of Thermodynamics". *A Manual of the Steam Engine and other Prime Movers*. London: Charles Griffin and Co. pp. 299–448.

[43] Pippard, A.B. (1957), p. 70.

[44] Partington, J.R. (1949), p. 615–621.

[45] Serrin, J. (1986). An outline of thermodynamical structure, Chapter 1, pp. 3–32 in Serrin, J. (1986).

[46] Callen, H.B. (1960/1985), Chapter 6, pages 131–152.

[47] Callen, H.B. (1960/1985), p. 13.

[48] Landsberg, P.T. (1978). *Thermodynamics and Statistical Mechanics*, Oxford University Press, Oxford UK, ISBN 0-19-851142-6, p. 1.

[49] Eu, B.C. (2002).

[50] Lebon, G., Jou, D., Casas-Vázquez, J. (2008).

[51] Grandy, W.T., Jr (2008), *passim* and p. 123.

[52] Callen, H.B. (1985), p. 26.

[53] Gibbs J.W. (1875), pp. 115–116.

[54] Bryan, G.H. (1907), p. 5.

[55] C. Carathéodory (1909). "Untersuchungen über die Grundlagen der Thermodynamik". *Mathematische Annalen* **67**: 355–386. A partly reliable translation is to be found at Kestin, J. (1976). *The Second Law of Thermodynamics*, Dowden, Hutchinson & Ross, Stroudsburg PA. doi:10.1007/BF01450409.

[56] Haase, R. (1971), p. 13.

[57] Bailyn, M. (1994), p. 145.

[58] Bailyn, M. (1994), Section 6.11.

[59] Planck, M. (1897/1903), passim.

[60] Partington, J.R. (1949), p. 129.

[61] Callen, H.B. (1960/1985), Section 4–2.

[62] Guggenheim, E.A. (1949/1967), §1.12.

[63] de Groot, S.R., Mazur, P., *Non-equilibrium thermodynamics*,1969, North-Holland Publishing Company, Amsterdam-London

[64] Moran, Michael J. and Howard N. Shapiro, 2008. *Fundamentals of Engineering Thermodynamics*. 6th ed. Wiley and Sons: 16.

[65] Planck, M. (1897/1903), p. 1.

[66] Rankine, W.J.M. (1953). *Proc. Roy. Soc. (Edin.)*, **20**(4).

[67] Maxwell, J.C. (1872), page 32.

[68] Maxwell, J.C. (1872), page 57.

[69] Planck, M. (1897/1903), pp. 1–2.

[70] Clausius, R. (1850). Ueber de bewegende Kraft der Wärme und die Gesetze, welche sich daraus für de Wärmelehre selbst ableiten lassen, *Annalen der Physik und Chemie*, **155** (3): 368–394.

[71] Rankine, W.J.M. (1850). On the mechanical action of heat, especially in gases and vapours. *Trans. Roy. Soc. Edinburgh*, **20**: 147–190.

[72] Helmholtz, H. von. (1897/1903). *Vorlesungen über Theorie der Wärme*, edited by F. Richarz, Press of Johann Ambrosius Barth, Leipzig, Section 46, pp. 176–182, in German.

[73] Planck, M. (1897/1903), p. 43.

[74] Guggenheim, E.A. (1949/1967), p. 10.

[75] Sommerfeld, A. (1952/1956), Section 4 A, pp. 13–16.

[76] Ilya Prigogine, I. & Defay, R., translated by D.H. Everett (1954). *Chemical Thermodynamics*. Longmans, Green & Co., London, p. 21.

[77] Lewis, G.N., Randall, M. (1961). *Thermodynamics*, second edition revised by K.S. Pitzer and L. Brewer, McGraw-Hill, New York, p. 35.

[78] Bailyn, M. (1994), page 79.

[79] Khanna, F.C., Malbouisson, A.P.C., Malbouisson, J.M.C., Santana, A.E. (2009). *Thermal Quantum Field Theory. Algebraic Aspects and Applications*, World Scientific, Singapore, ISBN 978-981-281-887-4, p. 6.

[80] Helmholtz, H. von, (1847). *Ueber die Erhaltung der Kraft*, G. Reimer, Berlin.

[81] Joule, J.P. (1847). On matter, living force, and heat, *Manchester Courier*, May 5 and May 12, 1847.

[82] Truesdell, C.A. (1980).

[83] Partington, J.R. (1949), page 150.

[84] Kondepudi & Prigogine (1998), pp. 31–32.

[85] Goody, R.M., Yung, Y.L. (1989). *Atmospheric Radiation. Theoretical Basis*, second edition, Oxford University Press, Oxford UK, ISBN 0-19-505134-3, p. 5

[86] Wallace, J.M., Hobbs, P.V. (2006). *Atmospheric Science. An Introductory Survey*, second edition, Elsevier, Amsterdam, ISBN 978-0-12-732951-2, p. 292.

[87] Partington, J.R. (1913). *A Text-book of Thermodynamics*, Van Nostrand, New York, page 37.

[88] Glansdorff, P., Prigogine, I., (1971). *Thermodynamic Theory of Structure, Stability and Fluctuations*, Wiley-Interscience, London, ISBN 0-471-30280-5, page 15.

[89] Haase, R., (1971), page 16.

[90] Eu, B.C. (2002), p. 13.

[91] Adkins, C.J. (1968/1975), pp. 46–49.

[92] Adkins, C.J. (1968/1975), p. 172.

[93] Lebon, G., Jou, D., Casas-Vázquez, J. (2008), pp. 37–38.

[94] Buchdahl, H.A. (1966). *The Concepts of Classical Thermodynamics*, Cambridge University Press, London, pp. 117–118.

[95] Guggenheim, E.A. (1949/1967), p. 6.

[96] Balescu, R. (1975). *Equilibrium and Non-equilibrium Statistical Mechanics*, Wiley-Interscience, New York, ISBN 0-471-04600-0, Section 3.2, pp. 64–72.

[97] Ilya Prigogine, I. & Defay, R., translated by D.H. Everett (1954). *Chemical Thermodynamics*. Longmans, Green & Co., London. pp. 1–6.

[98] Lavenda, B.H. (1978). *Thermodynamics of Irreversible Processes*, Macmillan, London, ISBN 0-333-21616-4, p. 12.

[99] Guggenheim, E.A. (1949/1967), p. 19.

[100] Guggenheim, E.A. (1949/1967), pp. 18–19.

[101] Grandy, W.T., Jr (2008), Chapter 5, pp. 59–68.

[102] Kondepudi & Prigogine (1998), pp. 116–118.

[103] Guggenheim, E.A. (1949/1967), Section 1.12, pp. 12–13.

[104] Planck, M. (1897/1903), p. 65.

[105] Planck, M. (1923/1926), Section 152A, pp. 121–123.

[106] Prigogine, I. Defay, R. (1950/1954). *Chemical Thermodynamics*, Longmans, Green & Co., London, p. 1.

[107] Adkins, pp. 43–46.

[108] Planck, M. (1897/1903), Section 70, pp. 48–50.

[109] Guggenheim, E.A. (1949/1967), Section 3.11, pp. 92–92.

[110] Sommerfeld, A. (1952/1956), Section 1.5 C, pp. 23–25.

[111] Callen, H.B. (1960/1985), Section 6.3.

[112] Adkins, pp. 164–168.

[113] Planck, M. (1897/1903), Section 236, pp. 211–212.

[114] Ilya Prigogine, I. & Defay, R., translated by D.H. Everett (1954). *Chemical Thermodynamics*. Longmans, Green & Co., London, Chapters 18–19.

[115] Truesdell, C.A. (1980), Section 11B, pp. 306–310.

[116] Truesdell, C.A. (1980), Sections 8G,8H, 9A, pp. 207–224.

[117] Ziegler, H., (1983). *An Introduction to Thermomechanics*, North-Holland, Amsterdam, ISBN 0-444-86503-9

[118] Ziegler, H. (1977). *An Introduction to Thermomechanics*, North-Holland, Amsterdam, ISBN 0-7204-0432-0.

[119] Planck M. (1922/1927).

[120] Guggenheim, E.A. (1949/1967).

[121] de Groot, S.R., Mazur, P. (1962). *Non-equilibrium Thermodynamics*, North Holland, Amsterdam.

[122] Gyarmati, I. (1970). *Non-equilibrium Thermodynamics*, translated into English by E. Gyarmati and W.F. Heinz, Springer, New York.

[123] Tro, N.J. (2008). *Chemistry. A Molecular Approach*, Pearson Prentice-Hall, Upper Saddle River NJ, ISBN 0-13-100065-9.

[124] Turner, L.A. (1962). Simplification of Carathéodory's treatment of thermodynamics, *Am. J. Phys. 30: 781–786.*

[125] Turner, L.A. (1962). Further remarks on the zeroth law, *Am. J. Phys.* **30**: 804–806.

[126] Thomsen, J.S., Hartka, T.J., (1962). Strange Carnot cycles; thermodynamics of a system with a density maximum, *Am. J. Phys.* **30**: 26–33, **30**: 388–389.

[127] C. Carathéodory (1909). "Untersuchungen über die Grundlagen der Thermodynamik". *Mathematische Annalen* **67**: 363. doi:10.1007/bf01450409. Axiom II: In jeder beliebigen Umgebung eines willkürlich vorgeschriebenen Anfangszustandes gibt es Zustände, die durch adiabatische Zustandsänderungen nicht beliebig approximiert werden können.

[128] Duhem, P. (1911). *Traité d'Energetique*, Gautier-Villars, Paris.

[129] Callen, H.B. (1960/1985).

[130] Truesdell, C., Bharatha, S. (1977). *The Concepts and Logic of Classical Thermodynamics as a Theory of Heat Engines, Rigorously Constructed upon the Foundation Laid by S. Carnot and F. Reech*, Springer, New York, ISBN 0-387-07971-8.

[131] Wright, P.G. (1980). Conceptually distinct types of thermodynamics, *Eur. J. Phys.* **1**: 81–84.

[132] Callen, H.B. (1960/1985), p. 14.

[133] Callen, H.B. (1960/1985), p. 16.

[134] Heisenberg, W. (1958). *Physics and Philosophy*, Harper & Row, New York, pp. 98–99.

[135] Gislason, E.A., Craig, N.C. (2005). Cementing the foundations of thermodynamics:comparison of system-based and surroundings-based definitions of work and heat, *J. Chem. Thermodynamics* **37**: 954–966.

[136] Kondepudi, D. (2008). *Introduction to Modern Thermodynamics*, Wiley, Chichester, ISBN 978-0-470-01598-8, p. 63.

[137] Planck, M. (1922/1927).

[138] Planck, M. (1926). Über die Begründung des zweiten Hauptsatzes der Thermodynamik, *Sitzungsberichte der Preußischen Akademie der Wissenschaften, physikalisch-mathematischen Klasse*, pp. 453–463.

[139] Münster, A. (1970). *Classical Thermodynamics*, translated by E.S. Halberstadt, Wiley–Interscience, London, ISBN 0-471-62430-6, p 41.

[140] Grandy, W.T., Jr (2008). *Entropy and the Time Evolution of Macroscopic Systems*, Oxford University Press, Oxford UK, ISBN 978-0-19-954617-6. p. 49.

[141] Iribarne, J.V., Godson, W.L. (1973/1989). *Atmospheric thermodynamics*, second edition, reprinted 1989, Kluwer Academic Publishers, Dordrecht, ISBN 90-277-1296-4.

[142] Peixoto, J.P., Oort, A.H. (1992). *Physics of climate*, American Institute of Physics, New York, ISBN 0-88318-712-4

[143] North, G.R., Erukhimova, T.L. (2009). *Atmospheric Thermodynamics. Elementary Physics and Chemistry*, Cambridge University Press, Cambridge UK, ISBN 978-0-521-89963-5.

[144] Holton, J.R. (2004). *An Introduction of Dynamic Meteorology*, fourth edition, Elsevier, Amsterdam, ISBN 978-0-12-354015-7.

[145] Mak, M. (2011). *Atmospheric Dynamics*, Cambridge University Press, Cambridge UK, ISBN 978-0-521-19573-7.

19.17   Cited bibliography

- Adkins, C.J. (1968/1975). *Equilibrium Thermodynamics*, second edition, McGraw-Hill, London, ISBN 0-07-084057-1.

- Bailyn, M. (1994). *A Survey of Thermodynamics*, American Institute of Physics Press, New York, ISBN 0-88318-797-3.

- Born, M. (1949). *Natural Philosophy of Cause and Chance*, Oxford University Press, London.

- Bryan, G.H. (1907). *Thermodynamics. An Introductory Treatise dealing mainly with First Principles and their Direct Applications*, B.G. Teubner, Leipzig.

- Callen, H.B. (1960/1985). *Thermodynamics and an Introduction to Thermostatistics*, (1st edition 1960) 2nd edition 1985, Wiley, New York, ISBN 0-471-86256-8.

- Eu, B.C. (2002). *Generalized Thermodynamics. The Thermodynamics of Irreversible Processes and Generalized Hydrodynamics*, Kluwer Academic Publishers, Dordrecht, ISBN 1-4020-0788-4.

- Fowler, R., Guggenheim, E.A. (1939). *Statistical Thermodynamics*, Cambridge University Press, Cambridge UK.

- Gibbs, J.W. (1875). On the equilibrium of heterogeneous substances, *Transactions of the Connecticut Academy of Arts and Sciences*, **3**: 108–248.

- Grandy, W.T., Jr (2008). *Entropy and the Time Evolution of Macroscopic Systems*, Oxford University Press, Oxford, ISBN 978-0-19-954617-6.

- Guggenheim, E.A. (1949/1967). *Thermodynamics. An Advanced Treatment for Chemists and Physicists*, (1st edition 1949) 5th edition 1967, North-Holland, Amsterdam.

- Haase, R. (1971). Survey of Fundamental Laws, chapter 1 of *Thermodynamics*, pages 1–97 of volume 1, ed. W. Jost, of *Physical Chemistry. An Advanced Treatise*, ed. H. Eyring, D. Henderson, W. Jost, Academic Press, New York, lcn 73–117081.

- Kondepudi, D., Prigogine, I. (1998). *Modern Thermodynamics. From Heat Engines to Dissipative Structures*, John Wiley & Sons, ISBN 0-471-97393-9.

- Lebon, G., Jou, D., Casas-Vázquez, J. (2008). *Understanding Non-equilibrium Thermodynamics*, Springer, Berlin, ISBN 978-3-540-74251-7.

- Partington, J.R. (1949). *An Advanced Treatise on Physical Chemistry*, volume 1, *Fundamental Principles. The Properties of Gases*, Longmans, Green and Co., London.

- Pippard, A.B. (1957). *The Elements of Classical Thermodynamics*, Cambridge University Press.

- Planck, M.(1897/1903). *Treatise on Thermodynamics*, translated by A. Ogg, Longmans, Green & Co., London.

- Planck, M. (1923/1926). *Treatise on Thermodynamics*, third English edition translated by A. Ogg from the seventh German edition, Longmans, Green & Co., London.

- Serrin, J. (1986). *New Perspectives in Thermodynamics*, edited by J. Serrin, Springer, Berlin, ISBN 3-540-15931-2.

- Sommerfeld, A. (1952/1956). *Thermodynamics and Statistical Mechanics*, Academic Press, New York.

- Tschoegl, N.W. (2000). *Fundamentals of Equilibrium and Steady-State Thermodynamics*, Elsevier, Amsterdam, ISBN 0-444-50426-5.

- Tisza, L. (1966). *Generalized Thermodynamics*, M.I.T Press, Cambridge MA.

- Truesdell, C.A. (1980). *The Tragicomical History of Thermodynamics, 1822–1854*, Springer, New York, ISBN 0-387-90403-4.

19.18   Further reading

- Goldstein, Martin, and Inge F. (1993). *The Refrigerator and the Universe*. Harvard University Press. ISBN 0-674-75325-9. OCLC 32826343. A nontechnical introduction, good on historical and interpretive matters.

- Kazakov, Andrei (July–August 2008). "Web Thermo Tables – an On-Line Version of the TRC Thermodynamic Tables" (PDF). *Journal of Research of the National Institutes of Standards and Technology* **113** (4): 209–220. doi:10.6028/jres.113.016.

The following titles are more technical:

- Cengel, Yunus A., & Boles, Michael A. (2002). *Thermodynamics – an Engineering Approach*. Mc-Graw Hill. ISBN 0-07-238332-1. OCLC 45791449 52263994 57548906.

- Fermi, E. (1956). *Thermodynamics*, Dover, New York.

- Kittel, Charles & Kroemer, Herbert (1980). *Thermal Physics*. W. H. Freeman Company. ISBN 0-7167-1088-9. OCLC 32932988 48236639 5171399.

19.19   External links

- Thermodynamics Data & Property Calculation Websites

- Thermodynamics OpenCourseWare from the University of Notre Dame

- Thermodynamics at *ScienceWorld*

- Biochemistry Thermodynamics

- Engineering Thermodynamics – A Graphical Approach

Chapter 20

Physical chemistry

Physical chemistry is the study of macroscopic, atomic, subatomic, and particulate phenomena in chemical systems in terms of laws and concepts of physics. It applies the principles, practices and concepts of physics such as motion, energy, force, time, thermodynamics, quantum chemistry, statistical mechanics and dynamics, equilibrium.

Physical chemistry, in contrast to chemical physics, is predominantly (but not always) a macroscopic or supramolecular science, as the majority of the principles on which physical chemistry was founded are concepts related to the bulk rather than on molecular/atomic structure alone. For example, chemical equilibrium, and colloids.

Some of the relationships that physical chemistry strives to resolve include the effects of:

1. Intermolecular forces that act upon the physical properties of materials (plasticity, tensile strength, surface tension in liquids).

2. Reaction kinetics on the rate of a reaction.

3. The identity of ions and the electrical conductivity of materials.

4. Surface chemistry and electrochemistry of membranes.[1]

5. Interaction of one body with another in terms of quantities of heat and work called thermodynamics.

6. Transfer of heat between a chemical system and its surroundings during change of phase or chemical reaction taking place called thermochemistry

7. Study of colligative properties of number of species present in solution.

8. Number of phases, number of components and degree of freedom (or variance) can be correlated with one another with help of phase rule.

9. Reactions of electrochemical cells.

20.1 Key concepts

The key concepts of physical chemistry are the ways in which pure physics is applied to chemical problems.

One of the key concepts in classical chemistry is that all chemical compounds can be described as groups of atoms bonded together and chemical reactions can be described as the making and breaking of those bonds. Predicting the properties of chemical compounds from a description of atoms and how they bond is one of the major goals of physical chemistry. To describe the atoms and bonds precisely, it is necessary to know both where the nuclei of the atoms are, and how electrons are distributed around them.[2]
Quantum chemistry, a subfield of physical chemistry especially concerned with the application of quantum mechanics to chemical problems, provides tools to determine how strong and what shape bonds are,[2] how nuclei move, and how light can be absorbed or emitted by a chemical compound.[3] Spectroscopy is the related sub-discipline of physical chemistry which is specifically concerned with the interaction of electromagnetic radiation with matter.

Another set of important questions in chemistry concerns what kind of reactions can happen spontaneously and which properties are possible for a given chemical mixture. This is studied in chemical thermodynamics, which sets limits on quantities like how far a reaction can proceed, or how much energy can be converted into work in an internal combustion engine, and which provides links between properties like the thermal expansion coefficient and rate of change of entropy with pressure for a gas or a liquid.[4] It can frequently be used to assess whether a reactor or engine design is feasible, or to check the validity of experimental data. To a limited extent, quasi-equilibrium and non-equilibrium thermodynamics can describe irreversible changes.[5] However, classical thermodynamics is mostly concerned with systems in equilibrium and reversible changes and not what actually does happen, or how fast, away from equilibrium.

Which reactions do occur and how fast is the subject of chemical kinetics, another branch of physical chemistry. A key idea in chemical kinetics is that for reactants to react and form products, most chemical species must go through transition states which are higher in energy than either the reactants or the products and serve as a barrier to reaction.[6] In general, the higher the barrier, the slower the reaction. A second is that most chemical reactions occur as a sequence of elementary reactions,[7] each with its own transition state. Key questions in kinetics include how the rate of reaction depends on temperature and on the concentrations of reactants and catalysts in the reaction mixture, as well as how catalysts and reaction conditions can be engineered to optimize the reaction rate.

The fact that how fast reactions occur can often be specified with just a few concentrations and a temperature, instead of needing to know all the positions and speeds of every molecule in a mixture, is a special case of another key concept in physical chemistry, which is that to the extent an engineer needs to know, everything going on in a mixture of very large numbers (perhaps of the order of the Avogadro constant, 6×10^{23}) of particles can often be described by just a few variables like pressure, temperature, and concentration. The precise reasons for this are described in statistical mechanics,[8] a specialty within physical chemistry which is also shared with physics. Statistical mechanics also provides ways to predict the properties we see in everyday life from molecular properties without relying on empirical correlations based on chemical similarities.[5]

20.2   History

See also: History of chemistry

The term "physical chemistry" was coined by Mikhail

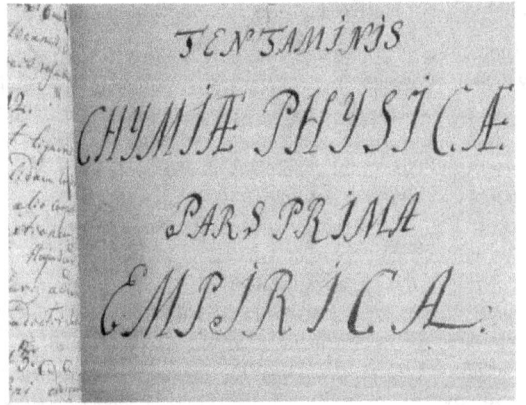

Fragment of M. Lomonosov's manuscript 'Physical Chemistry' (1752)

Lomonosov in 1752, when he presented a lecture course entitled "A Course in True Physical Chemistry" (Russian:

«Курс истинной физической химии») before the students of Petersburg University.[9] In the preamble to these lectures he gives definition: "Physical chemistry is the science that must explain under provisions of physical experiments the reason for what is happening in complex bodies through chemical operations".

Modern physical chemistry originated in the 1860s to 1880s with work on chemical thermodynamics, electrolytes in solutions, chemical kinetics and other subjects. One milestone was the publication in 1876 by Josiah Willard Gibbs of his paper, *On the Equilibrium of Heterogeneous Substances*. This paper introduced several of the cornerstones of physical chemistry, such as Gibbs energy, chemical potentials, Gibbs phase rule.[10] Other milestones include the subsequent naming and accreditation of enthalpy to Heike Kamerlingh Onnes and to macromolecular processes.

The first scientific journal specifically in the field of physical chemistry was the German journal, *Zeitschrift für Physikalische Chemie*, founded in 1887 by Wilhelm Ostwald and Jacobus Henricus van 't Hoff. Together with Svante August Arrhenius,[11] these were the leading figures in physical chemistry in the late 19th century and early 20th century. All three were awarded with the Nobel Prize in Chemistry between 1901-1909.

Developments in the following decades include the application of statistical mechanics to chemical systems and work on colloids and surface chemistry, where Irving Langmuir made many contributions. Another important step was the development of quantum mechanics into quantum chemistry from the 1930s, where Linus Pauling was one of the leading names. Theoretical developments have gone hand in hand with developments in experimental methods, where the use of different forms of spectroscopy, such as infrared spectroscopy, microwave spectroscopy, EPR spectroscopy and NMR spectroscopy, is probably the most important 20th century development.

Further development in physical chemistry may be attributed to discoveries in nuclear chemistry, especially in isotope separation (before and during World War II), more recent discoveries in astrochemistry,[12] as well as the development of calculation algorithms in the field of "additive physicochemical properties" (practically all physicochemical properties, such as boiling point, critical point, surface tension, vapor pressure, etc. - more than 20 in all - can be precisely calculated from chemical structure alone, even if the chemical molecule remains unsynthesized), and in this area is concentrated practical importance of contemporary physical chemistry.

See Group contribution method, Lydersen method, Joback method, Benson group increment theory, QSPR, QSAR

20.3 Journals

Main category: Physical chemistry journals

Some journals that deal with physical chemistry include:

- Zeitschrift für Physikalische Chemie (1887)
- Journal of Physical Chemistry A (from 1896 as *Journal of Physical Chemistry*, renamed in 1997)
- Physical Chemistry Chemical Physics (from 1999, formerly Faraday Transactions with a history dating back to 1905)
- Macromolecular Chemistry and Physics (1947)
- Annual Review of Physical Chemistry (1950)
- Molecular Physics (1957)
- Journal of Physical Organic Chemistry (1988)
- Journal of Physical Chemistry B (1997)
- ChemPhysChem (2000)
- Journal of Physical Chemistry C (2007)
- Journal of Physical Chemistry Letters (from 2010, combined letters previously published in the separate journals)

Historical journals that covered both chemistry and physics include Annales de chimie et de physique (started in 1789, published under the name given here from 1815–1914).

20.4 Branches and related topics

- Thermochemistry
- Chemical kinetics
- Quantum chemistry
- Electrochemistry
- Photochemistry
- Surface chemistry
- Solid-state chemistry
- Spectroscopy
- Biophysical chemistry
- Materials science
- Physical organic chemistry
- Micromeritics

20.5 See also

- List of important publications in chemistry#Physical chemistry
- List of unsolved problems in chemistry#Physical chemistry problems
- Physical biochemistry
- Category:Physical chemists

20.6 References

[1] Torben Smith Sørensen (1999). *Surface chemistry and electrochemistry of membranes*. CRC Press. p. 134. ISBN 0-8247-1922-0.

[2] Atkins, Peter and Friedman, Ronald (2005). *Molecular Quantum Mechanics*, p. 249. Oxford University Press, New York. ISBN 0-19-927498-3.

[3] Atkins, Peter and Friedman, Ronald (2005). *Molecular Quantum Mechanics*, p. 342. Oxford University Press, New York. ISBN 0-19-927498-3.

[4] Landau, L. D. and Lifshitz, E. M. (1980). *Statistical Physics*, 3rd Ed. p. 52. Elsevier Butterworth Heinemann, New York. ISBN 0-7506-3372-7.

[5] Hill, Terrell L. (1986). *Introduction to Statistical Thermodynamics*, p. 1. Dover Publications, New York. ISBN 0-486-65242-4.

[6] Schmidt, Lanny D. (2005). *The Engineering of Chemical Reactions*, 2nd Ed. p. 30. Oxford University Press, New York. ISBN 0-19-516925-5.

[7] Schmidt, Lanny D. (2005). *The Engineering of Chemical Reactions*, 2nd Ed. p. 25, 32. Oxford University Press, New York. ISBN 0-19-516925-5.

[8] Chandler, David (1987). *Introduction to Modern Statistical Mechanics*, p. 54. Oxford University Press, New York. ISBN 978-0-19-504277-1.

[9] Alexander Vucinich (1963). *Science in Russian culture*. Stanford University Press. p. 388. ISBN 0-8047-0738-3.

[10] Josiah Willard Gibbs, 1876, "On the Equilibrium of Heterogeneous Substances", *Transactions of the Connecticut Academy of Sciences*

[11] Laidler, Keith (1993). *The World of Physical Chemistry*. Oxford: Oxford University Press. p. 48. ISBN 0-19-855919-4.

[12] Herbst, Eric (May 12, 2005). "Chemistry of Star-Forming Regions". *Journal of Physical Chemistry A* **109** (18): 4017–4029. doi:10.1021/jp050461c. PMID 16833724.

20.7   External links

- Physical Chemistry (Keith J. Laidler, John H. Meiser and Bryan C. Sanctuary)

- The World of Physical Chemistry (Keith J. Laidler, 1993)

- Physical Chemistry from Ostwald to Pauling (John W. Servos, 1996)

- 100 Years of Physical Chemistry (Royal Society of Chemistry, 2004)

- Physical Chemistry: neither Fish nor Fowl? (Joachim Schummer, *The Autonomy of Chemistry*, Würzburg, Königshausen & Neumann, 1998, pp. 135–148)

- Cathedrals of Science (Patrick Coffey, 2008)

- The Cambridge History of Science: The modern physical and mathematical sciences (Mary Jo Nye, 2003)

Chapter 21

Noble gas

The **noble gases** make a group of chemical elements with similar properties. Under standard conditions, they are all odorless, colorless, monatomic gases with very low chemical reactivity. The six noble gases that occur naturally are helium (He), neon (Ne), argon (Ar), krypton (Kr), xenon (Xe), and the radioactive radon (Rn).

For the first six periods of the periodic table, the noble gases are exactly the members of **group 18** of the periodic table. It is possible that due to relativistic effects, the group 14 element flerovium exhibits some noble-gas-like properties,[1] instead of the group 18 element ununoctium.[2] Noble gases are typically highly unreactive except when under particular extreme conditions. The inertness of noble gases makes them very suitable in applications where reactions are not wanted. For example: argon is used in lightbulbs to prevent the hot tungsten filament from oxidizing; also, helium is breathed by deep-sea divers to prevent oxygen and nitrogen toxicity.

The properties of the noble gases can be well explained by modern theories of atomic structure: their outer shell of valence electrons is considered to be "full", giving them little tendency to participate in chemical reactions, and it has been possible to prepare only a few hundred noble gas compounds. The melting and boiling points for a given noble gas are close together, differing by less than 10 °C (18 °F); that is, they are liquids over only a small temperature range.

Neon, argon, krypton, and xenon are obtained from air in an air separation unit using the methods of liquefaction of gases and fractional distillation. Helium is sourced from natural gas fields which have high concentrations of helium in the natural gas, using cryogenic gas separation techniques, and radon is usually isolated from the radioactive decay of dissolved radium, thorium, or uranium compounds (since those compounds give off alpha particles). Noble gases have several important applications in industries such as lighting, welding, and space exploration. A helium-oxygen breathing gas is often used by deep-sea divers at depths of seawater over 55 m (180 ft) to keep the diver from experiencing oxygen toxemia, the lethal effect of high-pressure oxygen, and nitrogen narcosis, the distracting narcotic effect of the nitrogen in air beyond this partial-pressure threshold. After the risks caused by the flammability of hydrogen became apparent, it was replaced with helium in blimps and balloons.

21.1 History

Noble gas is translated from the German noun *Edelgas*, first used in 1898 by Hugo Erdmann[3] to indicate their extremely low level of reactivity. The name makes an analogy to the term "noble metals", which also have low reactivity. The noble gases have also been referred to as *inert gases*, but this label is deprecated as many noble gas compounds are now known.[4] *Rare gases* is another term that was used,[5] but this is also inaccurate because argon forms a fairly considerable part (0.94% by volume, 1.3% by mass) of the Earth's atmosphere due to decay of radioactive potassium-40.[6]

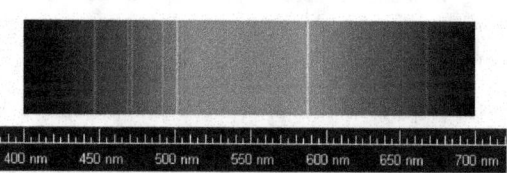

Helium was first detected in the Sun due to its characteristic spectral lines.

Pierre Janssen and Joseph Norman Lockyer discovered a new element on August 18, 1868 while looking at the chromosphere of the Sun, and named it helium after the Greek word for the Sun, ήλιος (*ílios* or *helios*).[7] No chemical analysis was possible at the time, but helium was later found to be a noble gas. Before them, in 1784, the English chemist and physicist Henry Cavendish had discovered that air contains a small proportion of a substance less reactive than nitrogen.[8] A century later, in 1895, Lord Rayleigh discovered that samples of nitrogen from the air were of a different density than nitrogen resulting from chemical reactions. Along with Scottish scientist William Ramsay at University College, London, Lord Rayleigh theorized that the nitrogen extracted from air was mixed with another gas, leading to an experiment that successfully isolated a new element, argon, from the Greek word αργός (*argós*, "inactive").[8] With this discovery, they realized an entire class of gases was missing from the periodic table. During his search for argon, Ramsay also managed to isolate helium for the first time while heating cleveite, a mineral. In 1902, having accepted the evidence for the elements helium and argon, Dmitri Mendeleev included these noble gases as group 0 in his arrangement of the elements, which would later become the periodic table.[9]

Ramsay continued to search for these gases using the method of fractional distillation to separate liquid air into several components. In 1898, he discovered the elements krypton, neon, and xenon, and named them after the Greek words κρυπτός (*kryptós*, "hidden"), νέος (*néos*, "new"), and ξένος (*xénos*, "stranger"), respectively. Radon was first identified in 1898 by Friedrich Ernst Dorn,[10] and was named *radium emanation*, but was not considered a noble gas until 1904 when its characteristics were found to be similar to those of other noble gases.[11] Rayleigh and Ramsay received the 1904 Nobel Prizes in Physics and in Chemistry, respectively, for their discovery of the noble gases;[12][13] in the words of J. E. Cederblom, then president of the Royal Swedish Academy of Sciences, "the discovery of an entirely new group of elements, of which no single representative had been known with any certainty, is something utterly unique in the history of chemistry, being intrinsically an advance in science of peculiar significance".[13]

The discovery of the noble gases aided in the development of a general understanding of atomic structure. In 1895, French chemist Henri Moissan attempted to form a reaction between fluorine, the most electronegative element, and argon, one of the noble gases, but failed. Scientists were unable to prepare compounds of argon until the end of the 20th century, but these attempts helped to develop new theories of atomic structure. Learning from these experiments, Danish physicist Niels Bohr proposed in 1913 that the electrons in atoms are arranged in shells surrounding the nucleus, and that for all noble gases except helium the outermost shell always contains eight electrons.[11] In 1916, Gilbert N. Lewis formulated the *octet rule*, which concluded an octet of electrons in the outer shell was the most stable arrangement for any atom; this arrangement caused them to be unreactive with other elements since they did not require any more electrons to complete their outer shell.[14]

In 1962, Neil Bartlett discovered the first chemical compound of a noble gas, xenon hexafluoroplatinate.[15] Compounds of other noble gases were discovered soon after: in 1962 for radon, radon difluoride,[16] and in 1963 for kryp-

ton, krypton difluoride (KrF 2).[17] The first stable compound of argon was reported in 2000 when argon fluorohydride (HArF) was formed at a temperature of 40 K (−233.2 °C; −387.7 °F).[18]

In December 1998, scientists at the Joint Institute for Nuclear Research working in Dubna, Russia bombarded plutonium (Pu) with calcium (Ca) to produce a single atom of element 114,[19] flerovium (Fl).[20] Preliminary chemistry experiments have indicated this element may be the first superheavy element to show abnormal noble-gas-like properties, even though it is a member of group 14 on the periodic table.[21] In October 2006, scientists from the Joint Institute for Nuclear Research and Lawrence Livermore National Laboratory successfully created synthetically ununoctium (Uuo), the seventh element in group 18,[22] by bombarding californium (Cf) with calcium (Ca).[23]

21.2 Physical and atomic properties

The noble gases have weak interatomic force, and consequently have very low melting and boiling points. They are all monatomic gases under standard conditions, including the elements with larger atomic masses than many normally solid elements.[11] Helium has several unique qualities when compared with other elements: its boiling and melting points are lower than those of any other known substance; it is the only element known to exhibit superfluidity; it is the only element that cannot be solidified by cooling under standard conditions—a pressure of 25 standard atmospheres (2,500 kPa; 370 psi) must be applied at a temperature of 0.95 K (−272.200 °C; −457.960 °F) to convert it to a solid.[26] The noble gases up to xenon have multiple stable isotopes. Radon has no stable isotopes; its longest-lived isotope, ^{222}Rn, has a half-life of 3.8 days and decays to form helium and polonium, which ultimately decays to lead.[11] Melting and boiling points generally increase going down the group.

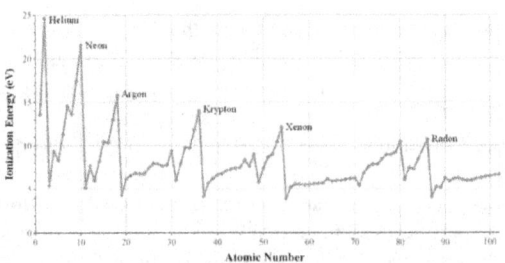

This is a plot of ionization potential versus atomic number. The noble gases, which are labeled, have the largest ionization potential for each period.

The noble gas atoms, like atoms in most groups, increase steadily in atomic radius from one period to the next due to the increasing number of electrons. The size of the atom is related to several properties. For example, the ionization potential decreases with an increasing radius because the valence electrons in the larger noble gases are farther away from the nucleus and are therefore not held as tightly together by the atom. Noble gases have the largest ionization potential among the elements of each period, which reflects the stability of their electron configuration and is related to their relative lack of chemical reactivity.[24] Some of the heavier noble gases, however, have ionization potentials small enough to be comparable to those of other elements and molecules. It was the insight that xenon has an ionization potential similar to that of the oxygen molecule that led Bartlett to attempt oxidizing xenon using platinum hexafluoride, an oxidizing agent known to be strong enough to react with oxygen.[15] Noble gases cannot accept an electron to form stable anions; that is, they have a negative electron affinity.[27]

The macroscopic physical properties of the noble gases are dominated by the weak van der Waals forces between the atoms. The attractive force increases with the size of the atom as a result of the increase in polarizability and the decrease in ionization potential. This results in systematic group trends: as one goes down group 18, the atomic radius, and with it the interatomic forces, increases, resulting in an increasing melting point, boiling point, enthalpy of vaporization, and solubility. The increase in density is due to the increase in atomic mass.[24]

The noble gases are nearly ideal gases under standard conditions, but their deviations from the ideal gas law provided important clues for the study of intermolecular interactions. The Lennard-Jones potential, often used to model intermolecular interactions, was deduced in 1924 by John Lennard-Jones from experimental data on argon before the development of quantum mechanics provided the tools for understanding intermolecular forces from first principles.[28] The theoretical analysis of these interactions became tractable because the noble gases are monatomic and the atoms spherical, which means that the interaction between the atoms is independent of direction, or isotropic.

21.3 Chemical properties

The noble gases are colorless, odorless, tasteless, and non-flammable under standard conditions. They were once labeled *group 0* in the periodic table because it was believed they had a valence of zero, meaning their atoms cannot combine with those of other elements to form compounds. However, it was later discovered some do indeed form compounds, causing this label to fall into disuse.[11]

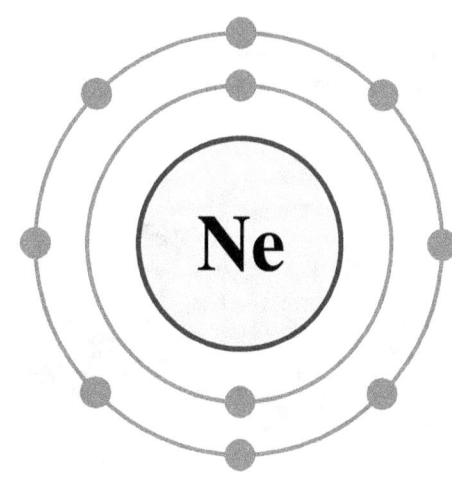

Neon, like all noble gases, has a full valence shell. Noble gases have eight electrons in their outermost shell, except in the case of helium, which has two.

21.3.1 Configuration

Main article: Noble gas configuration

Like other groups, the members of this family show patterns in its electron configuration, especially the outermost shells resulting in trends in chemical behavior:

The noble gases have full valence electron shells. Valence electrons are the outermost electrons of an atom and are normally the only electrons that participate in chemical bonding. Atoms with full valence electron shells are extremely stable and therefore do not tend to form chemical bonds and have little tendency to gain or lose electrons.[29] However, heavier noble gases such as radon are held less firmly together by electromagnetic force than lighter noble gases such as helium, making it easier to remove outer electrons from heavy noble gases.

As a result of a full shell, the noble gases can be used in conjunction with the electron configuration notation to form the *noble gas notation*. To do this, the nearest noble gas that precedes the element in question is written first, and then the electron configuration is continued from that point forward. For example, the electron notation of phosphorus is $1s^2 \, 2s^2 \, 2p^6 \, 3s^2 \, 3p^3$, while the noble gas notation is [Ne] $3s^2 \, 3p^3$. This more compact notation makes it easier to identify elements, and is shorter than writing out the full notation of atomic orbitals.[30]

21.3.2   Compounds

Main article: Noble gas compound
The noble gases show extremely low chemical reactivity;

Structure of XeF
4, one of the first noble gas compounds to be discovered

consequently, only a few hundred noble gas compounds have been formed. Neutral compounds in which helium and neon are involved in chemical bonds have not been formed (although there is some theoretical evidence for a few helium compounds), while xenon, krypton, and argon have shown only minor reactivity.[31] The reactivity follows the order Ne < He < Ar < Kr < Xe < Rn.

In 1933, Linus Pauling predicted that the heavier noble gases could form compounds with fluorine and oxygen. He predicted the existence of krypton hexafluoride (KrF
6) and xenon hexafluoride (XeF
6), speculated that XeF
8 might exist as an unstable compound, and suggested xenic acid could form perxenate salts.[32][33] These predictions were shown to be generally accurate, except that XeF
8 is now thought to be both thermodynamically and kinetically unstable.[34]

Xenon compounds are the most numerous of the noble gas compounds that have been formed.[35] Most of them have the xenon atom in the oxidation state of +2, +4, +6, or +8 bonded to highly electronegative atoms such as fluorine or oxygen, as in xenon difluoride (XeF
2), xenon tetrafluoride (XeF
4), xenon hexafluoride (XeF
6), xenon tetroxide (XeO
4), and sodium perxenate (Na
4XeO
6). Xenon reacts with fluorine to form numerous xenon fluorides according to the following equations:

$$Xe + F_2 \rightarrow XeF_2$$
$$Xe + 2F_2 \rightarrow XeF_4$$
$$Xe + 3F_2 \rightarrow XeF_6$$

Some of these compounds have found use in chemical synthesis as oxidizing agents; XeF
2, in particular, is commercially available and can be used as a fluorinating agent.[36] As of 2007, about five hundred compounds of xenon bonded to other elements have been identified, including organoxenon compounds (containing xenon bonded to carbon), and xenon bonded to nitrogen, chlorine, gold, mercury, and xenon itself.[31][37] Compounds of xenon bound to boron, hydrogen, bromine, iodine, beryllium, sulphur, titanium, copper, and silver have also been observed but only at low temperatures in noble gas matrices, or in supersonic noble gas jets.[31]

In theory, radon is more reactive than xenon, and therefore should form chemical bonds more easily than xenon does. However, due to the high radioactivity and short half-life of radon isotopes, only a few fluorides and oxides of radon have been formed in practice.[38]

Krypton is less reactive than xenon, but several compounds have been reported with krypton in the oxidation state of +2.[31] Krypton difluoride is the most notable and easily characterized. Under extreme conditions, krypton reacts with fluorine to form KrF_2 according to the following equation:

$$Kr + F_2 \rightarrow KrF_2$$

Compounds in which krypton forms a single bond to nitrogen and oxygen have also been characterized,[39] but are only stable below −60 °C (−76 °F) and −90 °C (−130 °F) respectively.[31]

Krypton atoms chemically bound to other nonmetals (hydrogen, chlorine, carbon) as well as some late transition metals (copper, silver, gold) have also been observed, but only either at low temperatures in noble gas matrices, or in supersonic noble gas jets.[31] Similar conditions were used to obtain the first few compounds of argon in 2000, such as argon fluorohydride (HArF), and some bound to the late transition metals copper, silver, and gold.[31] As of 2007, no stable neutral molecules involving covalently bound helium or neon are known.[31]

The noble gases—including helium—can form stable molecular ions in the gas phase. The simplest is the helium hydride molecular ion, HeH^+, discovered in 1925.[40] Because it is composed of the two most abundant elements in the universe, hydrogen and helium, it is believed to occur naturally in the interstellar medium, although it has not been detected yet.[41] In addition to these ions, there are many known neutral excimers of the noble gases. These are compounds such as ArF and KrF that are stable only when in an excited electronic state; some of them find application in excimer lasers.

In addition to the compounds where a noble gas atom is involved in a covalent bond, noble gases also form noncovalent compounds. The clathrates, first described in 1949,[42] consist of a noble gas atom trapped within cavities of crystal lattices of certain organic and inorganic substances. The essential condition for their formation is that the guest (noble gas) atoms must be of appropriate size to fit in the cavities of the host crystal lattice. For instance, argon, krypton, and xenon form clathrates with hydroquinone, but helium and neon do not because they are too small or insufficiently polarizable to be retained.[43] Neon, argon, krypton, and xenon also form clathrate hydrates, where the noble gas is trapped in ice.[44]

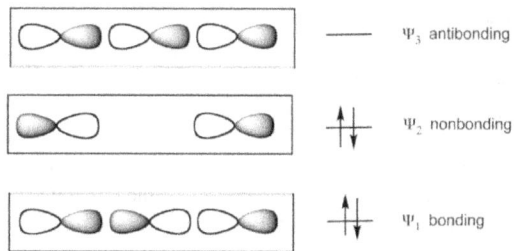

Bonding in XeF
2 according to the 3-center-4-electron bond model

Noble gas compounds such as xenon difluoride (XeF
2) are considered to be hypervalent because they violate the octet rule. Bonding in such compounds can be explained using a three-center four-electron bond model.[48][49] This model, first proposed in 1951, considers bonding of three collinear atoms. For example, bonding in XeF
2 is described by a set of three molecular orbitals (MOs) derived from p-orbitals on each atom. Bonding results from the combination of a filled p-orbital from Xe with one half-filled p-orbital from each F atom, resulting in a filled bonding orbital, a filled non-bonding orbital, and an empty antibonding orbital. The highest occupied molecular orbital is localized on the two terminal atoms. This represents a localization of charge which is facilitated by the high electronegativity of fluorine.[50]

The chemistry of heavier noble gases, krypton and xenon, are well established. The chemistry of the lighter ones, argon and helium, is still at an early stage, while a neon compound is yet to be identified.

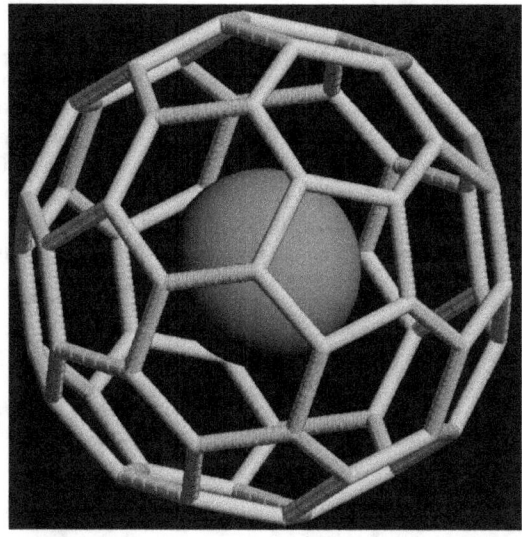

An endohedral fullerene compound containing a noble gas atom

Noble gases can form endohedral fullerene compounds, in which the noble gas atom is trapped inside a fullerene molecule. In 1993, it was discovered that when C
60, a spherical molecule consisting of 60 carbon atoms, is exposed to noble gases at high pressure, complexes such as He@C
60 can be formed (the @ notation indicates He is contained inside C
60 but not covalently bound to it).[45] As of 2008, endohedral complexes with helium, neon, argon, krypton, and xenon have been obtained.[46] These compounds have found use in the study of the structure and reactivity of fullerenes by means of the nuclear magnetic resonance of the noble gas atom.[47]

21.4 Occurrence and production

The abundances of the noble gases in the universe decrease as their atomic numbers increase. Helium is the most common element in the universe after hydrogen, with a mass fraction of about 24%. Most of the helium in the universe was formed during Big Bang nucleosynthesis, but the amount of helium is steadily increasing due to the fusion of hydrogen in stellar nucleosynthesis (and, to a very slight degree, the alpha decay of heavy elements).[51][52] Abundances on Earth follow different trends; for example, helium is only the third most abundant noble gas in the atmosphere. The reason is that there is no primordial helium in the atmosphere; due to the small mass of the atom, helium cannot be retained by the Earth's gravitational field.[53] Helium on Earth comes from the alpha decay of heavy elements such as uranium and thorium found in the Earth's crust, and tends to accumulate in natural gas deposits.[53] The abundance of argon, on the other hand, is increased as a result of the beta decay of potassium-40, also found in the Earth's crust, to form argon-40, which is the most abundant isotope of argon on Earth despite being relatively rare in the Solar System. This process is the base for the potassium-argon dating method.[54] Xenon has an unexpectedly low abundance in the atmosphere, in what has been called the *missing xenon problem*; one theory is that the missing xenon may be trapped in minerals inside the Earth's crust.[55] After the discovery of xenon dioxide, a research showed that Xe can substitute for Si in the quartz.[56] Radon is formed in the lithosphere as from the alpha decay of radium. It can seep into buildings through cracks in their foundation and accumulate in areas that are not well ventilated. Due to its high radioactivity, radon presents a significant health hazard; it is implicated in an estimated 21,000 lung cancer deaths per year in the United States alone.[57]

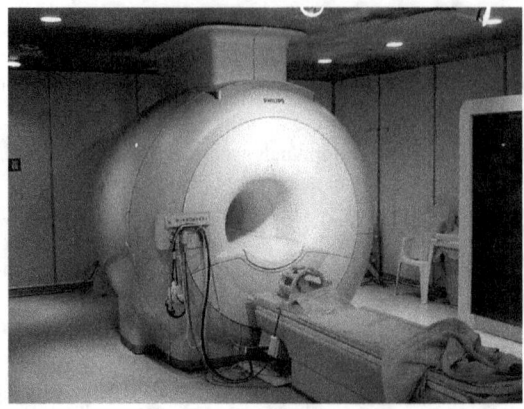

Liquid helium is used to cool the superconducting magnets in modern MRI scanners

For large-scale use, helium is extracted by fractional distillation from natural gas, which can contain up to 7% helium.[62]

Neon, argon, krypton, and xenon are obtained from air using the methods of liquefaction of gases, to convert elements to a liquid state, and fractional distillation, to separate mixtures into component parts. Helium is typically produced by separating it from natural gas, and radon is isolated from the radioactive decay of radium compounds.[11] The prices of the noble gases are influenced by their natural abundance, with argon being the cheapest and xenon the most expensive. As an example, the table to the right lists the 2004 prices in the United States for laboratory quantities of each gas.

21.5 Applications

Noble gases have very low boiling and melting points, which makes them useful as cryogenic refrigerants.[63] In particular, liquid helium, which boils at 4.2 K (−268.95 °C; −452.11 °F), is used for superconducting magnets, such as those needed in nuclear magnetic resonance imaging and nuclear magnetic resonance.[64] Liquid neon, although it does not reach temperatures as low as liquid helium, also finds use in cryogenics because it has over 40 times more refrigerating capacity than liquid helium and over three times more than liquid hydrogen.[60]

Helium is used as a component of breathing gases to replace nitrogen, due its low solubility in fluids, especially in lipids. Gases are absorbed by the blood and body tissues when under pressure like in scuba diving, which causes an anesthetic effect known as nitrogen narcosis.[65] Due to its reduced solubility, little helium is taken into cell membranes, and when helium is used to replace part of the breathing mixtures, such as in trimix or heliox, a decrease in the narcotic effect of the gas at depth is obtained.[66] Helium's reduced solubility offers further advantages for the condition known as decompression sickness, or *the bends*.[11][67] The reduced amount of dissolved gas in the body means that fewer gas bubbles form during the decrease in pressure of the ascent. Another noble gas, argon, is considered the best option for use as a drysuit inflation gas for scuba diving.[68] Helium is also used as filling gas in nuclear fuel rods for nuclear reactors.[69]

Goodyear Blimp

Since the *Hindenburg* disaster in 1937,[70] helium has replaced hydrogen as a lifting gas in blimps and balloons due to its lightness and incombustibility, despite an 8.6%[71] decrease in buoyancy.[11]

In many applications, the noble gases are used to provide an inert atmosphere. Argon is used in the synthesis of air-sensitive compounds that are sensitive to nitrogen. Solid argon is also used for the study of very unstable compounds, such as reactive intermediates, by trapping them in an inert matrix at very low temperatures.[72] Helium is used as the carrier medium in gas chromatography, as a filler gas for thermometers, and in devices for measuring radiation, such as the Geiger counter and the bubble chamber.[61] Helium and argon are both commonly used to shield welding arcs and the surrounding base metal from the atmosphere during welding and cutting, as well as in other metallurgical processes and in the production of silicon for the semiconductor industry.[60]

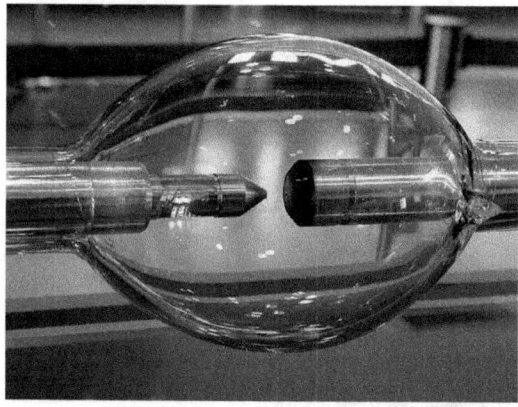

15,000-watt xenon short-arc lamp used in IMAX projectors

Noble gases are commonly used in lighting because of their lack of chemical reactivity. Argon, mixed with nitrogen, is used as a filler gas for incandescent light bulbs.[60] Krypton is used in high-performance light bulbs, which have higher color temperatures and greater efficiency, because it reduces the rate of evaporation of the filament more than argon; halogen lamps, in particular, use krypton mixed with small amounts of compounds of iodine or bromine.[60] The noble gases glow in distinctive colors when used inside gas-discharge lamps, such as "neon lights". These lights are called after neon but often contain other gases and phosphors, which add various hues to the orange-red color of neon. Xenon is commonly used in xenon arc lamps which, due to their nearly continuous spectrum that resembles daylight, find application in film projectors and as automobile headlamps.[60]

The noble gases are used in excimer lasers, which are based on short-lived electronically excited molecules known as excimers. The excimers used for lasers may be noble gas dimers such as Ar_2, Kr_2 or Xe_2, or more commonly, the noble gas is combined with a halogen in excimers such as ArF, KrF, XeF, or XeCl. These lasers produce ultraviolet light which, due to its short wavelength (193 nm for ArF and 248 nm for KrF), allows for high-precision imaging. Excimer lasers have many industrial, medical, and scientific applications. They are used for microlithography and microfabrication, which are essential for integrated circuit manufacture, and for laser surgery, including laser angioplasty and eye surgery.[73]

Some noble gases have direct application in medicine. Helium is sometimes used to improve the ease of breathing of asthma sufferers.[60] Xenon is used as an anesthetic because of its high solubility in lipids, which makes it more potent than the usual nitrous oxide, and because it is readily eliminated from the body, resulting in faster recovery.[74] Xenon finds application in medical imaging of the lungs through hyperpolarized MRI.[75] Radon, which is highly radioactive and is only available in minute amounts, is used in radiotherapy.[11]

21.6   Discharge color

The color of gas discharge emission depends on several factors, including the following:[76]

- discharge parameters (local value of current density and electric field, temperature, etc. – note the color variation along the discharge in the top row);

- gas purity (even small fraction of certain gases can affect color);

- material of the discharge tube envelope – note suppression of the UV and blue components in the bottom-row tubes made of thick household glass.

21.7   See also

- Noble gas (data page), for extended tables of physical properties.

- Noble metal, for metals that are resistant to corrosion or oxidation.

- Inert gas, for any gas that is not reactive under normal circumstances.

- Industrial gas

- Neutronium

- Noble gas configuration

21.8   Notes

[1] "Flerov laboratory of nuclear reactions" (PDF). JINR. Retrieved 2009-08-08.

[2] Nash, Clinton S. (2005). "Atomic and Molecular Properties of Elements 112, 114, and 118". *J. Phys. Chem. A* **109** (15): 3493–3500. doi:10.1021/jp050736o. PMID 16833687.

[3] Renouf, Edward (1901). "Noble gases". *Science* **13** (320): 268–270. Bibcode:1901Sci....13..268R. doi:10.1126/science.13.320.268.

[4] Ojima 2002, p. 30

[5] Ojima 2002, p. 4

[6] "argon". *Encyclopædia Britannica*. 2008.

[7] *Oxford English Dictionary* (1989), s.v. "helium". Retrieved December 16, 2006, from Oxford English Dictionary Online. Also, from quotation there: Thomson, W. (1872). *Rep. Brit. Assoc.* xcix: "Frankland and Lockyer find the yellow prominences to give a very decided bright line not far from D, but hitherto not identified with any terrestrial flame. It seems to indicate a new substance, which they propose to call Helium."

[8] Ojima 2002, p. 1

[9] Mendeleev 1903, p. 497

[10] Partington, J. R. (1957). "Discovery of Radon". *Nature* **179** (4566): 912. Bibcode:1957Natur.179..912P. doi:10.1038/179912a0.

[11] "Noble Gas". *Encyclopædia Britannica*. 2008.

[12] Cederblom, J. E. (1904). "The Nobel Prize in Physics 1904 Presentation Speech".

[13] Cederblom, J. E. (1904). "The Nobel Prize in Chemistry 1904 Presentation Speech".

[14] Gillespie, R. J.; Robinson, E. A. (2007). "Gilbert N. Lewis and the chemical bond: the electron pair and the octet rule from 1916 to the present day". *J Comput Chem* **28** (1): 87–97. doi:10.1002/jcc.20545. PMID 17109437.

[15] Bartlett, N. (1962). "Xenon hexafluoroplatinate $Xe^+[PtF_6]^-$". *Proceedings of the Chemical Society* (6): 218. doi:10.1039/PS9620000197.

[16] Fields, Paul R.; Stein, Lawrence; Zirin, Moshe H. (1962). "Radon Fluoride". *Journal of the American Chemical Society* **84** (21): 4164–4165. doi:10.1021/ja00880a048.

[17] Grosse, A. V.; Kirschenbaum, A. D.; Streng, A. G.; Streng, L. V. (1963). "Krypton Tetrafluoride: Preparation and Some Properties". *Science* **139** (3559): 1047–1048. Bibcode:1963Sci...139.1047G. doi:10.1126/science.139.3559.1047. PMID 17812982.

[18] Khriachtchev, Leonid; Pettersson, Mika; Runeberg, Nino; Lundell, Jan; Räsänen, Markku (2000). "A stable argon compound". *Nature* **406** (6798): 874–876. doi:10.1038/35022551. PMID 10972285.

[19] Oganessian, Yu. Ts.; Utyonkov, V.; Lobanov, Yu.; Abdullin, F.; Polyakov, A.; et al. (1999). "Synthesis of Superheavy Nuclei in the $^{48}Ca + ^{244}Pu$ Reaction". *Physical Review Letters* (American Physical Society) **83** (16): 3154–3157. Bibcode:1999PhRvL..83.3154O. doi:10.1103/PhysRevLett.83.3154.

[20] Woods, Michael (2003-05-06). "Chemical element No. 110 finally gets a name—darmstadtium". *Pittsburgh Post-Gazette*. Retrieved 2008-06-26.

[21] "Gas Phase Chemistry of Superheavy Elements" (PDF). Texas A&M University. Retrieved 2008-05-31.

[22] Robert C. Barber, Paul J. Karol, Hiromichi Nakahara, Emanuele Vardaci, and Erich W. Vogt (2011). "Discovery of the elements with atomic numbers greater than or equal to 113 (IUPAC Technical Report)*" (PDF). *Pure Appl. Chem.* (IUPAC) **83** (7). doi:10.1515/ci.2011.33.5.25b. Retrieved 2014-05-30.

[23] Oganessian, Yu. Ts.; Utyonkov, V.; Lobanov, Yu.; Abdullin, F.; Polyakov, A.,; et al. (2006). "Synthesis of the isotopes of elements 118 and 116 in the 249Cf and 245Cm + 48Ca fusion reactions". *Physical Review C* **74** (4): 44602. Bibcode:2006PhRvC..74d4602O. doi:10.1103/PhysRevC.74.044602.

[24] Greenwood 1997, p. 891

[25] Allen, Leland C. (1989). "Electronegativity is the average one-electron energy of the valence-shell electrons in ground-state free atoms". *Journal of the American Chemical Society* **111** (25): 9003–9014. doi:10.1021/ja00207a003.

[26] "Solid Helium". University of Alberta. Retrieved 2008-06-22.

[27] Wheeler, John C. (1997). "Electron Affinities of the Alkaline Earth Metals and the Sign Convention for Electron Affinity". *Journal of Chemical Education* **74**: 123–127. Bibcode:1997JChEd..74..123W. doi:10.1021/ed074p123.; Kalcher, Josef; Sax, Alexander F. (1994). "Gas Phase Stabilities of Small Anions: Theory and Experiment in Cooperation". *Chemical Reviews* **94** (8): 2291–2318. doi:10.1021/cr00032a004.

[28] Mott, N. F. (1955). "John Edward Lennard-Jones. 1894–1954". *Biographical Memoirs of Fellows of the Royal Society* **1**: 175–184. doi:10.1098/rsbm.1955.0013.

[29] Ojima 2002, p. 35

[30] CliffsNotes 2007, p. 15

[31] Grochala, Wojciech (2007). "Atypical compounds of gases, which have been called noble". *Chemical Society Reviews* **36** (10): 1632–1655. doi:10.1039/b702109g. PMID 17721587.

[32] Pauling, Linus (1933). "The Formulas of Antimonic Acid and the Antimonates". *Journal of the American Chemical Society* **55** (5): 1895–1900. doi:10.1021/ja01332a016.

[33] Holloway 1968

[34] Seppelt, Konrad (1979). "Recent developments in the Chemistry of Some Electronegative Elements". *Accounts of Chemical Research* **12** (6): 211–216. doi:10.1021/ar50138a004.

[35] Moody, G. J. (1974). "A Decade of Xenon Chemistry". *Journal of Chemical Education* **51** (10): 628–630. Bibcode:1974JChEd..51..628M. doi:10.1021/ed051p628. Retrieved 2007-10-16.

[36] Zupan, Marko; Iskra, Jernej; Stavber, Stojan (1998). "Fluorination with XeF$_2$. 44. Effect of Geometry and Heteroatom on the Regioselectivity of Fluorine Introduction into an Aromatic Ring". *J. Org. Chem* **63** (3): 878–880. doi:10.1021/jo971496e. PMID 11672087.

[37] Harding 2002, pp. 90–99

[38] .Avrorin, V. V.; Krasikova, R. N.; Nefedov, V. D.; Toropova, M. A. (1982). "The Chemistry of Radon". *Russian Chemical Review* **51** (1): 12–20. Bibcode:1982RuCRv..51...12A. doi:10.1070/RC1982v051n01ABEH002787.

[39] Lehmann, J (2002). "The chemistry of krypton". *Coordination Chemistry Reviews*. 233–234: 1–39. doi:10.1016/S0010-8545(02)00202-3.

[40] Hogness, T. R.; Lunn, E. G. (1925). "The Ionization of Hydrogen by Electron Impact as Interpreted by Positive Ray Analysis". *Physical Review* **26**: 44–55. Bibcode:1925PhRv...26...44H. doi:10.1103/PhysRev.26.44.

[41] Fernandez, J.; Martin, F. (2007). "Photoionization of the HeH$_2^+$ molecular ion". *J. Phys. B: At. Mol. Opt. Phys* **40** (12): 2471–2480. Bibcode:2007JPhB...40.2471F. doi:10.1088/0953-4075/40/12/020.

[42] H. M. Powell and M. Guter (1949). "An Inert Gas Compound". *Nature* **164** (4162): 240–241. Bibcode:1949Natur.164..240P. doi:10.1038/164240b0.

[43] Greenwood 1997, p. 893

[44] Dyadin, Yuri A.; et al. (1999). "Clathrate hydrates of hydrogen and neon". *Mendeleev Communications* **9** (5): 209–210. doi:10.1070/MC1999v009n05ABEH001104.

[45] Saunders, M.; Jiménez-Vázquez, H. A.; Cross, R. J.; Poreda, R. J. (1993). "Stable compounds of helium and neon. He@C60 and Ne@C60". *Science* **259** (5100): 1428–1430. Bibcode:1993Sci...259.1428S. doi:10.1126/science.259.5100.1428. PMID 17801275.

[46] Saunders, Martin; Jimenez-Vazquez, Hugo A.; Cross, R. James; Mroczkowski, Stanley; Gross, Michael L.; Giblin, Daryl E.; Poreda, Robert J. (1994). "Incorporation of helium, neon, argon, krypton, and xenon into fullerenes using high pressure". *J. Am. Chem. Soc.* **116** (5): 2193–2194. doi:10.1021/ja00084a089.

[47] Frunzi, Michael; Cross, R. Jame; Saunders, Martin (2007). "Effect of Xenon on Fullerene Reactions". *Journal of the American Chemical Society* **129** (43): 13343–6. doi:10.1021/ja075568n. PMID 17924634.

[48] Greenwood 1997, p. 897

[49] Weinhold 2005, pp. 275–306

[50] Pimentel, G. C. (1951). "The Bonding of Trihalide and Bifluoride Ions by the Molecular Orbital Method". *The Journal of Chemical Physics* **19** (4): 446–448. Bibcode:1951JChPh..19..446P. doi:10.1063/1.1748245.

[51] Weiss, Achim. "Elements of the past: Big Bang Nucleosynthesis and observation". Max Planck Institute for Gravitational Physics. Retrieved 2008-06-23.

[52] Coc, A.; et al. (2004). "Updated Big Bang Nucleosynthesis confronted to WMAP observations and to the Abundance of Light Elements". *Astrophysical Journal* **600** (2): 544–552. arXiv:astro-ph/0309480. Bibcode:2004ApJ...600..544C. doi:10.1086/380121.

[53] Morrison, P.; Pine, J. (1955). "Radiogenic Origin of the Helium Isotopes in Rock". *Annals of the New York Academy of Sciences* **62** (3): 71–92. Bibcode:1955NYASA..62...71M. doi:10.1111/j.1749-6632.1955.tb35366.x.

[54] Scherer, Alexandra (2007-01-16). "^{40}Ar/^{39}Ar dating and errors". Technische Universität Bergakademie Freiberg. Archived from the original on 2007-10-14. Retrieved 2008-06-26.

[55] Sanloup, Chrystèle; Schmidt, Burkhard C.; et al. (2005). "Retention of Xenon in Quartz and Earth's Missing Xenon". *Science* **310** (5751): 1174–1177. Bibcode:2005Sci...310.1174S. doi:10.1126/science.1119070. PMID 16293758.

[56] Tyler Irving (May 2011). "Xenon Dioxide May Solve One of Earth's Mysteries". L'Actualité chimique canadienne (Canadian Chemical News). Retrieved 2012-05-18.

[57] "A Citizen's Guide to Radon". U.S. Environmental Protection Agency. 2007-11-26. Retrieved 2008-06-26.

[58] Lodders, Katharina (July 10, 2003). "Solar System Abundances and Condensation Temperatures of the Elements" (PDF). *The Astrophysical Journal* (The American Astronomical Society) **591** (2): 1220–1247. Bibcode:2003ApJ...591.1220L. doi:10.1086/375492.

[59] "The Atmosphere". National Weather Service. Retrieved 2008-06-01.

[60] Häussinger, Peter; Glatthaar, Reinhard; Rhode, Wilhelm; Kick, Helmut; Benkmann, Christian; Weber, Josef; Wunschel, Hans-Jörg; Stenke, Viktor; Leicht, Edith; Stenger, Hermann (2002). "Noble gases". *Ullmann's Encyclopedia of Industrial Chemistry*. Wiley. doi:10.1002/14356007.a17_485.

[61] Hwang, Shuen-Chen; Lein, Robert D.; Morgan, Daniel A. (2005). "Noble Gases". *Kirk Othmer Encyclopedia of Chemical Technology*. Wiley. pp. 343–383. doi:10.1002/0471238961.0701190508230114.a01.

[62] Winter, Mark (2008). "Helium: the essentials". University of Sheffield. Retrieved 2008-07-14.

[63] "Neon". *Encarta*. 2008.

[64] Zhang, C. J.; Zhou, X. T.; Yang, L. (1992). "Demountable coaxial gas-cooled current leads for MRI superconducting magnets". *Magnetics, IEEE Transactions on* (IEEE) **28** (1): 957–959. Bibcode:1992ITM....28..957Z. doi:10.1109/20.120038.

[65] Fowler, B; Ackles, K. N.; Porlier, G. (1985). "Effects of inert gas narcosis on behavior—a critical review". *Undersea Biomed. Res.* **12** (4): 369–402. ISSN 0093-5387. OCLC 2068005. PMID 4082343. Retrieved 2008-04-08.

[66] Bennett 1998, p. 176

[67] Vann, R. D. (ed) (1989). "The Physiological Basis of Decompression". *38th Undersea and Hyperbaric Medical Society Workshop*. 75(Phys)6-1-89: 437. Retrieved 2008-05-31.

[68] Maiken, Eric (2004-08-01). "Why Argon?". Decompression. Retrieved 2008-06-26.

[69] Horhoianu, G; Ionescu, D.V; Olteanu, G (1999). "Thermal behaviour of CANDU type fuel rods during steady state and transient operating conditions". *Annals of Nuclear Energy* **26** (16): 1437–1445. doi:10.1016/S0306-4549(99)00022-5.

[70] "Disaster Ascribed to Gas by Experts". *The New York Times*. 1937-05-07. p. 1.

[71] Freudenrich, Craig (2008). "How Blimps Work". HowStuffWorks. Retrieved 2008-07-03.

[72] Dunkin, I. R. (1980). "The matrix isolation technique and its application to organic chemistry". *Chem. Soc. Rev.* **9**: 1–23. doi:10.1039/CS9800900001.

[73] Basting, Dirk; Marowsky, Gerd (2005). *Excimer Laser Technology*. Springer. ISBN 3-540-20056-8.

[74] Sanders, Robert D.; Ma, Daqing; Maze, Mervyn (2005). "Xenon: elemental anaesthesia in clinical practice". *British Medical Bulletin* **71** (1): 115–135. doi:10.1093/bmb/ldh034. PMID 15728132.

[75] Albert, M. S.; Balamore, D. (1998). "Development of hyperpolarized noble gas MRI". *Nuclear Instruments and Methods in Physics Research A* **402** (2–3): 441–453. Bibcode:1998NIMPA.402..441A. doi:10.1016/S0168-9002(97)00888-7. PMID 11543065.

[76] Ray, Sidney F. (1999). *Scientific photography and applied imaging*. Focal Press. pp. 383–384. ISBN 0-240-51323-1.

21.9 References

• Bennett, Peter B.; Elliott, David H. (1998). *The Physiology and Medicine of Diving*. SPCK Publishing. ISBN 0-7020-2410-4.

• Bobrow Test Preparation Services (2007-12-05). *CliffsAP Chemistry*. CliffsNotes. ISBN 0-470-13500-X.

• Greenwood, N. N.; Earnshaw, A. (1997). *Chemistry of the Elements* (2nd ed.). Oxford: Butterworth-Heinemann. ISBN 0-7506-3365-4.

• Harding, Charlie J.; Janes, Rob (2002). *Elements of the P Block*. Royal Society of Chemistry. ISBN 0-85404-690-9.

• Holloway, John H. (1968). *Noble-Gas Chemistry*. London: Methuen Publishing. ISBN 0-412-21100-9.

• Mendeleev, D. (1902–1903). *Osnovy Khimii (The Principles of Chemistry)* (in Russian) (7th ed.).

• Ojima, Minoru; Podosek, Frank A. (2002). *Noble Gas Geochemistry*. Cambridge University Press. ISBN 0-521-80366-7.

• Weinhold, F.; Landis, C. (2005). *Valency and bonding*. Cambridge University Press. ISBN 0-521-83128-8.

• Scerri, Eric R. (2007). *The Periodic Table, Its Story and Its Significance*. Oxford University Press. ISBN 0-19-530573-6.

Chapter 22

Niels Bohr

Niels Henrik David Bohr (Danish: [nels ˈb̥oɐ̯ˀ]; 7 October 1885 – 18 November 1962) was a Danish physicist who made foundational contributions to understanding atomic structure and quantum theory, for which he received the Nobel Prize in Physics in 1922. Bohr was also a philosopher and a promoter of scientific research.[1]

Bohr developed the Bohr model of the atom, in which he proposed that energy levels of electrons are discrete and that the electrons revolve in stable orbits around the atomic nucleus but can jump from one energy level (or orbit) to another. Although the Bohr model has been supplanted by other models, its underlying principles remain valid. He conceived the principle of complementarity: that items could be separately analysed in terms of contradictory properties, like behaving as a wave or a stream of particles. The notion of complementarity dominated Bohr's thinking in both science and philosophy.

Bohr founded the Institute of Theoretical Physics at the University of Copenhagen, now known as the Niels Bohr Institute, which opened in 1920. Bohr mentored and collaborated with physicists including Hans Kramers, Oskar Klein, George de Hevesy and Werner Heisenberg. He predicted the existence of a new zirconium-like element, which was named hafnium, after the Latin name for Copenhagen, where it was discovered. Later, the element bohrium was named after him.

During the 1930s, Bohr helped refugees from Nazism. After Denmark was occupied by the Germans, he had a famous meeting with Heisenberg, who had become the head of the German nuclear energy project. In September 1943, word reached Bohr that he was about to be arrested by the Germans, and he fled to Sweden. From there, he was flown to Britain, where he joined the British Tube Alloys nuclear weapons project, and was part of the British mission to the Manhattan Project. After the war, Bohr called for international cooperation on nuclear energy. He was involved with the establishment of CERN[2] and the Research Establishment Risø of the Danish Atomic Energy Commission, and became the first chairman of the Nordic Institute for Theoretical Physics in 1957.

22.1   Early years

Niels Bohr was born in Copenhagen, Denmark, on 7 October 1885, the second of three children of Christian Bohr,[3][4] a professor of physiology at the University of Copenhagen, and Ellen Adler Bohr, who came from a wealthy Danish Jewish family prominent in banking and parliamentary circles.[5] He had an elder sister, Jenny, and a younger brother Harald.[3] Jenny became a teacher,[4] while Harald became a mathematician and Olympic footballer who played for the Danish national team at the 1908 Summer Olympics in London. Niels was a passionate footballer as well, and the two brothers played several matches for the Copenhagen-based Akademisk Boldklub (Academic Football Club), with Niels as goalkeeper.[6]

Niels Bohr as a young man

Bohr was educated at Gammelholm Latin School, starting when he was seven.[7] In 1903, Bohr enrolled as an undergraduate at Copenhagen University. His major was physics, which he studied under Professor Christian Christiansen, the university's only professor of physics at that time. He also studied astronomy and mathematics under Professor Thorvald Thiele, and philosophy under Professor Harald Høffding, a friend of his father.[8][9]

In 1905, a gold medal competition was sponsored by the Royal Danish Academy of Sciences and Letters to investigate a method for measuring the surface tension of liquids that had been proposed by Lord Rayleigh in 1879. This involved measuring the frequency of oscillation of the radius of a water jet. Bohr conducted a series of experiments using his father's laboratory in the university; the university itself had no physics laboratory. To complete his experiments, he had to make his own glassware, creating test tubes with the required elliptical cross-sections. He went beyond the original task, incorporating improvements into both Rayleigh's theory and his method, by taking into account the viscosity of the water, and by working with finite amplitudes instead of just infinitesimal ones. His essay, which he submitted at the last minute, won the prize. He later submitted an improved version of the paper to the Royal Society in London for publication in the *Philosophical Transactions of the Royal Society*.[10][11][9][12]

Harald became the first of the two Bohr brothers to earn a master's degree, which he earned for mathematics in April 1909. Niels took another nine months to earn his. Students had to submit a thesis on a subject assigned by their supervisor. Bohr's supervisor was Christiansen, and the topic he chose was the electron theory of metals. Bohr subsequently elaborated his master's thesis into his much-larger Doctor of Philosophy (dr. phil.) thesis. He surveyed the literature on the subject, settling on a model postulated by Paul Drude and elaborated by Hendrik Lorentz, in which the electrons in a metal are considered to behave like a gas. Bohr extended Lorentz's model, but was still unable to account for phenomena like the Hall effect, and concluded that electron theory could not fully explain the magnetic properties of metals. The thesis was accepted in April 1911, and Bohr conducted his formal defence on 13 May. Harald had received his doctorate the previous year.[13] Bohr's thesis was groundbreaking, but attracted little interest outside Scandinavia because it was written in Danish, a Copenhagen University requirement at the time. In 1921, the Dutch physicist Hendrika Johanna van Leeuwen would independently derive a theorem from Bohr's thesis that is today known as the Bohr–van Leeuwen theorem.[14]

In 1910, Bohr met Margrethe Nørlund, the sister of the mathematician Niels Erik Nørlund.[15] Bohr resigned his membership in the Church of Denmark on 16 April 1912, and he and Margrethe were married in a civil ceremony

Niels Bohr and Margrethe Nørlund on their engagement in 1910.

at the town hall in Slagelse on 1 August. Years later, his brother Harald similarly left the church before getting married.[16] Niels and Margrethe had six sons.[17] The oldest, Christian, died in a boating accident in 1934,[18] and another, Harald, died from childhood meningitis.[17] Aage Bohr became a successful physicist, and in 1975 was awarded the Nobel Prize in physics, like his father. Hans became a physician; Erik, a chemical engineer; and Ernest, a lawyer.[19] Like his uncle Harald, Ernest Bohr became an Olympic athlete, playing field hockey for Denmark at the 1948 Summer Olympics in London.[20]

22.2   Physics

22.2.1   Bohr model

Main article: Bohr model

In 1911, Bohr travelled to England. At the time, it was where most of the theoretical work on the structure of atoms and molecules was being done.[21] He met J. J. Thomson of the Cavendish Laboratory and Trinity College, Cambridge. He attended lectures on electromagnetism given by

James Jeans and Joseph Larmor, and did some research on cathode rays, but failed to impress Thomson.[22][23] He had more success with younger physicists like the Australian William Lawrence Bragg,[24] and New Zealand's Ernest Rutherford, whose 1911 Rutherford model of the atom had challenged Thomson's 1904 plum pudding model.[25] Bohr received an invitation from Rutherford to conduct post-doctoral work at Victoria University of Manchester,[26] where Bohr met George de Hevesy and Charles Galton Darwin (whom Bohr referred to as "the grandson of the real Darwin").[27]

Bohr returned to Denmark in July 1912 for his wedding, and travelled around England and Scotland on his honeymoon. On his return, he became a *privatdocent* at the University of Copenhagen, giving lectures on thermodynamics. Martin Knudsen put Bohr's name forward for a *docent*, which was approved in July 1913, and Bohr then began teaching medical students.[28] His three papers, which later became famous as "the trilogy",[26] were published in *Philosophical Magazine* in July, September and November of that year.[29][30][31][32] He adapted Rutherford's nuclear structure to Max Planck's quantum theory and so created his Bohr model of the atom.[30]

Planetary models of atoms were not new, but Bohr's treatment was.[33] Taking the 1912 paper by Darwin on the role of electrons in the interaction of alpha particles with a nucleus as his starting point,[34][35] he advanced the theory of electrons travelling in orbits around the atom's nucleus, with the chemical properties of each element being largely determined by the number of electrons in the outer orbits of its atoms.[36] He introduced the idea that an electron could drop from a higher-energy orbit to a lower one, in the process emitting a quantum of discrete energy. This became a basis for what is now known as the old quantum theory.[37]

In 1885, Johann Balmer had come up with his Balmer series to describe the visible spectral lines of a hydrogen atoms:

$$\frac{1}{\lambda} = R_\text{H} \left(\frac{1}{2^2} - \frac{1}{n^2} \right) \quad \text{for } n = 3, 4, 5, \ldots$$

where λ is the wavelength of the absorbed or emitted light and RH is the Rydberg constant.[38] Balmer's formula was corroborated by the discovery of additional spectral lines, but for thirty years, no one could explain why it worked. In the first paper of his trilogy, Bohr was able to derive it from his model:

$$R_Z = \frac{2\pi^2 m_e Z^2 e^4}{h^3}$$

where m_e is the electron's mass, e is its charge, h is Planck's constant and Z is the atom's atomic number (1 for hydrogen).[39]

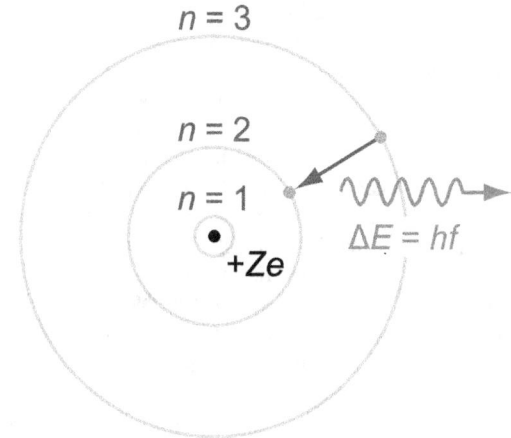

The **Bohr model** of the hydrogen atom. A negatively charged electron, confined to an atomic orbital, orbits a small, positively charged nucleus; a quantum jump between orbits is accompanied by an emitted or absorbed amount of electromagnetic radiation.

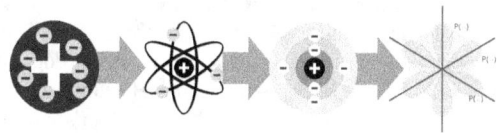

The evolution of atomic models in the 20th century: Thomson, Rutherford, Bohr, Heisenberg/Schrödinger

The model's first hurdle was the Pickering series, lines which did not fit Balmer's formula. When challenged on this by Alfred Fowler, Bohr replied that they were caused by ionised helium, helium atoms with only one electron. The Bohr model was found to work for such ions.[39] Many older physicists, like Thomson, Rayleigh and Hendrik Lorentz, did not like the trilogy, but the younger generation, including Rutherford, David Hilbert, Albert Einstein, Max Born and Arnold Sommerfeld saw it as a breakthrough.[40][41] The trilogy's acceptance was entirely due to its ability to explain phenomena which stymied other models, and to predict results that were subsequently verified by experiments.[42] Today, the Bohr model of the atom has been superseded, but is still the best known model of the atom, as it often appears in high school physics and chemistry texts.[43]

Bohr did not enjoy teaching medical students. He decided to return to Manchester, where Rutherford had offered him a job as a reader in place of Darwin, whose tenure had expired. Bohr accepted. He took a leave of absence from the University of Copenhagen, which he started by taking a holiday in Tyrol with his brother Harald and aunt Hanna

Adler. There, he visited the University of Göttingen and the Ludwig Maximilian University of Munich, where he met Sommerfeld and conducted seminars on the trilogy. The First World War broke out while they were in Tyrol, greatly complicating the trip back to Denmark and Bohr's subsequent voyage with Margrethe to England, where he arrived in October 1914. They stayed until July 1916, by which time he had been appointed to the Chair of Theoretical Physics at the University of Copenhagen, a position created especially for him. His docentship was abolished at the same time, so he still had to teach physics to medical students. New professors were formally introduced to King Christian X, who expressed his delight at meeting such a famous football player.[44]

22.2.2   Institute of Physics

In April 1917, Bohr began a campaign to establish an Institute of Theoretical Physics. He gained the support of the Danish government and the Carlsberg Foundation, and sizeable contributions were also made by industry and private donors, many of them Jewish. Legislation establishing the Institute was passed in November 1918. Now known as the Niels Bohr Institute, it opened its doors on 3 March 1921 with Bohr as its director. His family moved into an apartment on the first floor.[45][46] Bohr's institute served as a focal point for researchers into quantum mechanics and related subjects in the 1920s and 1930s, when most of the world's best known theoretical physicists spent some time in his company. Early arrivals included Hans Kramers from the Netherlands, Oskar Klein from Sweden, George de Hevesy from Hungary, Wojciech Rubinowicz from Poland and Svein Rosseland from Norway. Bohr became widely appreciated as their congenial host and eminent colleague.[47][48] Klein and Rosseland produced the Institute's first paper even before it opened.[46]

The Niels Bohr Institute

The Bohr model worked well for hydrogen, but could not explain more complex elements. By 1919, Bohr was moving away from the idea that electrons orbited the nucleus, and he developed heuristics to describe them. The rare earth elements posed a particular classification problem for chemists, because they were so chemically similar. An important development came in 1924 with Wolfgang Pauli's discovery of the Pauli exclusion principle, which put Bohr's models on a firm theoretical footing. Bohr was then able to declare that the as-yet-undiscovered element 72 was not a rare earth element, but an element with chemical properties similar to those of zirconium. He was immediately challenged by the French chemist Georges Urbain, who claimed to have discovered a rare earth element 72, which he called "celtium". At the Institute in Copenhagen, Dirk Coster and George de Hevesy took up the challenge of proving Bohr right and Urbain wrong. Starting with a clear idea of the chemical properties of the unknown element greatly simplified the search process. They went through samples from Copenhagen's Museum of Mineralogy looking for a zirconium-like element, and soon found it. The element, which they named hafnium, *Hafnia* being the Latin name for Copenhagen, turned out to be more common than gold.[49][50]

In 1922, Bohr was awarded the Nobel Prize in Physics "for his services in the investigation of the structure of atoms and of the radiation emanating from them".[51] The award thus recognised both the Trilogy and his early leading work in the emerging field of quantum mechanics. For his Nobel lecture, Bohr gave his audience a comprehensive survey of what was then known about the structure of the atom, including the correspondence principle, which he had formulated. This states that the behaviour of systems described by quantum theory reproduces classical physics in the limit of large quantum numbers.[52]

The discovery of Compton scattering by Arthur Holly Compton in 1923 convinced most physicists that light was composed of photons, and that energy and momentum were conserved in collisions between electrons and photons. In 1924, Bohr, Kramers and John C. Slater, an American physicist working at the Institute in Copenhagen, proposed the Bohr–Kramers–Slater theory (BKS). It was more a programme than a full physical theory, as the ideas it developed were not worked out quantitatively. BKS theory became the final attempt at understanding the interaction of matter and electromagnetic radiation on the basis of the old quantum theory, in which quantum phenomena were treated by imposing quantum restrictions on a classical wave description of the electromagnetic field.[53][54]

Modelling atomic behaviour under incident electromagnetic radiation using "virtual oscillators" at the absorption and emission frequencies, rather than the (different) apparent frequencies of the Bohr orbits, led Max Born, Werner Heisenberg and Kramers to explore different mathematical models. They led to the development of matrix mechanics, the first form of modern quantum mechanics. The BKS theory also generated discussion of, and renewed attention to, difficulties in the foundations of the old quantum theory.[55] The most provocative element of BKS – that momentum and energy would not necessarily be conserved in each interaction, but only statistically – was soon shown to be in conflict with experiments conducted by Walther Bothe and Hans Geiger.[56] In light of these results, Bohr informed Darwin, "there is nothing else to do than to give our revolutionary efforts as honourable a funeral as possible."[57]

22.2.3 Quantum mechanics

The introduction of spin by George Uhlenbeck and Samuel Goudsmit in November 1925 was a milestone. The next month, Bohr travelled to Leiden to attend celebrations of the 50th anniversary of Hendrick Lorentz receiving his doctorate. When his train stopped in Hamburg, he was met by Wolfgang Pauli and Otto Stern, who asked for his opinion of the spin theory. Bohr pointed out that he had concerns about the interaction between electrons and magnetic fields. When he arrived in Leiden, Paul Ehrenfest and Albert Einstein informed Bohr that Einstein had resolved this problem using relativity. Bohr then had Uhlenbeck and Goudsmit incorporate this into their paper. Thus, when he met Werner Heisenberg and Pascual Jordan in Göttingen on the way back, he had become, in his own words, "a prophet of the electron magnet gospel".[58]

1927 Solvay Conference in Brussels, October 1927. Bohr is on the right in the middle row, next to Max Born.

Heisenberg first came to Copenhagen in 1924, then re-turned to Göttingen in June 1925, shortly thereafter developing the mathematical foundations of quantum mechanics. When he showed his results to Max Born in Göttingen, Born realised that they could best be expressed using matrices. This work attracted the attention of the British physicist Paul Dirac,[59] who came to Copenhagen for six months in September 1926. Austrian physicist Erwin Schrödinger also visited in 1926. His attempt at explaining quantum physics in classical terms using wave mechanics impressed Bohr, who believed it contributed "so much to mathematical clarity and simplicity that it represents a gigantic advance over all previous forms of quantum mechanics".[60]

When Kramers left the Institute in 1926 to take up a chair as professor of theoretical physics at the Utrecht University, Bohr arranged for Heisenberg to return and take Kramers's place as a *lektor* at the University of Copenhagen.[61] Heisenberg worked in Copenhagen as a university lecturer and assistant to Bohr from 1926 to 1927.[62]

Bohr became convinced that light behaved like both waves and particles, and in 1927, experiments confirmed the de Broglie hypothesis that matter (like electrons) also behaved like waves.[63] He conceived the philosophical principle of complementarity: that items could have apparently mutually exclusive properties, such as being a wave or a stream of particles, depending on the experimental framework.[64] He felt that it was not fully understood by professional philosophers.[65]

In Copenhagen in 1927 Heisenberg developed his uncertainty principle,[66] which Bohr embraced. In a paper he presented at the Volta Conference at Como in September 1927, he demonstrated that the uncertainty principle could be derived from classical arguments, without quantum terminology or matrices.[66] Einstein preferred the determinism of classical physics over the probabilistic new quantum physics to which he himself had contributed. Philosophical issues that arose from the novel aspects of quantum mechanics became widely celebrated subjects of discussion. Einstein and Bohr had good-natured arguments over such issues throughout their lives.[67]

In 1914, Carl Jacobsen, the heir to Carlsberg breweries, bequeathed his mansion to be used for life by the Dane who had made the most prominent contribution to science, literature or the arts, as an honorary residence (Danish: *Æresbolig*). Harald Høffding had been the first occupant, and upon his death in July 1931, the Royal Danish Academy of Sciences and Letters gave Bohr occupancy. He and his family moved there in 1932.[68] He was elected president of the Academy on 17 March 1939.[69]

By 1929, the phenomenon of beta decay prompted Bohr to again suggest that the law of conservation of energy be abandoned, but Enrico Fermi's hypothetical neutrino and the subsequent 1932 discovery of the neutron provided another explanation. This prompted Bohr to create a new theory of the compound nucleus in 1936, which explained how neutrons could be captured by the nucleus. In this model, the nucleus could be deformed like a drop of liquid. He worked on this with a new collaborator, the Danish physicist Fritz Kalckar, who died suddenly in 1938.[70][71]

The discovery of nuclear fission by Otto Hahn in December 1938 (and its theoretical explanation by Lise Meitner) generated intense interest among physicists. Bohr brought the news to the United States where he opened the Fifth Washington Conference on Theoretical Physics with Fermi on 26 January 1939.[72] When Bohr told George Placzek that this resolved all the mysteries of transuranic elements, Placzek told him that one remained: the neutron capture energies of uranium did not match those of its decay. Bohr thought about it for a few minutes and then announced to Placzek, Léon Rosenfeld and John Wheeler that "I have understood everything."[73] Based on his liquid drop model of the nucleus, Bohr concluded that it was the uranium-235 isotope and not the more abundant uranium-238 that was primarily responsible for fission with thermal neutrons. In April 1940, John R. Dunning demonstrated that Bohr was correct.[72] In the meantime, Bohr and Wheeler developed a theoretical treatment which they published in a September 1939 paper on "The Mechanism of Nuclear Fission".[74]

22.3   Philosophy

Bohr read the 19th-century Danish Christian existentialist philosopher, Søren Kierkegaard. Richard Rhodes argued in *The Making of the Atomic Bomb* that Bohr was influenced by Kierkegaard through Høffding.[75] In 1909, Bohr sent his brother Kierkegaard's *Stages on Life's Way* as a birthday gift. In the enclosed letter, Bohr wrote, "It is the only thing I have to send home; but I do not believe that it would be very easy to find anything better ... I even think it is one of the most delightful things I have ever read." Bohr enjoyed Kierkegaard's language and literary style, but mentioned that he had some disagreement with Kierkegaard's philosophy.[76] Some of Bohr's biographers suggested that this disagreement stemmed from Kierkegaard's advocacy of Christianity, while Bohr was an atheist.[77][78][79]

There has been some dispute over the extent to which Kierkegaard influenced Bohr's philosophy and science. David Favrholdt argued that Kierkegaard had minimal influence over Bohr's work, taking Bohr's statement about disagreeing with Kierkegaard at face value,[80] while Jan Faye argued that one can disagree with the content of a theory while accepting its general premises and structure.[81][76]

22.4   Nazism and Second World War

The rise of Nazism in Germany prompted many scholars to flee their countries, either because they were Jewish or because they were political opponents of the Nazi regime. In 1933, the Rockefeller Foundation created a fund to help support refugee academics, and Bohr discussed this programme with the President of the Rockefeller Foundation, Max Mason, in May 1933 during a visit to the United States. Bohr offered the refugees temporary jobs at the Institute, provided them with financial support, arranged for them to be awarded fellowships from the Rockefeller Foundation, and ultimately found them places at institutions around the world. Those that he helped included Guido Beck, Felix Bloch, James Franck, George de Hevesy, Otto Frisch, Hilde Levi, Lise Meitner, George Placzek, Eugene Rabinowitch, Stefan Rozental, Erich Ernst Schneider, Edward Teller, Arthur von Hippel and Victor Weisskopf.[82]

In April 1940, early in the Second World War, Nazi Germany invaded and occupied Denmark.[83] To prevent the Germans from discovering Max von Laue's and James Franck's gold Nobel medals, Bohr had de Hevesy dissolve them in aqua regia. In this form, they were stored on a shelf at the Institute until after the war, when the gold was precipitated and the medals re-struck by the Nobel Foundation. Bohr kept the Institute running, but all the foreign scholars departed.[84]

22.4.1   Meeting with Heisenberg

Werner Heisenberg (left) with Bohr at the Copenhagen Conference in 1934

Bohr was aware of the possibility of using uranium-235 to construct an atomic bomb, referring to it in lectures in Britain and Denmark shortly before and after the war started, but he did not believe that it was technically feasible to extract a sufficient quantity of uranium-235.[85] In

September 1941, Heisenberg, who had become head of the German nuclear energy project, visited Bohr in Copenhagen. During this meeting the two men took a private moment outside, the content of which has caused much speculation, as both gave differing accounts. According to Heisenberg, he began to address nuclear energy, morality and the war, to which Bohr seems to have reacted by terminating the conversation abruptly while not giving Heisenberg hints about his own opinions.[86] Ivan Supek, one of Heisenberg's students and friends, claimed that the main subject of the meeting was Carl Friedrich von Weizsäcker, who had proposed trying to persuade Bohr to mediate peace between Britain and Germany.[87]

In 1957, Heisenberg wrote to Robert Jungk, who was then working on the book *Brighter than a Thousand Suns: A Personal History of the Atomic Scientists.* Heisenberg explained that he had visited Copenhagen to communicate to Bohr the views of several German scientists, that production of a nuclear weapon was possible with great efforts, and this raised enormous responsibilities on the world's scientists on both sides.[88] When Bohr saw Jungk's depiction in the Danish translation of the book, he drafted (but never sent) a letter to Heisenberg, stating that he never understood the purpose of Heisenberg's visit, was shocked by Heisenberg's opinion that Germany would win the war, and that atomic weapons could be decisive.[89]

Michael Frayn's 1998 play *Copenhagen* explores what might have happened at the 1941 meeting between Heisenberg and Bohr.[90] A BBC television film version of the play was first screened on 26 September 2002, with Stephen Rea as Bohr, and Daniel Craig as Heisenberg. The same meeting had previously been dramatised by the BBC's *Horizon* science documentary series in 1992, with Anthony Bate as Bohr, and Philip Anthony as Heisenberg.[91]

22.4.2 Manhattan Project

In September 1943, word reached Bohr and his brother Harald that the Nazis considered their family to be Jewish, since their mother, Ellen Adler Bohr, had been a Jew, and that they were therefore in danger of being arrested. The Danish resistance helped Bohr and his wife escape by sea to Sweden on 29 September.[92][93] The next day, Bohr persuaded King Gustaf V of Sweden to make public Sweden's willingness to provide asylum to Jewish refugees. On 2 October 1943, Swedish radio broadcast that Sweden was ready to offer asylum, and the mass rescue of the Danish Jews by their countrymen followed swiftly thereafter. Some historians claim that Bohr's actions led directly to the mass rescue, while others say that, though Bohr did all that he could for his countrymen, his actions were not a decisive influence on the wider events.[93][94][95][96] Eventually, over 7,000 Dan-

ish Jews escaped to Sweden.[97]

Bohr with James Franck, Albert Einstein and Isidor Isaac Rabi (LR)

When the news of Bohr's escape reached Britain, Lord Cherwell sent a telegram to Bohr asking him to come to Britain. Bohr arrived in Scotland on 6 October in a de Havilland Mosquito operated by the British Overseas Airways Corporation (BOAC). The Mosquitos were unarmed high-speed bomber aircraft that had been converted to carry small, valuable cargoes or important passengers. By flying at high speed and high altitude, they could cross German-occupied Norway, and yet avoid German fighters. Bohr, equipped with parachute, flying suit and oxygen mask, spent the three-hour flight lying on a mattress in the aircraft's bomb bay.[98] During the flight, Bohr did not wear his flying helmet as it was too small, and consequently did not hear the pilot's intercom instruction to turn on his oxygen supply when the aircraft climbed to high altitude to overfly Norway. He passed out from oxygen starvation and only revived when the aircraft descended to lower altitude over the North Sea.[99][100][101] Bohr's son Aage followed his father to Britain on another flight a week later, and became his personal assistant.[102][103]

Bohr was warmly received by James Chadwick and Sir John Anderson, but for security reasons Bohr was kept out of sight. He was given an apartment at St James's Palace and an office with the British Tube Alloys nuclear weapons development team. Bohr was astonished at the amount of progress that had been made.[102][103] Chadwick arranged for Bohr to visit the United States as a Tube Alloys consultant, with Aage as his assistant.[104] On 8 December 1943, Bohr arrived in Washington, D.C., where he met with the director of the Manhattan Project, Brigadier General Leslie R. Groves, Jr. He visited Einstein and Pauli at the Institute for Advanced Study in Princeton, New Jersey, and went to Los Alamos in New Mexico, where the nuclear weapons were being designed.[105] For security reasons, he went under the name of "Nicholas Baker" in the United States, while Aage became "James Baker".[106] In May 1944 the Danish resistance newspaper De frie Danske reported that they had learned that 'the famous son of Denmark Professor Niels Bohr' in October the previous year had fled his country via Sweden to London and from there travelled to Moscow from where he could be assumed to support the war effort.[107]

Bohr did not remain at Los Alamos, but paid a series of extended visits over the course of the next two years. Robert Oppenheimer credited Bohr with acting "as a scientific father figure to the younger men", most notably Richard Feynman.[108] Bohr is quoted as saying, "They didn't need my help in making the atom bomb."[109] Oppenheimer gave Bohr credit for an important contribution to the work on modulated neutron initiators. "This device remained a stubborn puzzle," Oppenheimer noted, "but in early February 1945 Niels Bohr clarified what had to be done."[108]

Bohr recognised early that nuclear weapons would change international relations. In April 1944, he received a letter from Peter Kapitza, written some months before when Bohr was in Sweden, inviting him to come to the Soviet Union. The letter convinced Bohr that the Soviets were aware of the Anglo-American project, and would strive to catch up. He sent Kapitza a non-committal response, which he showed to the authorities in Britain before posting.[110] Bohr met Churchill on 16 May 1944, but found that "we did not speak the same language".[111] Churchill disagreed with the idea of openness towards the Russians to the point that he wrote in a letter: "It seems to me Bohr ought to be confined or at any rate made to see that he is very near the edge of mortal crimes."[112]

Oppenheimer suggested that Bohr visit President Franklin D. Roosevelt to convince him that the Manhattan Project should be shared with the Soviets in the hope of speeding up its results. Bohr's friend, Supreme Court Justice Felix Frankfurter, informed President Roosevelt about Bohr's opinions, and a meeting between them took place on 26 August 1944. Roosevelt suggested that Bohr return to the United Kingdom to try to win British approval.[113][114] When Churchill and Roosevelt met at Hyde Park on 19 September 1944, they rejected the idea of informing the world about the project, and the aide-mémoire of their conversation contained a rider that "enquiries should be made regarding the activities of Professor Bohr and steps taken to ensure that he is responsible for no leakage of information, particularly to the Russians".[115]

In June 1950, Bohr addressed an "Open Letter" to the United Nations calling for international cooperation on nuclear energy.[116][117][118] In the 1950s, after the Soviet Union's first nuclear weapon test, the International Atomic Energy Agency was created along the lines of Bohr's

suggestion.[119] In 1957 he received the first ever Atoms for Peace Award.[120]

22.5   Later years

Niels Bohr's coat of arms

With the war now ended, Bohr returned to Copenhagen on 25 August 1945, and was re-elected President of the Royal Danish Academy of Arts and Sciences on 21 September.[121] At a memorial meeting of the Academy on 17 October 1947 for King Christian X, who had died in April, the new king, Frederick IX, announced that he was conferring the Order of the Elephant on Bohr. This award was normally awarded only to royalty and heads of state, but the king said that it honoured not just Bohr personally, but Danish science.[122][123] Bohr designed his own coat of arms which featured a taijitu (symbol of yin and yang) and a motto in Latin: *contraria sunt complementa*, "opposites are complementary".[124][123]

The Second World War demonstrated that science, and physics in particular, now required considerable financial and material resources. To avoid a brain drain to the United States, twelve European countries banded together to create CERN, a research organisation along the lines of the national laboratories in the United States, designed to undertake Big Science projects beyond the resources of any one of them alone. Questions soon arose regarding the best location for the facilities. Bohr and Kramers felt that the Institute in Copenhagen would be the ideal site. Pierre Auger, who organised the preliminary discussions, disagreed; he felt that both Bohr and his Institute were past their prime, and that Bohr's presence would overshadow others. After a long debate, Bohr pledged his support to CERN in February 1952, and Geneva was chosen as the site in October. The CERN Theory Group was based in Copenhagen until their new accommodation in Geneva was ready in 1957.[125] Victor Weisskopf, who later became the Director General of CERN, summed up Bohr's role, saying that "there were other personalities who started and conceived the idea of CERN. The enthusiasm and ideas of the other people would not have been enough, however, if a man of his stature had not supported it."[126][127]

Meanwhile, Scandinavian countries formed the Nordic Institute for Theoretical Physics in 1957, with Bohr as its chairman. He was also involved with the founding of the Research Establishment Risø of the Danish Atomic Energy Commission, and served as its first chairman from February 1956.[128]

Bohr died of heart failure at his home in Carlsberg on 18 November 1962.[129] He was cremated, and his ashes were buried in the family plot in the Assistens Cemetery in the Nørrebro section of Copenhagen, along with those of his parents, his brother Harald, and his son Christian. Years later, his wife's ashes were also interred there.[130] On 7 October 1965, on what would have been his 80th birthday, the Institute was officially renamed to what it had been called unofficially for many years: the Niels Bohr Institute.[131][132]

22.6   Accolades

See also: List of things named after Niels Bohr

Bohr received numerous honours and accolades. In addition to the Nobel Prize, he received the Hughes Medal in 1921, the Matteucci Medal in 1923,[133] the Franklin Medal in 1926,[134] the Copley Medal in 1938, the Order of the Elephant in 1947, the Atoms for Peace Award in 1957 and the Sonning Prize in 1961.[133] In 1923 he became foreign member of the Royal Netherlands Academy of Arts and Sciences.[135] The Bohr model's semicentennial was commemorated in Denmark on 21 November 1963 with a postage stamp depicting Bohr, the hydrogen atom and the formula for the difference of any two hydrogen en-

ergy levels: $h\nu = \epsilon_2 - \epsilon_1$. Several other countries have also issued postage stamps depicting Bohr.[136] In 1997, the Danish National Bank began circulating the 500-krone banknote with the portrait of Bohr smoking a pipe.[137][138] An asteroid, 3948 Bohr, was named after him,[139] as was the Bohr lunar crater,[133] and bohrium, the chemical element with atomic number 107.[140]

22.7   Bibliography

- Bohr, Niels (2008). Nielsen, J. Rud, ed. *Volume 1: Early Work (1905–1911)*. Niels Bohr Collected Works. Amsterdam: Elsevier. ISBN 978-0-444-53286-2. OCLC 272382249.

- —— (2008). Hoyer, Ulrich, ed. *Volume 2: Work on Atomic Physics (1912–1917)*. Niels Bohr Collected Works. Amsterdam: Elsevier. ISBN 978-0-444-53286-2. OCLC 272382249.

- —— (2008). Nielsen, J. Rud, ed. *Volume 3: The Correspondence Principle (1918–1923)*. Niels Bohr Collected Works. Amsterdam: Elsevier. ISBN 978-0-444-53286-2. OCLC 272382249.

- —— (2008). Nielsen, J. Rud, ed. *Volume 4: The Periodic System (1920–1923)*. Niels Bohr Collected Works. Amsterdam: Elsevier. ISBN 978-0-444-53286-2. OCLC 272382249.

- —— (2008). Stolzenburg, Klaus, ed. *Volume 5: The Emergence of Quantum Mechanics (mainly 1924–1926)*. Niels Bohr Collected Works. Amsterdam: Elsevier. ISBN 978-0-444-53286-2. OCLC 272382249.

- —— (2008). Kalckar, Jørgen, ed. *Volume 6: Foundations of Quantum Physics I (1926–1932)*. Niels Bohr Collected Works. Amsterdam: Elsevier. ISBN 978-0-444-53286-2. OCLC 272382249.

- —— (2008). Kalckar, Jørgen, ed. *Volume 7: Foundations of Quantum Physics I (1933–1958)*. Niels Bohr Collected Works. Amsterdam: Elsevier. ISBN 978-0-444-53286-2. OCLC 272382249.

- —— (2008). Thorsen, Jens, ed. *Volume 8: The Penetration of Charged Particles Through Matter (1912–1954)*. Niels Bohr Collected Works. Amsterdam: Elsevier. ISBN 978-0-444-53286-2. OCLC 272382249.

- —— (2008). Peierls, Rudolf, ed. *Volume 9: Nuclear Physics (1929–1952)*. Niels Bohr Collected Works. Amsterdam: Elsevier. ISBN 978-0-444-53286-2. OCLC 272382249.

- —— (2008). Favrholdt, David, ed. *Volume 10: Complementarity Beyond Physics (1928–1962)*. Niels Bohr Collected Works. Amsterdam: Elsevier. ISBN 978-0-444-53286-2. OCLC 272382249.

- —— (2008). Aaserud, Finn, ed. *Volume 11: The Political Arena (1934–1961)*. Niels Bohr Collected Works. Amsterdam: Elsevier. ISBN 978-0-444-53286-2. OCLC 272382249.

- —— (2008). Aaserud, Finn, ed. *Volume 12: Popularization and People (1911–1962)*. Niels Bohr Collected Works. Amsterdam: Elsevier. ISBN 978-0-444-53286-2. OCLC 272382249.

- —— (2008). Aaserud, Finn, ed. *Volume 13: Cumulative Subject Index*. Niels Bohr Collected Works. Amsterdam: Elsevier. ISBN 978-0-444-53286-2. OCLC 272382249.

22.8   Notes

[1] Cockcroft, J. D. (1963). "Niels Henrik David Bohr. 1885-1962". *Biographical Memoirs of Fellows of the Royal Society* **9**: 36–53. doi:10.1098/rsbm.1963.0002.

[2] "Niels Bohr centenary". *CERN Courier* **25** (10): 430–432. December 1985.

[3] *Politiets Registerblade* [*Register cards of the Police*] (in Danish). Copenhagen: Københavns Stadsarkiv. 7 June 1892. Station Dødeblade (indeholder afdøde i perioden). Filmrulle 0002. Registerblad 3341. ID 3308989.

[4] Pais 1991, pp. 44–45, 538–539.

[5] Pais 1991, pp. 35–39.

[6] There is no truth in the oft-repeated claim that Niels Bohr emulated his brother, Harald, by playing for the Danish national team. Dart, James (27 July 2005). "Bohr's footballing career". *The Guardian* (London). Retrieved 26 June 2011.

[7] "Niels Bohr's school years". Niels Bohr Institute. Retrieved 14 February 2013.

[8] Pais 1991, pp. 98–99.

[9] "Life as a Student". Niels Bohr Institute. Retrieved 14 February 2013.

[10] Rhodes 1986, pp. 62–63.

[11] Pais 1991, pp. 101–102.

[12] Aaserud & Heilbron 2013, p. 155.

[13] Pais 1991, pp. 107–109.

[14] Kragh 2012, pp. 43-45.

[15] Pais 1991, p. 112.

[16] Pais 1991, pp. 133–134.

[17] Pais 1991, pp. 226, 249.

[18] Stuewer 1985, p. 204.

[19] "Niels Bohr – Biography". Nobelprize.org. Retrieved 10 November 2011.

[20] "Ernest Bohr Biography and Olympic Results – Olympics". Sports-Reference.com. Retrieved 12 February 2013.

[21] Kragh 2012, p. 122.

[22] Kennedy 1985, p. 6.

[23] Pais 1991, pp. 117–121.

[24] Kragh 2012, p. 46.

[25] Pais 1991, pp. 121–125.

[26] Kennedy 1985, p. 7.

[27] Pais 1991, pp. 125–129.

[28] Pais 1991, pp. 134–135.

[29] Bohr, Niels (1913). "On the Constitution of Atoms and Molecules, Part I" (PDF). *Philosophical Magazine* **26** (151): 1–24. doi:10.1080/14786441308634955.

[30] Bohr, Niels (1913). "On the Constitution of Atoms and Molecules, Part II Systems Containing Only a Single Nucleus" (PDF). *Philosophical Magazine* **26** (153): 476–502. doi:10.1080/14786441308634993.

[31] Bohr, Niels (1913). "On the Constitution of Atoms and Molecules, Part III Systems containing several nuclei". *Philosophical Magazine* **26** (155): 857–875. doi:10.1080/14786441308635031.

[32] Pais 1991, p. 149.

[33] Kragh 2012, p. 22.

[34] Darwin, Charles Galton (1912). "A theory of the absorption and scattering of the alpha rays". *Philosophical Magazine* **23** (138): 901–920. doi:10.1080/14786440608637291. ISSN 1941-5982.

[35] Arabatzis, Theodore (2006). *Representing Electrons: A Biographical Approach to Theoretical Entities*. University of Chicago Press. p. 118. ISBN 978-0-226-02420-2.

[36] Kragh 1985, pp. 50–67, 385–391.

[37] Heilbron 1985, pp. 39–47.

[38] Heilbron 1985, p. 43.

[39] Pais 1991, pp. 146–149.

[40] Pais 1991, pp. 152–155.

[41] Kragh 2012, pp. 109–111.

[42] Kragh 2012, pp. 90–91.

[43] Kragh 2012, p. 39.

[44] Pais 1991, pp. 164–167.

[45] Aaserud, Finn. "History of the institute: The establishment of an institute". Niels Bohr Institute. Archived from the original on 5 April 2008. Retrieved 11 May 2008.

[46] Pais 1991, pp. 169–171.

[47] Kennedy 1985, pp. 9, 12, 13, 15.

[48] Hund 1985, pp. 71–73.

[49] Kragh 1985, pp. 61–64.

[50] Pais 1991, pp. 202–210.

[51] Pais 1991, p. 215.

[52] Bohr 1985, pp. 91–97.

[53] Bohr, N.; Kramers, H.A.; Slater, J.C. (1924). "The Quantum Theory of Radiation" (PDF). *Philosophical Magazine.* 6 **76** (287): 785–802. doi:10.1080/14786442408565262. Retrieved 18 February 2013.

[54] Pais 1991, pp. 232–239.

[55] Jammer 1989, p. 188.

[56] Pais 1991, p. 237.

[57] Pais 1991, p. 238.

[58] Pais 1991, p. 243.

[59] Pais 1991, pp. 275–279.

[60] Pais 1991, pp. 295–299.

[61] Pais 1991, p. 263.

[62] Pais 1991, pp. 272–275.

[63] Pais 1991, p. 301.

[64] MacKinnon 1985, pp. 112–113.

[65] MacKinnon 1985, p. 101.

[66] Pais 1991, pp. 304–309.

[67] Bohr 1985, pp. 121–140.

[68] Pais 1991, pp. 332–333.

[69] Pais 1991, pp. 464–465.

[70] Pais 1991, pp. 337–340, 368–370.

[71] Bohr, Niels (20 August 1937). "Transmutations of Atomic Nuclei". *Science* **86** (2225): 161–165. Bibcode:1937Sci....86..161B. doi:10.1126/science.86.2225.161.

[72] Stuewer 1985, pp. 211–214.

[73] Pais 1991, p. 456.

[74] Bohr, Niels; Wheeler, John Archibald (September 1939). "The Mechanism of Nuclear Fission" (PDF). *Physical Review* **56** (5): 426–450. Bibcode:1939PhRv...56..426B. doi:10.1103/PhysRev.56.426.

[75] Rhodes 1986, p. 60.

[76] Faye 1991, p. 37.

[77] Stewart 2010, p. 416.

[78] Aaserud & Heilbron 2013, pp. 159–160: "A statement about religion in the loose notes on Kierkegaard may throw light on the notion of wildness that appears in many of Bohr's letters. 'I, who do not feel in any way united with, and even less, bound to a God, and therefore am also much poorer [than Kierkegaard], would say that the good [is] the overall lofty goal, as only by being good [can one] judge according to worth and right.'"

[79] Aaserud & Heilbron 2013, p. 110: "Bohr's sort of humor, use of parables and stories, tolerance, dependence on family, feelings of indebtedness, obligation, and guilt, and his sense of responsibility for science, community, and, ultimately, humankind in general, are common traits of the Jewish intellectual. So too is a well-fortified atheism. Bohr ended with no religious belief and a dislike of all religions that claimed to base their teachings on revelations."

[80] Favrholdt 1992, pp. 42–63.

[81] Richardson & Wildman 1996, p. 289.

[82] Pais 1991, pp. 382–386.

[83] Pais 1991, p. 476.

[84] Pais 1991, pp. 480–481.

[85] Gowing 1985, pp. 267–268.

[86] Heisenberg 1984, p. 77.

[87] Portal Jutarnji.hr (19 March 2006). "Moj život s nobelovcima 20. stoljeća" [My Life with the 20th century Nobel Prizewinners]. *Jutarnji list* (in Croatian). Retrieved 13 August 2007. *Istinu sam saznao od Margrethe, Bohrove supruge. ... Ni Heisenberg ni Bohr nisu bili glavni junaci toga susreta nego Carl Friedrich von Weizsaecker. ... Von Weizsaeckerova ideja, za koju mislim da je bila zamisao njegova oca koji je bio Ribbentropov zamjenik, bila je nagovoriti Nielsa Bohra da posreduje za mir između Velike Britanije i Njemačke.* [I learned the truth from Margrethe, Bohr's wife. ... Neither Bohr nor Heisenberg were the main characters of this encounter, but Carl Friedrich von Weizsaecker. Von Weizsaecker's idea, which I think was the brainchild of his father who was Ribbentrop's deputy, was to persuade Niels Bohr to mediate for peace between Great Britain and Germany.] An interview with Ivan Supek relating to the 1941 Bohr – Heisenberg meeting.

[88] Heisenberg, Werner. "Letter From Werner Heisenberg to Author Robert Jungk". The Manhattan Project Heritage Preservation Association, Inc. Archived from the original on 17 October 2006. Retrieved 21 December 2006.

[89] Aaserud, Finn (6 February 2002). "Release of documents relating to 1941 Bohr-Heisenberg meeting". Niels Bohr Archive. Retrieved 4 June 2007.

[90] "Copenhagen – Michael Frayn". The Complete Review. Retrieved 27 February 2013.

[91] *Horizon: Hitler's Bomb*, BBC Two, 24 February 1992

[92] Rozental 1967, p. 168.

[93] Rhodes 1986, pp. 483–484.

[94] Hilberg 1961, p. 596.

[95] Kieler 2007, pp. 91–93.

[96] Stadtler, Morrison & Martin 1995, p. 136.

[97] Pais 1991, p. 479.

[98] Thirsk 2006, p. 374.

[99] Rife 1999, p. 242.

[100] Medawar & Pyke 2001, p. 65.

[101] Jones 1978, pp. 474–475.

[102] Jones 1985, pp. 280–282.

[103] Pais 1991, p. 491.

[104] Cockroft 1963, p. 46.

[105] Pais 1991, pp. 498–499.

[106] Gowing 1985, p. 269.

[107] "Professor Bohr ankommet til Moskva" [Professor Bohr arrived in Moscow]. *De frie Danske* (in Danish). May 1944. p. 7. Retrieved 18 November 2014.

[108] Pais 1991, p. 497.

[109] Pais 1991, p. 496.

[110] Gowing 1985, p. 270.

[111] Gowing 1985, p. 271.

[112] Aaserud 2006, p. 708.

[113] Rhodes 1986, pp. 528–538.

[114] Aaserud 2006, pp. 707–708.

[115] U.S. Government 1972, pp. 492–493.

[116] Aaserud 2006, pp. 708–709.

[117] Bohr, Niels (9 June 1950). "To the United Nations (open letter)". *Impact of Science on Society* I (2): 68. Retrieved 12 June 2012.
• Bohr, Niels (July 1950). "For An Open World". *Bulletin of the Atomic Scientists* 6 (7): 213–219. Retrieved 26 June 2011.

[118] Pais 1991, pp. 288–296.

[119] Gowing 1985, p. 276.

[120] Craig-McCormack, Elizabeth. "Guide to Atoms for Peace Awards Records" (PDF). Massachusetts Institute of Technology. Retrieved 28 February 2013.

[121] Pais 1991, p. 504.

[122] Pais 1991, pp. 166, 466–467.

[123] Wheeler 1985, p. 224.

[124] "Bohr crest". University of Copenhagen. 17 October 1947. Retrieved 16 March 2007.

[125] Pais 1991, pp. 519–522.

[126] Pais 1991, p. 521.

[127] Weisskopf, Victor (July 1963). "Tribute to Niels Bohr". *CERN Courier* 2 (11): 89.

[128] Pais 1991, pp. 523–525.

[129] "Niels Bohr". *CERN Courier* 2 (11): 10. November 1962.

[130] Pais 1991, p. 529.

[131] "History of the Niels Bohr Institute from 1921 to 1965". Niels Bohr Institute. Retrieved 28 February 2013.

[132] Reinhard, Stock (October 1998). "Niels Bohr and the 20th century". *CERN Courier* 38 (7): 19.

[133] "Niels Bohr". Soylent Communications. Retrieved 21 October 2013.

[134] "Niels Bohr – The Franklin Institute Awards – Laureate Database". Franklin Institute. Retrieved 21 October 2013.

[135] "N.H.D. Bohr (1885 - 1962)". Royal Netherlands Academy of Arts and Sciences. Retrieved 21 July 2015.

[136] Kennedy 1985, pp. 10–11.

[137] Danmarks Nationalbank 2005, pp. 20–21.

[138] "500-krone banknote, 1997 series". Danmarks Nationalbank. Retrieved 7 September 2010.

[139] Klinglesmith, Daniel A., III; Risley, Ethan; Turk, Janek; Vargas, Angelica; Warren, Curtis; Ferrero, Andera (January–March 2013). "Lightcurve Analysis Of 3948 Bohr and 4874 Burke: An International Collaboration" (PDF). *Minor Planet Bulletin* 40 (1): 15. Bibcode:2013MPBu...40...15K. Retrieved 28 February 2013.

[140] "Names and symbols of transfermium elements (IUPAC Recommendations 1997)". *Pure and Applied Chemistry* **69** (12): 2472. 1997. doi:10.1351/pac199769122471.

22.9 References

- Aaserud, Finn (2006). Kokowski, M., ed. *Niels Bohr's Mission for an 'Open World'* (PDF). Proceedings of the 2nd ICESHS. Cracow. pp. 706–709. Retrieved 26 June 2011.

- Aaserud, Finn; Heilbron, J. L. (2013). *Love, Literature and the Quantum Atom: Niels Bohr's 1913 Trilogy Revisited.* Oxford: Oxford University Press. ISBN 978-0-19-968028-3.

- Bohr, Niels (1985) [1922]. "Nobel Prize Lecture: The Structure of the Atom (excerpts)". In French, A. P.; Kennedy, P. J. *Niels Bohr: A Centenary Volume.* Cambridge, Massachusetts: Harvard University Press. pp. 91–97. ISBN 978-0-674-62415-3.

- Bohr, Niels (1985) [1949]. "The Bohr-Einstein Dialogue". In French, A. P.; Kennedy, P. J. *Niels Bohr: A Centenary Volume.* Cambridge, Massachusetts: Harvard University Press. pp. 121–140. ISBN 978-0-674-62415-3.
  - Excerpted from: Bohr, Niels (1949). "Discussions with Einstein on Epistemological Problems in Atomic Physics". In Paul Arthur Schilpp. *Albert Einstein: Philosopher-Scientist.* Evanston, Illinois: Library of Living Philosophers. pp. 208–241.

- Cockroft, John D. (1 November 1963). "Niels Henrik David Bohr. 1885–1962". *Biographical Memoirs of the Fellows of the Royal Society* **9**: 36–53. doi:10.1098/rsbm.1963.0002.

- Favrholdt, David (1992). *Niels Bohr's Philosophical Background.* Copenhagen: Munksgaard. ISBN 978-87-7304-228-1.

- Faye, January (1991). *Niels Bohr: His Heritage and Legacy.* Dordrecht: Kluwer Academic Publishers. ISBN 978-0-7923-1294-9.

- Gowing, Margaret (1985). "Niels Bohr and Nuclear Weapons". In French, A. P.; Kennedy, P. J. *Niels Bohr: A Centenary Volume.* Cambridge, Massachusetts: Harvard University Press. pp. 266–277. ISBN 978-0-674-62415-3.

- Heilbron, John L. (1985). "Bohr's First Theories of the Atom". In French, A. P.; Kennedy, P. J. *Niels Bohr: A Centenary Volume.* Cambridge, Massachusetts: Harvard University Press. pp. 33–49. ISBN 978-0-674-62415-3.

- Heisenberg, Elisabeth (1984). *Inner Exile: Recollections of a Life With Werner Heisenberg.* Boston: Birkhäuser. ISBN 978-0-8176-3146-8.

- Hilberg, Raul (1961). *The Destruction of the European Jews* **2**. New Haven, Connecticut: Yale University Press. ISBN 978-0-300-09557-9.

- Hund, Friedrich (1985). "Bohr, Göttingen, and Quantum Mechanics". In French, A. P.; Kennedy, P. J. *Niels Bohr: A Centenary Volume.* Cambridge, Massachusetts: Harvard University Press. pp. 71–75. ISBN 978-0-674-62415-3.

- Jammer, Max (1989). *The Conceptual Development of Quantum Mechanics.* Los Angeles: Tomash Publishers. ISBN 0-88318-617-9. OCLC 19517065.

- Jones, R . V. (1978). *Most Secret War.* London: Hamilton. ISBN 0-241-89746-7. OCLC 3717534.

- Jones, R. V. (1985). "Meetings in Wartime and After". In French, A. P.; Kennedy, P. J. *Niels Bohr: A Centenary Volume.* Cambridge, Massachusetts: Harvard University Press. pp. 278–287. ISBN 978-0-674-62415-3.

- Kennedy, P. J. (1985). "A Short Biography". In French, A. P.; Kennedy, P. J. *Niels Bohr: A Centenary Volume.* Cambridge, Massachusetts: Harvard University Press. pp. 3–15. ISBN 978-0-674-62415-3.

- Kieler, Jørgen (2007). *Resistance Fighter: A Personal History of the Danish Resistance.* Translated from the Danish by Eric Dickens. Jerusalem: Gefen Publishing House. ISBN 978-965-229-397-8.

- Kragh, Helge (1985). "The Theory of the Periodic System". In French, A. P.; Kennedy, P. J. *Niels Bohr: A Centenary Volume.* Cambridge, Massachusetts: Harvard University Press. pp. 50–67. ISBN 978-0-674-62415-3.

- Kragh, Helge (2012). *Niels Bohr and the quantum atom: the Bohr model of atomic structure, 1913–1925.* Oxford: Oxford University Press. ISBN 978-0-19-965498-7. OCLC 769989390.

- Medawar, Jean; Pyke, David (2001). *Hitler's Gift: The True Story of the Scientists Expelled by the Nazi Regime.* New York: Arcade Publishing. ISBN 1-55970-564-7.

- MacKinnon, Edward (1985). "Bohr on the Foundations of Quantum Theory". In French, A. P.;

Kennedy, P. J. *Niels Bohr: A Centenary Volume.* Cambridge, Massachusetts: Harvard University Press. pp. 101–120. ISBN 978-0-674-62415-3.

- Pais, Abraham (1991). *Niels Bohr's Times, In Physics, Philosophy and Polity.* Oxford: Clarendon Press. ISBN 978-0-19-852049-8.

- Rhodes, Richard (1986). *The Making of the Atomic Bomb.* New York: Simon and Schuster. ISBN 978-0-671-44133-3.

- Richardson, W. Mark; Wildman, Wesley J., eds. (1996). *Religion and Science: History, Method, Dialogue.* London, New York: Routledge. ISBN 978-0-415-91667-7.

- Rife, Patricia (1999). *Lise Meitner and the Dawn of the Nuclear Age.* Boston: Birkhäuser. ISBN 0-8176-3732-X.

- Rozental, Stefan (1967). *Niels Bohr: His Life and Work as Seen by his Friends and Colleagues.* Amsterdam: North-Holland. ISBN 978-0-444-86977-7. Previously published by John Wiley & Sons in 1964.

- Stadtler, Bea; Morrison, David Beal; Martin, David Stone (1995). *The Holocaust: A History of Courage and Resistance.* West Orange, New Jersey: Behrman House. ISBN 978-0-87441-578-0.

- Stewart, Melville Y. (2010). *Science and Religion in Dialogue, Two Volume Set.* Maiden, Massachusetts: John Wiley & Sons. ISBN 978-1-4051-8921-7.

- Stuewer, Roger H. (1985). "Niels Bohr and Nuclear Physics". In French, A. P.; Kennedy, P. J. *Niels Bohr: A Centenary Volume.* Cambridge, Massachusetts: Harvard University Press. pp. 197–220. ISBN 978-0-674-62415-3.

- Thirsk, Ian (2006). *De Havilland Mosquito: An Illustrated History, Volume 2.* Manchester: MBI Publishing Company. ISBN 0-85979-115-7.

- *The Conferences at Quebec 1944.* Foreign Relations of the United States. Washington, D.C.: U.S. Government Printing Office. 1972. OCLC 631921397.

- Wheeler, John A. (1985). "Physics in Copenhagen in 1934 and 1935". In French, A. P.; Kennedy, P. J. *Niels Bohr: A Centenary Volume.* Cambridge, Massachusetts: Harvard University Press. pp. 221–226. ISBN 978-0-674-62415-3.

- *The Coins and Banknotes of Denmark* (PDF). Danmarks Nationalbank. 2005. ISBN 978-87-87251-55-6. Archived from the original (PDF) on 23 May 2011. Retrieved 7 September 2010.

22.10 Further reading

- Aaserud, Finn (February 2002). "Release of documents relating to 1941 Bohr-Heisenberg meeting". Niels Bohr Archive. Retrieved 2 March 2013.

- Blaedel, Niels (1988). *Harmony and Unity: The Life of Niels Bohr.* Madison, Wisconsin: Science Tech. ISBN 0-910239-14-2. OCLC 17411890.

- Feilden, Tom (3 February 2010). "The Gunfighter's Dilemma". *news.bbc.co.uk.* Retrieved 2 March 2013. Bohr's researches on reaction times.

- Moore, Ruth (1966). *Niels Bohr: The Man, His Science, and the World They Changed.* New York: Knopf. ISBN 0-262-63101-6. OCLC 712016.

- Ottaviani, Jim; Purvis, Leland (2004). *Suspended In Language: Niels Bohr's Life, Discoveries, and the Century He Shaped.* Ann Arbor, Michigan: G.T. Labs. ISBN 0-9660106-5-5. OCLC 55739245.

- Frayn, Michael (2000). *Copenhagen.* New York: Anchor Books. ISBN 0-413-72490-5. OCLC 44467534.

- Segrè, Gino (2007). *Faust in Copenhagen: A Struggle for the Soul of Physics.* New York: Viking. ISBN 0-670-03858-X. OCLC 76416691.

- Vilhjálmsson, Vilhjálmur Örn; Blüdnikow, Bent (2006). "Rescue, Expulsion, and Collaboration: Denmark's Difficulties with its World War II Past". *Jewish Political Studies Review* **18**: 3–4. ISSN 0792-335X. Retrieved 29 June 2011.

22.11 External links

- Physics Tree: Niels Bohr Details

- "Niels Bohr Archive". Niels Bohr Archive. February 2002. Retrieved 2 March 2013.

- "The Bohr-Heisenberg meeting in September 1941". American Institute of Physics. Retrieved 2 March 2013.

- "Resources for Frayn's *Copenhagen*: Niels Bohr". Massachusetts Institute of Technology. Retrieved 9 October 2013.

- "Oral History interview transcript with Niels Bohr 31 October 1962". American Institute of Physics. Retrieved 2 March 2013.

- "Video – Niels Bohr (1962) : Atomic Physics and Human Knowledge". Lindau Nobel Laureate Meetings. Retrieved 9 July 2014.

- Author profile in the database zbMATH

Chapter 23

Bohr model

'Rutherford–Bohr model' and 'Bohr–Rutherford diagram' redirect to this page. 'Bohr model' is not to be confused with Bohr equation.

In atomic physics, the Rutherford–Bohr model or **Bohr**

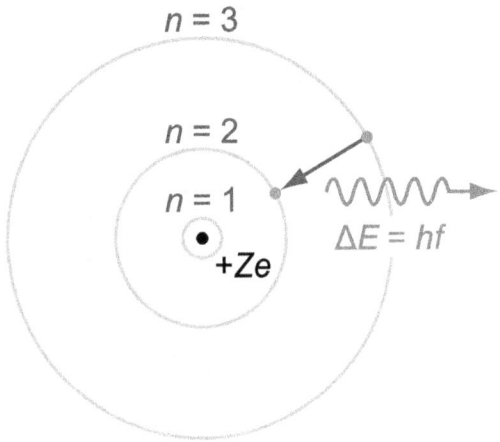

*The **Rutherford–Bohr model** of the hydrogen atom (Z = 1) or a hydrogen-like ion (Z > 1), where the negatively charged electron confined to an atomic shell encircles a small, positively charged atomic nucleus and where an electron jump between orbits is accompanied by an emitted or absorbed amount of electromagnetic energy (hν).[1] The orbits in which the electron may travel are shown as grey circles; their radius increases as n^2, where n is the principal quantum number. The 3 → 2 transition depicted here produces the first line of the Balmer series, and for hydrogen (Z = 1) it results in a photon of wavelength 656 nm (red light).*

model, introduced by Niels Bohr in 1913, depicts the atom as a small, positively charged nucleus surrounded by electrons that travel in circular orbits around the nucleus— similar in structure to the solar system, but with attraction provided by electrostatic forces rather than gravity. After the cubic model (1902), the plum-pudding model (1904), the Saturnian model (1904), and the Rutherford model (1911) came the **Rutherford–Bohr model** or just *Bohr model* for short (1913). The improvement to the Ruther-

ford model is mostly a quantum physical interpretation of it. The Bohr model has been superseded, but the quantum theory remains sound.

The model's key success lay in explaining the Rydberg formula for the spectral emission lines of atomic hydrogen. While the Rydberg formula had been known experimentally, it did not gain a theoretical underpinning until the Bohr model was introduced. Not only did the Bohr model explain the reason for the structure of the Rydberg formula, it also provided a justification for its empirical results in terms of fundamental physical constants.

The Bohr model is a relatively primitive model of the hydrogen atom, compared to the valence shell atom. As a theory, it can be derived as a first-order approximation of the hydrogen atom using the broader and much more accurate quantum mechanics and thus may be considered to be an obsolete scientific theory. However, because of its simplicity, and its correct results for selected systems (see below for application), the Bohr model is still commonly taught to introduce students to quantum mechanics or energy level diagrams before moving on to the more accurate, but more complex, valence shell atom. A related model was originally proposed by Arthur Erich Haas in 1910, but was rejected. The quantum theory of the period between Planck's discovery of the quantum (1900) and the advent of a full-blown quantum mechanics (1925) is often referred to as the old quantum theory.

23.1 Origin

In the early 20th century, experiments by Ernest Rutherford established that atoms consisted of a diffuse cloud of negatively charged electrons surrounding a small, dense, positively charged nucleus.[2] Given this experimental data, Rutherford naturally considered a planetary-model atom, the Rutherford model of 1911 – electrons orbiting a solar nucleus – however, said planetary-model atom has a technical difficulty. The laws of classical mechanics (i.e. the Larmor formula), predict that the electron will release electromagnetic radiation while orbiting a nucleus. Because the electron would lose energy, it would rapidly spiral inwards, collapsing into the nucleus on a timescale of around 16 picoseconds.[3] This atom model is disastrous, because it predicts that all atoms are unstable.[4]

Also, as the electron spirals inward, the emission would rapidly increase in frequency as the orbit got smaller and faster. This would produce a continuous smear, in frequency, of electromagnetic radiation. However, late 19th century experiments with electric discharges have shown that atoms will only emit light (that is, electromagnetic radiation) at certain discrete frequencies.

To overcome this difficulty, Niels Bohr proposed, in 1913, what is now called the *Bohr model of the atom*. He suggested that electrons could only have certain *classical* motions:

1. Electrons in atoms orbit the nucleus.

2. The electrons can only orbit stably, without radiating, in certain orbits (called by Bohr the "stationary orbits"[5]) at a certain discrete set of distances from the nucleus. These orbits are associated with definite energies and are also called energy shells or energy levels. In these orbits, the electron's acceleration does not result in radiation and energy loss as required by classical electromagnetics. The Bohr model of an atom was based upon Planck's quantum theory of radiation.

3. Electrons can only gain and lose energy by jumping from one allowed orbit to another, absorbing or emitting electromagnetic radiation with a frequency ν determined by the energy difference of the levels according to the Planck relation:

$$\Delta E = E_2 - E_1 = h\nu \, ,$$

where h is Planck's constant. The frequency of the radiation emitted at an orbit of period T is as it would be in classical mechanics; it is the reciprocal of the classical orbit period:

$$\nu = \tfrac{1}{T}.$$

The significance of the Bohr model is that the laws of classical mechanics apply to the motion of the electron about the nucleus *only when restricted by a quantum rule*. Although Rule 3 is not completely well defined for small orbits, because the emission process involves two orbits with two different periods, Bohr could determine the energy spacing between levels using Rule 3 and come to an exactly correct quantum rule: the angular momentum L is restricted to be an integer multiple of a fixed unit:

$$L = n\frac{h}{2\pi} = n\hbar$$

where $n = 1, 2, 3, ...$ is called the principal quantum number, and $\hbar = h/2\pi$. The lowest value of n is 1; this gives a smallest possible orbital radius of 0.0529 nm known as the Bohr radius. Once an electron is in this lowest orbit, it can get no closer to the proton. Starting from the angular momentum quantum rule, Bohr[2] was able to calculate the energies of the allowed orbits of the hydrogen atom and other hydrogen-like atoms and ions.

Other points are:

1. Like Einstein's theory of the Photoelectric effect, Bohr's formula assumes that during a quantum jump a *discrete* amount of energy is radiated. However, unlike Einstein, Bohr stuck to the *classical* Maxwell theory of the electromagnetic field. Quantization of the electromagnetic field was explained by the discreteness of the atomic energy levels; Bohr did not believe in the existence of photons.

2. According to the Maxwell theory the frequency ν of classical radiation is equal to the rotation frequency ν_{rot} of the electron in its orbit, with harmonics at integer multiples of this frequency. This result is obtained from the Bohr model for jumps between energy levels En and En_k when k is much smaller than n. These jumps reproduce the frequency of the k-th harmonic of orbit n. For sufficiently large values of n (so-called Rydberg states), the two orbits involved in the emission process have nearly the same rotation frequency, so that the classical orbital frequency is not ambiguous. But for small n (or large k), the radiation frequency has no unambiguous classical interpretation. This marks the birth of the correspondence principle, requiring quantum theory to agree with the classical theory only in the limit of large quantum numbers.

3. The Bohr-Kramers-Slater theory (BKS theory) is a failed attempt to extend the Bohr model, which violates the conservation of energy and momentum in quantum jumps, with the conservation laws only holding on average.

Bohr's condition, that the angular momentum is an integer multiple of $\hbar$ was later reinterpreted in 1924 by de Broglie as a standing wave condition: the electron is described by a wave and a whole number of wavelengths must fit along the circumference of the electron's orbit:

$$n\lambda = 2\pi r.$$

Substituting de Broglie's wavelength of $\lambda = h/p$ reproduces Bohr's rule. In 1913, however, Bohr justified his rule by appealing to the correspondence principle, without providing any sort of wave interpretation. In 1913, the wave behavior of matter particles such as the electron (i.e., matter waves) was not suspected.

In 1925 a new kind of mechanics was proposed, quantum mechanics, in which Bohr's model of electrons traveling in quantized orbits was extended into a more accurate model of electron motion. The new theory was proposed by Werner Heisenberg. Another form of the same theory, wave mechanics, was discovered by the Austrian physicist Erwin Schrödinger independently, and by different reasoning. Schrödinger employed de Broglie's matter waves, but sought wave solutions of a three-dimensional wave equation describing electrons that were constrained to move about the nucleus of a hydrogen-like atom, by being trapped by the potential of the positive nuclear charge.

23.2   Electron energy levels

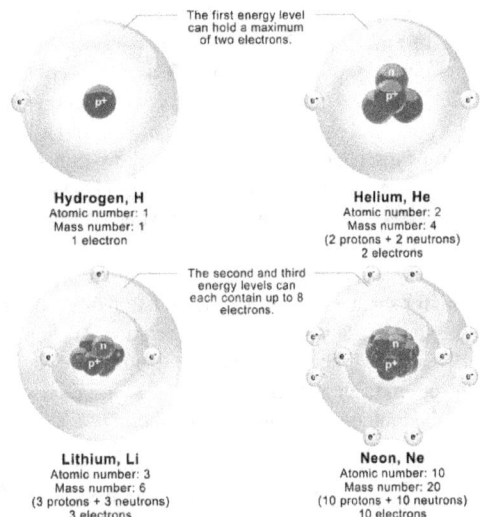

Models depicting electron energy levels in hydrogen, helium, lithium, and neon

The Bohr model gives almost exact results only for a system where two charged points orbit each other at speeds much less than that of light. This not only includes one-electron systems such as the hydrogen atom, singly ionized helium, doubly ionized lithium, but it includes positronium and Rydberg states of any atom where one electron is far away from everything else. It can be used for K-line X-ray transition calculations if other assumptions are added (see

Moseley's law below). In high energy physics, it can be used to calculate the masses of heavy quark mesons.

Calculation of the orbits requires two assumptions.

- **Classical mechanics**

  The electron is held in a circular orbit by electrostatic attraction. The centripetal force is equal to the Coulomb force.

  $$\frac{m_e v^2}{r} = \frac{Z k_e e^2}{r^2}$$

  where m_e is the electron's mass, e is the charge of the electron, k_e is Coulomb's constant and Z is the atom's atomic number. It is assumed here that the mass of the nucleus is much larger than the electron mass (which is a good assumption). This equation determines the electron's speed at any radius:

  $$v = \sqrt{\frac{Z k_e e^2}{m_e r}}.$$

  It also determines the electron's total energy at any radius:

  $$E = \frac{1}{2} m_e v^2 - \frac{Z k_e e^2}{r} = -\frac{Z k_e e^2}{2r}.$$

  The total energy is negative and inversely proportional to r. This means that it takes energy to pull the orbiting electron away from the proton. For infinite values of r, the energy is zero, corresponding to a motionless electron infinitely far from the proton. The total energy is half the potential energy, which is also true for noncircular orbits by the virial theorem.

- **A quantum rule**

  The angular momentum $L = m_e v r$ is an integer multiple of $\hbar$:

  $$m_e v r = n\hbar$$

  Substituting the expression for the velocity gives an equation for r in terms of n:

  $$m_e \sqrt{\frac{k_e Z e^2}{m_e r}} r = n\hbar$$

  so that the allowed orbit radius at any n is:

  $$r_n = \frac{n^2 \hbar^2}{Z k_e e^2 m_e}$$

The smallest possible value of r in the hydrogen atom ($Z=1$) is called the Bohr radius and is equal to:

$$r_1 = \frac{\hbar^2}{k_e e^2 m_e} \approx 5.29 \times 10^{-11} \text{m}$$

The energy of the n-th level for any atom is determined by the radius and quantum number:

$$E = -\frac{Z k_e e^2}{2 r_n} = -\frac{Z^2 (k_e e^2)^2 m_e}{2 \hbar^2 n^2} \approx \frac{-13.6 Z^2}{n^2} \text{eV}$$

An electron in the lowest energy level of hydrogen ($n = 1$) therefore has about 13.6 eV less energy than a motionless electron infinitely far from the nucleus. The next energy level ($n = 2$) is −3.4 eV. The third ($n = 3$) is −1.51 eV, and so on. For larger values of n, these are also the binding energies of a highly excited atom with one electron in a large circular orbit around the rest of the atom.

The combination of natural constants in the energy formula is called the Rydberg energy (RE):

$$R_E = \frac{(k_e e^2)^2 m_e}{2 \hbar^2}$$

This expression is clarified by interpreting it in combinations that form more natural units:

$m_e c^2$ is the rest mass energy of the electron (511 keV)

$\frac{k_e e^2}{\hbar c} = \alpha \approx \frac{1}{137}$ is the fine structure constant

$R_E = \frac{1}{2} (m_e c^2) \alpha^2$

Since this derivation is with the assumption that the nucleus is orbited by one electron, we can generalize this result by letting the nucleus have a charge $q = Z e$ where Z is the atomic number. This will now give us energy levels for hydrogenic atoms, which can serve as a rough order-of-magnitude approximation of the actual energy levels. So for nuclei with Z protons, the energy levels are (to a rough approximation):

$$E_n = -\frac{Z^2 R_E}{n^2}$$

The actual energy levels cannot be solved analytically for more than one electron (see n-body problem) because the electrons are not only affected by the nucleus but also interact with each other via the Coulomb Force.

When $Z = 1/\alpha$ ($Z \approx 137$), the motion becomes highly relativistic, and Z^2 cancels the α^2 in R; the orbit energy begins

to be comparable to rest energy. Sufficiently large nuclei, if they were stable, would reduce their charge by creating a bound electron from the vacuum, ejecting the positron to infinity. This is the theoretical phenomenon of electromagnetic charge screening which predicts a maximum nuclear charge. Emission of such positrons has been observed in the collisions of heavy ions to create temporary super-heavy nuclei.

The Bohr formula properly uses the reduced mass of electron and proton in all situations, instead of the mass of the electron: $m_{red} = \frac{m_e m_p}{m_e + m_p} = m_e \frac{1}{1 + m_e/m_p}$. However, these numbers are very nearly the same, due to the much larger mass of the proton, about 1836.1 times the mass of the electron, so that the reduced mass in the system is the mass of the electron multiplied by the constant $1836.1/(1+1836.1)$ = 0.99946. This fact was historically important in convincing Rutherford of the importance of Bohr's model, for it explained the fact that the frequencies of lines in the spectra for singly ionized helium do not differ from those of hydrogen by a factor of exactly 4, but rather by 4 times the ratio of the reduced mass for the hydrogen vs. the helium systems, which was much closer to the experimental ratio than exactly 4.

For positronium, the formula uses the reduced mass also, but in this case, it is exactly the electron mass divided by 2. For any value of the radius, the electron and the positron are each moving at half the speed around their common center of mass, and each has only one fourth the kinetic energy. The total kinetic energy is half what it would be for a single electron moving around a heavy nucleus.

$$E_n = \frac{R_E}{2 n^2}$$

23.3 Rydberg formula

The Rydberg formula, which was known empirically before Bohr's formula, is seen in Bohr's theory as describing the energies of transitions or quantum jumps between one orbital energy levels. Bohr's formula gives the numerical value of the already-known and measured Rydberg's constant, but in terms of more fundamental constants of nature, including the electron's charge and Planck's constant.

When the electron gets moved from its original energy level to a higher one, it then jumps back each level until it comes to the original position, which results in a photon being emitted. Using the derived formula for the different energy levels of hydrogen one may determine the wavelengths of light that a hydrogen atom can emit.

The energy of a photon emitted by a hydrogen atom is given by the difference of two hydrogen energy levels:

$$E = E_i - E_f = R_{\mathrm{E}} \left(\frac{1}{n_f^2} - \frac{1}{n_i^2} \right)$$

where nf is the final energy level, and ni is the initial energy level.

Since the energy of a photon is

$$E = \frac{hc}{\lambda},$$

the wavelength of the photon given off is given by

$$\frac{1}{\lambda} = R \left(\frac{1}{n_f^2} - \frac{1}{n_i^2} \right).$$

This is known as the Rydberg formula, and the Rydberg constant R is R_{E}/hc , or $R_{\mathrm{E}}/2\pi$ in natural units. This formula was known in the nineteenth century to scientists studying spectroscopy, but there was no theoretical explanation for this form or a theoretical prediction for the value of R, until Bohr. In fact, Bohr's derivation of the Rydberg constant, as well as the concomitant agreement of Bohr's formula with experimentally observed spectral lines of the Lyman ($n_f = 1$), Balmer ($n_f = 2$), and Paschen ($n_f = 3$) series, and successful theoretical prediction of other lines not yet observed, was one reason that his model was immediately accepted.

To apply to atoms with more than one electron, the Rydberg formula can be modified by replacing "Z" with "Z − b" or "n" with "n − b" where b is constant representing a screening effect due to the inner-shell and other electrons (see Electron shell and the later discussion of the "Shell Model of the Atom" below). This was established empirically before Bohr presented his model.

23.4   Shell model of heavier atoms

Bohr extended the model of hydrogen to give an approximate model for heavier atoms. This gave a physical picture that reproduced many known atomic properties for the first time.

Heavier atoms have more protons in the nucleus, and more electrons to cancel the charge. Bohr's idea was that each discrete orbit could only hold a certain number of electrons. After that orbit is full, the next level would have to be used.

This gives the atom a shell structure, in which each shell corresponds to a Bohr orbit.

This model is even more approximate than the model of hydrogen, because it treats the electrons in each shell as non-interacting. But the repulsions of electrons are taken into account somewhat by the phenomenon of screening. The electrons in outer orbits do not only orbit the nucleus, but they also move around the inner electrons, so the effective charge Z that they feel is reduced by the number of the electrons in the inner orbit.

For example, the lithium atom has two electrons in the lowest 1s orbit, and these orbit at Z=2. Each one sees the nuclear charge of Z=3 minus the screening effect of the other, which crudely reduces the nuclear charge by 1 unit. This means that the innermost electrons orbit at approximately 1/4 the Bohr radius. The outermost electron in lithium orbits at roughly Z=1, since the two inner electrons reduce the nuclear charge by 2. This outer electron should be at nearly one Bohr radius from the nucleus. Because the electrons strongly repel each other, the effective charge description is very approximate; the effective charge Z doesn't usually come out to be an integer. But Moseley's law experimentally probes the innermost pair of electrons, and shows that they do see a nuclear charge of approximately Z−1, while the outermost electron in an atom or ion with only one electron in the outermost shell orbits a core with effective charge Z−k where k is the total number of electrons in the inner shells.

The shell model was able to qualitatively explain many of the mysterious properties of atoms which became codified in the late 19th century in the periodic table of the elements. One property was the size of atoms, which could be determined approximately by measuring the viscosity of gases and density of pure crystalline solids. Atoms tend to get smaller toward the right in the periodic table, and become much larger at the next line of the table. Atoms to the right of the table tend to gain electrons, while atoms to the left tend to lose them. Every element on the last column of the table is chemically inert (noble gas).

In the shell model, this phenomenon is explained by shell-filling. Successive atoms become smaller because they are filling orbits of the same size, until the orbit is full, at which point the next atom in the table has a loosely bound outer electron, causing it to expand. The first Bohr orbit is filled when it has two electrons, which explains why helium is inert. The second orbit allows eight electrons, and when it is full the atom is neon, again inert. The third orbital contains eight again, except that in the more correct Sommerfeld treatment (reproduced in modern quantum mechanics) there are extra "d" electrons. The third orbit may hold an extra 10 d electrons, but these positions are not filled until a few more orbitals from the next level are filled (filling the n=3 d orbitals produces the 10 transition elements). The irregular filling pattern is an effect of interactions between electrons, which are not taken into account in either the Bohr or Sommerfeld models and which are difficult to calculate even in the modern treatment.

23.5 Moseley's law and calculation of K-alpha X-ray emission lines

Niels Bohr said in 1962, "You see actually the Rutherford work [the nuclear atom] was not taken seriously. We cannot understand today, but it was not taken seriously at all. There was no mention of it any place. The great change came from Moseley."

In 1913 Henry Moseley found an empirical relationship between the strongest X-ray line emitted by atoms under electron bombardment (then known as the K-alpha line), and their atomic number Z. Moseley's empiric formula was found to be derivable from Rydberg and Bohr's formula (Moseley actually mentions only Ernest Rutherford and Antonius Van den Broek in terms of models). The two additional assumptions that [1] this X-ray line came from a transition between energy levels with quantum numbers 1 and 2, and [2], that the atomic number Z when used in the formula for atoms heavier than hydrogen, should be diminished by 1, to $(Z-1)^2$.

Moseley wrote to Bohr, puzzled about his results, but Bohr was not able to help. At that time, he thought that the postulated innermost "K" shell of electrons should have at least four electrons, not the two which would have neatly explained the result. So Moseley published his results without a theoretical explanation.

Later, people realized that the effect was caused by charge screening, with an inner shell containing only 2 electrons. In the experiment, one of the innermost electrons in the atom is knocked out, leaving a vacancy in the lowest Bohr orbit, which contains a single remaining electron. This vacancy is then filled by an electron from the next orbit, which has n=2. But the n=2 electrons see an effective charge of $Z-1$, which is the value appropriate for the charge of the nucleus, when a single electron remains in the lowest Bohr orbit to screen the nuclear charge +Z, and lower it by −1 (due to the electron's negative charge screening the nuclear positive charge). The energy gained by an electron dropping from the second shell to the first gives Moseley's law for K-alpha lines:

$$E = h\nu = E_i - E_f = R_E(Z-1)^2 \left(\frac{1}{1^2} - \frac{1}{2^2} \right)$$

or

$$f = \nu = R_v \left(\frac{3}{4} \right) (Z-1)^2 = (2.46 \times 10^{15}\,\text{Hz})(Z-1)^2.$$

Here, $R_v = RE/h$ is the Rydberg constant, in terms of frequency equal to 3.28 x 10^{15} Hz. For values of Z between 11 and 31 this latter relationship had been empirically derived by Moseley, in a simple (linear) plot of the square root of X-ray frequency against atomic number (however, for silver, Z = 47, the experimentally obtained screening term should be replaced by 0.4). Notwithstanding its restricted validity,[6] Moseley's law not only established the objective meaning of atomic number (see Henry Moseley for detail) but, as Bohr noted, it also did more than the Rydberg derivation to establish the validity of the Rutherford/Van den Broek/Bohr nuclear model of the atom, with atomic number (place on the periodic table) standing for whole units of nuclear charge.

The K-alpha line of Moseley's time is now known to be a pair of close lines, written as ($K\alpha_1$ and $K\alpha_2$) in Siegbahn notation.

23.6 Shortcomings

The Bohr model gives an incorrect value $L=\hbar$ for the ground state orbital angular momentum. The angular momentum in the true ground state is known to be zero from experiment.[7] Although mental pictures fail somewhat at these levels of scale, an electron in the lowest modern "orbital" with no orbital momentum, may be thought of as not to rotate "around" the nucleus at all, but merely to go tightly around it in an ellipse with zero area (this may be pictured as "back and forth", without striking or interacting with the nucleus). This is only reproduced in a more sophisticated semiclassical treatment like Sommerfeld's. Still, even the most sophisticated semiclassical model fails to explain the fact that the lowest energy state is spherically symmetric - it doesn't point in any particular direction. Nevertheless, in the modern *fully quantum treatment in phase space*, the proper deformation (full extension) of the semi-classical result adjusts the angular momentum value to the correct effective one. As a consequence, the physical ground state expression is obtained through a shift of the vanishing quantum angular momentum expression, which corresponds to spherical symmetry.

In modern quantum mechanics, the electron in hydrogen is a spherical cloud of probability that grows denser near the nucleus. The rate-constant of probability-decay in hydrogen is equal to the inverse of the Bohr radius, but since Bohr worked with circular orbits, not zero area ellipses, the fact that these two numbers exactly agree is considered a "coincidence". (However, many such coincidental agreements are found between the semiclassical vs. full quantum mechanical treatment of the atom; these include identical energy levels in the hydrogen atom and the derivation of a fine structure constant, which arises from the relativistic Bohr–Sommerfeld model (see below) and which happens to be equal to an entirely different concept, in full modern quantum mechanics).

The Bohr model also has difficulty with, or else fails to explain:

- Much of the spectra of larger atoms. At best, it can make predictions about the K-alpha and some L-alpha X-ray emission spectra for larger atoms, if *two* additional ad hoc assumptions are made (see Moseley's law above). Emission spectra for atoms with a single outer-shell electron (atoms in the lithium group) can also be approximately predicted. Also, if the empiric electron–nuclear screening factors for many atoms are known, many other spectral lines can be deduced from the information, in similar atoms of differing elements, via the Ritz–Rydberg combination principles (see Rydberg formula). All these techniques essentially make use of Bohr's Newtonian energy-potential picture of the atom.

- the relative intensities of spectral lines; although in some simple cases, Bohr's formula or modifications of it, was able to provide reasonable estimates (for example, calculations by Kramers for the Stark effect).

- The existence of fine structure and hyperfine structure in spectral lines, which are known to be due to a variety of relativistic and subtle effects, as well as complications from electron spin.

- The Zeeman effect – changes in spectral lines due to external magnetic fields; these are also due to more complicated quantum principles interacting with electron spin and orbital magnetic fields.

- The model also violates the uncertainty principle in that it considers electrons to have known orbits and locations, two things which can not be measured simultaneously.

- Doublets and Triplets: Appear in the spectra of some atoms: Very close pairs of lines. Bohr's model cannot say why some energy levels should be very close together.

- Multi-electron Atoms: don't have energy levels predicted by the model. It doesn't work for (neutral) helium.

- A rotating charge, such as the electron classically orbiting around the nucleus, would constantly lose energy in form of electromagnetic radiation (via various mechanisms: dipole radiation, Bremsstrahlung,...). But such radiation is not observed.

23.7 Refinements

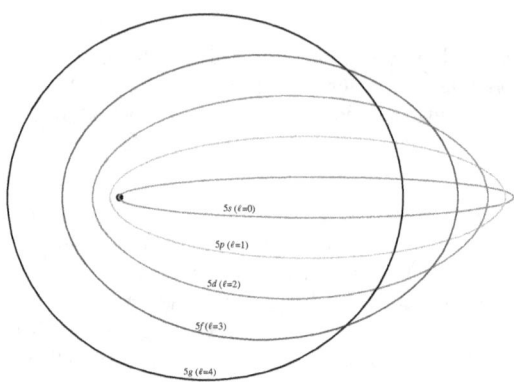

Elliptical orbits with the same energy and quantized angular momentum

Several enhancements to the Bohr model were proposed, most notably the **Sommerfeld model** or **Bohr–Sommerfeld model**, which suggested that electrons travel in elliptical orbits around a nucleus instead of the Bohr model's circular orbits.[1] This model supplemented the quantized angular momentum condition of the Bohr model with an additional radial quantization condition, the **Sommerfeld–Wilson quantization condition**[8][9]

$$\int_0^T p_r \, dq_r = nh$$

where *pr* is the radial momentum canonically conjugate to the coordinate *q* which is the radial position and *T* is one full orbital period. The integral is the action of action-angle coordinates. This condition, suggested by the correspondence principle, is the only one possible, since the quantum numbers are adiabatic invariants.

The Bohr–Sommerfeld model was fundamentally inconsistent and led to many paradoxes. The magnetic quantum number measured the tilt of the orbital plane relative to the *xy*-plane, and it could only take a few discrete values. This contradicted the obvious fact that an atom could be turned this way and that relative to the coordinates without restriction. The Sommerfeld quantization can be performed in different canonical coordinates and sometimes gives different answers. The incorporation of radiation corrections was difficult, because it required finding action-angle coordinates for a combined radiation/atom system, which is difficult when the radiation is allowed to escape. The whole theory did not extend to non-integrable motions, which meant that many systems could not be treated even in principle. In the end, the model was replaced by the modern quantum mechanical treatment of the hydrogen atom, which was first given by Wolfgang Pauli in 1925, using Heisenberg's matrix mechanics. The current picture of the hydrogen atom is based on the atomic orbitals of wave mechanics which Erwin Schrödinger developed in 1926.

However, this is not to say that the Bohr model was without its successes. Calculations based on the Bohr–Sommerfeld model were able to accurately explain a number of more complex atomic spectral effects. For example, up to first-order perturbations, the Bohr model and quantum mechanics make the same predictions for the spectral line splitting in the Stark effect. At higher-order perturbations, however, the Bohr model and quantum mechanics differ, and measurements of the Stark effect under high field strengths helped confirm the correctness of quantum mechanics over the Bohr model. The prevailing theory behind this difference lies in the shapes of the orbitals of the electrons, which vary according to the energy state of the electron.

The Bohr–Sommerfeld quantization conditions lead to questions in modern mathematics. Consistent semiclassical quantization condition requires a certain type of structure on the phase space, which places topological limitations on the types of symplectic manifolds which can be quantized. In particular, the symplectic form should be the curvature form of a connection of a Hermitian line bundle, which is called a prequantization.

23.8   See also

23.9   References

23.9.1   Footnotes

[1] Akhlesh Lakhtakia (Ed.); Salpeter, Edwin E. (1996). "Models and Modelers of Hydrogen". *American Journal of Physics* (World Scientific) **65** (9): 933. Bibcode:1997AmJPh..65..933L.   doi:10.1119/1.18691. ISBN 981-02-2302-1.

[2] Niels Bohr (1913). "On the Constitution of Atoms and Molecules, Part I" (PDF). *Philosophical Magazine* **26** (151): 1–24. doi:10.1080/14786441308634955.

[3] Olsen and McDonald 2005

[4] "CK12 – Chemistry Flexbook Second Edition – The Bohr Model of the Atom". Retrieved 30 September 2014.

[5] Niels Bohr (1913). "On the Constitution of Atoms and Molecules, Part II Systems Containing Only a Single Nucleus" (PDF). *Philosophical Magazine* **26** (153): 476–502. doi:10.1080/14786441308634993.

[6] M.A.B. Whitaker (1999).   "The Bohr–Moseley synthesis and a simple model for atomic x-ray energies". *European Journal of Physics* **20** (3): 213–220. Bibcode:1999EJPh...20..213W.   doi:10.1088/0143-0807/20/3/312.

[7] Smith, Brian. "Quantum Ideas: Week 2" Lecture Notes, p.17. University of Oxford. Retrieved Jan. 23, 2015.

[8] A.   Sommerfeld   (1916).   "Zur   Quantentheorie   der   Spektrallinien".   *Annalen   der   Physik* **51**   (17):   1.   Bibcode:1916AnP...356....1S. doi:10.1002/andp.19163561702.

[9] W. Wilson (1915). "The quantum theory of radiation and line spectra". *Philosophical Magazine* **29** (174): 795–802. doi:10.1080/14786440608635362.

23.9.2   Primary sources

• Niels Bohr (1913). "On the Constitution of Atoms and Molecules, Part I" (PDF). *Philosophical Magazine* **26** (151): 1–24. doi:10.1080/14786441308634955.

• Niels Bohr (1913). "On the Constitution of Atoms and Molecules, Part II Systems Containing Only a Single Nucleus" (PDF). *Philosophical Magazine* **26** (153): 476–502. doi:10.1080/14786441308634993.

• Niels Bohr (1913). "On the Constitution of Atoms and Molecules, Part III Systems containing several nuclei". *Philosophical Magazine* **26**: 857–875. doi:10.1080/14786441308635031.

• Niels Bohr (1914).   "The spectra of helium and hydrogen".   *Nature* **92** (2295): 231–232.   Bibcode:1913Natur..92..231B. doi:10.1038/092231d0.

• Niels Bohr (1921). "Atomic Structure". *Nature* **107** (2682):   104–107.   Bibcode:1921Natur.107..104B. doi:10.1038/107104a0.

• A. Einstein (1917). "Zum Quantensatz von Sommerfeld und Epstein". *Verhandlungen der Deutschen Physikalischen Gesellschaft* **19**: 82–92. Reprinted in *The Collected Papers of Albert Einstein*, A. Engel translator, (1997) Princeton University Press, Princeton. **6** p. 434. (provides an elegant reformulation of the Bohr–Sommerfeld quantization conditions, as well as an important insight into the quantization of non-integrable (chaotic) dynamical systems.)

23.10   Further reading

- Linus Carl Pauling (1970). "Chapter 5-1". *General Chemistry* (3rd ed.). San Francisco: W.H. Freeman & Co.

  - Reprint: Linus Pauling (1988). *General Chemistry*. New York: Dover Publications. ISBN 0-486-65622-5.

- George Gamow (1985). "Chapter 2". *Thirty Years That Shook Physics*. Dover Publications.

- Walter J. Lehmann (1972). "Chapter 18". *Atomic and Molecular Structure: the development of our concepts*. John Wiley and Sons.

- Paul Tipler and Ralph Llewellyn (2002). *Modern Physics* (4th ed.). W. H. Freeman. ISBN 0-7167-4345-0.

- Klaus Hentschel: Elektronenbahnen, Quantensprünge und Spektren, in: Charlotte Bigg & Jochen Hennig (eds.) Atombilder. Ikonografien des Atoms in Wissenschaft und Öffentlichkeit des 20. Jahrhunderts, Göttingen: Wallstein-Verlag 2009, pp. 51–61

- Steven and Susan Zumdahl (2010). "Chapter 7.4". *Chemistry* (8th ed.). Brooks/Cole. ISBN 978-0-495-82992-8.

- Helge Kragh (2011). "Conceptual objections to the Bohr atomic theory — do electrons have a "free will" ?". *European Physical Journal H* **36** (3): 327. Bibcode:2011EPJH...36..327K. doi:10.1140/epjh/e2011-20031-x.

23.11   External links

- Standing waves in Bohr's atomic model An interactive simulation to intuitively explain the quantization condition of standing waves in Bohr's atomic model

Chapter 24

Quantum chemistry

Quantum chemistry is a branch of chemistry whose primary focus is the application of quantum mechanics in physical models and experiments of chemical systems. It is also called molecular quantum mechanics.

24.1 Overview

It involves heavy interplay of experimental and theoretical methods:

- Experimental quantum chemists rely heavily on spectroscopy, through which information regarding the quantization of energy on a molecular scale can be obtained. Common methods are infra-red (IR) spectroscopy and nuclear magnetic resonance (NMR) spectroscopy.

- Theoretical quantum chemistry, the workings of which also tend to fall under the category of computational chemistry, seeks to calculate the predictions of quantum theory as atoms and molecules can only have discrete energies; as this task, when applied to polyatomic species, invokes the many-body problem, these calculations are performed using computers rather than by analytical "back of the envelope" methods, pen recorder or computerized data station with a VDU.

In these ways, quantum chemists investigate chemical phenomena.

- In reactions, quantum chemistry studies the ground state of individual atoms and molecules, the excited states, and the transition states that occur during chemical reactions.

- On the calculations: quantum chemical studies use also semi-empirical and other methods based on quantum mechanical principles, and deal with time dependent problems. Many quantum chemical studies assume the nuclei are at rest (Born–Oppenheimer approximation). Many calculations involve iterative methods that include self-consistent field methods. Major goals of quantum chemistry include increasing the accuracy of the results for small molecular systems, and increasing the size of large molecules that can be processed, which is limited by scaling considerations—the computation time increases as a power of the number of atoms.

24.2 History

Some view the birth of quantum chemistry in the discovery of the Schrödinger equation and its application to the hydrogen atom in 1926. However, the 1927 article of Walter Heitler and Fritz London is often recognised as the first milestone in the history of quantum chemistry. This is the first application of quantum mechanics to the diatomic hydrogen molecule, and thus to the phenomenon of the chemical bond. In the following years much progress was accomplished by Edward Teller, Robert S. Mulliken, Max Born, J. Robert Oppenheimer, Linus Pauling, Erich Hückel, Douglas Hartree, Vladimir Aleksandrovich Fock, to cite a few. The history of quantum chemistry also goes through the 1838 discovery of cathode rays by Michael Faraday, the 1859 statement of the black body radiation problem by Gustav Kirchhoff, the 1877 suggestion by Ludwig Boltzmann that the energy states of a physical system could be discrete, and the 1900 quantum hypothesis by Max Planck that any energy radiating atomic system can theoretically be divided into a number of discrete energy elements ε such that each of these energy elements is proportional to the frequency ν with which they each individually radiate energy and a numerical value called Planck's Constant. Then, in 1905, to explain the photoelectric effect (1839), i.e., that shining light on certain materials can function to eject electrons from the material, Albert Einstein postulated, based on Planck's quantum hypothesis, that light itself consists of individual quantum particles, which later came to be called photons (1926). In the years to follow, this theoretical basis slowly began to be applied to chemical structure, reactivity, and bonding. Probably the greatest contribution to the field was made by Linus Pauling.

24.3 Electronic structure

Main article: Computational chemistry § Electronic structure

The first step in solving a quantum chemical problem is usually solving the Schrödinger equation (or Dirac equation in relativistic quantum chemistry) with the electronic molecular Hamiltonian. This is called determining the **electronic structure** of the molecule. It can be said that the electronic structure of a molecule or crystal implies essentially its chemical properties. An exact solution for the Schrödinger equation can only be obtained for the hydrogen atom (though exact solutions for the bound state energies of the hydrogen molecular ion have been identified in terms of the generalized Lambert W function). Since all other atomic, or molecular systems, involve the motions of three or more "particles", their Schrödinger equations cannot be solved exactly and so approximate solutions must be sought.

24.3.1 Wave model

The foundation of quantum mechanics and quantum chemistry is the **wave model**, in which the atom is a small, dense, positively charged nucleus surrounded by electrons. The wave model is derived from the wavefunction, a set of possible equations derived from the time evolution of the Schrödinger equation which is applied to the wavelike probability distribution of subatomic particles. Unlike the earlier Bohr model of the atom, however, the wave model describes electrons as "clouds" moving in orbitals, and their positions are represented by probability distributions rather than discrete points. The strength of this model lies in its predictive power. Specifically, it predicts the pattern of chemically similar elements found in the periodic table. The wave model is so named because electrons exhibit properties (such as interference) traditionally associated with waves. See wave-particle duality. In this model, when we solve the Schrödinger Equation for an Hidrogenoid Atom, we obtain a solution that depends on some numbers, called quantum numbers, that describes the orbital, the most probable space where an electron can be. These are n, the principal quantum number, for the energy, l, or secondary quantum number, wich correlates to the angular momentum, ml, for the orientation, and ms the spin. This model can explain the new lines that appeared in the spec-

troscopy of atoms. For multielectron atoms we must introduce some rules as that the electrons fill orbitals in a way to minimize the energy of the atom, in order of increasing energy, the Pauli Exclusion Principle, the Hund's Rule, and the Aufbau Principle.

24.3.2 Valence bond

Main article: Valence bond theory

Although the mathematical basis of quantum chemistry had been laid by Schrödinger in 1926, it is generally accepted that the first true calculation in quantum chemistry was that of the German physicists Walter Heitler and Fritz London on the hydrogen (H_2) molecule in 1927. Heitler and London's method was extended by the American theoretical physicist John C. Slater and the American theoretical chemist Linus Pauling to become the **Valence-Bond (VB)** [or **Heitler–London–Slater–Pauling (HLSP)**] method. In this method, attention is primarily devoted to the pairwise interactions between atoms, and this method therefore correlates closely with classical chemists' drawings of bonds. It focuses on how the atomic orbitals of an atom combine to give individual chemical bonds when a molecule is formed.

24.3.3 Molecular orbital

Main article: Molecular orbital theory

An alternative approach was developed in 1929 by Friedrich Hund and Robert S. Mulliken, in which electrons are described by mathematical functions delocalized over an entire molecule. The **Hund–Mulliken** approach or **molecular orbital (MO) method** is less intuitive to chemists, but has turned out capable of predicting spectroscopic properties better than the VB method. This approach is the conceptional basis of the **Hartree–Fock method** and further post Hartree–Fock methods.

24.3.4 Density functional theory

Main article: Density functional theory

The **Thomas–Fermi model** was developed independently by Thomas and Fermi in 1927. This was the first attempt to describe many-electron systems on the basis of electronic density instead of wave functions, although it was not very successful in the treatment of entire molecules. The method did provide the basis for what is now known as **density functional theory**. Modern day DFT uses the **Kohn-Sham method**, where the density functional is split into four terms; the Kohn-Sham kinetic energy, an external potential, exchange and correlation energies. A large part of the focus on developing DFT is on improving the exchange and correlation terms. Though this method is less developed than post Hartree–Fock methods, its significantly lower computational requirements (scaling typically no worse than n^3 with respect to n basis functions, for the pure functionals) allow it to tackle larger polyatomic molecules and even macromolecules. This computational affordability and often comparable accuracy to MP2 and CCSD(T) (post-Hartree–Fock methods) has made it one of the most popular methods in computational chemistry at present.

24.4 Chemical dynamics

A further step can consist of solving the Schrödinger equation with the total molecular Hamiltonian in order to study the motion of molecules. Direct solution of the Schrödinger equation is called *quantum molecular dynamics*, within the semiclassical approximation *semiclassical molecular dynamics*, and within the classical mechanics framework *molecular dynamics (MD)*. Statistical approaches, using for example Monte Carlo methods, are also possible.

24.4.1 Adiabatic chemical dynamics

Main article: Adiabatic formalism or Born–Oppenheimer approximation

In **adiabatic dynamics**, interatomic interactions are represented by single scalar potentials called potential energy surfaces. This is the Born–Oppenheimer approximation introduced by Born and Oppenheimer in 1927. Pioneering applications of this in chemistry were performed by Rice and Ramsperger in 1927 and Kassel in 1928, and generalized into the RRKM theory in 1952 by Marcus who took the transition state theory developed by Eyring in 1935 into account. These methods enable simple estimates of unimolecular reaction rates from a few characteristics of the potential surface.

24.4.2 Non-adiabatic chemical dynamics

Main article: Vibronic coupling

Non-adiabatic dynamics consists of taking the interaction between several coupled potential energy surface (cor-

responding to different electronic quantum states of the molecule). The coupling terms are called **vibronic couplings**. The pioneering work in this field was done by Stueckelberg, Landau, and Zener in the 1930s, in their work on what is now known as the Landau–Zener transition. Their formula allows the transition probability between two diabatic potential curves in the neighborhood of an avoided crossing to be calculated.

24.5 See also

- Atomic physics
- Computational chemistry
- Condensed matter physics
- Electron localization function
- International Academy of Quantum Molecular Science
- Molecular modelling
- Physical chemistry
- List of quantum chemistry and solid-state physics software
- QMC@Home
- *Quantum Aspects of Life*
- Quantum chemistry computer programs
- Quantum electrochemistry
- Relativistic quantum chemistry
- Theoretical physics

24.6 References

- Atkins, P.W.; Friedman, R. (2005). *Molecular Quantum Mechanics* (4th ed.). Oxford University Press. ISBN 978-0-19-927498-7.

- Atkins, P.W. *Physical Chemistry*. Oxford University Press. ISBN 0-19-879285-9.

- Atkins, P.W.; Friedman, R. (2008). *Quanta, Matter and Change: A Molecular Approach to Physical Change*. ISBN 978-0-7167-6117-4.

- Pullman, Bernard; Pullman, Alberte (1963). *Quantum Biochemistry*. New York and London: Academic Press. ISBN 90-277-1830-X.

- Scerri, Eric R. (2006). *The Periodic Table: Its Story and Its Significance.* Oxford University Press. ISBN 0-19-530573-6. Considers the extent to which chemistry and especially the periodic system has been reduced to quantum mechanics.

- Kostas Gavroglu, Ana Simões: *NEITHER PHYSICS NOR CHEMISTRY. A History of Quantum Chemistry,* MIT Press, 2011, ISBN 0-262-01618-4

- McWeeny, R. *Coulson's Valence.* Oxford Science Publications. ISBN 0-19-855144-4.

- Karplus M., Porter R.N. (1971). *Atoms and Molecules. An introduction for students of physical chemistry,* Benjamin–Cummings Publishing Company, ISBN 978-0-8053-5218-4

- Szabo, Attila; Ostlund, Neil S. (1996). *Modern Quantum Chemistry: Introduction to Advanced Electronic Structure Theory.* Dover. ISBN 0-486-69186-1.

- Landau, L.D.; Lifshitz, E.M. *Quantum Mechanics: Non-relativistic Theory.* Course of Theoretical Physic **3**. Pergamon Press. ISBN 0-08-019012-X.

- Levine, I. (2008). *Physical Chemistry* (6th ed.). McGraw–Hill Science. ISBN 978-0-07-253862-5.

- Pauling, L. (1954). *General Chemistry.* Dover Publications. ISBN 0-486-65622-5.

- Pauling, L.; Wilson, E. B. (1963) [1935]. *Introduction to Quantum Mechanics with Applications to Chemistry.* Dover Publications. ISBN 0-486-64871-0.

- Simon, Z. (1976). *Quantum Biochemistry and Specific Interactions.* Taylor & Francis. ISBN 978-0-85626-087-2.

24.7 External links

- The Sherrill Group – Notes

- ChemViz Curriculum Support Resources

- Early ideas in the history of quantum chemistry

- The Particle Adventure

24.7.1 Nobel lectures by quantum chemists

- Walter Kohn's Nobel lecture

- Rudolph Marcus' Nobel lecture

- Robert Mulliken's Nobel lecture

- Linus Pauling's Nobel lecture

- John Pople's Nobel lecture

Chapter 25

History of electrochemistry

Electrochemistry, a branch of chemistry, went through several changes during its evolution from early principles related to magnets in the early 16th and 17th centuries, to complex theories involving conductivity, electric charge and mathematical methods. The term *electrochemistry* was used to describe electrical phenomena in the late 19th and 20th centuries. In recent decades, electrochemistry has become an area of current research, including research in batteries and fuel cells, preventing corrosion of metals, the use of electrochemical cells to remove refractory organics and similar contaminants in wastewater electrocoagulation and improving techniques in refining chemicals with electrolysis and electrophoresis.

25.1 Background and dawn of electrochemistry

The 16th century marked the beginning of scientific understanding of electricity and magnetism that culminated with the production of electric power and the industrial revolution in the late 19th century.

In the 1550s, English scientist William Gilbert spent 17 years experimenting with magnetism and, to a lesser extent, electricity. For his work on magnets, Gilbert became known as "The Father of Magnetism." His book *De Magnete* quickly became the standard work throughout Europe on electrical and magnetic phenomena. He made the first clear distinction between magnetism and what was then called the "amber effect" (static electricity).

In 1663, German physicist Otto von Guericke created the first electrostatic generator, which produced static electricity by applying friction. The generator was made of a large sulfur ball inside a glass globe, mounted on a shaft. The ball was rotated by means of a crank and a static electric spark was produced when a pad was rubbed against the ball as it rotated. The globe could be removed and used as an electrical source for experiments with electricity. Von Guericke used his generator to show that like charges repelled each

German physicist Otto von Guericke beside his electrical generator while conducting experiment.

other.

25.2 The 18th century and birth of electrochemistry

In 1709, Francis Hauksbee at the Royal Society in London discovered that by putting a small amount of mercury in the glass of Von Guericke's generator and evacuating the air from it, it would glow whenever the ball built up a charge and his hand was touching the globe. He had created the first gas-discharge lamp.

Between 1729 and 1736, two English scientists, Stephen Gray and Jean Desaguliers, performed a series of experiments which showed that a cork or other object as far away as 800 or 900 feet (245–275 m) could be electrified by connecting it via a charged glass tube to materials such as metal wires or hempen string. They found that other materials, such as silk, would not convey the effect.

By the mid-18th century, French chemist Charles François de Cisternay Du Fay had discovered two forms of static

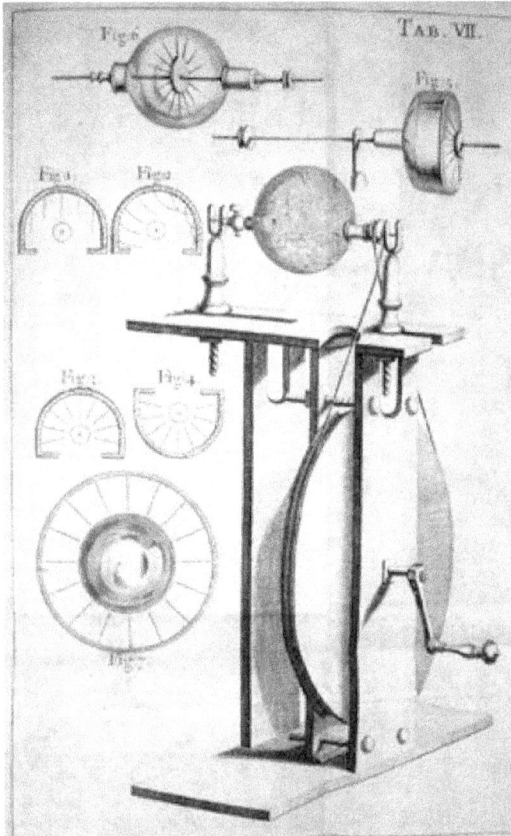

Francis Hauksbee's gas-discharge lamp

electricity, and that like charges repel each other while unlike charges attract. Du Fay announced that electricity consisted of two fluids: *vitreous* (from the Latin for "glass"), or positive, electricity; and *resinous*, or negative, electricity. This was the "two-fluid theory" of electricity, which was opposed by Benjamin Franklin's "one-fluid theory" later in the century.

In 1745, Jean-Antoine Nollet developed a theory of electrical attraction and repulsion that supposed the existence of a continuous flow of electrical matter between charged bodies. Nollet's theory at first gained wide acceptance, but met resistance in 1752 with the translation of Franklin's *Experiments and Observations on Electricity* into French. Franklin and Nollet debated the nature of electricity, with Franklin supporting action at a distance and two qualitatively opposing types of electricity, and Nollet advocating mechanical action and a single type of electrical fluid. Franklin's argument eventually won and Nollet's theory was abandoned.

In 1748, Nollet invented one of the first electrometers, the electroscope, which showed electric charge using

electrostatic attraction and repulsion. Nollet is reputed to be the first to apply the name "Leyden jar" to the first device for storing electricity. Nollet's invention was replaced by Horace-Bénédict de Saussure's electrometer in 1766.

By the 1740s, William Watson had conducted several experiments to determine the speed of electricity. The general belief at the time was that electricity was faster than sound, but no accurate test been devised to measure the velocity of a current. Watson, in the fields north of London, laid out a line of wire supported by dry sticks and silk which ran for 12,276 feet (3.7 km). Even at this length, the velocity of electricity seemed instantaneous. Resistance in the wire was also noticed but apparently not fully understood, as Watson related that "we observed again, that although the electrical compositions were very severe to those who held the wires, the report of the Explosion at the prime Conductor was little, in comparison of that which is heard when the Circuit is short." Watson eventually decided not to pursue his electrical experiments, concentrating instead upon his medical career.

By the 1750s, as the study of electricity became popular, efficient ways of producing electricity were sought. The generator developed by Jesse Ramsden was among the first electrostatic generators invented. Electricity produced by such generators was used to treat paralysis, muscle spasms, and to control heart rates. Other medical uses of electricity included filling the body with electricity, drawing sparks from the body, and applying sparks from the generator to the body.

Charles-Augustin de Coulomb developed the law of electrostatic attraction in 1781 as an outgrowth of his attempt to investigate the law of electrical repulsions as stated by Joseph Priestley in England. To this end, he invented a sensitive apparatus to measure the electrical forces involved in Priestley's law. He also established the inverse square law of attraction and repulsion magnetic poles, which became the basis for the mathematical theory of magnetic forces developed by Siméon Denis Poisson. Coulomb wrote seven important works on electricity and magnetism which he submitted to the Académie des Sciences between 1785 and 1791, in which he reported having developed a theory of attraction and repulsion between charged bodies, and went on to search for perfect conductors and dielectrics. He suggested that there was no perfect dielectric, proposing that every substance has a limit, above which it will conduct electricity. The SI unit of charge is called a coulomb in his honour.

In 1789, Franz Aepinus developed a device with the properties of a "condenser" (now known as a capacitor.) The Aepinus condenser was the first capacitor developed after the Leyden jar, and was used to demonstrate conduction and induction. The device was constructed so that the space between two plates could be adjusted, and the glass dielectric separating the two plates could be removed or replaced with other materials.

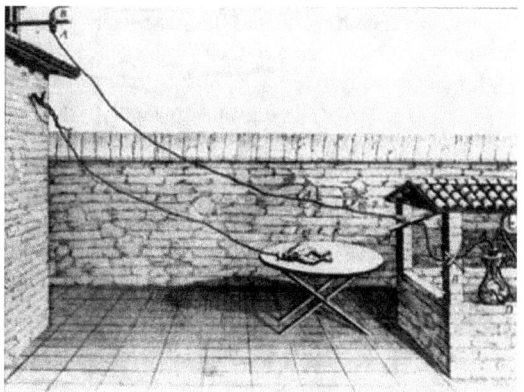

Late 1780s diagram of Galvani's experiment on frog legs.

Italian physicist Alessandro Volta showing his "battery" to French emperor Napoleon Bonaparte in the early 19th century.

Despite the gain in knowledge of electrical properties and the building of generators, it wasn't until the late 18th century that Italian physician and anatomist Luigi Galvani marked the birth of electrochemistry by establishing a bridge between muscular contractions and electricity with his 1791 essay *De Viribus Electricitatis in Motu Musculari Commentarius* (Commentary on the Effect of Electricity on Muscular Motion), where he proposed a "nerveo-electrical substance" in life forms.

In his essay, Galvani concluded that animal tissue contained a before-unknown innate, vital force, which he termed "animal electricity," which activated muscle when placed between two metal probes. He believed that this was evidence of a new form of electricity, separate from the "natural" form that is produced by lightning and the "artificial" form that is produced by friction (static electricity). He considered the brain to be the most important organ for the secretion of this "electric fluid" and that the nerves conducted the fluid to the muscles. He believed the tissues acted similarly to the outer and inner surfaces of Leyden jars. The flow of this electric fluid provided a stimulus to the muscle fibres.

Galvani's scientific colleagues generally accepted his views, but Alessandro Volta, the outstanding professor of physics at the University of Pavia, was not convinced by the analogy between muscles and Leyden jars. Deciding that the frogs' legs used in Galvani's experiments served only as an electroscope, he held that the contact of dissimilar metals was the true source of stimulation. He referred to the electricity so generated as "metallic electricity" and decided that the muscle, by contracting when touched by metal, resembled the action of an electroscope. Furthermore, Volta claimed that if two dissimilar metals in contact with each other also touched a muscle, agitation would also occur and increase

with the dissimilarity of the metals. Galvani refuted this by obtaining muscular action using two pieces of similar metal. Volta's name was later used for the unit of electrical potential, the volt.

25.3 Rise of electrochemistry as branch of chemistry

In 1800, English chemists William Nicholson and Johann Wilhelm Ritter succeeded in separating water into hydrogen and oxygen by electrolysis. Soon thereafter, Ritter discovered the process of electroplating. He also observed that the amount of metal deposited and the amount of oxygen produced during an electrolytic process depended on the distance between the electrodes. By 1801 Ritter had observed thermoelectric currents, which anticipated the discovery of thermoelectricity by Thomas Johann Seebeck.

In 1802, William Cruickshank designed the first electric battery capable of mass production. Like Volta, Cruickshank arranged square copper plates, which he soldered at their ends, together with plates of zinc of equal size. These plates were placed into a long rectangular wooden box which was sealed with cement. Grooves inside the box held the metal plates in position. The box was then filled with an electrolyte of brine, or watered down acid. This flooded design had the advantage of not drying out with use and provided more energy than Volta's arrangement, which used brine-soaked papers between the plates.

In the quest for a better production of platinum metals,

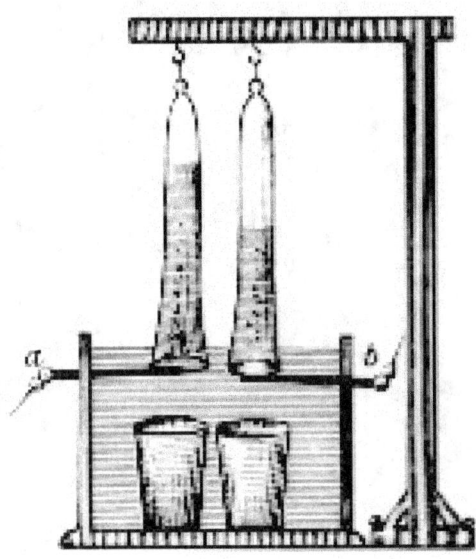

Scheme of Ritter's apparatus to separate water into hydrogen and oxygen by electrolysis

two scientists, William Hyde Wollaston and Smithson Tennant, worked together to design an efficient electrochemical technique to refine or purify platinum. Tennant ended up discovering the elements iridium and osmium. Wollaston's effort, in turn, led him to the discovery of the metals palladium in 1803 and rhodium in 1804.

Wollaston made improvements to the galvanic battery (named after Galvani) in the 1810s. In Wollaston's battery, the wooden box was replaced with an earthenware vessel, and a copper plate was bent into a U-shape, with a single plate of zinc placed in the center of the bent copper. The zinc plate was prevented from making contact with the copper by dowels (pieces) of cork or wood. In his single cell design, the U-shaped copper plate was welded to a horizontal handle for lifting the copper and zinc plates out of the electrolyte when the battery was not in use.

In 1809, Samuel Thomas von Soemmering developed the first telegraph. He used a device with 26 wires (1 wire for each letter of the German alphabet) terminating in a container of acid. At the sending station, a key, which completed a circuit with a battery, was connected as required to each of the line wires. The passage of current caused the acid to decompose chemically, and the message was read by observing at which of the terminals the bubbles of gas appeared. This is how he was able to send messages, one letter at a time.

Humphry Davy's work with electrolysis led to conclusion

that the production of electricity in simple electrolytic cells resulted from chemical reactions between the electrolyte and the metals, and occurred between substances of opposite charge. He reasoned that the interactions of electric currents with chemicals offered the most likely means of decomposing all substances to their basic elements. These views were explained in 1806 in his lecture *On Some Chemical Agencies of Electricity*, for which he received the Napoleon Prize from the Institut de France in 1807 (despite the fact that England and France were at war at the time). This work led directly to the isolation of sodium and potassium from their common compounds and of the alkaline earth metals from theirs in 1808.

Hans Christian Ørsted's discovery of the magnetic effect of electric currents in 1820 was immediately recognised as an important advance, although he left further work on electromagnetism to others. André-Marie Ampère quickly repeated Ørsted's experiment, and formulated them mathematically (which became Ampère's law) . Ørsted also discovered that not only is a magnetic needle deflected by the electric current, but that the live electric wire is also deflected in a magnetic field, thus laying the foundation for the construction of an electric motor. Ørsted's discovery of piperine, one of the pungent components of pepper, was an important contribution to chemistry, as was his preparation of aluminium in 1825.

During the 1820s, Robert Hare developed the Deflagrator, a form of voltaic battery having large plates used for producing rapid and powerful combustion. A modified form of this apparatus was employed in 1823 in volatilising and fusing carbon. It was with these batteries that the first use of voltaic electricity for blasting under water was made in 1831.

In 1821, the Estonian-German physicist, Thomas Johann Seebeck, demonstrated the electrical potential in the juncture points of two dissimilar metals when there is a temperature difference between the joints. He joined a copper wire with a bismuth wire to form a loop or circuit. Two junctions were formed by connecting the ends of the wires to each other. He then accidentally discovered that if he heated one junction to a high temperature, and the other junction remained at room temperature, a magnetic field was observed around the circuit.

He did not recognise that an electric current was being generated when heat was applied to a bi-metal junction. He used the term "thermomagnetic currents" or "thermomagnetism" to express his discovery. Over the following two years, he reported on his continuing observations to the Prussian Academy of Sciences, where he described his observation as "the magnetic polarization of metals and ores produced by a temperature difference." This Seebeck effect became the basis of the thermocouple, which is still considered the most accurate measurement of temperature today. The converse Peltier effect was seen over a decade later when a current was run through a circuit with two dissimilar metals, resulting in a temperature difference between the metals.

In 1827 German scientist Georg Ohm expressed his law in his famous book *Die galvanische Kette, mathematisch bearbeitet* (The Galvanic Circuit Investigated Mathematically) in which he gave his complete theory of electricity.

In 1829 Antoine-César Becquerel developed the "constant current" cell, forerunner of the well-known Daniell cell. When this acid-alkali cell was monitored by a galvanometer, current was found to be constant for an hour, the first instance of "constant current". He applied the results of his study of thermoelectricity to the construction of an electric thermometer, and measured the temperatures of the interior of animals, of the soil at different depths, and of the atmosphere at different heights. He helped validate Faraday's laws and conducted extensive investigations on the electroplating of metals with applications for metal finishing and metallurgy. Solar cell technology dates to 1839 when Becquerel observed that shining light on an electrode submerged in a conductive solution would create an electric current.

Michael Faraday began, in 1832, what promised to be a rather tedious attempt to prove that all electricities had precisely the same properties and caused precisely the same effects. The key effect was electrochemical decomposition. Voltaic and electromagnetic electricity posed no problems, but static electricity did. As Faraday delved deeper into the problem, he made two startling discoveries. First, electrical force did not, as had long been supposed, act at a distance upon molecules to cause them to dissociate. It was the passage of electricity through a conducting liquid medium that caused the molecules to dissociate, even when the electricity merely discharged into the air and did not pass through a "pole" or "center of action" in a voltaic cell. Second, the amount of the decomposition was found to be related directly to the amount of electricity passing through the solution.

These findings led Faraday to a new theory of electrochemistry. The electric force, he argued, threw the molecules of a solution into a state of tension. When the force was strong enough to distort the forces that held the molecules together so as to permit the interaction with neighbouring particles, the tension was relieved by the migration of particles along the lines of tension, the different parts of atoms migrating in opposite directions. The amount of electricity that passed, then, was clearly related to the chemical affinities of the substances in solution. These experiments led directly to Faraday's two laws of electrochemistry which state:

- The amount of a substance deposited on each electrode of an electrolytic cell is directly proportional to the amount of electricity passing through the cell.

- The quantities of different elements deposited by a given amount of electricity are in the ratio of their chemical equivalent weights.

William Sturgeon built an electric motor in 1832 and invented the commutator, a ring of metal-bristled brushes which allow the spinning armature to maintain contact with the electric current and changed the alternating current to a pulsating direct current. He also improved the voltaic battery and worked on the theory of thermoelectricity.

Hippolyte Pixii, a French instrument maker, constructed the first dynamo in 1832 and later built a direct current dynamo using the commutator. This was the first practical mechanical generator of electric current that used concepts demonstrated by Faraday.

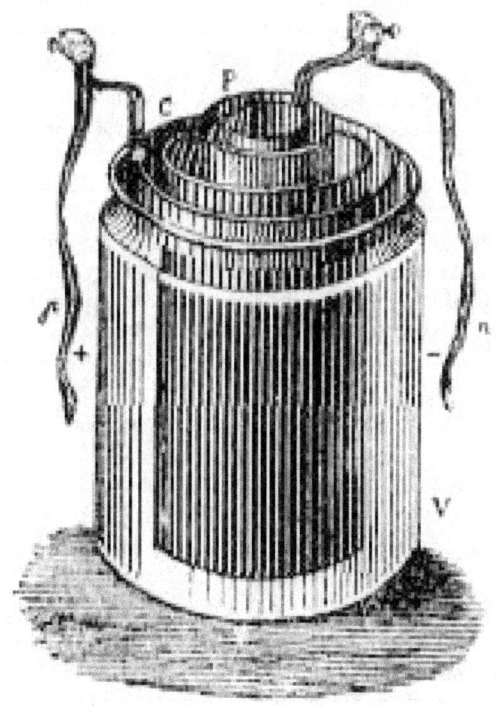

Daniell cell

John Daniell began experiments in 1835 in an attempt to improve the voltaic battery with its problems of being unsteady and a weak source of electric current. His experiments soon led to remarkable results. In 1836, he invented a primary cell in which hydrogen was eliminated in the generation of the electricity. Daniell had solved the problem of polarization. In his laboratory he had learned to alloy the amalgamated zinc of Sturgeon with mercury. His version was the first of the two-fluid class battery and the first battery that produced a constant reliable source of electric current over a long period of time.

William Grove produced the first fuel cell in 1839. He based his experiment on the fact that sending an electric current through water splits the water into its component parts of hydrogen and oxygen. So, Grove tried reversing the reaction—combining hydrogen and oxygen to produce electricity and water. Eventually the term *fuel cell* was coined in 1889 by Ludwig Mond and Charles Langer, who attempted to build the first practical device using air and industrial coal gas. He also introduced a powerful battery at the annual meeting of the British Association for the Advancement of Science in 1839. Grove's first cell consisted of zinc in diluted sulfuric acid and platinum in concentrated nitric acid, separated by a porous pot. The cell was able to generate about 12 amperes of current at about 1.8 volts. This cell had nearly double the voltage of the first Daniell cell. Grove's nitric acid cell was the favourite battery of the early American telegraph (1840–1860), because it offered strong current output.

As telegraph traffic increased, it was found that the Grove cell discharged poisonous nitrogen dioxide gas. As telegraphs became more complex, the need for a constant voltage became critical and the Grove device was limited (as the cell discharged, nitric acid was depleted and voltage was reduced). By the time of the American Civil War, Grove's battery had been replaced by the Daniell battery. In 1841 Robert Bunsen replaced the expensive platinum electrode used in Grove's battery with a carbon electrode. This led to large scale use of the "Bunsen battery" in the production of arc-lighting and in electroplating.

Wilhelm Weber developed, in 1846, the electrodynamometer, in which a current causes a coil suspended within another coil to turn when a current is passed through both. In 1852, Weber defined the absolute unit of electrical resistance (which was named the ohm after Georg Ohm). Weber's name is now used as a unit name to describe magnetic flux, the weber.

German physicist Johann Hittorf concluded that *ion movement* caused electric current. In 1853 Hittorf noticed that some ions traveled more rapidly than others. This observation led to the concept of transport number, the rate at which particular ions carried the electric current. Hittorf measured the changes in the concentration of electrolysed solutions, computed from these the transport numbers (relative carrying capacities) of many ions, and, in 1869, published his findings governing the migration of ions.

In 1866, Georges Leclanché patented a new battery system, which was immediately successful. Leclanché's original cell was assembled in a porous pot. The positive electrode (the

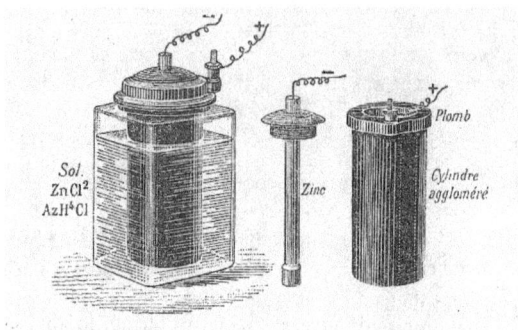

Fig. 293. — Élément Leclanché-Barbier.

Leclanché cell

cathode) consisted of crushed manganese dioxide with a little carbon mixed in. The negative pole (anode) was a zinc rod. The cathode was packed into the pot, and a carbon rod was inserted to act as a current collector. The anode and the pot were then immersed in an ammonium chloride solution. The liquid acted as the electrolyte, readily seeping through the porous pot and making contact with the cathode material. Leclanché's "wet" cell became the forerunner to the world's first widely used battery, the zinc-carbon cell.

25.4   Late 19th century advances and the advent of electrochemical societies

In 1869 Zénobe Gramme devised his first clean direct current dynamo. His generator featured a ring armature wound with many individual coils of wire.

Svante August Arrhenius published his thesis in 1884, *Recherches sur la conductibilité galvanique des électrolytes* (Investigations on the galvanic conductivity of electrolytes). From the results of his experiments, the author concluded that electrolytes, when dissolved in water, become to varying degrees split or dissociated into positive and negative ions. The degree to which this dissociation occurred depended above all on the nature of the substance and its concentration in the solution, being more developed the greater the dilution. The ions were supposed to be the carriers of not only the electric current, as in electrolysis, but also of the chemical activity. The relation between the actual number of ions and their number at great dilution (when all the molecules were dissociated) gave a quantity of special interest ("activity constant").

The race for the commercially viable production of aluminium was won in 1886 by Paul Héroult and Charles M. Hall. The problem many researchers had with extract-

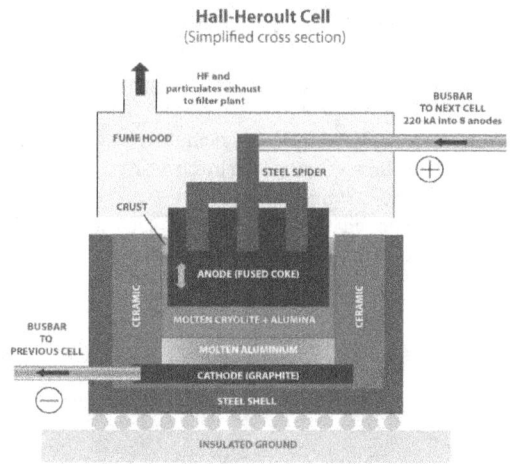

A Hall-Héroult industrial cell.

ing aluminium was that electrolysis of an aluminium salt dissolved in water yields aluminium hydroxide. Both Hall and Héroult avoided this problem by dissolving aluminium oxide in a new solvent— fused cryolite (Na_3AlF_6).

Wilhelm Ostwald, 1909 Nobel Laureate, started his experimental work in 1875, with an investigation on the law of mass action of water in relation to the problems of chemical affinity, with special emphasis on electrochemistry and chemical dynamics. In 1894 he gave the first modern definition of a catalyst and turned his attention to catalytic reactions. Ostwald is especially known for his contributions to the field of electrochemistry, including important studies of the electrical conductivity and electrolytic dissociation of organic acids.

Hermann Nernst developed the theory of the electromotive force of the voltaic cell in 1888. He developed methods for measuring dielectric constants and was the first to show that solvents of high dielectric constants promote the ionization of substances. Nernst's early studies in electrochemistry were inspired by Arrhenius' dissociation theory which first recognised the importance of ions in solution. In 1889, Nernst elucidated the theory of galvanic cells by assuming an "electrolytic pressure of dissolution," which forces ions from electrodes into solution and which was opposed to the osmotic pressure of the dissolved ions. He applied the principles of thermodynamics to the chemical reactions proceeding in a battery. In that same year he showed how the characteristics of the current produced could be used to calculate the free energy change in the chemical reaction producing the current. He constructed an equation, known as Nernst Equation, which describes the relation of a battery cell's voltage to its properties.

In 1898 Fritz Haber published his textbook, *Electrochem-*

istry: Grundriss der technischen Elektrochemie auf theoretischer Grundlage (The Theoretical Basis of Technical Electrochemistry), which was based on the lectures he gave at Karlsruhe. In the preface to his book he expressed his intention to relate chemical research to industrial processes and in the same year he reported the results of his work on electrolytic oxidation and reduction, in which he showed that definite reduction products can result if the voltage at the cathode is kept constant. In 1898 he explained the reduction of nitrobenzene in stages at the cathode and this became the model for other similar reduction processes.

In 1909, Robert Andrews Millikan began a series of experiments to determine the electric charge carried by a single electron. He began by measuring the course of charged water droplets in an electrical field. The results suggested that the charge on the droplets is a multiple of the elementary electric charge, but the experiment was not accurate enough to be convincing. He obtained more precise results in 1910 with his famous oil-drop experiment in which he replaced water (which tended to evaporate too quickly) with oil.

Jaroslav Heyrovský, a Nobel laureate, eliminated the tedious weighing required by previous analytical techniques, which used the differential precipitation of mercury by measuring drop-time. In the previous method, a voltage was applied to a dropping mercury electrode and a reference electrode was immersed in a test solution. After 50 drops of mercury were collected, they were dried and weighed. The applied voltage was varied and the experiment repeated. Measured weight was plotted versus applied voltage to obtain the curve. In 1921, Heyrovský had the idea of measuring the current flowing through the cell instead of just studying drop-time.

On February 10, 1922, the "polarograph" was born as Heyrovský recorded the current-voltage curve for a solution of 1 mol/L NaOH. Heyrovský correctly interpreted the current increase between -1.9 and -2.0 V as being due to the deposit of Na^+ ions, forming an amalgam. Shortly thereafter, with his Japanese colleague Masuzo Shikata, he constructed the first instrument for the automatic recording of polarographic curves, which became world famous later as the polarograph.

In 1923, Johannes Nicolaus Brønsted and Thomas Martin Lowry published essentially the same theory about how acids and bases behave using electrochemical basis.

The International Society of Electrochemistry (ISE) was founded in 1949, and some years later the first sophisticated electrophoretic apparatus was developed in 1937 by Arne Tiselius, who was awarded the 1948 Nobel prize for his work in protein electrophoresis. He developed the "moving boundary," which later would become known as *zone electrophoresis*, and used it to separate serum proteins in solution. Electrophoresis became widely developed in

Heyrovský's Polarograph

Nobelist in Chemistry, Wilhelm Ostwald: Elektrochemie: Ihre Geschichte und Lehre, Wilhelm Ostwald, Veit, Leipzig, 1896. (http://www.archive.org/details/elektrochemieih00ostwgoog). An English version is available as "Electrochemistry: history and theory" (2 volumes), translated by N. P. Date. It was published for the Smithsonian Institution and the National Science Foundation, Washington, DC, by Amerind Publ. Co., New Delhi, 1980.

the 1940s and 1950s when the technique was applied to molecules ranging from the largest proteins to amino acids and even inorganic ions.

During the 1960s and 1970s quantum electrochemistry was developed by Revaz Dogonadze and his pupils.

25.5   See also

- Electrochemistry

- History of the battery

- Karpen Pile

25.6   References

- "Physician-described use of electricity in medicine". *T.Gale's Electricity, or Ethereal Fire, Considered, 1802*. Retrieved March 10, 2008.

- Corrosion-Doctors.org

- A classic and knowledgeable - but dated - reference on the history of electrochemistry is by 1909

Chapter 26

Timeline of chemistry

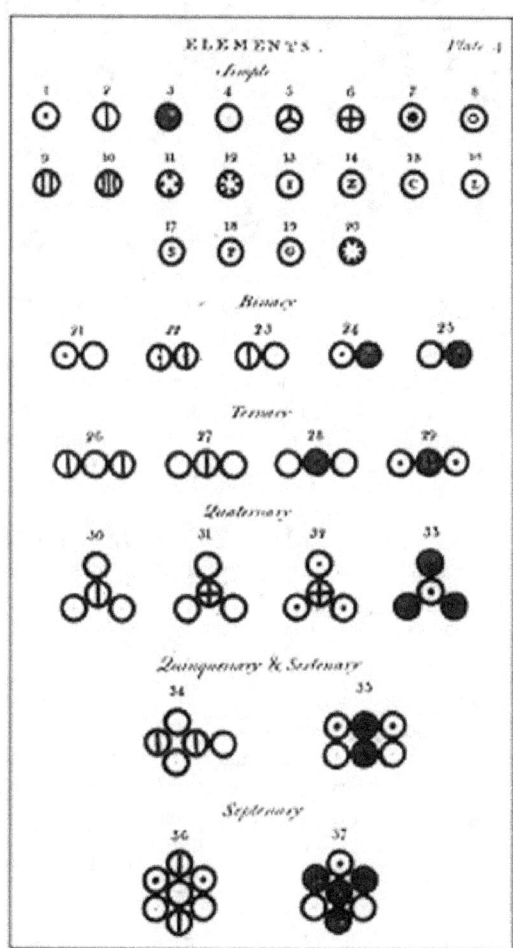

An image from John Dalton's A New System of Chemical Philosophy, *the first modern explanation of atomic theory.*

The **timeline of chemistry** lists important works, discoveries, ideas, inventions, and experiments that significantly changed humanity's understanding of the modern science known as chemistry, defined as the scientific study of the composition of matter and of its interactions. The history of chemistry in its modern form arguably began with the Irish scientist Robert Boyle, though its roots can be traced back to the earliest recorded history.

Early ideas that later became incorporated into the modern science of chemistry come from two main sources. Natural philosophers (such as Aristotle and Democritus) used deductive reasoning in an attempt to explain the behavior of the world around them. Alchemists (such as Geber and Rhazes) were people who used experimental techniques in an attempt to extend the life or perform material conversions, such as turning base metals into gold.

In the 17th century, a synthesis of the ideas of these two disciplines, that is the *deductive* and the *experimental*, leads to the development of a process of thinking known as the scientific method. With the introduction of the scientific method, the modern science of chemistry was born.

Known as "the central science", the study of chemistry is strongly influenced by, and exerts a strong influence on, many other scientific and technological fields. Many events considered central to our modern understanding of chemistry are also considered key discoveries in such fields as physics, biology, astronomy, geology, and materials science to name a few.[1]

26.1 Pre-17th century

Prior to the acceptance of the scientific method and its application to the field of chemistry, it is somewhat controversial to consider many of the people listed below as "chemists" in the modern sense of the word. However, the ideas of certain great thinkers, either for their prescience, or for their wide and long-term acceptance, bear listing here.

c. 3000 BC Egyptians formulate the theory of the Ogdoad, or the "primordial forces", from which all was formed. These were the elements of chaos, numbered in eight, that existed before the creation of the sun.[2]

235

Aristotle (384–322 BCE)

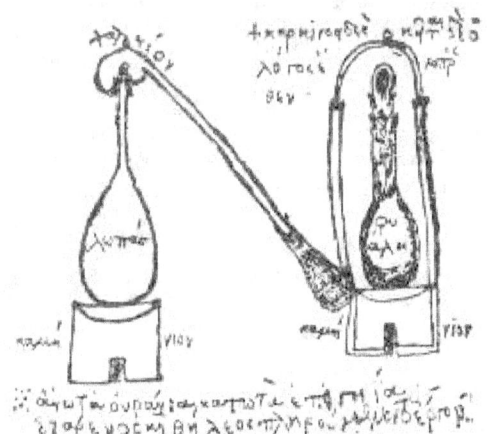

Ambix, cucurbit and retort, the alchemical implements of Zosimus c. 300, from Marcelin Berthelot, Collection des anciens alchimistes grecs *(3 vol., Paris, 1887–88)*

c. 1200 BC Tapputi-Belatikallim, a perfume-maker and early chemist, was mentioned in a cuneiform tablet in Mesopotamia.[3]

c. 450 BC Empedocles asserts that all things are composed of four primal elements: earth, air, fire, and

Geber (d. 815) is considered by some to be the "father of chemistry".

water, whereby two active and opposing forces, love and hate, or affinity and antipathy, act upon these elements, combining and separating them into infinitely varied forms.[4]

c. 440 BC Leucippus and Democritus propose the idea of the atom, an indivisible particle that all matter is made of. This idea is largely rejected by natural philosophers in favor of the Aristotlean view (see below).[5][6]

c. 360 BC Plato coins term 'elements' (*stoicheia*) and in his dialogue Timaeus, which includes a discussion of the composition of inorganic and organic bodies and is a rudimentary treatise on chemistry, assumes that the minute particle of each element had a special geometric shape: tetrahedron (fire), octahedron (air), icosahedron (water), and cube (earth).[7]

c. 350 BC Aristotle, expanding on Empedocles, proposes idea of a substance as a combination of *matter* and *form*. Describes theory of the Five Elements, fire, water, earth, air, and aether. This theory is largely accepted throughout the western world for over 1000 years.[8]

c. 50 BC Lucretius publishes *De Rerum Natura*, a poetic description of the ideas of atomism.[9]

c. 300 Zosimos of Panopolis writes some of the oldest known books on alchemy, which he defines as the study of the composition of waters, movement, growth, embodying and disembodying, drawing the spirits from bodies and bonding the spirits within bodies.[10]

c. 770 Abu Musa Jabir ibn Hayyan (aka Geber), an Arab/Persian alchemist who is "considered by many to be the father of chemistry",[11][12][13] develops an early experimental method for chemistry, and isolates numerous acids, including hydrochloric acid, nitric acid, citric acid, acetic acid, tartaric acid, and aqua regia.[14]

c. 1000 Abū al-Rayhān al-Bīrūnī[15] and Avicenna,[16] both Persian chemists, refute the practice of alchemy and the theory of the transmutation of metals.

c. 1167 Magister Salernus of the School of Salerno makes the first references to the distillation of wine.[17]

c. 1220 Robert Grosseteste publishes several Aristotelian commentaries where he lays out an early framework for the scientific method.[18]

c 1250 Tadeo Alderotti develops fractional distillation, which is much more effective than its predecessors.[19]

c 1260 St Albertus Magnus discovers arsenic[20] and silver nitrate.[21] He also made one of the first references to sulfuric acid.[22]

c. 1267 Roger Bacon publishes *Opus Maius*, which among other things, proposes an early form of the scientific method, and contains results of his experiments with gunpowder.[23]

c. 1310 Pseudo-Geber, an anonymous Spanish alchemist who wrote under the name of Geber, publishes several books that establish the long-held theory that all metals were composed of various proportions of sulfur and mercury.[24] He is one of the first to describe nitric acid, aqua regia, and aqua fortis.[25]

c. 1530 Paracelsus develops the study of iatrochemistry, a subdiscipline of alchemy dedicated to extending life, thus being the roots of the modern pharmaceutical industry. It is also claimed that he is the first to use the word "chemistry".[10]

1597 Andreas Libavius publishes *Alchemia*, a prototype chemistry textbook.[26]

26.2 17th and 18th centuries

1605 Sir Francis Bacon publishes *The Proficience and Advancement of Learning*, which contains a description of what would later be known as the scientific method.[27]

1605 Michal Sedziwój publishes the alchemical treatise *A New Light of Alchemy* which proposed the existence of the "food of life" within air, much later recognized as oxygen.[28]

1615 Jean Beguin publishes the *Tyrocinium Chymicum*, an early chemistry textbook, and in it draws the first-ever chemical equation.[29]

1637 René Descartes publishes *Discours de la méthode*, which contains an outline of the scientific method.[30]

1648 Posthumous publication of the book *Ortus medicinae* by Jan Baptist van Helmont, which is cited by some as a major transitional work between alchemy and chemistry, and as an important influence on Robert Boyle. The book contains the results of numerous experiments and establishes an early version of the law of conservation of mass.[31]

1661 Robert Boyle publishes *The Sceptical Chymist*, a treatise on the distinction between chemistry and alchemy. It contains some of the earliest modern ideas of atoms, molecules, and chemical reaction, and marks the beginning of the history of modern chemistry.[32]

1662 Robert Boyle proposes Boyle's law, an experimentally based description of the behavior of gases, specifically the relationship between pressure and volume.[32]

1735 Swedish chemist Georg Brandt analyzes a dark blue pigment found in copper ore. Brandt demonstrated that the pigment contained a new element, later named cobalt.[33][34]

1754 Joseph Black isolates carbon dioxide, which he called "fixed air".[35]

1757 Louis Claude Cadet de Gassicourt, while investigating arsenic compounds, creates Cadet's fuming liquid, later discovered to be cacodyl oxide, considered to be the first synthetic organometallic compound.[36]

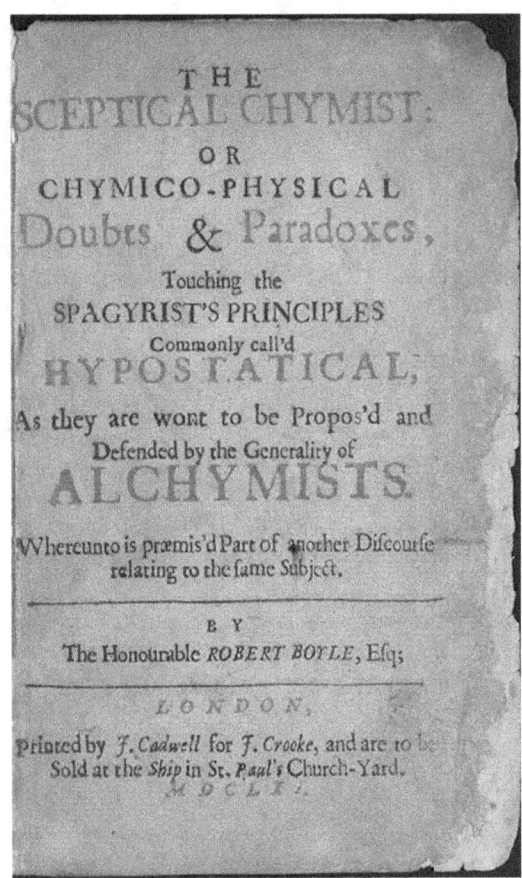

Title page of The Sceptical Chymist *by Robert Boyle (1627–91)*

A typical chemical laboratory of the 18th century

1758 Joseph Black formulates the concept of latent heat to explain the thermochemistry of phase changes.[37]

1766 Henry Cavendish discovers hydrogen as a colorless, odourless gas that burns and can form an explosive mixture with air.[38]

1773–1774 Carl Wilhelm Scheele and Joseph Priestley independently isolate oxygen, called by Priestley "dephlogisticated air" and Scheele "fire air".[39][40]

Antoine-Laurent de Lavoisier (1743–94) is considered the "Father of Modern Chemistry".

1778 Antoine Lavoisier, considered "The father of modern chemistry",[41] recognizes and names oxygen, and recognizes its importance and role in combustion.[42]

1787 Antoine Lavoisier publishes *Méthode de nomenclature chimique*, the first modern system of chemical nomenclature.[42]

1787 Jacques Charles proposes Charles's law, a corollary of Boyle's law, describes relationship between temperature and volume of a gas.[43]

1789 Antoine Lavoisier publishes *Traité Élémentaire de Chimie*, the first modern chemistry textbook. It is a complete survey of (at that time) modern chemistry, including the first concise definition of the law of conservation of mass, and thus also represents the founding of the discipline of stoichiometry or quantitative chemical analysis.[42][44]

1797 Joseph Proust proposes the law of definite proportions, which states that elements always combine in small, whole number ratios to form compounds.[45]

1800 Alessandro Volta devises the first chemical battery, thereby founding the discipline of electrochemistry.[46]

26.3 19th century

John Dalton (1766–1844)

1801 John Dalton proposes Dalton's law, which describes relationship between the components in a mixture of gases and the relative pressure each contributes to that of the overall mixture.[47]

1805 Joseph Louis Gay-Lussac discovers that water is composed of two parts hydrogen and one part oxygen by volume.[48]

1808 Joseph Louis Gay-Lussac collects and discovers several chemical and physical properties of air and of other gases, including experimental proofs of Boyle's and Charles's laws, and of relationships between density and composition of gases.[49]

1808 John Dalton publishes *New System of Chemical Philosophy*, which contains first modern scientific description of the atomic theory, and clear description of the law of multiple proportions.[47]

1808 Jöns Jakob Berzelius publishes *Lärbok i Kemien* in which he proposes modern chemical symbols and notation, and of the concept of relative atomic weight.[50]

1811 Amedeo Avogadro proposes Avogadro's law, that equal volumes of gases under constant temperature and pressure contain equal number of molecules.[51]

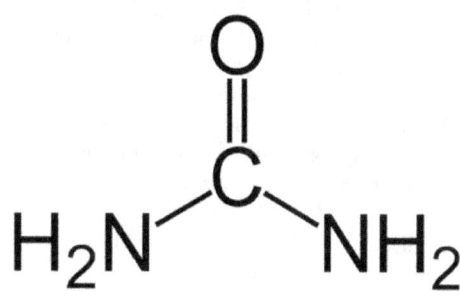

Structural formula of urea

1825 Friedrich Wöhler and Justus von Liebig perform the first confirmed discovery and explanation of isomers, earlier named by Berzelius. Working with cyanic acid and fulminic acid, they correctly deduce that isomerism was caused by differing arrangements of atoms within a molecular structure.[52]

1827 William Prout classifies biomolecules into their modern groupings: carbohydrates, proteins and lipids.[53]

1828 Friedrich Wöhler synthesizes urea, thereby establishing that organic compounds could be produced from inorganic starting materials, disproving the theory of vitalism.[52]

1832 Friedrich Wöhler and Justus von Liebig discover and explain functional groups and radicals in relation to organic chemistry.[52]

1840 Germain Hess proposes Hess's law, an early statement of the law of conservation of energy, which establishes that energy changes in a chemical process depend only on the states of the starting and product materials and not on the specific pathway taken between the two states.[54]

1847 Hermann Kolbe obtains acetic acid from completely inorganic sources, further disproving vitalism.[55]

1848 Lord Kelvin establishes concept of absolute zero, the temperature at which all molecular motion ceases.[56]

1849 Louis Pasteur discovers that the racemic form of tartaric acid is a mixture of the levorotatory and dextrotatory forms, thus clarifying the nature of optical rotation and advancing the field of stereochemistry.[57]

1852 August Beer proposes Beer's law, which explains the relationship between the composition of a mixture and the amount of light it will absorb. Based partly on earlier work by Pierre Bouguer and Johann Heinrich Lambert, it establishes the analytical technique known as spectrophotometry.[58]

1855 Benjamin Silliman, Jr. pioneers methods of petroleum cracking, which makes the entire modern petrochemical industry possible.[59]

1856 William Henry Perkin synthesizes Perkin's mauve, the first synthetic dye. Created as an accidental byproduct of an attempt to create quinine from coal tar. This discovery is the foundation of the dye synthesis industry, one of the earliest successful chemical industries.[60]

1857 Friedrich August Kekulé von Stradonitz proposes that carbon is tetravalent, or forms exactly four chemical bonds.[61]

1859–1860 Gustav Kirchhoff and Robert Bunsen lay the foundations of spectroscopy as a means of chemical analysis, which lead them to the discovery of caesium and rubidium. Other workers soon used the same technique to discover indium, thallium, and helium.[62]

1860 Stanislao Cannizzaro, resurrecting Avogadro's ideas regarding diatomic molecules, compiles a table of atomic weights and presents it at the 1860 Karlsruhe Congress, ending decades of conflicting atomic weights and molecular formulas, and leading to Mendeleev's discovery of the periodic law.[63]

1862 Alexander Parkes exhibits Parkesine, one of the earliest synthetic polymers, at the International Exhibition in London. This discovery formed the foundation of the modern plastics industry.[64]

1862 Alexandre-Emile Béguyer de Chancourtois publishes the telluric helix, an early, three-dimensional version of the periodic table of the elements.[65]

1864 John Newlands proposes the law of octaves, a precursor to the periodic law.[65]

1864 Lothar Meyer develops an early version of the periodic table, with 28 elements organized by valence.[66]

1864 Cato Maximilian Guldberg and Peter Waage, building on Claude Louis Berthollet's ideas, proposed the law of mass action.[67][68][69]

1865 Johann Josef Loschmidt determines exact number of molecules in a mole, later named Avogadro's number.[70]

1865 Friedrich August Kekulé von Stradonitz, based partially on the work of Loschmidt and others, establishes structure of benzene as a six carbon ring with alternating single and double bonds.[61]

1865 Adolf von Baeyer begins work on indigo dye, a milestone in modern industrial organic chemistry which revolutionizes the dye industry.[71]

ОПЫТЪ СИСТЕМЫ ЭЛЕМЕНТОВЪ.

ОСНОВАННОЙ НА ИХЪ АТОМНОМЪ ВѢСѢ И ХИМИЧЕСКОМЪ СХОДСТВѢ.

	Ti = 50	Zr = 90	? = 180.
	V = 51	Nb = 94	Ta = 182.
	Cr = 52	Mo = 96	W = 186.
	Mn = 55	Rh = 104,4	Pt = 197,1.
	Fe = 56	Rn = 104,4	Ir = 198.
	Ni = Co = 59	Pl = 106,6	O = 199.
H = 1	Cu = 63,4	Ag = 108	Hg = 200.
Be = 9,4 Mg = 24	Zn = 65,2	Cd = 112	
B = 11 Al = 27,4	? = 68	Ur = 116	Au = 197?
C = 12 Si = 28	? = 70	Sn = 118	
N = 14 P = 31	As = 75	Sb = 122	Bi = 210?
O = 16 S = 32	Se = 79,4	Te = 128?	
F = 19 Cl = 35,6	Br = 80	I = 127	
Li = 7 Na = 23 K = 39	Rb = 85,4	Cs = 133	Tl = 204.
Ca = 40	Sr = 87,6	Ba = 137	Pb = 207.
? = 45 Ce = 92			
?Er = 56 La = 94			
?Yt = 60 Di = 95			
?In = 75,6 Th = 118?			

Д. Менделѣевъ

Mendeleev's 1869 Periodic table

1869 Dmitri Mendeleev publishes the first modern periodic table, with the 66 known elements organized by atomic weights. The strength of his table was its ability to accurately predict the properties of as-yet unknown elements.[65][66]

1873 Jacobus Henricus van 't Hoff and Joseph Achille Le Bel, working independently, develop a model of chemical bonding that explains the chirality experiments of Pasteur and provides a physical cause for optical activity in chiral compounds.[72]

1876 Josiah Willard Gibbs publishes *On the Equilibrium of Heterogeneous Substances*, a compilation of his work on thermodynamics and physical chemistry which lays out the concept of free energy to explain the physical basis of chemical equilibria.[73]

1877 Ludwig Boltzmann establishes statistical derivations of many important physical and chemical concepts, including entropy, and distributions of molecular velocities in the gas phase.[74]

1883 Svante Arrhenius develops ion theory to explain conductivity in electrolytes.[75]

1884 Jacobus Henricus van 't Hoff publishes *Études de Dynamique chimique*, a seminal study on chemical kinetics.[76]

1884 Hermann Emil Fischer proposes structure of purine, a key structure in many biomolecules, which he later synthesized in 1898. Also begins work on the chemistry of glucose and related sugars.[77]

1884 Henry Louis Le Chatelier develops Le Chatelier's principle, which explains the response of dynamic chemical equilibria to external stresses.[78]

1885 Eugene Goldstein names the cathode ray, later discovered to be composed of electrons, and the canal ray, later discovered to be positive hydrogen ions that had been stripped of their electrons in a cathode ray tube. These would later be named protons.[79]

1893 Alfred Werner discovers the octahedral structure of cobalt complexes, thus establishing the field of coordination chemistry.[80]

1894–1898 William Ramsay discovers the noble gases, which fill a large and unexpected gap in the periodic table and led to models of chemical bonding.[81]

1897 J. J. Thomson discovers the electron using the cathode ray tube.[82]

1898 Wilhelm Wien demonstrates that canal rays (streams of positive ions) can be deflected by magnetic fields, and that the amount of deflection is proportional to the mass-to-charge ratio. This discovery would lead to the analytical technique known as mass spectrometry.[83]

1898 Maria Sklodowska-Curie and Pierre Curie isolate radium and polonium from pitchblende.[84]

c. 1900 Ernest Rutherford discovers the source of radioactivity as decaying atoms; coins terms for various types of radiation.[85]

26.4 20th century

1903 Mikhail Semyonovich Tsvet invents chromatography, an important analytic technique.[86]

1904 Hantaro Nagaoka proposes an early nuclear model of the atom, where electrons orbit a dense massive nucleus.[87]

1905 Fritz Haber and Carl Bosch develop the Haber process for making ammonia from its elements, a milestone in industrial chemistry with deep consequences in agriculture.[88]

1905 Albert Einstein explains Brownian motion in a way that definitively proves atomic theory.[89]

1907 Leo Hendrik Baekeland invents bakelite, one of the first commercially successful plastics.[90]

1909 Robert Millikan measures the charge of individual electrons with unprecedented accuracy through the oil drop experiment, confirming that all electrons have the same charge and mass.[91]

1909 S. P. L. Sørensen invents the pH concept and develops methods for measuring acidity.[92]

1911 Antonius van den Broek proposes the idea that the elements on the periodic table are more properly organized by positive nuclear charge rather than atomic weight.[93]

1911 The first Solvay Conference is held in Brussels, bringing together most of the most prominent scientists of the day. Conferences in physics and chemistry continue to be held periodically to this day.[94]

1911 Ernest Rutherford, Hans Geiger, and Ernest Marsden perform the gold foil experiment, which proves the nuclear model of the atom, with a small, dense, positive nucleus surrounded by a diffuse electron cloud.[85]

Robert A. Millikan performed the oil drop experiment.

1912 William Henry Bragg and William Lawrence Bragg propose Bragg's law and establish the field of X-ray crystallography, an important tool for elucidating the crystal structure of substances.[95]

1912 Peter Debye develops the concept of molecular dipole to describe asymmetric charge distribution in some molecules.[96]

1913 Niels Bohr introduces concepts of quantum mechanics to atomic structure by proposing what is now known as the Bohr model of the atom, where electrons exist only in strictly defined orbitals.[97]

1913 Henry Moseley, working from Van den Broek's earlier idea, introduces concept of atomic number to fix inadequacies of Mendeleev's periodic table, which had been based on atomic weight.[98]

1913 Frederick Soddy proposes the concept of isotopes, that elements with the same chemical properties may have differing atomic weights.[99]

1913 J. J. Thomson expanding on the work of Wien, shows that charged subatomic particles can be separated by

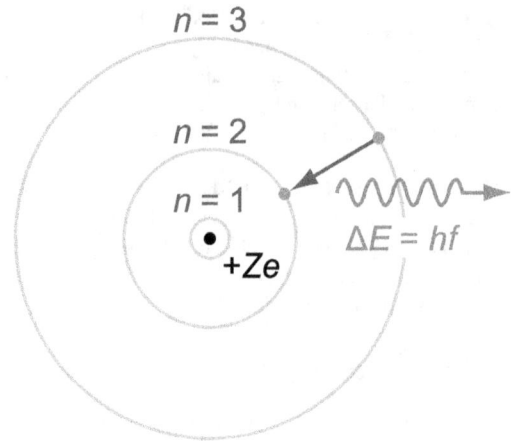

The Bohr model of the atom

their mass-to-charge ratio, a technique known as mass spectrometry.[100]

1916 Gilbert N. Lewis publishes "The Atom and the Molecule", the foundation of valence bond theory.[101]

1921 Otto Stern and Walther Gerlach establish concept of quantum mechanical spin in subatomic particles.[102]

1923 Gilbert N. Lewis and Merle Randall publish *Thermodynamics and the Free Energy of Chemical Substances*, first modern treatise on chemical thermodynamics.[103]

1923 Gilbert N. Lewis develops the electron pair theory of acid/base reactions.[101]

1924 Louis de Broglie introduces the wave-model of atomic structure, based on the ideas of wave–particle duality.[104]

1925 Wolfgang Pauli develops the exclusion principle, which states that no two electrons around a single nucleus may have the same quantum state, as described by four quantum numbers.[105]

1926 Erwin Schrödinger proposes the Schrödinger equation, which provides a mathematical basis for the wave model of atomic structure.[106]

1927 Werner Heisenberg develops the uncertainty principle which, among other things, explains the mechanics of electron motion around the nucleus.[107]

$$H(t) \mid \psi(t)\rangle = i\hbar \frac{d}{dt} \mid \psi(t)\rangle$$

The Schrödinger equation

1927 Fritz London and Walter Heitler apply quantum mechanics to explain covalent bonding in the hydrogen molecule,[108] which marked the birth of quantum chemistry.[109]

1929 Linus Pauling publishes Pauling's rules, which are key principles for the use of X-ray crystallography to deduce molecular structure.[110]

1931 Erich Hückel proposes Hückel's rule, which explains when a planar ring molecule will have aromatic properties.[111]

1931 Harold Urey discovers deuterium by fractionally distilling liquid hydrogen.[112]

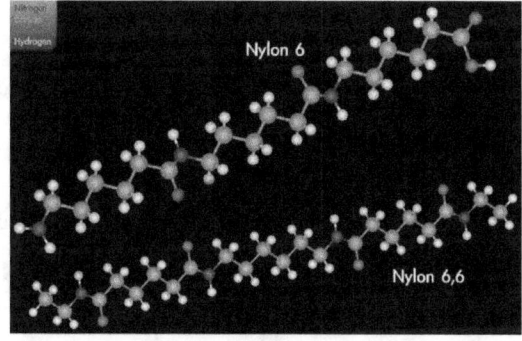

Model of two common forms of nylon

1932 James Chadwick discovers the neutron.[113]

1932–1934 Linus Pauling and Robert Mulliken quantify electronegativity, devising the scales that now bear their names.[114]

1935 Wallace Carothers leads a team of chemists at DuPont who invent nylon, one of the most commercially successful synthetic polymers in history.[115]

1937 Carlo Perrier and Emilio Segrè perform the first confirmed synthesis of technetium-97, the first artificially

produced element, filling a gap in the periodic table. Though disputed, the element may have been synthesized as early as 1925 by Walter Noddack and others.[116]

1937 Eugene Houdry develops a method of industrial scale catalytic cracking of petroleum, leading to the development of the first modern oil refinery.[117]

1937 Pyotr Kapitsa, John Allen and Don Misener produce supercooled helium-4, the first zero-viscosity superfluid, a substance that displays quantum mechanical properties on a macroscopic scale.[118]

1938 Otto Hahn discovers the process of nuclear fission in uranium and thorium.[119]

1939 Linus Pauling publishes *The Nature of the Chemical Bond*, a compilation of a decades worth of work on chemical bonding. It is one of the most important modern chemical texts. It explains hybridization theory, covalent bonding and ionic bonding as explained through electronegativity, and resonance as a means to explain, among other things, the structure of benzene.[110]

1940 Edwin McMillan and Philip H. Abelson identify neptunium, the lightest and first synthesized transuranium element, found in the products of uranium fission. McMillan would found a lab at Berkeley that would be involved in the discovery of many new elements and isotopes.[120]

1941 Glenn T. Seaborg takes over McMillan's work creating new atomic nuclei. Pioneers method of neutron capture and later through other nuclear reactions. Would become the principal or co-discoverer of nine new chemical elements, and dozens of new isotopes of existing elements.[120]

1945 Jacob A. Marinsky, Lawrence E. Glendenin, and Charles D. Coryell perform the first confirmed synthesis of Promethium, filling in the last "gap" in the periodic table.[121]

1945–1946 Felix Bloch and Edward Mills Purcell develop the process of nuclear magnetic resonance, an analytical technique important in elucidating structures of molecules, especially in organic chemistry.[122]

1951 Linus Pauling uses X-ray crystallography to deduce the secondary structure of proteins.[110]

1952 Alan Walsh pioneers the field of atomic absorption spectroscopy, an important quantitative spectroscopy method that allows one to measure specific concentrations of a material in a mixture.[123]

1952 Robert Burns Woodward, Geoffrey Wilkinson, and Ernst Otto Fischer discover the structure of ferrocene, one of the founding discoveries of the field of organometallic chemistry.[124]

1953 James D. Watson and Francis Crick propose the structure of DNA, opening the door to the field of molecular biology.[125]

1957 Jens Skou discovers Na^+/K^+-ATPase, the first ion-transporting enzyme.[126]

1958 Max Perutz and John Kendrew use X-ray crystallography to elucidate a protein structure, specifically sperm whale myoglobin.[127]

1962 Neil Bartlett synthesizes xenon hexafluoroplatinate, showing for the first time that the noble gases can form chemical compounds.[128]

1962 George Olah observes carbocations via superacid reactions.[129]

1964 Richard R. Ernst performs experiments that will lead to the development of the technique of Fourier transform NMR. This would greatly increase the sensitivity of the technique, and open the door for magnetic resonance imaging or MRI.[130]

1965 Robert Burns Woodward and Roald Hoffmann propose the Woodward–Hoffmann rules, which use the symmetry of molecular orbitals to explain the stereochemistry of chemical reactions.[124]

1966 Hotosi Nozaki and Ryōji Noyori discovered the first example of asymmetric catalysis (hydrogenation) using a structurally well-defined chiral transition metal complex.[131][132]

1970 John Pople develops the Gaussian program greatly easing computational chemistry calculations.[133]

1971 Yves Chauvin offered an explanation of the reaction mechanism of olefin metathesis reactions.[134]

1975 Karl Barry Sharpless and group discover a stereoselective oxidation reactions including Sharpless epoxidation,[135][136] Sharpless asymmetric dihydroxylation,[137][138][139] and Sharpless oxyamination.[140][141][142]

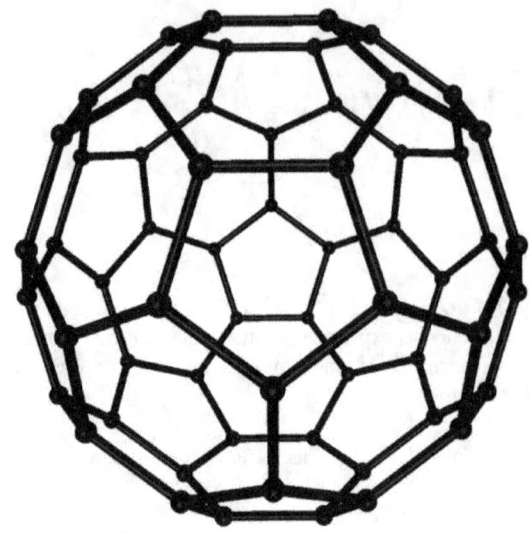

Buckminsterfullerene, C_{60}

1985 Harold Kroto, Robert Curl and Richard Smalley discover fullerenes, a class of large carbon molecules superficially resembling the geodesic dome designed by architect R. Buckminster Fuller.[143]

1991 Sumio Iijima uses electron microscopy to discover a type of cylindrical fullerene known as a carbon nanotube, though earlier work had been done in the field as early as 1951. This material is an important component in the field of nanotechnology.[144]

1994 First total synthesis of Taxol by Robert A. Holton and his group.[145][146][147]

1995 Eric Cornell and Carl Wieman produce the first Bose–Einstein condensate, a substance that displays quantum mechanical properties on the macroscopic scale.[148]

26.5   See also

- History of chemistry

- Nobel Prize in chemistry

- List of Nobel laureates in Chemistry

- Timeline of chemical elements discoveries

26.6   References

[1] "Chemistry – The Central Science". *The Chemistry Hall of Fame*. York University. Retrieved 2006-09-12.

[2] J. Gwyn Griffiths (1955). "The Orders of Gods in Greece and Egypt (According to Herodotus)". *The Journal of Hellenic Studies* (The Society for the Promotion of Hellenic Studies) **75**: 21–23. doi:10.2307/629164. JSTOR 629164.

[3] Giese, Patsy Ann. "Women in Science: 5000 Years of Obstacles and Achievements". SHiPS Resource Center for Sociology, History and Philosophy in Science Teaching. Retrieved 2007-03-11.

[4] Parry, Richard (2005-03-04). "Empedocles". *Stanford Encyclopedia of Philosophy*. Metaphysics Research Lab, CSLI, Stanford University. Retrieved 2007-03-11.

[5] Berryman, Sylvia (2004-08-14). "Leucippus". *Stanford Encyclopedia of Philosophy*. Metaphysics Research Lab, CSLI, Stanford University. Retrieved 2007-03-11.

[6] Berryman, Sylvia (2004-08-15). "Democritus". *Stanford Encyclopedia of Philosophy*. Metaphysics Research Lab, CSLI, Stanford University. Retrieved 2007-03-11.

[7] Hillar, Marian (2004). "The Problem of the Soul in Aristotle's De anima". NASA WMAP. Retrieved 2006-08-10.

[8] "HISTORY/CHRONOLOGY OF THE ELEMENTS". Retrieved 2007-03-12.

[9] Sedley, David (2004-08-04). "Lucretius". *Stanford Encyclopedia of Philosophy*. Metaphysics Research Lab, CSLI, Stanford University. Retrieved 2007-03-11.

[10] Strathern, Paul (2000). *Mendeleyev's Dream – The Quest for the Elements*. Berkley Books. ISBN 0-425-18467-6.

[11] Derewenda, Zygmunt S. (2007), "On wine, chirality and crystallography", *Acta Crystallographica Section A: Foundations of Crystallography* **64**: 246–258 [247], Bibcode:2008AcCrA..64..246D, doi:10.1107/S0108767307054293, PMID 18156689

[12] John Warren (2005). "War and the Cultural Heritage of Iraq: a sadly mismanaged affair", *Third World Quarterly*, Volume 26, Issue 4 & 5, p. 815-830.

[13] Dr. A. Zahoor (1997). "JABIR IBN HAIYAN (Geber)". University of Indonesia. Archived from the original on 2008-06-30. Retrieved 2011-2009-30. Check date values in: |access-date= (help)

[14] "Father of Chemistry: Jabir Ibn Haiyan". *Famous Muslism*. Famousmuslims.com. 2003. Retrieved 2007-03-12.

[15] Marmura, Michael E. (1965). "*An Introduction to Islamic Cosmological Doctrines. Conceptions of Nature and Methods Used for Its Study by the Ikhwan Al-Safa'an, Al-Biruni, and Ibn Sina* by Seyyed Hossein Nasr". *Speculum* **40** (4): 744–746. doi:10.2307/2851429.

[16] Robert Briffault (1938). *The Making of Humanity*, p. 196-197.

[17] Forbes, Robert James (1970). *A short history of the art of distillation: from the beginnings up to the death of Cellier Blumenthal*. BRILL. p. 88. ISBN 978-90-04-00617-1. Retrieved 29 June 2010.

[18] "Robert Grosseteste" in the 1913 *Catholic Encyclopedia*.

[19] Holmyard, Eric John (1990). *Alchemy*. Courier Dover Publications. p. 288. ISBN 0-486-26298-7.

[20] Emsley, John (2001). *Nature's Building Blocks: An A-Z Guide to the Elements*. Oxford: Oxford University Press. pp. 43,513,529. ISBN 0-19-850341-5.

[21] Davidson, Michael W.; National High Magnetic Field Laboratory at The Florida State University (2003-08-01). "Molecular Expressions: Science, Optics and You — Timeline — Albertus Magnus". The Florida State University. Retrieved 2009-11-28.

[22] Vladimir Karpenko, John A. Norris(2001), Vitriol in the history of Chemistry, Charles University

[23] O'Connor, J. J.; Robertson, E. F. (2003). "Roger Bacon". *MacTutor*. School of Mathematics and Statistics University of St Andrews, Scotland. Retrieved 2007-03-12.

[24] Zdravkovski, Zoran; Stojanoski, Kiro (1997-03-09). "GEBER". Institute of Chemistry, Skopje, Macedonia. Retrieved 2007-03-12.

[25] Encyclopædia Britannica 1911, *Alchemy*

[26] "From liquid to vapor and back: origins". *Special Collections Department*. University of Delaware Library. Retrieved 2007-03-12.

[27] Asarnow, Herman (2005-08-08). "Sir Francis Bacon: Empiricism". *An Image-Oriented Introduction to Backgrounds for English Renaissance Literature*. University of Portland. Retrieved 2007-02-22.

[28] "Sedziwój, Michal". *infopoland: Poland on the Web*. University at Buffalo. Retrieved 2007-02-22.

[29] Crosland, M.P. (1959). "The use of diagrams as chemical 'equations' in the lectures of William Cullen and Joseph Black". *Annals of Science* **15** (2): 75–90. doi:10.1080/00033795900200088.

[30] "René Descartes" in the 1913 *Catholic Encyclopedia*.

[31] "Johann Baptista van Helmont". *History of Gas Chemistry*. Center for Microscale Gas Chemistry, Creighton University. 2005-09-25. Retrieved 2007-02-23.

[32] "Robert Boyle". *Chemical Achievers: The Human Face of Chemical Sciences*. Chemical Heritage Foundation. 2005. Retrieved 2007-02-22.

[33] Georg Brandt first showed cobalt to be a new metal in: G. Brandt (1735) "Dissertatio de semimetallis" (Dissertation on semi-metals), *Acta Literaria et Scientiarum Sveciae* (Journal of Swedish literature and sciences), vol. 4, pages 1–10. See also: (1) G. Brandt (1746) "Rön och anmärkningar angäende en synnerlig färg — cobolt" (Observations and remarks concerning an extraordinary pigment — cobalt), *Kongliga Svenska vetenskapsakademiens handlingar* (Transactions of the Royal Swedish Academy of Science), vol.7, pages 119–130; (2) G. Brandt (1748) "Cobalti nova species examinata et descripta" (Cobalt, a new element examined and described), *Acta Regiae Societatis Scientiarum Upsaliensis* (Journal of the Royal Scientific Society of Uppsala), 1st series, vol. 3, pages 33–41; (3) James L. Marshall and Virginia R. Marshall (Spring 2003) "Rediscovery of the Elements: Riddarhyttan, Sweden," *The Hexagon* (official journal of the Alpha Chi Sigma fraternity of chemists), vol. 94, no. 1, pages 3–8.

[34] Wang, Shijie (2006). "Cobalt—Its recovery, recycling, and application". *Journal of the Minerals, Metals and Materials Society* **58** (10): 47–50. Bibcode:2006JOM....58j..47W. doi:10.1007/s11837-006-0201-y.

[35] Cooper, Alan (1999). "Joseph Black". *History of Glasgow University Chemistry Department*. University of Glasgow Department of Chemistry. Archived from the original on 2006-04-10. Retrieved 2006-02-23.

[36] Seyferth, Dietmar (2001). "Cadet's Fuming Arsenical Liquid and the Cacodyl Compounds of Bunsen". *Organometallics* **20** (8): 1488–1498. doi:10.1021/om0101947.

[37] Partington, J.R. (1989). *A Short History of Chemistry*. Dover Publications, Inc. ISBN 0-486-65977-1.

[38] Cavendish, Henry (1766). "Three Papers Containing Experiments on Factitious Air, by the Hon. Henry Cavendish". *Philosophical Transactions* (The University Press) **56**: 141–184. doi:10.1098/rstl.1766.0019. Retrieved 6 November 2007.

[39] "Joseph Priestley". *Chemical Achievers: The Human Face of Chemical Sciences*. Chemical Heritage Foundation. 2005. Retrieved 2007-02-22.

[40] "Carl Wilhelm Scheele". *History of Gas Chemistry*. Center for Microscale Gas Chemistry, Creighton University. 2005-09-11. Retrieved 2007-02-23.

[41] "Lavoisier, Antoine." Encyclopædia Britannica. 2007. Encyclopædia Britannica Online. 24 July 2007 <http://www.britannica.com/eb/article-9369846>.

[42] Weisstein, Eric W. (1996). "Lavoisier, Antoine (1743–1794)". *Eric Weisstein's World of Scientific Biography*. Wolfram Research Products. Retrieved 2007-02-23.

[43] "Jacques Alexandre César Charles". *Centennial of Flight*. U.S. Centennial of Flight Commission. 2001. Retrieved 2007-02-23.

[44] Burns, Ralph A. (1999). *Fundamentals of Chemistry*. Prentice Hall. p. 32. ISBN 0-02-317351-3.

[45] "Proust, Joseph Louis (1754–1826)". *100 Distinguished Chemists*. European Association for Chemical and Molecular Science. 2005. Archived from the original on 2008-05-15. Retrieved 2007-02-23.

[46] "Inventor Alessandro Volta Biography". *The Great Idea Finder*. The Great Idea Finder. 2005. Retrieved 2007-02-23.

[47] "John Dalton". *Chemical Achievers: The Human Face of Chemical Sciences*. Chemical Heritage Foundation. 2005. Retrieved 2007-02-22.

[48] "The Human Face of Chemical Sciences". Chemical Heritage Foundation. 2005. Retrieved 2007-02-22.

[49] "December 6 Births". *Today in Science History*. Today in Science History. 2007. Retrieved 2007-03-12.

[50] "Jöns Jakob Berzelius". *Chemical Achievers: The Human Face of Chemical Sciences*. Chemical Heritage Foundation. 2005. Retrieved 2007-02-22.

[51] "Michael Faraday". *Famous Physicists and Astronomers*. Retrieved 2007-03-12.

[52] "Justus von Liebig and Friedrich Wöhler". *Chemical Achievers: The Human Face of Chemical Sciences*. Chemical Heritage Foundation. 2005. Retrieved 2007-02-22.

[53] "William Prout". Retrieved 2007-03-12.

[54] "Hess, Germain Henri". Retrieved 2007-03-12.

[55] "Kolbe, Adolph Wilhelm Hermann". *100 Distinguished European Chemists*. European Association for Chemical and Molecular Sciences. 2005. Archived from the original on 2008-10-11. Retrieved 2007-03-12.

[56] Weisstein, Eric W. (1996). "Kelvin, Lord William Thomson (1824–1907)". *Eric Weisstein's World of Scientific Biography*. Wolfram Research Products. Retrieved 2007-03-12.

[57] "History of Chirality". Stheno Corporation. 2006. Archived from the original on 2007-03-07. Retrieved 2007-03-12.

[58] "Lambert-Beer Law". Sigrist-Photometer AG. 2007-03-07. Retrieved 2007-03-12.

[59] "Benjamin Silliman, Jr. (1816–1885)". *Picture History*. Picture History LLC. 2003. Retrieved 2007-03-24.

[60] "William Henry Perkin". *Chemical Achievers: The Human Face of Chemical Sciences*. Chemical Heritage Foundation. 2005. Retrieved 2007-03-24.

[61] "Archibald Scott Couper and August Kekulé von Stradonitz". *Chemical Achievers: The Human Face of Chemical Sciences*. Chemical Heritage Foundation. 2005. Retrieved 2007-02-22.

[62] O'Connor, J. J.; Robertson, E.F. (2002). "Gustav Robert Kirchhoff". *MacTutor*. School of Mathematics and Statistics University of St Andrews, Scotland. Retrieved 2007-03-24.

[63] Eric R. Scerri, *The Periodic Table: Its Story and Its Significance*, Oxford University Press, 2006.

[64] "Alexander Parkes (1813–1890)". *People & Polymers*. Plastics Historical Society. Archived from the original on 2007-03-15. Retrieved 2007-03-24.

[65] "The Periodic Table". The Third Millennium Online. Retrieved 2007-03-24.

[66] "Julius Lothar Meyer and Dmitri Ivanovich Mendeleev". *Chemical Achievers: The Human Face of Chemical Sciences*. Chemical Heritage Foundation. 2005. Retrieved 2007-02-22.

[67] C.M. Guldberg and P. Waage,"Studies Concerning Affinity" *C. M. Forhandlinger: Videnskabs-Selskabet i Christiana* (1864), 35

[68] P. Waage, "Experiments for Determining the Affinity Law" ,*Forhandlinger i Videnskabs-Selskabet i Christiania*, (1864) 92.

[69] C.M. Guldberg, "Concerning the Laws of Chemical Affinity", *C. M. Forhandlinger i Videnskabs-Selskabet i Christiania* (1864) 111

[70] John H. Lienhard (2003). "Johann Josef Loschmidt". *The Engines of Our Ingenuity*. Episode 1858http://www.uh.edu/engines/epi1858.htm |transcripturl= missing title (help). NPR. KUHF-FM Houston.

[71] "Adolf von Baeyer: The Nobel Prize in Chemistry 1905". *Nobel Lectures, Chemistry 1901–1921*. Elsevier Publishing Company. 1966. Retrieved 2007-02-28.

[72] "Jacobus Henricus van't Hoff". *Chemical Achievers: The Human Face of Chemical Sciences*. Chemical Heritage Foundation. 2005. Retrieved 2007-02-22.

[73] O'Connor, J. J.; Robertson, E.F. (1997). "Josiah Willard Gibbs". *MacTutor*. School of Mathematics and Statistics University of St Andrews, Scotland. Retrieved 2007-03-24.

[74] Weisstein, Eric W. (1996). "Boltzmann, Ludwig (1844–1906)". *Eric Weisstein's World of Scientific Biography*. Wolfram Research Products. Retrieved 2007-03-24.

[75] "Svante August Arrhenius". *Chemical Achievers: The Human Face of Chemical Sciences*. Chemical Heritage Foundation. 2005. Retrieved 2007-02-22.

[76] "Jacobus H. van 't Hoff: The Nobel Prize in Chemistry 1901". *Nobel Lectures, Chemistry 1901–1921*. Elsevier Publishing Company. 1966. Retrieved 2007-02-28.

[77] "Emil Fischer: The Nobel Prize in Chemistry 1902". *Nobel Lectures, Chemistry 1901–1921*. Elsevier Publishing Company. 1966. Retrieved 2007-02-28.

[78] "Henry Louis Le Châtelier". *World of Scientific Discovery*. Thomson Gale. 2005. Retrieved 2007-03-24.

[79] "History of Chemistry". *Intensive General Chemistry*. Columbia University Department of Chemistry Undergraduate Program. Retrieved 2007-03-24.

[80] "Alfred Werner: The Nobel Prize in Chemistry 1913". *Nobel Lectures, Chemistry 1901–1921*. Elsevier Publishing Company. 1966. Retrieved 2007-03-24.

[81] "William Ramsay: The Nobel Prize in Chemistry 1904". *Nobel Lectures, Chemistry 1901–1921*. Elsevier Publishing Company. 1966. Retrieved 2007-03-20.

[82] "Joseph John Thomson". *Chemical Achievers: The Human Face of Chemical Sciences*. Chemical Heritage Foundation. 2005. Retrieved 2007-02-22.

[83] "Alfred Werner: The Nobel Prize in Physics 1911". *Nobel Lectures, Physics 1901–1921*. Elsevier Publishing Company. 1967. Retrieved 2007-03-24.

[84] "Marie Sklodowska Curie". *Chemical Achievers: The Human Face of Chemical Sciences*. Chemical Heritage Foundation. 2005. Retrieved 2007-02-22.

[85] "Ernest Rutherford: The Nobel Prize in Chemistry 1908". *Nobel Lectures, Chemistry 1901–1921*. Elsevier Publishing Company. 1966. Retrieved 2007-02-28.

[86] "Tsvet, Mikhail (Semyonovich)". *Compton's Desk Reference*. Encyclopædia Britannica. 2007. Retrieved 2007-03-24.

[87] "Physics Time-Line 1900 to 1949". Weburbia.com. Retrieved 2007-03-25.

[88] "Fritz Haber". *Chemical Achievers: The Human Face of Chemical Sciences*. Chemical Heritage Foundation. 2005. Retrieved 2007-02-22.

[89] Cassidy, David (1996). "Einstein on Brownian Motion". The Center for History of Physics. Retrieved 2007-03-25.

[90] "Leo Hendrik Baekeland". *Chemical Achievers: The Human Face of Chemical Sciences*. Chemical Heritage Foundation. 2005. Retrieved 2007-02-22.

[91] "Robert A. Millikan: The Nobel Prize in Physics 1923". *Nobel Lectures, Physics 1922–1941*. Elsevier Publishing Company. 1965. Retrieved 2007-07-17.

[92] "Søren Sørensen". *Chemical Achievers: The Human Face of Chemical Sciences*. Chemical Heritage Foundation. 2005. Retrieved 2007-02-22.

[93] Parker, David. "Nuclear Twins: The Discovery of the Proton and Neutron". *Electron Centennial Page*. Retrieved 2007-03-25.

[94] "Solvay Conference". Einstein Symposium. 2005. Retrieved 2007-03-28.

[95] "The Nobel Prize in Physics 1915". *Nobelprize.org*. The Nobel Foundation. Retrieved 2007-02-28.

[96] "Peter Debye: The Nobel Prize in Chemistry 1936". *Nobel Lectures, Chemistry 1922–1941*. Elsevier Publishing Company. 1966. Retrieved 2007-02-28.

[97] "Niels Bohr: The Nobel Prize in Physics 1922". *Nobel Lectures, Chemistry 1922–1941*. Elsevier Publishing Company. 1966. Retrieved 2007-03-25.

[98] Weisstein, Eric W. (1996). "Moseley, Henry (1887–1915)". *Eric Weisstein's World of Scientific Biography*. Wolfram Research Products. Retrieved 2007-03-25.

[99] "Frederick Soddy The Nobel Prize in Chemistry 1921". *Nobel Lectures, Chemistry 1901–1921*. Elsevier Publishing Company. 1966. Retrieved 2007-03-25.

[100] "Early Mass Spectrometry". *A History of Mass Spectrometry*. Scripps Center for Mass Spectrometry. 2005. Archived from the original on 2007-03-03. Retrieved 2007-03-26.

[101] "Gilbert Newton Lewis and Irving Langmuir". *Chemical Achievers: The Human Face of Chemical Sciences*. Chemical Heritage Foundation. 2005. Retrieved 2007-02-22.

[102] "Electron Spin". Retrieved 2007-03-26.

[103] LeMaster, Nancy; McGann, Diane (1992). "GILBERT NEWTON LEWIS: AMERICAN CHEMIST (1875–1946)". *Woodrow Wilson Leadership Program in Chemistry*. The Woodrow Wilson National Fellowship Foundation. Retrieved 2007-03-25.

[104] "Louis de Broglie: The Nobel Prize in Physics 1929". *Nobel Lectures, Physics 1922–1941*. Elsevier Publishing Company. 1965. Retrieved 2007-02-28.

[105] "Wolfgang Pauli: The Nobel Prize in Physics 1945". *Nobel Lectures, Physics 1942–1962*. Elsevier Publishing Company. 1964. Retrieved 2007-02-28.

[106] "Erwin Schrödinger: The Nobel Prize in Physics 1933". *Nobel Lectures, Physics 1922–1941*. Elsevier Publishing Company. 1965. Retrieved 2007-02-28.

[107] "Werner Heisenberg: The Nobel Prize in Physics 1932". *Nobel Lectures, Physics 1922–1941*. Elsevier Publishing Company. 1965. Retrieved 2007-02-28.

[108] Heitler, Walter; London, Fritz (1927). "Wechselwirkung neutraler Atome und homöopolare Bindung nach der Quantenmechanik". *Zeitschrift für Physik* **44**: 455–472. Bibcode:1927ZPhy...44..455H. doi:10.1007/BF01397394.

[109] Ivor Grattan-Guinness. *Companion Encyclopedia of the History and Philosophy of the Mathematical Sciences*. Johns Hopkins University Press, 2003, p. 1266.; Jagdish Mehra, Helmut Rechenberg. *The Historical Development of Quantum Theory*. Springer, 2001, p. 540.

[110] "Linus Pauling: The Nobel Prize in Chemistry 1954". *Nobel Lectures, Chemistry 1942–1962*. Elsevier. 1964. Retrieved 2007-02-28.

[111] Rzepa, Henry S. "The aromaticity of Pericyclic reaction transition states". Department of Chemistry, Imperial College London. Retrieved 2007-03-26.

[112] "Harold C. Urey: The Nobel Prize in Chemistry 1934". *Nobel Lectures, Chemistry 1922–1941*. Elsevier Publishing Company. 1965. Retrieved 2007-03-26.

[113] "James Chadwick: The Nobel Prize in Physics 1935". *Nobel Lectures, Physics 1922–1941*. Elsevier Publishing Company. 1965. Retrieved 2007-02-28.

[114] William B. Jensen (2003). "Electronegativity from Avogadro to Pauling: II. Late Nineteenth- and Early Twentieth-Century Developments". *Journal of Chemical Education* **80** (3): 279. Bibcode:2003JChEd..80..279J. doi:10.1021/ed080p279.

[115] "Wallace Hume Carothers". *Chemical Achievers: The Human Face of Chemical Sciences*. Chemical Heritage Foundation. 2005. Retrieved 2007-02-22.

[116] "Emilio Segrè: The Nobel Prize in Physics 1959". *Nobel Lectures, Physics 1942–1962*. Elsevier Publishing Company. 1965. Retrieved 2007-02-28.

[117] "Eugene Houdry". *Chemical Achievers: The Human Face of Chemical Sciences*. Chemical Heritage Foundation. 2005. Retrieved 2007-02-22.

[118] "Pyotr Kapitsa: The Nobel Prize in Physics 1978". *Les Prix Nobel, The Nobel Prizes 1991*. Nobel Foundation. 1979. Retrieved 2007-03-26.

[119] "Otto Hahn: The Nobel Prize in Chemistry 1944". *Nobel Lectures, Chemistry 1942–1962*. Elsevier Publishing Company. 1964. Retrieved 2007-04-07.

[120] "Glenn Theodore Seaborg". *Chemical Achievers: The Human Face of Chemical Sciences*. Chemical Heritage Foundation. 2005. Retrieved 2007-02-22.

[121] "History of the Elements of the Periodic Table". AUS-e-TUTE. Retrieved 2007-03-26.

[122] "The Nobel Prize in Physics 1952". *Nobelprize.org*. The Nobel Foundation. Retrieved 2007-02-28.

[123] Hannaford, Peter. "Alan Walsh 1916–1998". *AAS Biographical Memoirs*. Australian Academy of Science. Archived from the original on 2007-02-24. Retrieved 2007-03-26.

[124] Cornforth, Lord Todd, John; Cornforth, J.; T., A. R.; C., J. W. (November 1981). "Robert Burns Woodward. 10 April 1917-8 July 1979". *Biographical Memoirs of Fellows of the Royal Society* (JSTOR) **27** (Nov., 1981): pp. 628–695. doi:10.1098/rsbm.1981.0025. JSTOR 198111. *note: authorization required for web access.*

[125] "The Nobel Prize in Medicine 1962". *Nobelprize.org*. The Nobel Foundation. Retrieved 2007-02-28.

[126] Skou J (1957). "The influence of some cations on an adenosine triphosphatase from peripheral nerves.". *Biochim Biophys Acta* **23** (2): 394–401. doi:10.1016/0006-3002(57)90343-8. PMID 13412736.

[127] "The Nobel Prize in Chemistry 1962". *Nobelprize.org*. The Nobel Foundation. Retrieved 2007-02-28.

[128] "Neil Bartlett and the Reactive Noble Gases". American Chemical Society. Retrieved June 5, 2012.

[129] G. A. Olah, S. J. Kuhn, W. S. Tolgyesi, E. B. Baker, J. Am. Chem. Soc. 1962, 84, 2733; G. A. Olah, lieu. Chim. (Buchrest), 1962, 7, 1139 (Nenitzescu issue); G. A. Olah, W. S. Tolgyesi, S. J. Kuhn, M. E. Moffatt, I. J. Bastien, E. B. Baker, J. Am. Chem. Soc. 1963, 85, 1328.

[130] "Richard R. Ernst The Nobel Prize in Chemistry 1991". *Les Prix Nobel, The Nobel Prizes 1991*. Nobel Foundation. 1992. Retrieved 2007-03-27.

[131] H. Nozaki, S. Moriuti, H. Takaya, R. Noyori, Tetrahedron Lett. 1966, 5239;

[132] H. Nozaki, H. Takaya, S. Moriuti, R. Noyori, Tetrahedron 1968, 24, 3655.

[133] W. J. Hehre, W. A. Lathan, R. Ditchfield, M. D. Newton, and J. A. Pople, Gaussian 70 (Quantum Chemistry Program Exchange, Program No. 237, 1970).

[134] *Catalyse de transformation des oléfines par les complexes du tungstène. II. Télomérisation des oléfines cycliques en présence d'oléfines acycliques* Die Makromolekulare Chemie Volume 141, Issue 1, Date: 9 February **1971**, Pages: 161–176 Par Jean-Louis Hérisson, Yves Chauvin doi:10.1002/macp.1971.021410112

[135] Katsuki, T.; Sharpless, K. B. *J. Am. Chem. Soc.* **1980**, *102*, 5974. (doi:10.1021/ja00538a077)

[136] Hill, J. G.; Sharpless, K. B.; Exon, C. M.; Regenye, R. *Org. Syn.*, Coll. Vol. 7, p.461 (1990); Vol. 63, p.66 (1985). (Article)

[137] Jacobsen, E. N.; Marko, I.; Mungall, W. S.; Schroeder, G.; Sharpless, K. B. *J. Am. Chem. Soc.* **1988**, *110*, 1968. (doi:10.1021/ja00214a053)

[138] Kolb, H. C.; Van Nieuwenhze, M. S.; Sharpless, K. B. *Chem. Rev.* **1994**, *94*, 2483–2547. (Review) (doi:10.1021/cr00032a009)

[139] Gonzalez, J.; Aurigemma, C.; Truesdale, L. *Org. Syn.*, Coll. Vol. 10, p.603 (2004); Vol. 79, p.93 (2002). (Article)

[140] Sharpless, K. B.; Patrick, D. W.; Truesdale, L. K.; Biller, S. A. *J. Am. Chem. Soc.* **1975**, 97, 2305. (doi:10.1021/ja00841a071)

[141] Herranz, E.; Biller, S. A.; Sharpless, K. B. *J. Am. Chem. Soc.* **1978**, *100*, 3596–3598. (doi:10.1021/ja00479a051)

[142] Herranz, E.; Sharpless, K. B. *Org. Syn.*, Coll. Vol. 7, p.375 (1990); Vol. 61, p.85 (1983). (Article)

[143] "The Nobel Prize in Chemistry 1996". *Nobelprize.org*. The Nobel Foundation. Retrieved 2007-02-28.

[144] "Benjamin Franklin Medal awarded to Dr. Sumio Iijima, Director of the Research Center for Advanced Carbon Materials, AIST". National Institute of Advanced Industrial Science and Technology. 2002. Retrieved 2007-03-27.

[145] *First total synthesis of taxol 1*. Functionalization of the B ring Robert A. Holton, Carmen Somoza, Hyeong Baik Kim, Feng Liang, Ronald J. Biediger, P. Douglas Boatman, Mitsuru Shindo, Chase C. Smith, Soekchan Kim, et al.; J. Am. Chem. Soc.; **1994**; 116(4); 1597–1598. DOI Abstract

[146] *First total synthesis of taxol. 2*. Completion of the C and D rings Robert A. Holton, Hyeong Baik Kim, Carmen Somoza, Feng Liang, Ronald J. Biediger, P. Douglas Boatman, Mitsuru Shindo, Chase C. Smith, Soekchan Kim, and et al. J. Am. Chem. Soc.; **1994**; 116(4) pp 1599–1600 DOI Abstract

[147] *A synthesis of taxusin* Robert A. Holton, R. R. Juo, Hyeong B. Kim, Andrew D. Williams, Shinya Harusawa, Richard E. Lowenthal, Sadamu Yogai J. Am. Chem. Soc.; **1988**; 110(19); 6558–6560. Abstract

[148] "Cornell and Wieman Share 2001 Nobel Prize in Physics". *NIST News Release*. National Institute of Standards and Technology. 2001. Retrieved 2007-03-27.

26.7 Further reading

- Servos, John W., *Physical chemistry from Ostwald to Pauling : the making of a science in America*, Princeton, N.J. : Princeton University Press, 1990. ISBN 0-691-08566-8

26.8 External links

- Chemical Achievers: The Human Face of the Chemical Sciences

- Eric Weisstein's World of Scientific Biography

- History of Gas Chemistry

- list of all Nobel Prize laureates

- History of Elements of the Periodic Table

- Chemsoc timeline

Chapter 27

Timeline of chemical element discoveries

The discovery of the elements known to exist today is presented here in chronological order. The elements are listed generally in the order in which each was first defined as the pure element, as the exact date of discovery of most elements cannot be accurately defined.

Given is each element's name, atomic number, year of first report, name of the discoverer, and some notes related to the discovery.

27.1 Table

27.2 Unrecorded discoveries

27.3 Recorded discoveries

27.4 Unconfirmed discoveries

27.5 Graphics

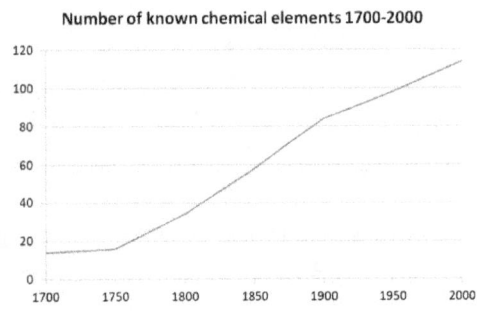

Development in discovery

27.6 See also

- History of the periodic table
- Periodic table
- The Mystery of Matter: Search for the Elements (2015 PBS film)

27.7 References

[1] "Copper History". Rameria.com. Retrieved 2008-09-12.

[2] CSA – Discovery Guides, A Brief History of Copper

[3] "The History of Lead – Part 3". Lead.org.au. Retrieved 2008-09-12.

[4] 47 Silver

[5] "Silver Facts – Periodic Table of the Elements". Chemistry.about.com. Retrieved 2008-09-12.

[6] "26 Iron". Elements.vanderkrogt.net. Retrieved 2008-09-12.

[7] Weeks, Mary Elvira; Leichester, Henry M. (1968). "Elements Known to the Ancients". *Discovery of the Elements*. Easton, PA: Journal of Chemical Education. pp. 29–40. ISBN 0-7661-3872-0. LCCCN 68-15217.

[8] "Notes on the Significance of the First Persian Empire in World History". Courses.wcupa.edu. Retrieved 2008-09-12.

[9] "History of Carbon and Carbon Materials – Center for Applied Energy Research – University of Kentucky". Caer.uky.edu. Retrieved 2008-09-12.

[10] "Chinese made first use of diamond". BBC News. 17 May 2005. Retrieved 2007-03-21.

[11] Ferchault de Réaumur, R-A (1722). *L'art de convertir le fer forgé en acier, et l'art d'adoucir le fer fondu, ou de faire des ouvrages de fer fondu aussi finis que le fer forgé (English translation from 1956)*. Paris, Chicago.

[12] Senese, Fred (September 9, 2009). "Who discovered carbon?". Frostburg State University. Retrieved 2007-11-24.

[13] "50 Tin". Elements.vanderkrogt.net. Retrieved 2008-09-12.

[14] "History of Metals". Neon.mems.cmu.edu. Retrieved 2008-09-12.

[15] "Sulfur History". Georgiagulfsulfur.com. Retrieved 2008-09-12.

[16] "Mercury and the environment — Basic facts". *Environment Canada, Federal Government of Canada*. 2004. Retrieved 2008-03-27.

[17] Craddock, P. T. et al. (1983), "Zinc production in medieval India", *World Archaeology* **15** (2), Industrial Archaeology, p. 13

[18] "30 Zinc". Elements.vanderkrogt.net. Retrieved 2008-09-12.

[19] Weeks, Mary Elvira (1933). "III. Some Eighteenth-Century Metals". *The Discovery of the Elements*. Easton, PA: Journal of Chemical Education. p. 21. ISBN 0-7661-3872-0.

[20] "Arsenic". Los Alamos National Laboratory. Retrieved 3 March 2013.

[21] SHORTLAND, A. J. (2006-11-01). "APPLICATION OF LEAD ISOTOPE ANALYSIS TO A WIDE RANGE OF LATE BRONZE AGE EGYPTIAN MATERIALS". *Archaeometry* **48** (4): 657–669. doi:10.1111/j.1475-4754.2006.00279.x.

[22] "15 Phosphorus". Elements.vanderkrogt.net. Retrieved 2008-09-12.

[23] "27 Cobalt". Elements.vanderkrogt.net. Retrieved 2008-09-12.

[24] "78 Platinum". Elements.vanderkrogt.net. Retrieved 2008-09-12.

[25] "28 Nickel". Elements.vanderkrogt.net. Retrieved 2008-09-12.

[26] "Bismuth". Los Alamos National Laboratory. Retrieved 3 March 2013.

[27] "12 Magnesium". Elements.vanderkrogt.net. Retrieved 2008-09-12.

[28] "01 Hydrogen". Elements.vanderkrogt.net. Retrieved 2008-09-12.

[29] Andrews, A. C. (1968). "Oxygen". In Clifford A. Hampel. *The Encyclopedia of the Chemical Elements*. New York: Reinhold Book Corporation. p. 272. LCCN 68-29938.

[30] "08 Oxygen". Elements.vanderkrogt.net. Retrieved 2008-09-12.

[31] Cook, Gerhard A.; Lauer, Carol M. (1968). "Oxygen". In Clifford A. Hampel. *The Encyclopedia of the Chemical Elements*. New York: Reinhold Book Corporation. pp. 499–500. LCCN 68-29938.

[32] Roza, Greg (2010). *The Nitrogen Elements: Nitrogen, Phosphorus, Arsenic, Antimony, Bismuth*. p. 7. ISBN 9781435853355.

[33] "07 Nitrogen". Elements.vanderkrogt.net. Retrieved 2008-09-12.

[34] "17 Chlorine". Elements.vanderkrogt.net. Retrieved 2008-09-12.

[35] "25 Manganese". Elements.vanderkrogt.net. Retrieved 2008-09-12.

[36] "56 Barium". Elements.vanderkrogt.net. Retrieved 2008-09-12.

[37] "42 Molybdenum". Elements.vanderkrogt.net. Retrieved 2008-09-12.

[38] "52 Tellurium". Elements.vanderkrogt.net. Retrieved 2008-09-12.

[39] IUPAC. "74 Tungsten". Elements.vanderkrogt.net. Retrieved 2008-09-12.

[40] "38 Strontium". Elements.vanderkrogt.net. Retrieved 2008-09-12.

[41] "Lavoisier". Homepage.mac.com. Retrieved 2008-09-12.

[42] "Chronology – Elementymology". Elements.vanderkrogt.net. Retrieved 2008-09-12.

[43] Lide, David R., ed. (2007–2008). "CRC Handbook of Chemistry and Physics" **4**. New York: CRC Press. p. 42. 978-0-8493-0488-0. |contribution= ignored (help)

[44] M. H. Klaproth (1789). "Chemische Untersuchung des Uranits, einer neuentdeckten metallischen Substanz". *Chemische Annalen* **2**: 387–403.

[45] E.-M. Péligot (1842). "Recherches Sur L'Uranium". *Annales de chimie et de physique* **5** (5): 5–47.

[46] "Titanium". Los Alamos National Laboratory. 2004. Retrieved 2006-12-29.

[47] Barksdale, Jelks (1968). The Encyclopedia of the Chemical Elements. Skokie, Illinois: Reinhold Book Corporation. pp. 732–38 "Titanium". LCCCN 68-29938.

[48] Browning, Philip Embury (1917). "Introduction to the Rarer Elements". *Kongl. Vet. Acad. Handl.* **XV**: 137.

[49] *Crell Anal.* **I**: 313. 1796. Missing or empty |title= (help)

[50] Vauquelin, Louis Nicolas (1798). "Memoir on a New Metallic Acid which exists in the Red Lead of Sibiria". *Journal of Natural Philosophy, Chemistry, and the Art* **3**: 146.

[51] "04 Beryllium". Elements.vanderkrogt.net. Retrieved 2008-09-12.

[52] "23 Vanadium". Elements.vanderkrogt.net. Retrieved 2008-09-12.

[53] "41 Niobium". Elements.vanderkrogt.net. Retrieved 2008-09-12.

[54] "73 Tantalum". Elements.vanderkrogt.net. Retrieved 2008-09-12.

[55] "46 Palladium". Elements.vanderkrogt.net. Retrieved 2008-09-12.

[56] "58 Cerium". Elements.vanderkrogt.net. Retrieved 2008-09-12.

[57] "76 Osmium". Elements.vanderkrogt.net. Retrieved 2008-09-12.

[58] "77 Iridium". Elements.vanderkrogt.net. Retrieved 2008-09-12.

[59] "45 Rhodium". Elements.vanderkrogt.net. Retrieved 2008-09-12.

[60] "19 Potassium". Elements.vanderkrogt.net. Retrieved 2008-09-12.

[61] "11 Sodium". Elements.vanderkrogt.net. Retrieved 2008-09-12.

[62] "05 Boron". Elements.vanderkrogt.net. Retrieved 2008-09-12.

[63] "09 Fluorine". Elements.vanderkrogt.net. Retrieved 2008-09-12.

[64] "53 Iodine". Elements.vanderkrogt.net. Retrieved 2008-09-12.

[65] "03 Lithium". Elements.vanderkrogt.net. Retrieved 2008-09-12.

[66] "48 Cadmium". Elements.vanderkrogt.net. Retrieved 2008-09-12.

[67] "34 Selenium". Elements.vanderkrogt.net. Retrieved 2008-09-12.

[68] "14 Silicon". Elements.vanderkrogt.net. Retrieved 2008-09-12.

[69] "13 Aluminium". Elements.vanderkrogt.net. Retrieved 2008-09-12.

[70] "35 Bromine". Elements.vanderkrogt.net. Retrieved 2008-09-12.

[71] "90 Thorium". Elements.vanderkrogt.net. Retrieved 2008-09-12.

[72] "57 Lanthanum". Elements.vanderkrogt.net. Retrieved 2008-09-12.

[73] "68 Erbium". Elements.vanderkrogt.net. Retrieved 2008-09-12.

[74] "65 Terbium". Elements.vanderkrogt.net. Retrieved 2008-09-12.

[75] "44 Ruthenium". Elements.vanderkrogt.net. Retrieved 2008-09-12.

[76] "55 Caesium". Elements.vanderkrogt.net. Retrieved 2008-09-12.

[77] Caesium

[78] "37 Rubidium". Elements.vanderkrogt.net. Retrieved 2008-09-12.

[79] "81 Thallium". Elements.vanderkrogt.net. Retrieved 2008-09-12.

[80] "49 Indium". Elements.vanderkrogt.net. Retrieved 2008-09-12.

[81] "02 Helium". Elements.vanderkrogt.net. Retrieved 2008-09-12.

[82] "31 Gallium". Elements.vanderkrogt.net. Retrieved 2008-09-12.

[83] "70 Ytterbium". Elements.vanderkrogt.net. Retrieved 2008-09-12.

[84] "67 Holmium". Elements.vanderkrogt.net. Retrieved 2008-09-12.

[85] "69 Thulium". Elements.vanderkrogt.net. Retrieved 2008-09-12.

[86] "21 Scandium". Elements.vanderkrogt.net. Retrieved 2008-09-12.

[87] "62 Samarium". Elements.vanderkrogt.net. Retrieved 2008-09-12.

[88] "64 Gadolinium". Elements.vanderkrogt.net. Retrieved 2008-09-12.

[89] "59 Praseodymium". Elements.vanderkrogt.net. Retrieved 2008-09-12.

[90] "60 Neodymium". Elements.vanderkrogt.net. Retrieved 2008-09-12.

[91] "32 Germanium". Elements.vanderkrogt.net. Retrieved 2008-09-12.

[92] "18 Argon". Elements.vanderkrogt.net. Retrieved 2008-09-12.

[93] "10 Neon". Elements.vanderkrogt.net. Retrieved 2008-09-12.

[94] "54 Xenon". Elements.vanderkrogt.net. Retrieved 2008-09-12.

[95] "84 Polonium". Elements.vanderkrogt.net. Retrieved 2008-09-12.

[96] "88 Radium". Elements.vanderkrogt.net. Retrieved 2008-09-12.

[97] Partington, J. R. (May 1957). "Discovery of Radon". *Nature* **179** (4566): 912. Bibcode:1957Natur.179..912P. doi:10.1038/179912a0.

[98] Ramsay, W.; Gray, R. W. (1910). "La densité de l'emanation du radium". *Comptes rendus hebdomadaires des séances de l'Académie des sciences* **151**: 126–128.

[99] "89 Actinium". Elements.vanderkrogt.net. Retrieved 2008-09-12.

[100] "63 Europium". Elements.vanderkrogt.net. Retrieved 2008-09-12.

[101] "71 Lutetium". Elements.vanderkrogt.net. Retrieved 2008-09-12.

[102] http://www.maik.ru/abstract/radchem/0/radchem0535_abstract.pdf

[103] "72 Hafnium". Elements.vanderkrogt.net. Retrieved 2008-09-12.

[104] Noddack, W.; Tacke, I.; Berg, O (1925). "Die Ekamangane". *Naturwissenschaften* **13** (26): 567. Bibcode:1925NW.....13..567.. doi:10.1007/BF01558746.

[105] "91 Protactinium". Elements.vanderkrogt.net. Retrieved 2008-09-12.

[106] Emsley, John (2001). *Nature's Building Blocks* ((Hardcover, First Edition) ed.). Oxford University Press. p. 347. ISBN 0-19-850340-7.

[107] "43 Technetium". Elements.vanderkrogt.net. Retrieved 2008-09-12.

[108] *History of the Origin of the Chemical Elements and Their Discoverers*, Individual Element Names and History, "Technetium"

[109] "87 Francium". Elements.vanderkrogt.net. Retrieved 2008-09-12.

[110] Adloff, Jean-Pierre; Kaufman, George B. (2005-09-25). Francium (Atomic Number 87), the Last Discovered Natural Element. *The Chemical Educator* **10** (5). [2007-03-26]

[111] "85 Astatine". Elements.vanderkrogt.net. Retrieved 2008-09-12.

[112] Close, Frank E. (2004). *Particle Physics: A Very Short Introduction*. Oxford University Press. p. 2. ISBN 978-0-19-280434-1.

[113] "93 Neptunium". Elements.vanderkrogt.net. Retrieved 2008-09-12.

[114] "94 Plutonium". Elements.vanderkrogt.net. Retrieved 2008-09-12.

[115] "95 Americium". Elements.vanderkrogt.net. Retrieved 2008-09-12.

[116] "96 Curium". Elements.vanderkrogt.net. Retrieved 2008-09-12.

[117] "97 Berkelium". Elements.vanderkrogt.net. Retrieved 2008-09-12.

[118] "98 Californium". Elements.vanderkrogt.net. Retrieved 2008-09-12.

[119] "99 Einsteinium". Elements.vanderkrogt.net. Retrieved 2008-09-12.

[120] "100 Fermium". Elements.vanderkrogt.net. Retrieved 2008-09-12.

[121] "101 Mendelevium". Elements.vanderkrogt.net. Retrieved 2008-09-12.

[122] "102 Nobelium". Elements.vanderkrogt.net. Retrieved 2008-09-12.

[123] "103 Lawrencium". Elements.vanderkrogt.net. Retrieved 2008-09-12.

[124] "104 Rutherfordium". Elements.vanderkrogt.net. Retrieved 2008-09-12.

[125] "105 Dubnium". Elements.vanderkrogt.net. Retrieved 2008-09-12.

[126] "106 Seaborgium". Elements.vanderkrogt.net. Retrieved 2008-09-12.

[127] "107 Bohrium". Elements.vanderkrogt.net. Retrieved 2008-09-12.

[128] "109 Meitnerium". Elements.vanderkrogt.net. Retrieved 2008-09-12.

[129] "108 Hassium". Elements.vanderkrogt.net. Retrieved 2008-09-12.

[130] "110 Darmstadtium". Elements.vanderkrogt.net. Retrieved 2008-09-12.

[131] "111 Roentgenium". Elements.vanderkrogt.net. Retrieved 2008-09-12.

[132] "112 Copernicium". Elements.vanderkrogt.net. Retrieved 2009-07-17.

[133] "Discovery of the Element with Atomic Number 112". www.iupac.org. 2009-06-26. Retrieved 2009-07-17.

[134] Oganessian, Yu. Ts.; Utyonkov, V. K.; Lobanov, Yu. V.; Abdullin, F. Sh.; Polyakov, A. N.; Shirokovsky, I. V.; Tsyganov, Yu. S.; Gulbekian, G. G.; Bogomolov, S. L.; Gikal, B.; Mezentsev, A.; Iliev, S.; Subbotin, V.; Sukhov, A.; Buklanov, G.; Subotic, K.; Itkis, M.; Moody, K.; Wild, J.; Stoyer, N.; Stoyer, M.; Lougheed, R. (October 1999). "Synthesis of Superheavy Nuclei in the ^{48}Ca + ^{244}Pu Reaction". *Physical Review Letters* **83** (16): 3154. Bibcode:1999PhRvL..83.3154O. doi:10.1103/PhysRevLett.83.3154.

[135] Oganessian, Yu. Ts.; Utyonkov, V. K.; Lobanov, Yu. V.; Abdullin, F. Sh.; Polyakov, A. N.; Shirokovsky, I. V.; Tsyganov, Yu. S.; Gulbekian, G. G.; Bogomolov, S. L.; Gikal, B.; Mezentsev, A.; Iliev, S.; Subbotin, V.; Sukhov, A.; Ivanov, O.; Buklanov, G.; Subotic, K.; Itkis, M.; Moody, K.; Wild, J.; Stoyer, N.; Stoyer, M.; Lougheed, R.; Laue, C.; Karelin, Ye.; Tatarinov, A. (2000). "Observation of the decay of ^{292}116". *Physical Review C* **63**: 011301. Bibcode:2001PhRvC..63a1301O. doi:10.1103/PhysRevC.63.011301.

[136] Oganessian, Yu. Ts.; Utyonkov, V. K.; Lobanov, Yu. V.; Abdullin, F. Sh.; Polyakov, A. N.; Sagaidak, R. N.; Shirokovsky, I. V.; Tsyganov, Yu. S.; Voinov, A. A.; Gulbekian, G.; Bogomolov, S.; Gikal, B.; Mezentsev, A.; Iliev, S.; Subbotin, V.; Sukhov, A.; Subotic, K.; Zagrebaev, V.; Vostokin, G.; Itkis, M.; Moody, K.; Patin, J.; Shaughnessy, D.; Stoyer, M.; Stoyer, N.; Wilk, P.; Kenneally, J.; Landrum, J.; Wild, J.; Lougheed, R. (2006). "Synthesis of the isotopes of elements 118 and 116 in the ^{249}Cf and ^{245}Cm+^{48}Ca fusion reactions". *Physical Review C* **74** (4): 044602. Bibcode:2006PhRvC..74d4602O. doi:10.1103/PhysRevC.74.044602.

[137] Oganessian, Yu. Ts.; Utyonkov, V. K.; Dmitriev, S. N.; Lobanov, Yu. V.; Itkis, M. G.; Polyakov, A. N.; Tsyganov, Yu. S.; Mezentsev, A. N.; Yeremin, A. V.; Voinov, A. A.; Sokol, E.; Gulbekian, G.; Bogomolov, S.; Iliev, S.; Subbotin, V.; Sukhov, A.; Buklanov, G.; Shishkin, S.; Chepygin, V.; Vostokin, G.; Aksenov, N.; Hussonnois, M.; Subotic, K.; Zagrebaev, V.; Moody, K.; Patin, J.; Wild, J.; Stoyer, M.; Stoyer, N.; et al. (2005). "Synthesis of elements 115 and 113 in the reaction ^{243}Am + ^{48}Ca". *Physical Review C* **72** (3): 034611. Bibcode:2005PhRvC..72c4611O. doi:10.1103/PhysRevC.72.034611.

[138] Oganessian, Yu. Ts.; Abdullin, F. Sh.; Bailey, P. D.; Benker, D. E.; Bennett, M. E.; Dmitriev, S. N.; Ezold, J. G.; Hamilton, J. H.; Henderson, R. A.; Itkis, M. G.; Lobanov, Yu. V.; Mezentsev, A. N.; Moody, K. J.; Nelson, S. L.; Polyakov, A. N.; Porter, C. E.; Ramayya, A. V.; Riley, F. D.; Roberto, J. B.; Ryabinin, M. A.; Rykaczewski, K. P.; Sagaidak, R. N.; Shaughnessy, D. A.; Shirokovsky, I. V.; Stoyer, M. A.; Subbotin, V. G.; Sudowe, R.; Sukhov, A. M.; Tsyganov, Yu. S.; et al. (April 2010). "Synthesis of a New Element with Atomic Number Z=117". *Physical Review Letters* **104** (14): 142502. Bibcode:2010PhRvL.104n2502O. doi:10.1103/PhysRevLett.104.142502. PMID 20481935.

27.8 External links

• History of the Origin of the Chemical Elements and Their Discoverers Last updated by Boris Pritychenko on March 30, 2004

• History of Elements of the Periodic Table

• Timeline of Element Discoveries

• Discovery of the Elements - The Movie - YouTube (1:18)

• The History Of Metals Timeline. A timeline showing the discovery of metals and the development of metallurgy.

27.9 Text and image sources, contributors, and licenses

27.9.1 Text

- **History of chemistry** *Source:* https://en.wikipedia.org/wiki/History_of_chemistry?oldid=687200177 *Contributors:* Rmhermen, William Avery, D, Polimerek, Wnissen, CBDunkerson, Fibonacci, Jerzy, Shantavira, Fredrik, Arkuat, Gandalf61, Blainster, Michael Snow, Alan Liefting, Ancheta Wis, Alexwcovington, Quadell, Ukexpat, Grstain, Jkl, Discospinster, Wadewitz, Jaberwocky6669, Violetriga, Eric Forste, Aranel, Chairboy, John Vandenberg, Nsaa, AnnaP, Grenavitar, Falcorian, Megan1967, Linas, StradivariusTV, Carcharoth, Ruud Koot, Wikiklrsc, Paughsw, Allen3, Nanite, Drbogdan, Rjwilmsi, Ctdunstan, AllanHainey, Vegaswikian, DonSiano, Yamamoto Ichiro, ACrush, RexNL, Gurch, Srleffler, Igordebraga, Mushin, Deeptrivia, RussBot, Fabartus, MarcK, RazorICE, Ragesoss, Osu8907, Aleichem, Mysid, Davidc, Tetracube, Whitejay251, LeonardoRob0t, Katieh5584, ChemGardener, Itub, Sardanaphalus, SmackBot, TestPilot, Pavlovič, Jagged 85, Arniep, Edgar181, Gilliam, Schmiteye, Chris the speller, MalafayaBot, Akanemoto, Sbharris, Colonies Chris, Hallenrm, Addshore, DMacks, Sadi Carnot, The undertow, Heimstern, Maziar fayaz, Stefan2, Peterlewis, Aleenf1, IronGargoyle, Ckatz, Rinnenadtrosc, Shoeofdeath, StephenBuxton, Ivan-Lanin, Penbat, Glenn4pr, Myasuda, Warhorus, Christian75, Epbr123, N5iln, Escarbot, Tariqhada, BokicaK, Opelio, Jayron32, Mary Mark Ockerbloom, Dane 1981, The Transhumanist, Matthew Fennell, OM, VoABot II, JamesBWatson, KConWiki, 28421u2232nfenfcenc, David Eppstein, Adventurer, JaGa, MartinBot, STBot, David J Wilson, R'n'B, CommonsDelinker, FANSTARbot, Lanternix, Entropy, Natl1, Signalhead, Pking123, Paul EJ King, Rei-bot, JayC, Sintaku, Monty845, Sfmammamia, Deconstructhis, Thw1309, SieBot, YourEyesOnly, Caltas, Happysailor, Oxymoron83, Nuttycoconut, Escape Orbit, Mr. Granger, Tomasz Prochownik, ClueBot, Matdrodes, Uncle Milty, J8079s, Niceguyedc, Bbb2007, DragonBot, Alexbot, LaosLos, Iohannes Animosus, P1415926535, Thingg, Aitias, SoxBot III, Shayno123, Dthomsen8, Addbot, Brumski, Vero.Verite, Favonian, LinkFA-Bot, Tassedethe, Sylvania w, Tide rolls, Osado, Gail, Luckas-bot, AnomieBOT, HairyPerry, Materialscientist, Citation bot, Quebec99, Poetaris, Famizban~enwiki, Marciag, Ignoranteconomist, Shadowjams, Methcub, Magnagr, Zuckyd1, Citation bot 1, PigFlu Oink, DrilBot, Pinethicket, Tanweer Morshed, Trappist the monk, Vrenator, Diannaa, John of Reading, Juanita Saenz S, WikitanvirBot, Cas6039, Dewritech, Syncategoremata, Barackobama808, GoingBatty, ZxxZxxZ, AsceticRose, JSquish, Trinidade, Jpvandijk, H3llBot, Snap1234, Hccc, Dagko, Donner60, ClueBot NG, Hcrunyon, Nolabob, Helpful Pixie Bot, HMSSolent, Bibcode Bot, DBigXray, Car Henkel, AvocatoBot, CitationCleanerBot, Ricardokramer, Harizotoh9, Aisteco, Johnny7794, Dexbot, Mogism, Faizan, Mrm7171, Polybacon, JaconaFrere, ThatRusskiiGuy, Federico Leva (BEIC), Supdiop, CZarcula, Clinamental and Anonymous: 202

- **History of thermodynamics** *Source:* https://en.wikipedia.org/wiki/History_of_thermodynamics?oldid=680385706 *Contributors:* Collabi, Lumos3, Arkuat, Gandalf61, Cutler, Karol Langner, Eric Forste, PAR, Marianika~enwiki, Carcharoth, Benbest, Rjwilmsi, Ligulem, Srleffler, Chobot, Gaius Cornelius, CambridgeBayWeather, Ragesoss, Dhollm, Moe Epsilon, Rayc, Tropylium, SmackBot, Jagged 85, TimBentley, Colonies Chris, A.R., DMacks, Ligulembot, Mion, Sadi Carnot, Pilotguy, JzG, JorisvS, Peterlewis, Special-T, AdultSwim, Lottamiata, Myasuda, FilipeS, Gtxfrance, Doug Weller, M karzarj, Barticus88, D.H, Greg L, EdJogg, VoABot II, Cardamon, Jtir, Inwind, ElinorD, Riick, Enviroboy, Radagast3, Natox, SieBot, I Love Pi, Anchor Link Bot, Tomasz Prochownik, MCCRogers, Taroaldo, J8079s, Djr32, CohesionBot, Eeekster, XLinkBot, Saeed.Veradi, Ariconte, Kwjbot, Addbot, Lightbot, Wikkidd, Luckas-bot, Yobot, Ptbotgourou, Ajh16, AnomieBOT, Citation bot, ArthurBot, Xqbot, J04n, GrouchoBot, SassoBot, Geraldo61, Fortdj33, Machine Elf 1735, Citation bot 1, TobeBot, Marie Poise, Syncategoremata, ClueBot NG, Helpful Pixie Bot, Bibcode Bot, Ludi Romani, Bfong2828, Belief action, Nerlost, CleanEnergyPundit and Anonymous: 32

- **Ferrous metallurgy** *Source:* https://en.wikipedia.org/wiki/Ferrous_metallurgy?oldid=681589998 *Contributors:* Dwmyers, Michael Hardy, Paul A, Genie, Joy, Robbot, PBS, Auric, DocWatson42, Luis Dantas, Jorge Stolfi, Utcursch, Burschik, Eyrian, Rich Farmbrough, Vsmith, Morten Blaabjerg, Dbachmann, Smalljim, QuantumEleven, Ricky81682, Svartalf, Snowolf, Woohookitty, Benbest, Mandarax, BD2412, Josh Parris, Rjwilmsi, Samkass, Bgwhite, Wester, Deeptrivia, RussBot, Peterkingiron, Scott5834, Madcoverboy, Dialectric, Grafen, Rjensen, Malangali, Poppy, E Wing, Petri Krohn, WIN, Yvwv, SmackBot, Jagged 85, HeartofaDog, Gilliam, Hmains, Chris the speller, Full Shunyata, Hibernian, MaxSem, Kelvin Case, X@x, BillFlis, Magere Hein, Peter Horn, Wwagner, Dan Gluck, Wizard191, CmdrObot, ShelfSkewed, Odie5533, Shirulashem, Barticus88, Noclevername, Ela112, Kbthompson, Omeganian, Trey314159, Severo, Majorarcanum, JaGa, Gun Powder Ma, B9 hummingbird hovering, EyeSerene, CommonsDelinker, Verdatum, DrKay, Balthazarduju, DadaNeem, Achillobator, Squids and Chips, Lieutenant pepper, Signalhead, VolkovBot, Th.Rehren, Jimthedigger, Aymatth2, Mzmadmike, Phirosiberia, Andy Dingley, Taharqa, Fleela, Phmoreno, PericlesofAthens, Undead warrior, Sir Reginald Aylmer Ranfurly Plunkett-Ernle-Erle-Drax, Steorra, Shaheenjim, Wilson44691, Randy Kryn, Zakaz~enwiki, Alastair J. Campbell, EoGuy, Aaa3-other, J8079s, Auntof6, DragonBot, Audaciter, Lx 121, DumZiBoT, Mpondoporido, Dthomsen8, Addbot, DOI bot, MartinezMD, Leszek Jańczuk, Proxima Centauri, Sardur, Ptbotgourou, AnomieBOT, Theseeker4, Citation bot, Teeninvestor, Bob Burkhardt, LilHelpa, Geskenk, Br77rino, Tulocci, Bruneauinfo, Riventree, Citation bot 1, Full-date unlinking bot, Tobias1984, DASHBot, EmausBot, Laszlovszky András, ZéroBot, Rcsprinter123, ChuispastonBot, ClueBot NG, Morgan Riley, Rurik the Varangian, Mmarre, Helpful Pixie Bot, Bibcode Bot, BG19bot, TeamRocketPikachu, Bryanpiczon, Solomon7968, Zedshort, Nelg, BattyBot, Simeondahl, ChrisGualtieri, YFdyh-bot, Khazar2, Dexbot, Hmainsbot1, Numbermaniac, Reatlas, Bodha2, HIST406-13Petesg, EditorialChance, Writingalltheway, Akanady5, Monkbot, Purple Phoenician, Kautilya3, Julietdeltalima, R2d2m2, Lappspira and Anonymous: 71

- **History of metallurgy in South Asia** *Source:* https://en.wikipedia.org/wiki/History_of_metallurgy_in_South_Asia?oldid=681773664 *Contributors:* Goethean, Auric, Dina, Burschik, Rich Farmbrough, Kwamikagami, Foobaz, Wiki-uk, Riana, SteinbDJ, Ceyockey, Machaon, Josh Parris, Bgwhite, Avecit, Epolk, Gaius Cornelius, NawlinWiki, Karl Meier, Deepak~enwiki, Attilios, SmackBot, Jagged 85, Edgar181, Hmains, Amarrg, Shyamsunder, JHunterJ, BillFlis, Noah Salzman, Wizard191, Rayfield, Bharatveer, CmdrObot, ShelfSkewed, Timtrent, After Midnight, Nick Number, JEH, Arman Padaryan, KConWiki, JaGa, Ekotkie, EyeSerene, CommonsDelinker, Fconaway, Polenth, Katharineamy, Morinae, TopGun, Davecrosby uk, GrahamHardy, Malinaccier, Kww, Wilson44691, Philly jawn, ImageRemovalBot, Niceguyedc, SchreiberBike, Belasd, DumZiBoT, WikHead, addbot, JSR, AnomieBOT, Materialscientist, LilHelpa, Capricorn42, Thehelpfulbot, FrescoBot, RedBot, Armando-Martin, Generalboss3, BCtl, John of Reading, Look2See1, Mar4d, SporkBot, ClueBot NG, Frietjes, J1776, संतोष दहिवळ, Marcocapelle, Khazar2, AK456, Hmainsbot1, ASharma23, JaconaFrere, Chitain deep singh and Anonymous: 29

- **Atomism** *Source:* https://en.wikipedia.org/wiki/Atomism?oldid=686082451 *Contributors:* Eloquence, SimonP, Maury Markowitz, Michael Hardy, Alan Peakall, Tregoweth, Ellywa, Ahoerstemeier, JWSchmidt, Charles Matthews, Rednblu, Wik, Hr oskar, Manika, Wetman, Banno, Carlossuarez46, Robbot, Dittaeva, Psychonaut, Ashley Y, Nilmerg, Wikibot, Wile E. Heresiarch, Alan Liefting, Fabiform, Wolfkeeper, Xerxes314, Daniel Brockman, Eequor, Antandrus, Kusunose, Mukerjee, Deglr6328, Cacycle, Xezbeth, Dbachmann, Bender235, Silentlight, El C, Laurascudder, Bobo192, Truthflux, John Vandenberg, Kjkolb, Ral315, Merope, Jumbuck, Ungtss, ABCD, Samohyl Jan, Evil Monkey, Gene Nygaard,

Oleg Alexandrov, Woohookitty, Linas, Kurzon, Rjwilmsi, Vary, Kajmal, HappyCamper, FlaBot, SchuminWeb, Godlord2, Aethralis, Uriah923, YurikBot, Spacepotato, Pigman, SpuriousQ, Bachrach44, Ragesoss, Amakuha, Tomisti, Ninly, LeonardoRob0t, RG2, Finell, Sardanaphalus, Intangible, SmackBot, Jagged 85, JimmyBlackwing, Gilliam, Hmains, GwydionM, Mhss, Fabiolucas~enwiki, SonOfNothing, Funper, Sbharris, Atomist, Kcordina, LoveMonkey, DMacks, SashatoBot, Petr Kopač, Kashmiri, Rinnenadtrosc, Dr Greg, Special-T, Dhp1080, Brandizzi, Twas Now, Tdmg, CmdrObot, Sir Vicious, Eiorgiomugini, Comrade42, GalliasM, Myasuda, Gregbard, Shanoman, Cydebot, Peterdjones, Dfritter4, Hkyriazi, Strom, Odie5533, Doug Weller, SteveMcCluskey, Thijs!bot, Kubanczyk, Oerjan, Headbomb, Ujm, Tellyaddict, Escarbot, Igorwindsor~enwiki, Stannered, AntiVandalBot, Danger, Danny lost, Indian Chronicles, Sluzzelin, Husond, The Transhumanist, Paper.plane.pilot, Leolaursen, .anacondabot, Meredyth, K95, NoychoH, Mdsats, Yensin, Rettetast, R'n'B, CommonsDelinker, Erkan Yilmaz, AlphaEta, Bogey97, Ianmathwiz7, Benio76, Cadby Waydell Bainbrydge, Ontoraul, JhsBot, Thanatos666, Seraphita~enwiki, Karriaagzh, SieBot, Flyer22 Reborn, Anchor Link Bot, 3rdAlcove, Athenean, ClueBot, Roshraw, J8079s, Singinglemon~enwiki, Huzzam, Andriolo, Gbertoli3, SchreiberBike, Editor2020, Saeed.Veradi, Avoided, Addbot, Sa9097, Guoguo12, Mathew Rammer, Favonian, Ginosbot, Alcove, Lightbot, Zorrobot, Luckas-bot, Yobot, Reindra, AnomieBOT, Jim1138, Materialscientist, Citation bot, DirlBot, LilHelpa, Xqbot, JimVC3, Leucipp3, Quixotex, GrouchoBot, Omnipaedista, Exxoo, FrescoBot, Isospin, Fortdj33, Machine Elf 1735, Diremarc, Buybuydandavis, Citation bot 1, Jean-François Clet, MastiBot, Prophetvcn, Jauhienij, Yunshui, Cowlibob, Omkargokhale, DASHBot, EmausBot, Acather96, Syncategoremata, IncognitoErgoSum, G upadhyay, Janus549, ZéroBot, ChuispastonBot, ClueBot NG, Helpful Pixie Bot, Snow Rise, WithSelet, Barnaculus, JimRenge, Monkbot, Srini Kolla, Kimberley Anne Davey, Strongjam, Shyskirwaiktcb, KasparBot and Anonymous: 139

- **Minima naturalia** *Source:* https://en.wikipedia.org/wiki/Minima_naturalia?oldid=671322496 *Contributors:* Sardanaphalus, Rinnenadtrosc, Hebrides, Machine Elf 1735 and Trappist the monk

- **Alchemy** *Source:* https://en.wikipedia.org/wiki/Alchemy?oldid=685787230 *Contributors:* MichaelTinkler, Lee Daniel Crocker, BF, Vicki Rosenzweig, Bryan Derksen, The Anome, Stephen Gilbert, Manning Bartlett, Sjc, RK, Andre Engels, Eclecticology, Matusz, Fubar Obfusco, Apollia, Mswake, Heron, BryceHarrington, Edward, Patrick, Tim Starling, Kwertii, DopefishJustin, Nixdorf, Liftarn, Wapcaplet, Ixfd64, Sannse, Tregoweth, Looxix~enwiki, Ihcoyc, Ronz, William M. Connolley, Theresa knott, Angela, Jdforrester, Александър, Nikai, Susurrus, Rotem Dan, Jouster, Ghewgill, Norwikian, Heidimo, Charles Matthews, Timwi, RickK, Stone, Slathering, Malcohol, Andrewman327, Zoicon5, Markhurd, Evan~enwiki, Big Bob the Finder, Maximus Rex, SEWilco, Omegatron, Wernher, Elwoz, Wetman, Flockmeal, David.Monniaux, Donarreiskoffer, Robbot, Vardion, Mazin07, Chrism, Fredrik, Ly, Romanm, Mirv, Chiramabi, Flauto Dolce, Blainster, Humus sapiens, Timrollpickering, Hadal, Jsonitsac, Wereon, TPK, Wayland, Timvasquez, Smjg, DocWatson42, Christopher Parham, Gtrmp, Sj, Dr spork, Lupin, Timpo, Monedula, Bradeos Graphon, Xerxes314, Everyking, Anville, LarryGilbert, Duncharris, Gilgamesh~enwiki, Jorge Stolfi, Mboverload, Luigi30, Solipsist, Darrien, Chameleon, SWAdair, Bobblewik, Tagishsimon, Wmahan, Bacchiad, Isidore, Chowbok, R. fiend, Jonel, GeneralPatton, Pcarbonn, Quadell, Antandrus, Zaha, Phe, MisfitToys, Piotrus, Jossi, Rdsmith4, Oneiros, Tothebarricades.tk, Bodnotbod, Kuralyov, Icairns, Gscshoyru, LHOON, Neutrality, Urhixidur, Hilarleo, Joyous!, Syvanen, Fenrir~enwiki, Adashiel, Esperant, ProjeX, Ashami, PRiis, Oskar Sigvardsson, Mr Bound, Jayjg, Freakofnurture, DanielCD, Mercurius~enwiki, Moverton, Discospinster, Steve Farrell, Rich Farmbrough, Guanabot, JBradHicks, Vsmith, Silence, Bishonen, MeltBanana, Dbachmann, Paul August, SpookyMulder, Indrian, Stbalbach, Bender235, ESkog, Metaquasi, Kbh3rd, Fenice, Crux Ansata, Brian0918, El C, Huntster, Bletch, Edward Z. Yang, Chairboy, RoyBoy, AlexTheMartian, Thuresson, Bobo192, 23skidoo, Ray Dassen, Htmlism, Sivaraj, Viriditas, Jericho4.0, Markryherd, Idban, Forteanajones, Tiresias BC, Microtony, Sam Korn, Fugg, Jumbuck, Patsw, Alansohn, Anthony Appleyard, Gerweck, CyberSkull, Wiki-uk, Keenan Pepper, Jet57, Andrew Gray, Riana, Primalchaos, SlimVirgin, Fourthgeek, Alex '05, Malo, Avenue, House of Shin, DreamGuy, Snowolf, Homo universalis, Fasten, Velella, Here, Knowledge Seeker, Suruena, Wimvandorst, Rhialto, LFaraone, BDD, Sleigh, Ghirlandajo, Zereshk, Oleg Alexandrov, Marasmusine, Bloodsorr0w, Richard Arthur Norton (1958-), Woohookitty, Linas, Onari, T. Baphomet, Uncle G, Kzollman, Nefertum17, WadeSimMiser, JeremyA, QuetschJL, Jeff3000, LeaMaimone, Twthmoses, Kmg90, Wikiklrsc, Damicatz, TotoBaggins, Adam Field, Tickle me, Macaddct1984, Rchamberlain, GalaazV, Crucis, MarcoTolo, Sweetfreek, V8rik, Cuchullain, BD2412, Galwhaa, Kbdank71, Jclemens, Josh Parris, Sjö, Drbogdan, Sjakkalle, Rjwilmsi, Nightscream, Matt.whitby, Phileas, Tawker, Oblivious, Ligulem, NeonMerlin, ElKevbo, Mjsedgwick, Brighterorange, TheGWO, Sango123, Yamamoto Ichiro, SlaunchaMan, FlaBot, RobertG, Windchaser, AJR, Gparker, RexNL, Gurch, Whateley23, Mitsukai, Robert Prummel, Mehrshad123, Codex Sinaiticus, Nick81, Bihzad, Samuel Levine, Benjwong, Jidan, DaGizza, Aethralis, Korg, Digitalme, Gwernol, EamonnPKeane, Satanael, YurikBot, Wavelength, Spacepotato, RobotE, JJB, PowerGamer6, Sceptre, Hairy Dude, Deeptrivia, Rtkat3, Hillman, Brandmeister (old), 999~enwiki, Pip2andahalf, Lighterside, The Storm Surfer, Hornplease, Pigman, Chris Capoccia, Sasuke Sarutobi, Jtbandes, GG Crono, Akamad, Stephenb, Manop, Shell Kinney, Gaius Cornelius, Theelf29, Bisqwit, Wimt, EnakoNosaj, NawlinWiki, Nahallac Silverwinds, Leutha, Onias, Joshdboz, Henleydude, Janarius, Justin Eiler, Ragesoss, Rubaphilos, Aaron Brenneman, PhilipC, Dr Debug, Waqas1987, Dayana Hashim, Misza13, Semperf, Srammij, Syrthiss, Mkill, Mishalak, Mysid, Kortoso, Karl Meier, PS2pcGAMER, Wujastyk, T-rex, FestivalOfSouls, Phenz, Vaisnavi, Nlu, Wknight94, Protozoid~enwiki, Cheese Sandwich, Ms2ger, Bomkia~enwiki, Wiqi55, NickD, Zzuuzz, Ninly, Theda, Jwissick, Reyk, Brina700, Nothlit, Hound Doggie, LeonardoRob0t, Cjwright79, Fram, TopGear, Curpsbot-unicodify, Nightscrawler, Staxringold, Kramden, Ephilei, Kungfuadam, JDspeeder1, Cookiedog, Payneos, CIreland, NickelShoe, Boss1000, That Guy, From That Show!, Abramul, Luk, ChemGardener, MaeseLeon, Itub, Yvwv, Edenbeast, Attilios, Crystallina, Joshbuddy, SmackBot, Moeron, Oxford Comma, Reedy, KnowledgeOfSelf, VigilancePrime, Lagalag, Ze miguel, Pgk, MeiStone, Bomac, Jagged 85, Thunderboltz, Chairman S., Delldot, Sleevies, J.J.Sagnella, Teiler Köregäten, Vassyana, Aksi great, Gilliam, Monkeytail39, Hmains, Skizzik, Kevinalewis, Chris the speller, Kurykh, IMacThere4iAm, Jnelson09, Jprg1966, Aro888, MalafayaBot, Honey bee155, Kashami, Nozzleman, Bazonka, Jerome Charles Potts, Kungming2, Gracenotes, Hotwiki, Royboycrashfan, Zsinj, Rogermw, Quaque, Can't sleep, clown will eat me, Vanished user llkd8wtiuawfhiuweuhncu3tr, TheGerm, Benten, Kr5t, Dan Ferrario, Rrburke, Kittybrewster, Addshore, Tlusťa, Dali, Huon, Khoikhoi, King Vegita, Brogersoc, Makemi, Theodore7, Kntrabssi, Mistamagic28, EVula, D J L, Harvestman, Localzuk, Wirbelwind, Drphilharmonic, LordHoborgXVII, Fuzzypeg, Henrydms, SuperDT, Sadi Carnot, Josellis, Pilotguy, Kukini, Yoshiko-Chan, Deepred6502, Doug Miller, Ken M., Rory096, Harryboyles, Axem Titanium, NormalGoddess, Kuru, John, STemplar, DavidCooke, DRaGZ, Heimstern, SilkTork, Alcumista, Sir Nicholas de Mimsy-Porpington, Pthag, Tktktk, Linnell, JoshuaZ, Maziar fayaz, ManiF, Peterlewis, IronGargoyle, Pennyforth, Asdfv, Drork, A. Parrot, Slakr, Piercetp, Alethiophile, DragoonWraith, Dicklyon, Xiaphias, Aarktica, Mets501, Doczilla, Ryulong, Sharnak, Citicat, PSUMark2006, WindOwl, John1014, Gotnoglory, Keahapana, Iridescent, Sarastro777, Thenobleageofsteam, Colinjl, JoeBot, Twas Now, Somaterc, Courcelles, Heliomance, Tawkerbot2, Daniel5127, Filelakeshoe, Hirtenfeuer, Xcentaur, Planktonbot, CmdrObot, Sir Vicious, Groovysoul, Comrade42, Ninetyone, Anakata, KyraVixen, Leevanjackson, Dgw, DanielRigal, NickW557, Casper2k3, Ken Gallager, John S Moore, Pratikthakore, Speedy342315, Cydebot, Astrochemist, Goldfritha, Synergy, Frater5, Odie5533, Karafias, Doug Weller, Christian75, DumbBOT, Aintsemic, Chrislk02, Dooly00000, SteveMcCluskey, JodyB, Satori Son, .:Debil:., Thijs!bot, Epbr123, Barticus88, Rsage, RevolverOcelotX, Marek69, Esowteric, Tapir Terrific, Msbluerasp, GideonF,

Nick Number, ThePeg, Natalie Erin, Mentifisto, WikiSlasher, AntiVandalBot, G.o.narada, Majorly, Phuff, Luna Santin, Yomangani, Fire3500, The Templar, Prolog, Dr. Blofeld, Julia Rossi, Goldenrowley, Mary Mark Ockerbloom, Cinnamon42, Fayenatic london, Ereiyo, Bakabaka, Modernist, Dylan Lake, Glacierfairy, Leevclarke, Rico402, Myanw, Defordj, Gökhan, Mightywayne, Avani patel, Davewho2, Aurumsolis, Laboratorio.Ricerche.Evolutive, OllyG, Hut 8.5, Chevellefan11, Flashinpon, Acroterion, I80and, Caffeinepuppy, Magioladitis, Hroðulf, Bongwarrior, VoABot II, Jeff Dahl, Professor marginalia, Juanradeliz, Ling.Nut, Sodabottle, Vikrant A Phadkay, Nyttend, Balloonguy, Samjohnston, Nick Cooper, NotACow, FrF, Animum, Upholder, IkonicDeath, Practical123, Deepinthenight, MetsBot, Deanostrodamus, Hamiltonstone, Allstarecho, Frotz, Vssun, Ksvaughan2, DerHexer, Edward321, Thyroidpsychic, Teardrop onthefire, DukeTwicep, War wizard90, Gun Powder Ma, Ztobor, B9 hummingbird hovering, FisherQueen, Paul Gard, Hdt83, Mmoneypenny, CliffC, Equisis, Dr. TaO, Artemis-Arethusa, C0nsumer, Naohiro19, Anaxial, Vincentrijlaarsdam, David J Wilson, R'n'B, Rh64815, WelshMatt, Smokizzy, LedgendGamer, Amt1018, J.delanoy, M.m.a, PCock, Owl320, Bitethesilverbullet, Bogey97, Uncle Dick, Maurice Carbonaro, HiddenTreasures06, Ishamid, Polenth, TomS TDotO, Ian.thomson, Exploravisionx2, Jkaplan, Sssuuuzzzaaannn, Minderbinder~enwiki, Xyzt1234, Bot-Schafter, L'Aquatique, AdamBMorgan, Gurchzilla, WebHamster, 2wingo, AntiSpamBot, Plasticup, NewEnglandYankee, Vermeer1, SJP, Alnokta, Fjbfour, Doug4, KylieTastic, STBotD, Ot Manu, FuegoFish, Jamesontai, Skryinv~enwiki, XKiichigo, DrunkinRoxtar, Gtg204y, Bonadea, Levcampbell, Jones9999, Sixiki, Jakeman2005, CardinalDan, Idioma-bot, Sesshoumaru-sama~enwiki, Chromancer, Alchemy999, VolkovBot, Metal.lunchbox, Alienlifeformz, Xeltifon, ABF, Beatnik Party, DOHC Holiday, Macedonian, TallNapoleon, Shinju, AlnoktaBOT, LeilaniLad, QuackGuru, CinderAlchemist, Philip Trueman, Ken.Dickey, Saziel, TXiKiBoT, Oshwah, GimmeBot, BuickCenturyDriver, Keefer.t, Hobe, Apepch7, Atomcoeur, IPSOS, Qxz, Goldisgood, Zmod101, DennyColt, Martin451, Ripepette, Psyche825, Vgranucci, Natg 19, Flyingw, Fishhook, CO, RadiantRay, Isaacrob, Chickenpower, Joelwyland, Staka, Vritti, AnnekeBart, Merkabaman, Nmhall, TravelingCat, Blaze Flame, Spinningspark, Mikrofone877, Ajrocke, Why Not A Duck, Anton H, HiDrNick, AlleborgoBot, LuigiManiac, Yaksar, MrChupon, W4chris, Wikiajmail, Steven Weston, D. Recorder, LOTRrules, Ponyo, Gabe777, SieBot, Paul20070, Anyep, Alabaster Crow, Scarian, BotMultichill, Dawn Bard, YBorg, Caltas, Racro 16, Kageskull, Ralecourtze, Calabraxthis, Mrpearcee, SuzanneIAM, Typritc, Nummer29, Mar(c), Keilana, Flyer22 Reborn, Radon210, Perspicacite, Larek, Oda Mari, Momo san, Doctorfluffy, Oxymoron83, Treehappy, KoshVorlon, Historicus800, Lightmouse, Smartkid112358, Nancy, Dogbeast, Shadygrove2007, Spartan-James, StaticGull, Dcattell, Sxp151, Centralplexus, Hamiltondaniel, Vanished User 8902317830, Dust Filter, Ptr123, Wahrmund, Simsin1, Denisarona, Eriador~enwiki, WickerGuy, Troy 07, Explicit, BHenry1969, JiggeryPokery, Onemillionthmonkey, Faithlessthewonderboy, Secfile, Felipe Aira, Loren.wilton, Martarius, Sfan00 IMG, ClueBot, Hawks vial er-u-m vail, Alchemic-Paladin, Thirteen O' Clock, The Thing That Should Not Be, EoGuy, Tseavey1, Csarami, AeliusHadrianus, J8079s, Niceguyedc, Rotational, Trivialist, Neverquick, ACHKC, John J. Bulten, SpencerJordan, XofWiki, Sirius85, DragonBot, Excirial, Jusdafax, Waiwai933, Alejandrocaro35, Sun Creator, Creatatron, ZuluPapa5, Ernobe, AmontonarPapeles, Arjayay, Btre2007, Bremerenator, JamieS93, Tnxman307, CowboySpartan, Andreas Groß, Frozen4322, Curious Blue, SchreiberBike, Joshua Arent, Orathaic, Shikyomaru, 1ForTheMoney, Mczack26, Versus22, Painus69, LiuMasters, Phynicen, SoxBot III, Apparition11, +u3)u!^ 7!3N, Oore, DumZiBoT, Zodiacdog, Alchemist Jack, Wednesday Next, XLinkBot, Spitfire, Pichpich, BodhisattvaBot, Rror, Mavigogun, Pgallert, Cozmo131, WikHead, Yoshi thomas, Jd027, Will in China, Noctibus, NHJG, Navy Blue, The full metal alchemist, Boocah305, Xp54321, Proofreader77, Some jerk on the Internet, Guoguo12, Mabdul, Non-dropframe, GSMR, PicklesofDOOM, Ronhjones, CanadianLinuxUser, Leszek Jańczuk, Abbiejoice, MrOllie, Download, Redheylin, Theneogon, AndersBot, Favonian, Zi Yi Quan, Doniago, LinkFA-Bot, Kidsrule15893, Dtrain8211, Kisbesbot, AgadaUrbanit, Tassedethe, Tide rolls, Bfigura's puppy, Isis2197, MuZemike, ReginaGoesMoo, LuK3, Angrysockhop, Legobot, Luckas-bot, Nyanatusita, Yobot, Pink!Teen, Billaknz43, Fraggle81, Nutfortuna, Karanne, Dsowner, Ningauble, Call Me K, W-dueck, AnomieBOT, DemocraticLuntz, IRP, Aeirom1, Piano non troppo, Ularevalo98, AdjustShift, Kingpin13, Cyanidethistles, Mahmudmasri, Materialscientist, Citation bot, Kasaalan, Andraiw, Arthur-Bot, Quebec99, LilHelpa, Vulcan Hephaestus, Xqbot, Kpind6916, Sudha, Capricorn42, Dont watch dat, Sellyme, Timmyshin, RedKiteUK, بطل علي حسن, Buleleader, Fordee11, -), Khruner, Petropoxy (Lithoderm Proxy), J04n, GrouchoBot, Xashaiar, Webwat, Omnipaedista, Anoot7, Ssarti, Etgel, Vinnyintrousers, TboneMN, Sophus Bie, Calcinations, Shadowjams, FrescoBot, Looplog, Vishnu2011, Machine Elf 1735, Garfgarfgarf, Tetraedycal, Citation bot 1, Gdje je nestala duša svijeta, Grannieweatherwax, Pinethicket, I dream of horses, Parabombastus, Elockid, Demon Lord 302, GrailTyger, Skyerise, Petermopar, Annbrepols, Archangels9, Kibi78704, Tea with toast, FoxBot, Cmdahler, TobeBot, Trappist the monk, Layth Sbaihat, Vrenator, January, Kildwyke, Dddiiii, Seahorseruler, Jeffrd10, Unrulyevil, Reach Out to the Truth, Lord of the Pit, RjwilmsiBot, Lung salad, Rocko945, NameIsRon, Act Up, Blueinc123, Jamol96, Enauspeaker, DASHBot, EmausBot, John of Reading, Acather96, Juanita Saenz S, Avenue X at Cicero, Barkert89, Gfoley4, Facesmasher69, Syncategoremata, Active Banana, RenamedUser01302013, Mjazena, Vanished user zq46pw21, CrimsonBlue, Tommy2010, Wikipelli, Selfchosen, Werieth, AvicBot, ZéroBot, Luzo-Graal, Ida Shaw, Entiex, Creepy geek, Érico Júnior Wouters, Zloyvolsheb, Deaburnham, Fagetboy, Wayne Slam, Tolly4bolly, Staszek Lem, EricWesBrown, THEGREATKRAMER, Sbmeirow, Jacobisq, AIM1796, Brandmeister, L Kensington, Donner60, Inka 888, Xiaoyu of Yuxi, Homeaccount, Orange Suede Sofa, Dansalignatious, ChuispastonBot, Herk1955, Ladnadruk, GrayFullbuster, Ordibehesht22, DASHBotAV, Spicemix, 28bot, Michael Bailes, WMC, Will Beback Auto, ClueBot NG, Gareth Griffith-Jones, Proclus27, IKill-Animation, Eonsword, Angelcook, J151e, Movses-bot, Alchemyalchemy, Encycloshave, Rahence, Wonderingraven, Thexiii, HazelAB, Dream of Nyx, Rurik the Varangian, Bidwoud, Helpful Pixie Bot, Bidwoud13, Baghermolavi, HMSSolent, Mdavid9, Bibcode Bot, 2001:db8, WNYY98, Jeraphine Gryphon, Will Timony, Ph.D, Car Henkel, Rijinatwiki, Draven099, Captainmighty, Highandseeking, Surya Neeraj, GrammarFascist, Hurricanefan24, MrBill3, Glacialfox, Dr. Remo F. Roth, Lieutenant of Melkor, Janus945, BattyBot, Matthawk0328, EgillSchollogrim, TheCascadian, ChrisGualtieri, ZappaOMati, Torvalu4, Khazar2, Iry-Hor, JYBot, Dexbot, Alchemy 1997, Webclient101, Mogism, Barnaculus, Luna Fire, Lugia2453, UnsourcedBlanker222, Tawnyaninjacat, Blamestars, Lerr, King jakob c, Jinx1002, JustAMuggle, Silverbrooks100, BurritoBazooka, Sunlight1361, Ngoodnow, Madisongouin, Ilacin, Demgiraffes, Harshal123456, Lukekfreeman, Shrikarsan, DavidLeighEllis, Metadox, Ginsuloft, Jackmcbarn, AhBengI, Rons corner, StevenD99, Theseus001, Stamptrader, Fidasty, Politicalanthropology, Eisborne, JaconaFrere, Alexis1102, Kintastic hair, Naelahkiin, Overthemountains222, Great Escape Hero, Monkbot, Chrisbrooks59, Renganwa, Howrde1, Jeremiah90moon, Egg151, Amjertyu, ChamithN, Monkutis, Pixal Storm, Inazuma261, Wafflesareprettyflippincool, Orduin, Obasha1, Chris degs, TheEditorOnline, Scroom84, KasparBot, Totochemist, Gmr1994, ChildsC24, Slashingfear, Emperorclothes, KSFT, Di Serra, Bob5525, Ziyahashmi, Fornax7 and Anonymous: 1718

- **Alchemy and chemistry in medieval Islam** *Source:* https://en.wikipedia.org/wiki/Alchemy_and_chemistry_in_medieval_Islam?oldid= 684939343 *Contributors:* William M. Connolley, Rursus, Icairns, Rich Farmbrough, Bender235, Kwamikagami, Wtmitchell, Woohookitty, Ruud Koot, Kralizec!, BD2412, Rjwilmsi, Koavf, Nihiltres, Mehrshad123, DVdm, Spacepotato, RussBot, Pigman, Reyk, Boss1000, SmackBot, Jagged 85, Mscuthbert, Delldot, Colonies Chris, Aboudaqn, Msmolik, Harryboyles, FairuseBot, Drinibot, Hemlock Martinis, Doug Weller, DumbBOT, Alaibot, Mbell, Tiamut, Mentifisto, Once in a Blue Moon, Sluzzelin, MER-C, Yahel Guhan, Wasell, JamesBWatson, باس م, Gabriel Kielland, David Eppstein, Frotz, JaGa, Gun Powder Ma, Indamoney, R'n'B, J.delanoy, Belovedfreak, Knulclunk, Alnokta, Squids and Chips,

Signalhead, Philip Trueman, Davehi1, Aymatth2, Enviroboy, Insanity Incarnate, LOTRrules, Malcolmxl5, Oxymoron83, Dcattell, Altzinn, ClueBot, AlchemicPaladin, Plastikspork, J8079s, SamuelTheGhost, Seanwal111111, Al-Andalusi, Error −128, Johnuniq, Editor2020, Xiquet, Addbot, Element16, Jncraton, CanadianLinuxUser, Redheylin, LarryJeff, Yobot, Aboalbiss, ReasOFF, AnomieBOT, Jim1138, SusBot~enwiki, Samythekid, Citation bot, Cavila, FaleBot, Siavashk, D'ohBot, Citation bot 1, Plasticspork, Full-date unlinking bot, IJBall, Smithers946, Jingleheimer Smith, River6969us, Seahorseruler, RjwilmsiBot, Syncategoremata, Aquib American Muslim, ZxxZxxZ, ZéroBot, Kaka Mughal, Someone65, Δ, EdoBot, WMC, Techbotic, ClueBot NG, Joefromrandb, Gpcfox, Widr, Helpful Pixie Bot, Car Henkel, بپارس آملی, NotWith, Ariaveeg, Zh84, Dexbot, Schwalltj, Alyssasordo, Romipes, Y410123, 468SM and Anonymous: 79

- **Chemical revolution** *Source:* https://en.wikipedia.org/wiki/Chemical_revolution?oldid=673302349 *Contributors:* Edward, Julesd, Stone, Joyous!, Rich Farmbrough, Espoo, Anthony Appleyard, GrouchyDan, Graham87, Rjwilmsi, Tdoune, YurikBot, Okedem, Ragesoss, Petri Krohn, Itub, Jagged 85, XenoNeon, Kurykh, OrphanBot, Sadi Carnot, Kensor, RekishiEJ, Cydebot, Herd of Swine, Daniel, Matthew Fennell, Cardamon, The Real Marauder, DH85868993, Amaher, AlleborgoBot, SieBot, SamuelTheGhost, Alexbot, Addbot, Luckas-bot, Citation bot, Xqbot, Raafael, Pinethicket, Nmillerche, Syncategoremata, John of Lancaster, ZéroBot, KLindblom, Alex Nico, Helpful Pixie Bot, Gjue, AvocatoBot, ASCIIn2Bme, Fkeeling and Anonymous: 18

- **Atomic theory** *Source:* https://en.wikipedia.org/wiki/Atomic_theory?oldid=686888082 *Contributors:* Mav, Ubiquity, Tim Starling, Tannin, Ixfd64, Ahoerstemeier, Smack, David Shay, HarryHenryGebel, Bloodshedder, Sverdrup, Rursus, Blainster, Hadal, Michael Snow, Rho~enwiki, Giftlite, Awolf002, Bensaccount, Beland, Karol Langner, H Padleckas, Gscshoyru, Neutrality, Engleman, Karl Dickman, Deglr6328, Discospinster, Qutezuce, Pjacobi, Vsmith, Paul August, Eric Forste, RJHall, El C, Laurascudder, Grick, Bobo192, DanielNuyu, AnyFile, Wipe, BrokenSegue, Maurreen, Joe Jarvis, Hooperbloob, Nsaa, Alansohn, Iothiania, Riana, BryanD, CJ, Malo, Wtmitchell, Velella, CloudNine, Bsadowski1, Falcorian, Feezo, Kelly Martin, Linas, Mindmatrix, Kurzon, MONGO, Terence, Mathewtse, Wayward, MarcoTolo, Mandarax, Edison, Rjwilmsi, Ctdunstan, R.e.b., Ems57fcva, Bubba73, Brighterorange, Krash, Yamamoto Ichiro, Jameshfisher, Alphachimp, Physchim62, King of Hearts, Chobot, Moocha, Cactus.man, Hall Monitor, Gwernol, Banaticus, YurikBot, Sceptre, Brandmeister (old), Petiatil, DanMS, ColoradoZ, Shell Kinney, Gaius Cornelius, NawlinWiki, Wiki alf, Grafen, Ragesoss, Mlouns, Syrthiss, Derek.cashman, Oysteinp, Petri Krohn, Tevildo, Archer7, TLSuda, Moomoomoo, Banus, Zvika, Elliskev, Finell, Luk, Itub, SmackBot, Tarret, Prodego, Vald, Jagged 85, Delldot, Eskimbot, Gaff, Betacommand, Kurykh, Audacity, NCurse, Bduke, Jprg1966, Miquonranger03, Complexica, Dustimagic, Sbharris, Colonies Chris, Hallenrm, Darth Panda, Sct72, Can't sleep, clown will eat me, ApolloCreed, Voyajer, Thrane, SnappingTurtle, TrogdorPolitiks, DMacks, Salamurai, Pkeets, SashatoBot, Nishkid64, ALUOPline2, Mathboy965, Ckatz, Rinnenadtrosc, Noah Salzman, KNBDunlop, Timmy2, Ryanjrr, Waggers, SandyGeorgia, Mets501, This is so strange, Ryulong, LaMenta3, Sifaka, Iridescent, Shoeofdeath, Francl, Igoldste, Blehfu, Courcelles, Tawkerbot2, Kurtan~enwiki, JForget, CmdrObot, KyraVixen, JohnCD, NickW557, MrFish, Jac16888, Bddmagic, Misterhay, Bellerophon5685, Synergy, Tawkerbot4, Carstensen, Doug Weller, Christian75, Codetiger, DumbBOT, Kozuch, SteveMcCluskey, Vanished User jdksfajlasd, Casliber, Thijs!bot, Epbr123, Qwyrxian, Headbomb, Marek69, JustAGal, Natalie Erin, CTZMSC3, Mentifisto, AntiVandalBot, KP Botany, Danger, Indian Chronicles, Etr52, JAnDbot, MER-C, Arch dude, MSBOT, Bongwarrior, VoABot II, Ling.Nut, Catgut, Froman77, Nposs, Allstarecho, J Hill, User A1, Adventurer, DerHexer, Hbent, Yensin, Slippknotryan, S3000, MartinBot, Kbrewer, Rettetast, R'n'B, AlexiusHoratius, J.delanoy, Adavidb, Bogey97, Mthibault, Rhinestone K, Yonidebot, Choihei, Hellomoto021, L337 kybldmstr, P.wormer, Belovedfreak, King turtal, Sunderland06, TottyBot, Idioma-bot, Bottobbot, Moblinmaniac, Deor, TreasuryTag, Jeff G., Rtrace, Hilarious Bookbinder, Kyle the bot, Philip Trueman, Zarcusian, TXiKiBoT, Oshwah, Kriak, Hqb, Dchall1, Z.E.R.O., Crohnie, JayC, Billy1223billy1223, Sintaku, Dgiroux, The Wilschon, Martin451, Sanfranman59, TehZorroness, Bean915, Herindes, Enigmaman, SallyBoseman, Falcon8765, RaseaC, Superhockeyman, Ajrocke, Sfmammamia, Aanankhurma, Hrafn, SieBot, Ivan Štambuk, Caltas, Dzexon, Keilana, Happysailor, Flyer22 Reborn, Oda Mari, Lightmouse, Techman224, KathrynLybarger, Pappapippa, Sunrise, Maxxamaxx, Anchor Link Bot, Mygerardromance, Mr. Stradivarius, Athenean, Tomasz Prochownik, Elassint, ClueBot, Trojancowboy, Snigbrook, Fyyer, The Thing That Should Not Be, Gaia Octavia Agrippa, WDavis1911, J8079s, CounterVandalismBot, LizardJr8, Ratreuser, Anthon01, Djr32, Excirial, Jusdafax, Mrmoney426, Runningamok19, Praveen khm, Vivio Testarossa, Eagleman343, NuclearWarfare, Hans Adler, Mynameisages, J99867, Thingg, Aitias, BVBede, PCHS-NJROTC, Egmontaz, BarretB, Pippin254, Proofreader77, Non-dropframe, DaughterofSun, Mathew Rammer, Ronhjones, TutterMouse, CanadianLinuxUser, WFPM, Download, CarsracBot, LAAFan, DFS454, Glane23, Bassbonerocks, 5 albert square, Barak Sh, Tassedethe, Tide rolls, OlEnglish, Jan eissfeldt, Krano, QuadrivialMind, Gail, Loupeter, Zorrobot, Suiseiseki, Ben Ben, Legobot, Yobot, Tohd8BohaithuGh1, Anypodetos, Eric-Wester, Tempodivalse, Synchronism, AnomieBOT, Killiondude, Piano non troppo, Flewis, Materialscientist, Citation bot, OllieFury, Addihockey10, Capricorn42, Ibanezer23, Samiam2312, SassoBot, Spellage, SlavaPhD, Dougofborg, ⁇⁇, Chjoaygame, Wupop, Happy1npink13, Legion23, Priyul2020, Steve Quinn, Machine Elf 1735, Diremarc, DrilBot, Pinethicket, I dream of horses, Elockid, MJ94, Calmer Waters, 124Nick, FoxBot, Trappist the monk, Sheogorath, Aarmentapalacios, Tbhotch, DARTH SIDIOUS 2, RjwilmsiBot, NameIsRon, Noommos, Bloopityblab, DASHBot, John of Reading, Immunize, *devunt, Super48paul, Da500063, Dewritech, Syncategoremata, Itsamee, Exmadxman2, K6ka, AsceticRose, Hhhippo, CanonLawJunkie, Amrodrgz, Bollyjeff, Traxs7, Editingallmistakes22, Access Denied, Esaethan, Gz33, Wayne Slam, Smurf03, PaulRobertson1, Nicoaslolbooze, L Kensington, Mayur, Sunshine4921, Dazmansimmons, DASHBotAV, Bbourne20, Shadowbloodzzz, ClueBot NG, Cwmhiraeth, Hiperfelix, Photoemission, Gartxoak, O.Koslowski, Widr, Harkalelibre, MerlIwBot, Helpful Pixie Bot, Bobherry, Calabe1992, Bibcode Bot, Lowercase sigmabot, ElamTheStiffPole, MusikAnimal, Mark Arsten, CitationCleanerBot, Account.ka.naam, Duxwing, BattyBot, ChrisGualtieri, EuroCarGT, Douglas2012, Dexbot, Jscher18, Webclient101, TwoTwoHello, LlamaDude78, Syum90, Jamesx12345, SassyLilNugget, Joeinwiki, MercurianTerr, Lovebug2013, Beuracrat, Wikichman, Zenibus, Dansantos303, JustBerry, JaconaFrere, Mahusha, SantiLak, Michaeldmoreno, Liance, L.exhc, Tayahgirlxo, Tymon.r, Bose skipper, PossiblyLying, Bhatde, Harsh mahesheka, Rennzomataastumalon, 221science, Dickwads and Anonymous: 782

- **Chemical formula** *Source:* https://en.wikipedia.org/wiki/Chemical_formula?oldid=686819593 *Contributors:* Marj Tiefert, Tarquin, Patrick, Ixfd64, Card~enwiki, Loooix~enwiki, Bdonlan, Ellywa, Ahoerstemeier, Mac, Theresa knott, Cherkash, Schneelocke, Timwi, 4lex, Maximus Rex, Gentgeen, Robbot, Stewartadcock, Hadal, HaeB, Tobias Bergemann, Giftlite, Lee J Haywood, Bensaccount, Unconcerned, Guanaco, Beland, DragonflySixtyseven, Icairns, Popadopolis, ⁇⁇, Grunt, Mike Rosoft, Vesta~enwiki, Random contributor, Discospinster, Guanabot, Cacycle, Cnwb, Wrp103, Vsmith, Shadow demon, Jpgordon, Bobo192, Smalljim, Elipongo, Tomgally, Kjkolb, Hagerman, Jumbuck, Alansohn, Mo0, Benjah-bmm27, Sl, Walkerma, Snowolf, Projoe~enwiki, Evil Monkey, RainbowOfLight, Shimeru, Pol098, Mpatel, Eleassar777, Dah31, JRHorse, Tom W.M., Palica, MassGalactusUniversum, Graham87, V8rik, DePiep, Jclemens, Drbogdan, The wub, FlaBot, RexNL, Gurch, Krishnavedala, Dj Capricorn, ColdFeet, Roboto de Ajvol, YurikBot, Wavelength, Sputnikcccp, Quinlan Vos~enwiki, GLaDOS, Hellbus, Stephenb, Chaos, Wimt, Seegoon, TimK MSI, Wknight94, Tetracube, Donald Albury, Pb30, KGasso, Summersilk, Itub, Sardanaphalus, SmackBot, Lagalag, Kopaka649, TheDoctor10, Edgar181, Likeitsmyjob, Gilliam, Skizzik, Kurykh, SlimJim, Deli nk, Sbharris, Mladifilozof,

Hgrosser, NYKevin, AeroSpace, RedHillian, Aldaron, Flyguy649, Tinctorius, Ian01, DMacks, Edwy, Chodorkovskiy, Mgiganteus1, Bjanku-
loski06en~enwiki, Ekrub-ntyh, Beetstra, Meco, Mets501, Dreftymac, IvanLanin, Mr Chuckles, Tawkerbot2, Astirmays, Rambam rashi, CBM,
Raz1el, Nick Wilson, Chasingsol, Shirulashem, Christian75, Marqmike2, Thijs!bot, Barticus88, Wikid77, Danielle dk, Pjvpjv, Natalie Erin,
Escarbot, Adam B, AntiVandalBot, Seaphoto, Voyaging, Jj137, Farosdaughter, Qwerty Binary, Canadian-Bacon, JAnDbot, Leuko, Garda40,
Couchpotato99, Bakilas, Bongwarrior, VoABot II, Nikevich, Daarznieks, Dirac66, Loonymonkey, Thompson.matthew, MartinBot, Anaxial,
Leyo, J.delanoy, Pharaoh of the Wizards, Trusilver, Marcusmax, Katalaveno, Mikael Häggström, Gurchzilla, NewEnglandYankee, Touch Of
Light, Mufka, KylieTastic, Juliancolton, Remember the dot, TheNewPhobia, Ischemia, Regicollis, Cezn, Wikieditor06, VolkovBot, Barneca,
Philip Trueman, TXiKiBoT, MrZhuKeeper, Vipinhari, Anna Lincoln, Vnesh, AlleborgoBot, GoddersUK, Winchelsea, Flyer22 Reborn, Oiws,
Oxymoron83, Byrialbot, Faradayplank, OKBot, StaticGull, Anchor Link Bot, Mygerardromance, Denisarona, ClueBot, The Thing That Should
Not Be, Tomas e, Drmies, Super propane, SpikeToronto, Rhododendrites, Fassitude, Razorflame, Aitias, Lincmaster, DerBorg, SoxBot III, Ed-
itorofthewiki, AlanM1, Doc9871, Eric Baer, Addbot, Some jerk on the Internet, Wickey-nl, Tanhabot, Ronhjones, KorinoChikara, Vishnava,
Skyezx, Tide rolls, Luckas-bot, Yobot, Andreasmperu, MinorProphet, AnomieBOT, Itssanjith, Kristen Eriksen, 90, Kingpin13, Law, Materi-
alscientist, The High Fin Sperm Whale, Citation bot, Pepsimax120, Maxis ftw, Xqbot, Capricorn42, Bihco, Weeeeeeeee1234, GrouchoBot,
Polargeo, Alpha77a, Thehelpfulbot, FrescoBot, Finalius, Pinethicket, I dream of horses, V.narsikar, Mikespedia, Gamewizard71, FoxBot,
عقیل فشاک, Andrewharold, ThinkEnemies, DARTH SIDIOUS 2, Armando-Martin, DASHBot, EmausBot, Orphan Wiki, WikitanvirBot,
Syncategoremata, K6ka, JSquish, ZéroBot, Kaipai94, Dffgd, Deniska.G, Wayne Slam, TyA, Morgankevinj, Inka 888, Carmichael, Matthewr-
bowker, MacStep, DASHBotAV, E. Fokker, Mikhail Ryazanov, ClueBot NG, Tiger11199, Vibhijain, Helpful Pixie Bot, Zacharyledford, Trnhg-
duoc2222, MusikAnimal, Mark Arsten, IraChesterfield, Klilidiplomus, Acratta, Shaun, Anbu121, Njaohnt, ChrisGualtieri, JYBot, Cmdavis34,
TwoTwoHello, Lugia2453, Kevin12xd, Vamsi manikanta, Valites, DudeWithAFeud, Atomdgbgfhffvgxhgdggdvhfc, Trackteur, Knowingevery-
thing101, Saddas123sad, Ivana Pupu, KH-1, Toriwood72, KasparBot, The Quixotic Potato and Anonymous: 554

• **Electrolysis** *Source:* https://en.wikipedia.org/wiki/Electrolysis?oldid=686058634 *Contributors:* Bryan Derksen, Timo Honkasalo, The Anome,
Mark, Mirwin, Peterlin~enwiki, Heron, Ubiquity, D, Michael Hardy, Tim Starling, Fred Bauder, Gabbe, Delirium, Mdebets, Ahoerstemeier,
Mac, Bogdangiusca, Netsnipe, Nohat, Stone, Maximus Rex, Nv8200pa, Rm, Omegatron, Toreau, Jni, Pcollison, Branddobbe, Gentgeen, Rob-
bot, Diberri, Cyrius, Tobias Bergemann, Giftlite, Tom harrison, Everyking, Kandar, Gadfium, Utcursch, Pcarbonn, Antandrus, Aulis Eskola,
Togo~enwiki, Jokestress, Icairns, Karl-Henner, Richardb43, PeR, Neutrality, Muijz, Adashiel, D6, Freakofnurture, Limeheadnyc, Cacycle,
Pjacobi, Vsmith, Sgeo, FirstPrinciples, Kwamikagami, Shanes, Femto, Devil Master, Enric Naval, Cohesion, Rainbird, Deryck Chan, Axyjo,
MPerel, Sam Korn, Hooperbloob, HasharBot~enwiki, Oolong, Alansohn, CountdownCrispy, Nuclear man, Jnothman, BRW, Wtshymanski,
Shoefly, Apblum, DV8 2XL, Windsok, Kelisi, Bkwillwm, HappyApple, Palica, Galwhaa, Rjwilmsi, FutureNJGov, HappyCamper, Nigosh,
THE KING, Yamamoto Ichiro, FlaBot, Eubot, RexNL, SteveBaker, Srleffler, Chobot, Banaticus, YurikBot, Wavelength, RobotE, Brandmeis-
ter (old), Sillybilly, Limulus, CanadianCaesar, Russoc4, Salsb, Wimt, NawlinWiki, Grafen, Zwobot, Tmallon, Sandstein, Pb30, Wanopanog,
Peter, Anclation~enwiki, Asterion, Mejor Los Indios, One, That Guy, From That Show!, ChemGardener, Itub, Danielsavoiu, SmackBot, Un-
school, CarbonCopy, Brianyoumans, Bomac, Delldot, Xaosflux, Gilliam, Betacommand, Master Jay, Bluebot, Ciacchi, Quinsareth, Persian
Poet Gal, MalafayaBot, Metacomet, DHN-bot~enwiki, Zven, VirtualSteve, MaxSem, Shadow7789, Can't sleep, clown will eat me, Shalom
Yechiel, Frap, Saitanmayi, Rrburke, Flyguy649, Nakon, Smokefoot, Polonium, Drphilharmonic, Where, Mion, FelisLeo, Ohconfucius, Sashato-
Bot, Tdw1203, Kuru, Peterlewis, 16@r, Stwalkerster, Greenbeans, Galactor213, Hu12, Iridescent, Walther Atkinson, Astral highway, Quodfui,
Courcelles, Tawkerbot2, Binks, Orangutan, Tar7arus, Thatperson, Nunquam Dormio, Mike.mcdevitt, Avillia, Rudjek, Gogo Dodo, Omicron-
persei8, Rocodoco, Slysa, CrusherEAGLE, Thijs!bot, Epbr123, VKemyss, Michagal, MrXow, Nonagonal Spider, CTZMSC3, Escarbot, Anti-
VandalBot, BokicaK, Luna Santin, Quintote, Kainino, Deadbeef, D99figge, DuncanHill, MER-C, Arch dude, Attarparn, Beaumont, Fishhooky,
Karlhahn, Bongwarrior, VoABot II, Nulled, JamesBWatson, Engineman, Dirac66, Excuseu5000, Allstarecho, Just James, Glen, DerHexer,
0612, Hdt83, BetBot~enwiki, Youghurt, Pruthvi.Vallabh, Bfesser, CommonsDelinker, Leyo, Tgeairn, T.vanschaik, J.delanoy, Aussie111, Lord-
ofthe9, Eliz81, Ppreeper, Kenwinston, Cadwaladr, SJP, KylieTastic, Bob, STBotD, Jkeohane, Stevekemsley, Idioma-bot, VolkovBot, Jeff G.,
Bsroiaadn, Barneca, Philip Trueman, TXiKiBoT, 99DBSIMLR, Mybadluck22, Anna Lincoln, Cmreigrut, Nafordham, Capper01, Axiosaurus,
Jackfork, LeaveSleaves, Peacekills, Mishlai, Blalalalala, Random54321, Dirkbb, Hanjabba, Gorank4, Rhopkins8, Sylent, PGWG, GreaterWik-
iholic, SieBot, OMCV, Scarian, NB-NB, Jauerback, Vanished User 8a9b4725f8376, WildWildBil, PookeyMaster, RucasHost, Flyer22 Reborn,
Master Belgarath, Volubile25, Clive long, Shkedi, ImageRemovalBot, ClueBot, Daniel73480, The Thing That Should Not Be, Playtimee, Swathi-
nath, Spadesp, VandalCruncher, Etalli, Skihatboatbike, Excirial, Quercus basaseachicensis, Dave 710, Zaharous, Mr45acp, Saebjorn, Katanada,
SoxBot III, XLinkBot, Ivan Akira, Rror, P30Carl, WikHead, Nadavs, Addbot, Some jerk on the Internet, Laasworld, TutterMouse, Vishnava,
Asphatasawhale, 2corner, Favonian, Tide rolls, Lightbot, Vintagerave, Therussianlife, Ben Ben, Luckas-bot, Mazoues, Yobot, Crispmuncher,
Eric-Wester, AnomieBOT, Ciphers, Jim1138, Sz-iwbot, Crystal whacker, Materialscientist, Poepzak, Aleg 2311, Xqbot, TheAMmollusc, Capri-
corn42, Jmundo, PhysicsR, GrouchoBot, Amaury, Shadowjams, BoomerAB, Databytecorp, LonelyMountain, DigbyDalton, Pinethicket, Jone-
sey95, Xcvista, Serols, Jujutacular, TobeBot, Time9, Dinamik-bot, Extra999, Imagepixel, Robbin' Knowledge, DARTH SIDIOUS 2, John of
Reading, RA0808, Smerdakas, Wikipelli, Felixh238, Makecat, JuTa, Puffin, DaGr8N8, 28bot, Petrb, ClueBot NG, CocuBot, Rainbowwrasse,
Cntras, Widr, Lewis027, BG19bot, AnnaBennett, Davidiad, Toshiki, Snow Blizzard, Coeco, Jimwater, Mdann52, Hebert Peró, Chrisvel-
net, Olly314, Silviacalvotome, Jeffson odiete, Kaleb.LT, Ugog Nizdast, Spyglasses, Marcelocarmo1, Cypherquest, Kind Tennis Fan, Crow,
TrekkieSpeller, Arunimasd, Mahusha, Monkbot, Xappiens, Teddyktchan, IiKkEe, Bravocheese3, Lavransg, Monofdday, Mudit Balooja, Dr
Ozzzzz, Sudhselva, KasparBot, Juself, The Quixotic Potato, Hermionedidallthework and Anonymous: 677

• **Avogadro's law** *Source:* https://en.wikipedia.org/wiki/Avogadro'{}s_law?oldid=678984059 *Contributors:* Bryan Derksen, Tarquin, Heron,
Wshun, Delirium, Ddoherty, Alex S, Andrewman327, Robbot, Romanm, Mervyn, Christopher Parham, Bensaccount, Rpyle731, Antandrus,
Karol Langner, Icairns, Kelson, Sam, Clemwang, Vsmith, Murtasa, Andrew Maiman, Semifamous, Elwikipedista~enwiki, Duk, Alansohn,
Velella, RJFJR, Gene Nygaard, Jetru, Yousaf465, Benbest, Mpatel, Knuckles, Isnow, Magister Mathematicae, V8rik, BD2412, The wub,
FlaBot, Nivix, TheDJ, Physchim62, Chobot, YurikBot, RobotE, Quinlan Vos~enwiki, Lepidoptera, Nick C, Zythe, BOT-Superzerocool, Lep-
tictidium, Dspradau, CWenger, Itub, SmackBot, Bluebot, Richard001, Vina-iwbot~enwiki, C.jeynes, Myasuda, Kanags, Oosoom, Epbr123,
Escarbot, Mentifisto, Myanw, TuvicBot, JAnDbot, Stellmach, Gaeddal, JamesBWatson, Edmundwoods, Cpl Syx, CommonsDelinker, Tgeairn,
J.delanoy, Rhinestone K, Radar33, Largoplazo, DorganBot, SoCalSuperEagle, Thedjatclubrock, Al.locke, TXiKiBoT, Abc135246, Ann Stouter,
KyleRGiggs, Mptoks, Remilo, Kbrose, SieBot, Arkwatem, VVVBot, Cwkmail, Flyer22 Reborn, KoshVorlon, Spitfire19, Efe, A.C. Norman,
ClueBot, Binksternet, Excirial, Johnuniq, Addbot, Jncraton, Chamberlain2007, Download, Glane23, Luckas-bot, TaBOT-zerem, Eric-Wester,
Jim1138, Brane.Blokar, GrouchoBot, Olthebol, Pinethicket, 5Celcious, Jschnur, RedBot, FoxBot, RjwilmsiBot, ModWilson, K6ka, JSquish,

ZéroBot, Mutomana, Bollyjeff, Donner60, ChuispastonBot, Rmashhadi, ClueBot NG, O.Koslowski, SuperJedi224, 149AFK, Widr, Northamerica1000, Coolbones15, LHcheM, Ulupoi, MJDoughty, King jakob c, N.masoud, Mevo957, ZeAnonimous, Laborerchemist, Chemistry74, Blank xray, Jaffles, Davisbr01, Rajdeep parmar and Anonymous: 147

- **Vitalism** *Source:* https://en.wikipedia.org/wiki/Vitalism?oldid=684412904 *Contributors:* AdamRetchless, Chas zzz brown, Michael Hardy, Paul Barlow, Lexor, Nixdorf, MartinHarper, Wapcaplet, Delirium, Angela, JWSchmidt, Pratyeka, AugPi, Markhurd, Robbot, Sam Spade, Mr-Natural-Health, Aetheling, Pengo, Alan Liefting, Art Carlson, Wighson, Electric goat, Michael Devore, Bensaccount, Eequor, Christofurio, Andycjp, Beland, Sharavanabhava, Rich Farmbrough, Cacycle, Pjacobi, Vsmith, Bishonen, Carlon, Kwamikagami, I9Q79oL78KiL0QTFHgyc, NickSchweitzer, Orangemarlin, Pinar, Sjschen, Sgtpepper6344, Deacon of Pndapetzim, Jheald, Alkarex, Palica, Bunchofgrapes, Sjö, Rjwilmsi, Helvetius, Smithfarm, FlaBot, AED, Backin72, Travis.Thurston, Le Anh-Huy, Random user 39849958, B.~enwiki, YurikBot, Borgx, RobotE, Michael Slone, Luis Fernández García, Pigman, Anomalocaris, Semolo75, Grafen, Ragesoss, Crasshopper, Bota47, 2over0, RDF, Chase me ladies, I'm the Cavalry, Fram, Flowersofnight, JDspeeder1, NetRolller 3D, SmackBot, InverseHypercube, Melchoir, Jim62sch, Kintetsubuffalo, Hmains, Jushi, Isaac Dupree, Dingar, Chris the speller, DoctorW, Jefffire, Tschwenn, Ne0Freedom, Richard001, BullRangifer, Mariawiki~enwiki, Bdiscoe, Salamurai, Sadi Carnot, JzG, Khazar, Gleng, AB, Extremophile, Steth, Newone, Tawkerbot2, Gregbard, Cydebot, Ttiotsw, Dematt, Hughgr, KrishnaVindaloo, Lindsay658, Barticus88, Headbomb, Edhubbard, Grandin, BrightonRock101, StringRay, TimVickers, Tjmayerinsf, Freebytes, JAnDbot, Narssarssuaq, DuncanHill, Skomorokh, .anacondabot, Harristweed, Nyttend, WhatamIdoing, CuteLittleFaery, Cgingold, Gomm, WLU, B9 hummingbird hovering, MartinBot, RealDefender, R'n'B, KTo288, Earthdenizen, MistyMorn, Tmtoulouse, Tarotcards, Belovedfreak, Robertgreer, Wfaze, Megscherer, Inwind, Sacramentis, Andrewaskew, Brainmuncher, David Marjanović, Tiddly Tom, Alexbrn, Sunrise, Wahrmund, Vanyo, Romit3, Martarius, Hafspajen, MARKELLOS, Rhododendrites, SchreiberBike, Editor2020, DumZiBoT, LittleVeryLittle, Roxy the dog, Little Mountain 5, Addbot, Rmsydiana, Landon1980, Bennó, Redheylin, Hogart, AndersBot, Nolelover, Cesiumfrog, Luckas-bot, Yobot, Amirobot, AnomieBOT, Rubinbot, Choij, Jim1138, Neptune5000, Citation bot, LilHelpa, Dr Oldekop, Omnipaedista, Lpetrich, FrescoBot, Machine Elf 1735, RedBot, Orenburg1, Alph Bot, EmausBot, WikitanvirBot, Jorgesca, Dcirovic, Solomonfromfinland, AvicBot, ZéroBot, Wingman4l7, Jacobisq, Just granpa, Herk1955, ByronoryB, ClueBot NG, SheenShin, Liveintheforests, Helpful Pixie Bot, Curb Chain, Ucalegon2011, Frze, Modern Familyy, Latelier.mb, John.kineman, MrBill3, GreenUniverse, BattyBot, La marts boys, Nusaybah, Mogism, The Vintage Feminist, StillStanding-247, Joolzzt, Kevin12xd, François Robere, Itc editor2, YiFeiBot, Kjphill1977, Subtendant, Gronk Oz, Jorge Guerra Pires, Sharp-shinned.hawk, Jerodlycett, Gerardeux and Anonymous: 103

- **Dmitri Mendeleev** *Source:* https://en.wikipedia.org/wiki/Dmitri_Mendeleev?oldid=686512212 *Contributors:* Magnus Manske, Mav, Tarquin, Koyaanis Qatsi, Malcolm Farmer, Css, Fnielsen, Danny, XJaM, William Avery, Drbug, Heron, Fonzy, Ewen, Erik Zachte, Ezra Wax, Wapcaplet, Ixfd64, Dcljr, Nine Tail Fox, Ahoerstemeier, Suisui, Angela, Александър, Lupinoid, Error, Kwekubo, Rob Hooft, Hashar, Reddi, Malcohol, DJ Clayworth, Tpbradbury, Maximus Rex, Taxman, Paul-L~enwiki, Rnbc, Jose Ramos, Bevo, Xyb, Jerzy, Gentgeen, Robbot, Vardion, Hankwang, Fredrik, Baldhur, Altenmann, Henrygb, Flauto Dolce, Auric, Wikibot, JackofOz, NeoThe1, Guy Peters, Jooler, Xyzzyva, Centrx, Giftlite, Pmerriam, Sj, Nunh-huh, Tom harrison, Fastfission, Monedula, No Guru, Curps, Alison, Pashute, Mellum, Solipsist, Bobblewik, Explendido Rocha, Chowbok, Alexf, Toytoy, Chirlu, Blankfaze, MarkSweep, Piotrus, Untifler, Gene s, Sebbe, Icairns, Lumidek, Joyous!, Picapica, Adashiel, Grstain, D6, Atrian, Freakofnurture, Discospinster, Rich Farmbrough, Wikiwide, Vsmith, Slipstream, Zazou, SpookyMulder, Kenb215, DcoetzeeBot~enwiki, Tgies, Janderk, Mashford, Eric Forste, MyNameIsNotBob, RJHall, Miraceti, RoyBoy, Adambro, Bobo192, Smalljim, John Vandenberg, BrokenSegue, Enric Naval, Shenme, Larsie, Juzeris, SpeedyGonsales, Pschemp, Sam Korn, Nsaa, Lysdexia, Ranveig, Jumbuck, Alansohn, Ben davison, Jeltz, Cjthellama, Riana, Lectonar, Ayeroxor, SidP, Helixblue, Docboat, RainbowOfLight, Cmapm, Ghirlandajo, Redvers, KTC, Nuno Tavares, Benji2~enwiki, Jeffrey O. Gustafson, OwenX, Woohookitty, Camw, Benbest, Scjessey, Lenar, Dionyziz, Karmosin, Kralizec!, MarcoTolo, Marudubshinki, Emerson7, Mandarax, Graham87, Deltabeignet, FreplySpang, DePiep, Edison, Canderson7, Ketiltrout, Sjö, Rjwilmsi, Lordkinbote, Ligulem, Peripatetic, Olessi, Matt Deres, Ecelan, Titoxd, Jwkpiano1, FlaBot, SchuminWeb, RobertG, Nihiltres, Itinerant1, Kmorozov, Anzelm, JYOuyang, RexNL, Gurch, Wars, RasputinAXP, GreyCat, Russavia, Physchim62, Mallocks, Imnotminkus, Introvert, Chobot, Bgwhite, Cactus.man, Gwernol, EamonnPKeane, The Rambling Man, YurikBot, Wavelength, Vuvar1, Brandmeister (old), Pip2andahalf, Rylz, Cyberherbalist, NAveryW, Conscious, Splash, Akamad, Alex Bakharev, Wimt, Lusanaherandraton, Ornilnas, NawlinWiki, Wiki alf, Bachrach44, Aeusoes1, LaszloWalrus, Howcheng, Arima, Tomburbine, Ragesoss, Anetode, Ravedave, Raven4x4x, Sfnhltb, Zwobot, Aaron Schulz, BOT-Superzerocool, DeadEyeArrow, Private Butcher, 2over0, Lt-wiki-bot, Closedmouth, Donald Albury, GraemeL, Dr U, JoanneB, Bandurist, Ordinary Person, T. Anthony, Mikus, Nixer, Kungfuadam, Thomas Blomberg, GrinBot~enwiki, Lunch, Itub, Attilios, SmackBot, Monkeyblue, Haverpopper, Bobet, Brianyoumans, InverseHypercube, Royalguard11, Kitchka, Delldot, Eskimbot, Kintetsubuffalo, Gilliam, Ohnoitsjamie, Master Jay, Bluebot, Keegan, F382d56d7a18630cf764a5b576ea1b4810467238, MalafayaBot, BrendelSignature, H i-c h-a M~enwiki, DHN-bot~enwiki, Colonies Chris, Darth Panda, Marblefluss, DTR, Can't sleep, clown will eat me, Scott3, Eschbaumer, Writtenright, Chlewbot, Rrburke, GeorgeMoney, Addshore, Nuklear, Khoikhoi, Jmlk17, Emact, BostonMA, Dirk gently~enwiki, Khukri, Nakon, Coolag12345, Derek R Bullamore, Badgerpatrol, Dantadd, DMacks, Er Komandante, Jóna Pórunn, Springnuts, Vina-iwbot~enwiki, RossF18, Drmaik, Ohconfucius, Pinktulip, SashatoBot, Swatjester, JzG, John, Kipala, Breno, This user has left wikipedia, Barry Kent~enwiki, Thraxas, IronGargoyle, Frokor, Anatopism, Ryulong, Manifestation, Dr.K., MTSbot~enwiki, Fromeout11, Angryxpeh, Hu12, Joseph Solis in Australia, Suresh K. Sheth, Domitori, Blehfu, Murf661, Esn, Tawkerbot2, Chelydra, Lbr123, Emote, JForget, Mellery, Meisam.fa, Unionhawk, Scohoust, KyraVixen, Nczempin, Orayzio, Kylu, Moreschi, Chicheley, ElPoojmar, Cydebot, Kanags, Galassi, Steel, Scottiscool, Doomed Rasher, Gogo Dodo, JFreeman, Corpx, Booty3535, Tkynerd, Christian75, Codetiger, DumbBOT, Omicronpersei8, Danielil, Lordhatrus, Epbr123, Divyangmithaiwala, CopperKettle, Pampas Cat, Marek69, John254, NorwegianBlue, James86, Yettie0711, Jonny-mt, Zachary, Lithpiperpilot, SusanLesch, Natalie Erin, Hempfel, Escarbot, Ilion2, Thadius856, AntiVandalBot, RobotG, Fedayee, Eamezaga, Richiel101, Jj137, NSH001, Modernist, Malcolm, Spartaz, Phanerozoic, Caper13, Ioeth, JAnDbot, Husond, MER-C, Matthew Fennell, Plm209, Andonic, Connormah, VoABot II, JNW, Kajasudhakarababu, Think outside the box, Caroldermoid, CTF83!, Waacstats, Eldumpo, Dirac66, 28421u2232nfenfcenc, Rebecca777, Glen, DerHexer, JaGa, Edward321, Lelkesa, Hbent, MartinBot, BetBot~enwiki, Arjun01, Jaystar11, Rettetast, Kateshortforbob, CommonsDelinker, LittleOldMe old, John Duncan, Tgeairn, J.delanoy, Greenflower89, Nev1, Filll, DrKay, Trusilver, Skeptic2, Hans Dunkelberg, Maurice Carbonaro, Yonidebot, Derwig, Smeira, Petersec, Ephebi, Mikhail Dvorkin, NewEnglandYankee, Mufka, ObseloV, Entropy, Burzmali, Spitacular, MishaPan, CardinalDan, Pqwo, Matty202033, To my arse, Sumo su, Wikieditor06, Lights, Timotab, VolkovBot, Mp3boy3239, Jeff G., AlnoktaBOT, Philip Trueman, Muut21, TXiKiBoT, Slvrstn, Dale134, Kovbasa, Rei-bot, Miguel Chong, Qxz, Minol, Anna Lincoln, Lradrama, Seungfire, Broadbot, Farever, LeaveSleaves, Xelda, Maroonedsorrow, Luuva, Duncan.Hull, Maxim, Plazak, Ashnard, David Marjanović, Rhopkins8, T0nyM0ntana 420, Jamiemouse90, BrianY, Jjdon, AlleborgoBot, Symane, Carnelain, Tvinh, Pediaknowledge, Billytrousers, Demmy, Sergwiki, Rozmysl, SieBot, Ttonyb1, PlanetStar, Tiddly Tom, Scarian, WereSpielChequers, BotMultichill,

Gerakibot, Seeyardee, LeadSongDog, GrooveDog, Arbor to SJ, Dwiakigle, Monegasque, Oxymoron83, Ddxc, KoshVorlon, Seth Whales, Vojvodaen, MadmanBot, Gb-oh6, Mygerardromance, Denisarona, Kanonkas, Samcristiano7, SallyForth123, Mr. Granger, Cjc15153, Atif.t2, Loren.wilton, Martarius, ClueBot, Snigbrook, The Thing That Should Not Be, All Hallow's Wraith, Techdawg667, FileMaster, Tomas e, Drmies, Uncle Milty, Polyamorph, CounterVandalismBot, Megam~enwiki, Polker03, Kgandrews, Mindmelter, DragonBot, Ktr101, Excirial, Christine1107, Rohbat, Vivio Testarossa, Eemyaj, Jotterbot, Njardarlogar, Diehard4.0, Lenary, SchreiberBike, Thehelpfulone, Thingg, G3421hi, Aitias, 05bysstern, DumZiBoT, XLinkBot, Fastily, Nathan Johnson, BodhisattvaBot, Dthomsen8, Nicki.The.Random.Chick, Little Mountain 5, Kwjbot, Good Olfactory, Airplaneman, Kbdankbot, HexaChord, Decanc, Addbot, Sarah jamieson, Jojhutton, Ronhjones, KorinoChikara, CanadianLinuxUser, NjardarBot, Cst17, Mentisock, CarsracBot, Tangoed1whiskey, Cheesepieman, Wurk a kurk, Debresser, Favonian, Tubesidiom, Numbo3-bot, Tide rolls, Tressor, David0811, Greyhood, Legobot, Luckas-bot, Yobot, Veraladeramanera, TaBOT-zerem, Heidas, II MusLiM HyBRiD II, Tavy08, Tilnakk, Widey, Jobroluver98, IW.HG, Tempodivalse, DiverDave, AnomieBOT, Rubinbot, Götz, Jim1138, IRP, AdjustShift, Kingpin13, Seanaloisi, Flewis, Citation bot, Geregen2, Weirdskateguy, Neurolysis, ArthurBot, PhilAnG, Stewart96, Parthian Scribe, Xqbot, Mick10793, I have a cool name, Agleeson, Capricorn42, Drilnoth, Renaissancee, Notmax, Maxximus514, HUZZAH123, Omnipaedista, Ducki17, RibotBOT, SassoBot, Chris.urs-o, Nedim Ardoğa, GhalyBot, Ninja Scaley, The Sceptical Chymist, Thermokarst, FrescoBot, Wikipe-tan, Lagelspeil, VS6507, Happykg, BenzolBot, Jamesooders, GrayScaleRainbow, Vlp92, Louperibot, Citation bot 1, Eightofnine, Ntse, RMN1390, Bryant james, Traleo, Abductive, Rameshngbot, Plucas58, Tomcat7, SpaceFlight89, Fixer88, Tlhslobus, PrinceRegentLuitpold, Merlion444, FoxBot, Double sharp, TobeBot, نوری ناراس,ئ Lotje, Javierito92, Dinamik-bot, Vrenator, Actoreng1, Earthandmoon, Brambleclawx, Kidkaos707, DARTH SIDIOUS 2, RjwilmsiBot, Ripchip Bot, DASHBot, EmausBot, GregZak, Wikipelli, HiW-Bot, ZéroBot, StringTheory11, Plotfeat, Battoe19, Zloyvolsheb, Masteratc, Kafern, Fanofnaruto2, Alfio66, JeanneMish, Mayur, Orange Suede Sofa, Negovori, Chuispaston-Bot, Xero200, UDHAUEIFD, Xanchester, ClueBot NG, Jnorton7558, Goose friend, Mpaa, ساجد امجد ساجد, Helpful Pixie Bot, Sceptic1954, ଓଡ଼ିଆ କାଠିଆ, Athkalany, Wald, Calabe1992, Vagobot, Sergeispb-10, OttawaAC, The Almighty Drill, TheEditor12345, Writ Keeper, Snow Blizzard, Uikmi, Lizzyah, Unknownkarma, Wrath X, Vassto, Ninmacer20, JYBot, Ukrained2012, Jamesx12345, Passengerpigeon, HobbyGD, Monkbot, Acagastya, FalconJackson, Jonarnold1985, Ghiutun, Josjos69, KasparBot and Anonymous: 805

• **Periodic table** *Source:* https://en.wikipedia.org/wiki/Periodic_table?oldid=686835311 *Contributors:* AxelBoldt, Dreamyshade, Chuck Smith, Lee Daniel Crocker, Mav, Bryan Derksen, Timo Honkasalo, The Anome, Tarquin, DanKeshet, Rjstott, Andre Engels, XJaM, Christian List, PierreAbbat, Heron, Fonzy, Youandme, Olivier, Someone else, Bob Jonkman, Patrick, Infrogmation, Michael Hardy, Erik Zachte, TMC, Kwertii, Dan Koehl, Shellreef, Taras, Wapcaplet, Ixfd64, Dcljr, Tomi, Eric119, Kosebamse, Egil, Mdebets, Ahoerstemeier, Stan Shebs, Ronz, Jpatokal, Theresa knott, Snoyes, Suisui, Den fjättrade ankan~enwiki, Kragen, Salsa Shark, Cyan, Stefan-S, Poor Yorick, Kwekubo, Jiang, Eirik (usurped), Mxn, BRG, Smack, Schneelocke, Jengod, Okome~enwiki, Emperorbma, EL Willy, Eszett, Adam Bishop, Reddi, Stone, Piolinfax, Dtgm, Selket, Tpbradbury, Rarb, Maximus Rex, Nv8200pa, Tempshill, Bevo, Traroth, Shizhao, Stormie, Dpbsmith, Bcorr, Secretlondon, Jusjih, Just another user 2, Darthchaos, Jeffq, Lumos3, Denelson83, Jni, Nofutureuk, Gromlakh, Gentgeen, Robbot, Phisite, Juve82, Fredrik, Chris 73, WormRunner, Altenmann, Romanm, Naddy, Lowellian, WebElements, Yosri, Rfc1394, Texture, Hippietrail, Caknuck, Bkell, David Edgar, Borislav, Eliashedberg, Radagast, David Gerard, Giftlite, DocWatson42, Haeleth, Ævar Arnfjörð Bjarmason, Tom harrison, Lupin, Everyking, Bkonrad, No Guru, NeoJustin, Bensaccount, Zaphod Beeblebrox, AJim, Avsa, Yekrats, Dmmaus, Archenzo, Brockert, Darrien, SWAdair, Bobblewik, Deus Ex, Edcolins, Lucky 6.9, Peter Ellis, Gadfium, Zed0, Ran, Antandrus, Ctachme, PDH, Jossi, Exigentsky, Kesac, Vbs, Icairns, Sam Hocevar, Clemwang, Karl Dickman, Adashiel, Iwilcox, EagleOne, Mike Rosoft, Alkivar, D6, Andrew11, Poccil, Zarxos, EugeneZelenko, Felix Wan, A-giau, Noisy, Discospinster, Rich Farmbrough, KarlaQat, Cacycle, Inkypaws, Vsmith, Samboy, Joeclark, SpookyMulder, Bender235, TerraFrost, Sunborn, Klenje, RJHall, El C, Kwamikagami, Shanes, Briséis~enwiki, RoyBoy, Femto, Semper discens, Grick, Bobo192, AlHalawi, Whosyourjudas, Nyenyec, Reinyday, Clawson, Cwolfsheep, Dbchip, Giraffedata, SpeedyGonsales, Jojit fb, Nk, Eddideigel, Conget~enwiki, Jhd, Conny, Stephen G. Brown, Danski14, Honeycake, Orzetto, Alansohn, Mo0, Atlant, Keenan Pepper, Plumbago, Sl, Damnreds, AzaToth, Mac Davis, Caesura, Blobglob, Wtmitchell, ClockworkSoul, Unconventional, Helixblue, Stephan Leeds, Harej, RJFJR, Skatebiker, Computerjoe, GabrielF, Ghirlandajo, HGB, Feline1, Weyes, Lucent, Philthecow, Cimex, TigerShark, Benbest, Mpatel, Schzmo, U10ajf, Bluemoose, CharlesC, Waldir, SeventyThree, EarthmatriX, MarcoTolo, Cataclysm, V8rik, Qwertyus, Kbdank71, FreplySpang, DePiep, Dwaipayanc, Canderson7, Drbogdan, Saperaud~enwiki, Angusmclellan, Joe Decker, Koavf, Oblivious, SeanMack, Shalmanese, Sango123, Ptdecker, Yamamoto Ichiro, RobertG, Pumeleon, Nivix, Pathoschild, RexNL, Gurch, Kolbasz, Brendan Moody, Scerri, Alphachimp, Kri, Dalta~enwiki, Glenn L, Physchim62, Imnotminkus, Chobot, Visor, Jared Preston, DVdm, Bgwhite, Gwernol, EamonnPKeane, Roboto de Ajvol, Mercury McKinnon, YurikBot, Wavelength, Hairy Dude, Deeptrivia, Phantomsteve, RussBot, Vlad4599, Fabartus, SpuriousQ, IanManka, Stephenb, Rintrah, Alvinrune, Schoen, Rsrikanth05, Bovineone, Wimt, Stassats, Anomalocaris, EngineerScotty, NawlinWiki, Wiki alf, E123, Test-tools~enwiki, Jaxl, Terfili, Yahya Abdal-Aziz, Mkouklis, Nick, Ragesoss, Dhollm, Cholmes75, Dmoss, Matticus78, RUL3R, AdiJapan, Ryanminier, Juanpdp, Hv, Misza13, Beanyk, Aaron Schulz, Bota47, CorbieVreccan, Derek.cashman, DRosenbach, Elkman, Phaedrus86, Smaines, Wknight94, Tetracube, FF2010, Ageekgal, Closedmouth, Jwissick, Ketsuekigata, Sean Whitton, Petri Krohn, DGaw, CWenger, Smurrayinchester, Kungfuadam, Junglecat, RG2, NeilN, DVD R W, Itub, Thecroman, SmackBot, Android 93, Bobet, Reedy, InverseHypercube, KnowledgeOfSelf, TestPilot, Melchoir, Unyoyega, KocjoBot~enwiki, Davewild, Thunderboltz, Milesnfowler, Anastrophe, Delldot, J0lt C0la, Knowhow, Elk Salmon, Edgar181, Half-Shadow, Eupedia, Srnec, Gilliam, Ohnoitsjamie, Skizzik, Carbon-16, JRSP, Chris the speller, Keegan, Iskander32, RDBrown, Thumperward, Fuzzform, Lollerskates, EncMstr, MalafayaBot, OrangeDog, Roscelese, Bonaparte, Xoyorkie13, Metacomet, Dustimagic, DHN-bot~enwiki, DNAmaster, Darth Panda, Suicidalhamster, Can't sleep, clown will eat me, Onorem, Clorox, Konczewski, Squadoosh, Andy120290, Ddon, DR04, UU, Grover cleveland, Jachapo, PiMaster3, TotalSpaceshipGuy3, Savidan, Dreadstar, Pwjb, Aco47, Peterwhy, Jklin, DMacks, BrotherFlounder, Suidafrikaan, Sadi Carnot, The undertow, SashatoBot, Mchavez, Nishkid64, Tarantola, LtPowers, Archimedead, Khazar, Vitall, Scientizzle, Gobonobo, Btg2290, Anoop.m, Olin, ManiF, JohnWittle, Moop stick, Jaywubba1887, Ckatz, Dale101usa, Chrisch, Garudabd, Digger3000, Slakr, Rainwarrior, Beetstra, Mr Stephen, AxG, Arkrishna, Mets501, Ambuj.Saxena, Ryulong, RichardF, Jose77, DGtal, WOWGeek, Sifaka, Asyndeton, Ramuman, Dead3y3, Michaelbusch, Walton One, Tabfugnic, David Little, J Di, CapitalR, DavidOaks, Supertigerman, Pearson3372, Az1568, Courcelles, Túrelio, Ziusudra, Dpeters11, Tawkerbot2, Bobby131313, VinceB, Cryptic C62, Kaischwartz, Lincmad, TranClan, JForget, Betaeleven, Deon, Eli84, Van helsing, NullAshton, CBM, Rawling, DSachan, GHe, Fork me, Egmonster, Black and White, FlyingToaster, Wikiman7~enwiki, WeggeBot, Logical2u, Pi Guy 31415, Johnlogic, MrFish, Bill Sayre, Dmsc893, Rudjek, Nebular110, Reywas92, Grahamec, MC10, Rasmus vendelboe, Vanished user vjhsduheuiui4t5hjri, Rifleman 82, Corpx, GeorgeTopouria, Islander, Mycroft.Holmes, Methyl~enwiki, Fifo, Christian75, DumbBOT, Chrislk02, Shrikethestalker, Taylor4452, Ndufour, Memorymike, JodyB, Rowlaj01, Calvero JP, Satori Son, Casliber, Thijs!bot, Full On, Barticus88, ShayneRyan, Opabinia regalis, Kiwi137, Corsair18, Dagrimdialer619, Headbomb, Sobreira, Marek69, Ydoommas, John254, Racantrell, Dmitri Lytov, Philippe, Nezzington, Nemti, Escarbot, Mentifisto, Lani123, Tom dl, AntiVandalBot, Luna Santin, Michael phan, Bigtimepeace, Random user 8384993, Chill doubt, LegitimateAndEvenCompelling, Myanw, Figma, Tomertomer,

JAnDbot, Barek, MER-C, Zerotjon, Sanchom, Hut 8.5, Kirrages, Kerotan, Maurakt, Magioladitis, Canjth, Bongwarrior, VoABot II, JNW, Kinston eagle, Redaktor, Aa35te, SparrowsWing, Avicennasis, Superworms, Ahecht, Nposs, Wikiak, Dirac66, Adrian J. Hunter, Allstarecho, ChrisSmol, StuFifeScotland, User A1, Musicloudball, Cpl Syx, Vssun, Just James, DerHexer, Khalid Mahmood, TheRanger, DancingPenguin, FisherQueen, Hdt83, MartinBot, HLewis, PostScript, Rettetast, Roastytoast, 1993 lol, Glrx, Kateshortforbob, CommonsDelinker, Supia, PrestonH, Thomasrive, Exodecai101, Slash, J.delanoy, Pharaoh of the Wizards, Ilovestars89, ChickenMarengo, Hans Dunkelberg, Psycho Kirby, Smartweb, I2yu, Sergeibernstein, Tempnegro, Extransit, WarthogDemon, Munkimunki, Richard777, Thom.fynn, Tdadamemd, Yvonr, Adamsbriand, EH74DK, Rescorbic, McSly, Aonrotar, Ryan Postlethwaite, Ephebi, Notapotato, Gurchzilla, Wasitgood69, Wasitgood, Paulbkirk, Monkeybutt5423, AntiSpamBot, Gffootball58, Coin945, Bigsnake 19, NewEnglandYankee, Creator58, Joka1991, SJP, C0RNF1AK35, Shoessss, Bob, Vanished user 39948282, BrianScanlan, Nat682, Darklama, Hyuuganeji0123, Arjun Rana, AnjuX, Suuperturtle, Idioma-bot, Johnnieblue, Deor, 28bytes, VolkovBot, Thedjatclubrock, Iosef, Jmocenigo, Christophenstein, Jeff G., JohnBlackburne, TheOtherJesse, RemoteCar, Barneca, Philip Trueman, Af648, Drunkenmonkey, Sweetness46, TXiKiBoT, TheVault, A4bot, Quilbert, Caster23, GDonato, Miranda, Chrisk12, Sankalpdravid, Qxz, Littlealien182, Anna Lincoln, Corvus cornix, Martin451, Jackfork, LeaveSleaves, Andrewrost3241981, DBragagnolo, Luuva, Quindraco, Gona.eu, RadiantRay, Madhero88, Jinglesmells999, Finngall, Aciddoll, Superjustinbros., Deeryh01, Synthebot, Enviroboy, Chengyq19942007, Insanity Incarnate, Everybody's Got One, Why Not A Duck, Brianga, Jaybo007, HiDrNick, LuigiManiac, Petergans, ConnTorrodon, NHRHS2010, SieBot, Coffee, Mikemoral, TJRC, Rihanij, PlanetStar, Jmwwiki, Borgdylan, Gprince007, Tiddly Tom, Scarian, WereSpielChequers, Jauerback, Jack Merridew, Gerakibot, Dawn Bard, Viskonsas, Caltas, Kragenz, The way, the truth, and the light, Michfg, Sat84, Whiteghost.ink, Til Eulenspiegel, Purbo T, Tiptoety, Exert, Elcobbola, Nopetro, Hamilton hogs, Hiddenfromview, Segalsegal, SquirrelMonkeySpiderFace, Oxymoron83, Nuttycoconut, Lightmouse, Hwn tls, Hak-kâ-ngìn, BenoniBot~enwiki, Jack the Stripper, Sunrise, Dillard421, Werldwayd, Pappapasd, Maelgwnbot, DixonD, Nergaal, Precious Roy, Escape Orbit, Into The Fray, Jimmy Slade, Kanonkas, Georgedriver, Mr. Granger, Twinsday, ClueBot, PipepBot, Dinamik, The Thing That Should Not Be, Apastrophe, Techdawg667, Gawaxay, Syhon, Thompsontm, Drmies, Polyamorph, Elsweyn, Ryoutou, CounterVandalismBot, Ansh666, Blanchardb, LizardJr8, Dylan620, Timex987, Wifiless, Puchiko, Natasha.fielding, Dlorang, Robert Skyhawk, Excirial, Alexbot, BirgerH, Omgosh2, Jazjaz92, NuclearWarfare, Pearrari, Kaeso Dio, Jotterbot, LarryMorseDCOhio, Psinu, Realm up, DeltaQuad, Kaiba, Dekisugi, Pwntskater, NolanRichard, Thehelpfulone, La Pianista, Another Believer, Kiran the great, Viper275, Aitias, Blargblarg89, Versus22, Sch00l3r, SoxBot III, MairAW, JDT1991, BalkanFever, Vanished user uih38riiw4hjlsd, Indopug, TimothyRias, Jean-claude perez, Neuralwarp, XLinkBot, Shpakovich, Gonzonoir, Nsimya, TheSickBehemoth, Ryuken14, Nsim, WikHead, Noctibus, JinJian, ZooFari, MystBot, RyanCross, Thatguyflint, Roentgenium111, Tomilee0001, Yoenit, OmgItsTheSmartGuy, NicholasSThompson, JenR32, Vchorozopoulos, WFPM, Cst17, Download, SoSaysChappy, EconoPhysicist, Mathmarker, Glane23, Plutonium55, Debresser, LinkFA-Bot, Drova, LiveAgain, PoliteCarbide, Numbo3-bot, Tide rolls, BrianKnez, Luckas Blade, Greyhood, Arbitrarily0, Angrysockhop, Jack who built the house, Luckas-bot, Yobot, Essam Sharaf, Azylber, KamikazeBot, Jobroluver98, EnDaLeCoMpLeX, MacTire02, Tempodivalse, AnomieBOT, Zhieaanm, Helixer, Rubinbot, Jake Fuersturm, Degg444, Daniele Pugliesi, Piano non troppo, Collieuk, AdjustShift, Kingpin13, Ulric1313, Materialscientist, RadioBroadcast, The High Fin Sperm Whale, Citation bot, ArthurBot, Carturo222, Xqbot, Ziaix, Timir2, Sionus, Gopal81, Vidshow, Tfts, YBG, Grim23, Srich32977, WingedSkiCap, Pirateer, GrouchoBot, RibotBOT, Ian Fraser at Temple Newsam House, Spesh531, FaTony, AlimanRuna, A. di M., Nolimits5017, Dougofborg, Thehelpfulbot, R8R Gtrs, FrescoBot, Ylime715, Dogposter, StaticVision, Michael93555, Car132, Zach112233, Weetoddid, Strongbadmanofme, Maxus96, Grandiose, Xhaoz, Dellacomp, Citation bot 1, Pezzells, Pshent, Redrose64, ArnaudContet, Pinethicket, I dream of horses, M.pois, Hamtechperson, BTolli, Gemmi3, RedBot, Fishekad, NarSakSasLee, Kangxi emperor6868, Hardwigg, Abc518, Tjlafave, Lightlowemon, Gamewizard71, FoxBot, Double sharp, Adult Swim Addict, TobeBot, Graniggo, Jaeger Lotno, Pitcroft, Sillyboy67, Hughbert512369, Dinamik-bot, BZRatfink, Fyandcena, Kelvin35, Mattvirajrenaudbrandon, Turn off 2, Imawsome 09, Mass09, Tbhotch, Pldx1, Naughtysriram, DARTH SIDIOUS 2, Luhar1997, Twonernator, Pickweed, Dancojocari, EmausBot, Sir Arthur Williams, WikitanvirBot, GA bot, Franjklogos, Tf1321, Tommy2010, P. S. F. Freitas, Kitrkatr, 以前以前以前以前, Zainiadragon10000, JSquish, Jakers69, StringTheory11, Dffgd, Cobaltcigs, Шуруян, Joshlepaknpsa, Hzb pangus, Elly4web, Arman Cagle, KotVa, Sven nestle2, Hh73wiki, Ego White Tray, Negovori, ChuispastonBot, GermanJoe, Sunshine4921, Rmashhadi, Chaotic iak, Whoop whoop pull up, Heidslovesearl, Xanchester, ClueBot NG, Ihakeycakeyabreak, Wd930, Ozkithar Salas, Manubot, CocuBot, Lanthanum-138, Frietjes, Hazhk, Moneya, Parcly Taxel, Metaknowledge, Willzuk, Helpful Pixie Bot, RobertGustafson, ಪಳ್ಳಿ ಕಾವಿ‌ಪ್ರೆ, Curb Chain, Bibcode Bot, BG19bot, MKar, Vagobot, Nagasturg, Sandbh, Mark Arsten, IraChesterfield, Soerfm, Zedshort, FeralOink, Razzat99, SoylentPurple, BattyBot, Justincheng12345-bot, Judiakok1985, Ziggypowe, Ushau97, Maxronnersjo, ChrisGualtieri, JYBot, Dexbot, LightandDark2000, Aditya Mahar, Mogism, Burzuchius, Carpelogos, Jtrevor99, Leitoxx, Kazim5294, AmericanLemming, WikiEditor2563, 08adamsm, Michel Djerzinski, The Herald, Shearflyer, Quenhitran, Asadwarraich, Kind Tennis Fan, Monkbot, HiYahhFriend, Deepak harshal nagle, Trackteur, Encyclopedia Lu, IiKkEe, Shane Stachwick, Mogie Bear, R. Portela F., Forscienceonly, KasparBot, Alistairgray42, Equinox, Lexi sioz and Anonymous: 1391

- **History of the periodic table** *Source:* https://en.wikipedia.org/wiki/History_of_the_periodic_table?oldid=687172027 *Contributors:* AxelBoldt, The Anome, William Avery, Ewlloyd, D, Dcljr, Eric119, Ellywa, Ahoerstemeier, Stone, Jerzy, Gentgeen, Fredrik, Korath, Chris 73, Altenmann, Romanm, Arkuat, SoLando, Diberri, Everyking, Tweenk, Pne, Rrw, Alexf, Goog, Fredcondo, Ruzulo, DragonflySixtyseven, RetiredUser2, PFHLai, Icairns, Deglr6328, Mike Rosoft, Discospinster, Vsmith, Bishonen, SpookyMulder, Rubicon, Bletch, Bobo192, Fir0002, Smalljim, Honeycake, Wtmitchell, Helixblue, RainbowOfLight, Dominic, Eztli, Feline1, Kazvorpal, Angr, Linas, DonPMitchell, Camw, MarcoTolo, Allen3, RuM, DePiep, -DjD-, Drbogdan, HappyCamper, Ems57fcva, Bubba73, The wub, Bhadani, Dracontes, Titoxd, TheMidnighters, Gurch, Scerri, Physchim62, DVdm, Cactus.man, Jimp, Kordas, Conscious, Wimt, NawlinWiki, Shreshth91, Grafen, Darkmeerkat, DGJM, Lockesdonkey, Kkmurray, Fabiob~enwiki, Wknight94, Zzuuzz, Closedmouth, Dspradau, Petri Krohn, Kevin, Itub, SmackBot, MattieTK, Unschool, Reedy, KnowledgeOfSelf, TestPilot, VigilancePrime, Hydrogen Iodide, Davewild, Cool3, Gilliam, Skizzik, Persian Poet Gal, Bduke, Jprg1966, SchfiftyThree, Darth Panda, Jahiegel, Eliyahu S, Neo139, Pieter1, Konczewski, RedHillian, AiOlorWile, Edwtie, RJN, DMacks, Kuru, John, Microchip08, Jaganath, Shlomke, JoshuaZ, Noah Salzman, Arkrishna, DabMachine, Mikehelms, Tawkerbot2, Dgw, Black and White, Glenn4pr, Equendil, Krauss, Christian75, DumbBOT, Danogo, Thijs!bot, Mojo Hand, Purple Paint, Headbomb, Marek69, Kathovo, Majorly, Luna Santin, RapidR, Shift6, Wan nni, Gökhan, TuvicBot, JAnDbot, MER-C, Kerotan, .anacondabot, Bongwarrior, VoABot II, Romtobbi, Dirac66, Thibbs, DerHexer, Davidbws23, Amitchell125, SquidSK, Rikpotts, Rettetast, J.delanoy, Rhinestone K, Uncle Dick, Tdadamemd, DoubleParadox, Ephebi, Juliancolton, IceDragon64, DarkNiGHTs, X!, Jeff G., JohnBlackburne, Philip Trueman, Gigo12, Anna Lincoln, Gekritzl, Leafyplant, One half 3544, BigDunc, Vector Potential, Insanity Incarnate, HeirloomGardener, AlleborgoBot, Logan, LuigiManiac, Jehorn, Graham Beards, Caltas, Matthew Brandon Yeager, Keilana, Aillema, Happysailor, Cat1993127, Oxymoron83, Aflumpire, AnonGuy, OKBot, Anchor Link Bot, Nergaal, Pmigdal, Elassint, ClueBot, The Thing That Should Not Be, S Levchenkov, Pi zero, CounterVandalismBot, Delta1989, Pointillist, Djr32, Excirial, Pumpmeup, Attheweaneiffer, PixelBot, Drsrisenthil, Razorflame, SchreiberBike, BOTarate, Versus22, Epiclolz, Romaine, PCHS-

NJROTC, SoxBot III, Jean-claude perez, Dark Mage, Avoided, Airplaneman, Kinokaru, Mootros, Laurinavicius, Vishnava, Fluffernutter, Download, Favonian, Drova, Ehrenkater, Tide rolls, Narayan, MissAlyx, Finbob83, Cflm001, Cottonshirt, Paul Siebert, Eric-Wester, Synchronism, AnomieBOT, Rubinbot, Ginyild, Piano non troppo, Aditya, Materialscientist, The High Fin Sperm Whale, Srinivas, Alexphudson, LilHelpa, Xqbot, Zad68, S h i v a (Visnu), Mattstead93, 78.26, Ignoranteconomist, Clementisbad, Amrosabra, A.amitkumar, Dougofborg, Eyea0012, Liamhaha, Paine Ellsworth, VS6507, Pinethicket, I dream of horses, Serols, Merlion444, Utility Monster, Cnwilliams, Tim1357, Double sharp, Tubby23, Vrenator, TBloemink, Oenrhysbiggs, Extra999, Allen4names, Aoidh, Laurielaurielaurie, Hornlitz, DARTH SIDIOUS 2, Onel5969, NerdyScienceDude, DASHBot, Orphan Wiki, Acather96, Immunize, Sponk, RenamedUser01302013, Ilikefod, Tommy2010, Wikipelli, K6ka, Anirudh Emani, PS., Dfern22, Fæ, StringTheory11, Kieran Nash, Openstrings, Flightx52, MonoAV, Donner60, Negovori, DASHBotAV, Petrb, ClueBot NG, Satellizer, Alex Nico, Widr, Antiqueight, TORNELLcello, BG19bot, Betty Noire, Wiki13, Soerfm, Paolo Raneses, Sni56996, Klilidiplomus, Razzat99, Simeondahl, Hghyux, Mrt3366, BlaBlaBaberBabe, Mediran, MadGuy7023, TBBT Chase, Dexbot, Sam 365, Lugia2453, SFK2, Little green rosetta, Kevin12xd, Bangladesh News, Epicgenius, Sarsarpow, I am One of Many, Tentinator, George8211, Ameer Hasan Khan, JaconaFrere, BillyTanjung, Polymathica, Filedelinkerbot, Ash.flowers.palad, ArdentWhiteraven, Dude128123, Jacobo95, Zondaj, ChemWarfare, KSFT, 420BlazeIt69Sex, Thefalsehistorybuff and Anonymous: 586

- **Statistical mechanics** *Source:* https://en.wikipedia.org/wiki/Statistical_mechanics?oldid=683967942 *Contributors:* The Cunctator, Derek Ross, Bryan Derksen, The Anome, Ap, Miguel~enwiki, Peterlin~enwiki, Edward, Patrick, Michael Hardy, Tim Starling, Den fjättrade ankan~enwiki, Bogdangiusca, Mxn, Charles Matthews, Phys, Nnh, Eman, Fuelbottle, Isopropyl, Cordell, Ancheta Wis, Giftlite, Andries, Mikez, Monedula, Alison, Tweenk, John Palkovic, Karol Langner, APH, Karl-Henner, Edsanville, Michael L. Kaufman, Chris Howard, Brianjd, Bender235, El-wikipedista~enwiki, Linuxlad, Jumbuck, Ryanmcdaniel, BryanD, PAR, Jheald, Woohookitty, Linas, StradivariusTV, Kzollman, Pol098, Mpatel, SDC, DaveApter, Nanite, Rjwilmsi, HappyCamper, FlaBot, Margosbot~enwiki, Gurch, Fephisto, GangofOne, Sanpaz, YurikBot, Wavelength, The.orpheus, DiceDiceBaby, JabberWok, Mary blackwell, Dhollm, E2mb0t~enwiki, Aleksas, Teply, That Guy, From That Show!, SmackBot, Pavlovič, Charele, Jyoshimi, Weiguxp, David Woolley, Edgar181, Drttm, Steve Omohundro, Skizzik, DMTagatac, ThorinMuglindir, Kmarinas86, Bluebot, MK8, Complexica, Sbharris, Wiki me, Phudga, Radagast83, RandomP, G716, Sadi Carnot, Yevgeny Kats, Lambiam, Chrisch, Frokor, Mets501, Politepunk, Iridescent, IvanLanin, Daniel5127, Van helsing, Djus, Mct mht, Cydebot, Forthommel, Boardhead, Dancter, Joyradost, Christian75, Abtract, Thijs!bot, Headbomb, Spud Gun, Samkung, Alphachimpbot, Perelaar, Chandraveer, JAnDbot, Yill577, Magioladitis, WolfmanSF, VoABot II, Dirac66, Jorgenumata, Peabeejay, SimpsonDG, Lantonov, Sheliak, Gerrit C. Groenenboum, VolkovBot, Scorwin, LokiClock, The Original Wildbear, Agricola44, Moondarkx, Locke9k, PhysPhD, Anoko moonlight, Kbrose, SieBot, Damorbel, LeadSongDog, Melcombe, StewartMH, Apuldram, Plastikspork, Razimantv, Mild Bill Hiccup, Davennmarr, Vql, Lyonspen, Djr32, CohesionBot, Brews ohare, Mlys~enwiki, Doprendek, SchreiberBike, Thingg, Edkarpov, Qwfp, JKeck, Koumz, TravisAF, Truthnlove, Addbot, Xp54321, DOI bot, Wickey-nl, Looie496, Netzwerkerin, ⸮, SPat, Gail, Loupeter, Yobot, Ht686rg90, TaBOT-zerem, ^musaz, Xqbot, P99am, Charvest, Hlfhjwlrdglsp, Baz.77.243.99.32, Anterior1, RjwilmsiBot, Pullister, EmausBot, Michael assis, JSquish, ZéroBot, Wikfr, AManWithNoPlan, Kyucasio, Hpubliclibrary, Keulian, Rashhypothesis, IBensone, RockMagnetist, EdoBot, Amviotd, ClueBot NG, CocuBot, Landregn, Frietjes, Theopolisme, Helpful Pixie Bot, Robwf, PhnomPencil, Op47, Acmedogs, F=q(E+v^B), JZCL, Roshan220195, Egm4313.s12, Illia Connell, Dexbot, Mogism, Mark viking, Alefbenedetti, W. P. Uzer, KeithFratus, Michael Lee Baker, PhilippeTilly, Ա₂ᴨᴜᴨՏᴸᴨ, Udus97, Scientific Adviser, Alakzi, VexorAbVikipædia, KasparBot and Anonymous: 160

- **Thermodynamics** *Source:* https://en.wikipedia.org/wiki/Thermodynamics?oldid=683858986 *Contributors:* Bryan Derksen, Stokerm, Andre Engels, Danny, Miguel~enwiki, Roadrunner, Jdpipe, Heron, Arj, Olivier, Ram-Man, Michael Hardy, Tim Starling, Kku, Menchi, Jedimike, TakuyaMurata, Dgrant, Looxix~enwiki, Ahoerstemeier, CatherineMunro, Glenn, Victor Gijsbers, Jeff Relf, Mxn, Smack, Ehn, Tantalate, Reddi, Lfh, Peregrine981, Eadric, Miterdale, Phys, Fvw, Raul654, Seherr, Mjmcb1, Lumos3, RadicalBender, Rogper~enwiki, Robbot, R3m0t, Babbage, Moink, Hadal, Fuelbottle, Quadalpha, Seth Ilys, Diberri, Ancheta Wis, Giftlite, Mshonle~enwiki, N12345n, Lee J Haywood, Monedula, Wwoods, Dratman, Curps, Michael Devore, Bensaccount, Abqwildcat, Macrakis, Foobar, Physicist, Louis Labrèche, Daen, Antandrus, BozMo, OverlordQ, Karol Langner, APH, H Padleckas, Icairns, Monn0016, Sam Hocevar, MulderX, Agro r, Edsanville, Klemen Kocjancic, Mike Rosoft, Poccil, CALR, EugeneZelenko, Masudr, Llh, Vsmith, Jpk, Pavel Vozenilek, Dmr2, Bender235, Eric Forste, Pmetzger, El C, Hayabusa future, Femto, CDN99, Bobo192, Jung dalglish, SpeedyGonsales, Sasquatch, MPerel, Helix84, Haham hanuka, Pearle, Jumbuck, Ixfalia, Alansohn, Gary, Dbeardsl, Atlant, PAR, Cdc, Malo, Cortonin, Wtmitchell, NAshbery, Docboat, Jheald, Gene Nygaard, Falcorian, Zntrip, Alyblaith, Miaow Miaow, Uncle G, Plek, Carcharoth, Kzollman, Jwulsin, Sympleko, Pkeck, Tylerni7, Jwanders, Keta, Mido, CBdorsett, Dzordzm, Frankie1969, Prashanthns, Mandarax, Graham87, Jclemens, Melesse, Rjwilmsi, DrTorstenHenning, SMC, Ligulem, Dar-Ape, JohnnoShadbolt, Sango123, Dyolf Knip, Titoxd, FlaBot, MacRusgail, RexNL, Jrtayloriv, Lynxara, Thecurran, Srleffler, Chobot, DVdm, Bgwhite, Roboto de Ajvol, The Rambling Man, Siddhant, RobotE, Pip2andahalf, Sillybilly, Anonymous editor, Anubis1975, JabberWok, Casey56, Wavesmikey, Stephenb, Okedem, The1physicist, CambridgeBayWeather, Rsrikanth05, Wiki alf, Hagiographer, UDScott, Nick, Dhollm, Abb3w, DeadEyeArrow, Ms2ger, Spinkysam, Enormousdude, Lt-wiki-bot, Arthur Rubin, Pb30, KGasso, MaNeMeBasat, Banus, RG2, Bo Jacoby, DVD R W, That Guy, From That Show!, Quadpus, Luk, ChemGardener, Vanka5, A13ean, SmackBot, Aim Here, Bobet, C J Cowie, Sounny, Bomac, Jagged 85, Onebravemonkey, Sundaryourfriend, Gilliam, Hmains, Skizzik, ThorinMuglindir, Saros136, Bluebot, Bduke, Silly rabbit, SchfiftyThree, Complexica, DHN-bot~enwiki, Antonrojo, Stedder, Sholto Maud, EvelinaB, HGS, Nakon, Lagrangian, Dreadstar, Richard001, Hammer1980, BryanG, Jklin, DMacks, Sadi Carnot, Kukini, SashatoBot, Ocee, ML5, CatastrophicToad~enwiki, JoseREMY, Nonsuch, Pflatau, Ben Moore, CyrilB, Frokor, Tasc, Beetstra, Waggers, Zεύς, Funnybunny, Negrulio, Peyre, Ejw50, Lottamiata, Shoeofdeath, Mattmaccourt, Ivy mike, Moocowisi, Tawkerbot2, Dlohcierekim, Daniel5127, Deathcrap, Spudcrazy, Meisam.fa, CRGreathouse, Dycedarg, Scohoust, Albert.white, TVC 15, Ruslik0, Dgw, McVities, MarsRover, Freedumb, Casper2k3, Grj23, Cydebot, Gtxfrance, Rifleman 82, Bazzargh, Miketwardos, Shirulashem, Tpot2688, Omicronpersei8, Freak in the bunnysuit, Thijs!bot, MuTau, Barticus88, Bill Nye the wheelin' guy, Coelacan, Knakts, Kablammo, Headbomb, Pjvpjv, Gerry Ashton, James086, D.H, Stannered, Spud Gun, Austin Maxwell, AntiVandalBot, Gioto, Luna Santin, Jnyanydts, FrankLambert, Dylan Lake, JAnDbot, MER-C, Matthew Fennell, Acroterion, Lidnariq, Bongwarrior, VoABot II, JNW, Indon, Loonymonkey, User A1, Pax:Vobiscum, Oneileri, A666666, Jtir, BetBot~enwiki, Mermaid from the Baltic Sea, NAHID, Rettetast, Ravichandar84, R'n'B, LittleOldMe old, Mausy5043, Ludatha, Rhinestone K, Uncle Dick, Maurice Carbonaro, Yonidebot, Brien Clark, Ian.thomson, Dispenser, Katalaveno, MikeEagling, Notreallydavid, AntiSpamBot, Wariner, Nwbeeson, Ontarioboy, Rumpelstiltskin223, WilfriedC, KylieTastic, Bob, Joshmt, Lyctc, Vagr7, Biff Laserfire, CA387, Idioma-bot, Funandtrvl, VolkovBot, Macedonian, Orthologist, Philip Trueman, TXiKiBoT, Reibot, Anonymous Dissident, Sankalpdravid, Baatarchuluun~enwiki, Qxz, Anna Lincoln, CaptinJohn, Sillygoosemo, JhsBot, Leafyplant, Jackfork, Psyche825, Nny12345, Zion biñas, Appieters, Whbstare, Enigmaman, Sploonie, Synthebot, Falcon8765, Enviroboy, Phmoreno, A Raider Like Indiana, Furious.baz, SvNH, Jianni, EmxBot, Kbrose, Arjun024, SieBot, Damorbel, Paradoctor, Jason Patton, LeadSongDog, JerrySteal, Hoax user, Ddsmartie, Bentogoa, Happysailor, Flyer22 Reborn, Dhatfield, BrianGregory86, Oxymoron83, Antonio Lopez, CultureShock582, OK-Bot, Correogsk, Mygerardromance, Hamiltondaniel, JL-Bot, Tomasz Prochownik, Loren.wilton, ClueBot, Namasi, The Thing That Should Not Be, DesertAngel, Taroaldo, Therealmilton, Pak umrfrq, Kdruhl, LizardJr8, Whoever101, ChandlerMapBot, Notburnt, GrapeSmuckers, Aua, Djr32, Jusdafax, LaosLos, Chrisban0314, Pmronchi, Eeekster, Lartoven, Brews ohare, NuclearWarfare, Jotterbot, PhySusie, Scog, Sidsawsome, SoxBot, Razorflame, DEMOLISHOR, CheddarMan, Aitias, Dank, MagDude101, Galor612, Cableman1112, SoxBot III, RexxS, Faulcon DeLacy, Spitfire, Shres58tha, Avoided, Snapperman2, Thatguyflint, Mls1492, Thebestofall007, Addbot, Power.corrupts, DOI bot, Morri028, DougsTech, Patrosnoopy, Glane23, Bob K31416, Numbo3-bot, Landofthedead2, Lightbot, OlEnglish, Gatewayofintrigue, Ben Ben, Luckasbot, Yobot, THEN WHO WAS PHONE?, Bos7, QueenCake, IW.HG, Magog the Ogre, AnomieBOT, Paranoidhuman, IncidentalPoint, Daniele Pugliesi, Jim1138, Flewis, Materialscientist, Celtis123, Citation bot, Fredde 99, LilHelpa, Xqbot, Addihockey10, Capricorn42, Fireballxyz, -), Almabot, GrouchoBot, Omnipaedista, Jezhotwells, Waleswatcher, Logger9, Twested, Chjoaygame, FrescoBot, VS6507, Wallyau, Petr10, Galorr, WikiCatalogEdit701, Sae1962, Denello, Neutiquam, HamburgerRadio, Citation bot 1, Pinethicket, HRoestBot, 10metreh, Calmer Waters, Thermo771, RedBot, MastiBot, Serols, TobeBot, Yunshui, CathySc, Thermodynoman, Thomas85127, Myleneo, Schmei, Brian the Editor, Unbitwise, Sundareshan, DARTH SIDIOUS 2, TjBot, Beyond My Ken, EmausBot, Orphan Wiki, Domesticenginerd, WikitanvirBot, Obamafan70, AriasFco, Helptry, Racerx11, GoingBatty, Nag 08, Your Lord and Master, Weleepoxypoo, Wikipelli, K6ka, John of Lancaster, Hhhippo, Traxs7, Shannon1, Azuris, H3llBot, Libb Thims, Wayne Slam, Tolly4bolly, Vanished user fois8fhow3iqf9hsrlgkjw4tus, EricWesBrown, Mayur, Donner60, Jbergste, DennisIsMe, Haiti333, Hazard-Bot, ChuispastonBot, Levis ken, Matewis1, LaurentRDC, 28bot, Sonicyouth86, Anshul173, ClueBot NG, Coverman6, Piast93, Chester Markel, Andreas.Persson, Chronic21, Jj1236, Duciswrong1234, Suresh 5, Widr, The Troll lolololololololol, NuclearEnergy, Helpful Pixie Bot, Calabe1992, DBigXray, Nomi12892, Necatikaval, BG19bot, Xonein, Krenair, BeRo999, Fedor Babkin, PTJoshua, Balajits93, Defladamouse, MusikAnimal, Metricopolus, Ushakaron, Mariano Blasi, CitationCleanerBot, Hollycliff, Zedshort, Asaydjari, Blodslav, Nascar90210, DarafshBot, Adwaele, Dexbot, Duncanpark, Joeljoeljoel12345, Czforest, Josophie, Miyangoo, Beans098, Reatlas, Rejnej, Nerlost, Epicgenius, Georgegeorge127, Deadmau8****, I am One of Many, Harlem Baker Hughes, Dakkagon, DavidLeighEllis, Vinodhchennu, Ugog Nizdast, Prokaryotes, Eff John Wayne, Ginsuloft, Bubba58, Nanapanners, Hknaik1307, Monkbot, Codebreaker1999, BTHB2010, Bunlip, Zirus101, Xmlhttp.readystate, Qpdatabase, Jayashree1203, Youlikeman, JellyPatotie, Loveusujeet, Isambard Kingdom, Nashrudin13l, Supdiop, The Collapsation of The Sensation, KasparBot, CabbagePotato, Amangautam1995, Лагічна рэвалюцыйны, Ravi.dhami.234, Bishwajeet Panda, HenryGroupman, Dctfgijkm, Paragnar and Anonymous: 872

- **Physical chemistry** *Source:* https://en.wikipedia.org/wiki/Physical_chemistry?oldid=680577510 *Contributors:* Mav, Andre Engels, XJaM, Rmhermen, DavidLevinson, Edward, Ubiquity, SebastianHelm, Looxix~enwiki, Ahoerstemeier, Suisui, Mxn, Lfh, Gentgeen, Robbot, Altenmann, Mayooranathan, Holeung, Giftlite, Bensaccount, LiDaobing, Karol Langner, APH, Tothebarricades.tk, Icairns, D6, Discospinster, Dr. Strangelove~enwiki, Vsmith, Ivan Bajlo, El C, Bobo192, Directorstratton, Maureen, Linuxlad, Walkerma, 3light, HenryLi, Woohookitty, Carcharoth, Palica, Rjwilmsi, Mayumashu, FlaBot, Chobot, Dj Capricorn, Adoniscik, YurikBot, TimNelson, Wavesmikey, Million Little Gods, Ithacagorges, Nick, Jpbowen, Sguzior, Kkmurray, Max Schwarz, Arthur Rubin, GraemeL, JeramieHicks, Pifvyubjwm, Willtron, Itub, Smack-Bot, Unyoyega, Edgar181, M stone, Hmains, JSpudeman, Hugo-cs, Bduke, Fadeev, DHN-bot~enwiki, Hallenrm, Stedder, Suicidalhamster, Smokefoot, DMacks, Sadi Carnot, Mathboy965, RekishiEJ, AlsatianRain, Sir Vicious, Van helsing, BeenAroundAWhile, Black and White, WeggeBot, Neelix, Jowan2005, Yaris678, Rifleman 82, B, Thijs!bot, Publicola, Marek69, Big Bird, JAnDbot, Avaya1, Matthew Fennell, BenB4, LittleOldMe, Magioladitis, Midgrid, Dirac66, Adventurer, Auctor~enwiki, JaGa, MartinBot, Trusilver, Maurice Carbonaro, Srs144, Bob, Squids and Chips, VolkovBot, TXiKiBoT, Littlealien182, Gekritzl, CapJ15, Logan, Adeez, Karlc1980, SieBot, Coffee, Hollis9210, LeadSongDog, Oxymoron83, Diego Grez-Cañete, Sean.hoyland, ClueBot, PipepBot, The Thing That Should Not Be, Tomas e, ChandlerMapBot, Aitias, AC+79 3888, BodhisattvaBot, FTGHSmith, Addbot, Some jerk on the Internet, Fgnievinski, SamatBot, LinkFA-Bot, Longbowman, Lightbot, Luckas-bot, Yobot, عالم محبوب‎, AnomieBOT, PolskiOwl, Materialscientist, Brane.Blokar, Xqbot, Tasudrty, AL3X TH3 GR8, RibotBOT, Logger9, Membraner, FrescoBot, Proepro, Steve Quinn, Bvilleog, FearXtheXfro, HRoestBot, Jujutacular, Jauhienij, Gamewizard71, Marie Poise, Difu Wu, Qbaqbas, EmausBot, Orphan Wiki, K6ka, ZéroBot, Nicolas Eynaud, Rashhypothesis, Donner60, ChuispastonBot, ClueBot NG, Rtucker913, CaroleHenson, Helpful Pixie Bot, Yapatel, Snaevar-bot, Bryan Sanctuary, AvocatoBot, ChE Fundamentalist, AdventurousSquirrel, Cpgdaarfob, Aisteco, APerson, SFK2, Gdaniel111, Aavika, Ugog Nizdast, Sam Sailor, Wikicology, KasparBot and Anonymous: 147

- **Noble gas** *Source:* https://en.wikipedia.org/wiki/Noble_gas?oldid=687136225 *Contributors:* AxelBoldt, Carey Evans, Derek Ross, LC~enwiki, Mav, Tarquin, Andre Engels, Christian List, DrBob, Fonzy, Hephaestos, Olivier, Edward, Shellreef, Ahoerstemeier, Jimfbleak, Rlandmann, Salsa Shark, Nikai, GCarty, Smack, Lenaic, Stone, Jake Nelson, Tero~enwiki, Paul-L~enwiki, Taoster, Betterworld, Fvw, Shantavira, Donarreiskoffer, Gentgeen, Robbot, Sander123, Jakohn, Romanm, Merovingian, Rfc1394, Flauto Dolce, Meelar, Mervyn, Hadal, JackofOz, Robinh, Lupo, Dina, Giftlite, DocWatson42, Herbee, Monedula, Karn, Everyking, Slyguy, Kandar, Quackor, Andycjp, Antandrus, 1297, Icairns, Sam Hocevar, Gscshoyru, Deglr6328, Adashiel, Mike Rosoft, DanielCD, Discospinster, Rich Farmbrough, Vsmith, Ponder, Paul August, NeilTarrant, Geoking66, RJHall, El C, Dnwq, Shanes, Remember, Sietse Snel, Art LaPella, Femto, Marco Polo, Shenme, Viriditas, SpeedyGonsales, Severious, Obradovic Goran, Haham hanuka, Nsaa, Eddideigel, Orangemarlin, Ranveig, Jumbuck, Alansohn, Gary, Jared81, Keenan Pepper, Plumbago, Cjthellama, InShaneee, Suruena, Vuo, Gene Nygaard, Feline1, Kay Dekker, Boothy443, Woohookitty, Cimex, TarmoK, LOL, Pol098, WadeSimMiser, Mpatel, Wayward, Shanedidona, Palica, Stevey7788, Paxsimius, Graham87, David Levy, DePiep, Canderson7, Rjwilmsi, Mfwills, Vary, HappyCamper, Matjlav, Vuong Ngan Ha, RobertG, Latka, Nihiltres, Strangnet, RexNL, Mjp797, DevastatorIIC, Goudzovski, Scerri, Kri, King of Hearts, Chobot, DVdm, Bgwhite, EamonnPKeane, YurikBot, Chaser, Ollie holton, Stephenb, Yyy, NawlinWiki, Bachrach44, Jaxl, Adamn, Semperf, Zirland, Bota47, T-rex, Thetoaster3, Wknight94, Tetracube, Leptictidium, Phgao, Zzuuzz, Adilch, Scoutersig, Keepiru, HereToHelp, Katieh5584, NeilN, GrinBot~enwiki, DVD R W, Tom Morris, That Guy, From That Show!, Itub, Anthony Duff, SmackBot, FocalPoint, Tarret, KnowledgeOfSelf, Shoy, Kilo-Lima, Edgar181, Gaff, Aksi great, Gilliam, Isaac Dupree, Pslawinski, Durova, Bluebot, Bduke, SchfiftyThree, Moshe Constantine Hassan Al-Silverburg, CSWarren, Robth, DHN-bot~enwiki, Darth Panda, Gyrobo, Tsca.bot, Eric Olson, MJCdetroit, Rrburke, Aldaron, Nakon, Eganev, Pwjb, Smokefoot, DMacks, Wizardman, The undertow, SashatoBot, Lambiam, Krashlandon, Titus III, John, Zaphraud, Jaganath, Breno, Anoop.m, IronGargoyle, JHunterJ, Slakr, Beetstra, Noah Salzman, Optimale, Kpengboy, SandyGeorgia, AdultSwim, MTSbot~enwiki, BranStark, Iridescent, StephenBuxton, Jaksmata, Tawkerbot2, Swampgas, JForget, Irwangatot, Ruslik0, CompRhetoric, Dorothybaez, Bentleymrk, Gogo Dodo, A Softer Answer, Hibou8, SteveMcCluskey, Mattisse, Thijs!bot, Epbr123, Kablammo, Headbomb, JaimeAnnaMoore, Straussian, Werdnanoslen, Dantheman531, Mentifisto, AntiVandalBot, Luna Santin, Opelio, Cinnamon42, Scepia, LibLord, Xnuiem, Gökhan, IanOsgood, Nicholas Tan, Easchiff, Animaly2k2, Magioladitis, Bongwarrior, VoABot II, Hasek is the best, Ling.Nut, LorenzoB, Thibbs, Vssun, DerHexer, JaGa, Awolnetdiva, Mattinbgn, Hdt83, ChemNerd, Polartsang, CommonsDelinker,

AlexiusHoratius, Leyo, J.delanoy, Pharaoh of the Wizards, Phillip.northfield, Hans Dunkelberg, Dhruv17singhal, Uncle Dick, Jeri Aulurtve, Extransit, Cpiral, Bombhead, Wandering Ghost, Shay Guy, Coppertwig, Plasticup, Andraaide, Belovedfreak, Acey365, NewEnglandYankee, Najlepszy, Numerjeden, Matthardingu, KChiu7, WinterSpw, Brvman, Wilhelm meis, Squids and Chips, Idioma-bot, Wikieditor06, Deor, VolkovBot, Eakka, JGHowes, Jeff G., Chris Dybala, AlnoktaBOT, Nousernamesleft, Philip Trueman, Photonikonman, TXiKiBoT, Tavix, GimmeBot, Muro de Aguas, Rei-bot, Slysplace, Enigmaman, Davidmwhite, CephasE, Sylent, TinribsAndy, Owainbut, AlleborgoBot, Surfrat60793, SieBot, Calliopejen1, OTAVIO1981, Graham Beards, BotMultichill, ToePeu.bot, Jauerback, Nathan, Triwbe, Agesworth, Keilana, Flyer22 Reborn, The Evil Spartan, Arbor to SJ, Sohelpme, Scorpion451, Enok Walker, Lightmouse, OKBot, Nielg, Nimbusania, Nergaal, Escape Orbit, PerpetualSX, Runtishpaladin, UKe-CH, Martarius, ClueBot, Artichoker, The Thing That Should Not Be, Cygnis insignis, Manbearpig4, Franamax, Blanchardb, Piledhigheranddeeper, ChandlerMapBot, Puchiko, DragonBot, Excirial, Sidias300, GngstrMNKY, Jusdafax, Finch-HIMself, Estirabot, Poigol5043, Cenarium, Zomno, Jotterbot, Bellax22, Chaser (away), Werson, Boatcolour, SeanFarris, Thingg, Aitias, Dank, Versus22, RexxS, Boleyn, Neuralwarp, Feinoha, Little Mountain 5, Skarebo, Frood, Freestyle-69, CalumH93, Addbot, Mr0t1633, Roentgenium111, DOI bot, Theleftorium, Popopee, Ronhjones, Jncraton, Moosehadley, CanadianLinuxUser, AnnaFrance, Jasper Deng, Alchemist-hp, Numbo3-bot, Tide rolls, Zorrobot, Angrysockhop, Arimareiji, Legobot, Seresin.public, Luckas-bot, ZX81, Yobot, IsFari, TaBOT-zerem, Rsquire3, Bloody Mary (folklore), KamikazeBot, Widey, Synchronism, Andme2, AnomieBOT, Lolcopter666, Jcsdude, Navneethmohan, Jim1138, IRP, Law, Materialscientist, Citation bot, E2eamon, Maxis ftw, ArthurBot, Xqbot, Capricorn42, Nickkid5, Tad Lincoln, Turk oğlan, NocturneNoir, Lop242438, Pmlineditor, GrouchoBot, Doulos Christos, Antonjad, Jilkmarine, Smot94, Robo37, OgreBot, Citation bot 1, AstaBOTh15, Pinethicket, HRoestBot, Calmer Waters, I own in the bed, Marine79, Double sharp, TobeBot, Yopure, 777sms, Navy101, Reach Out to the Truth, Minimac, DARTH SIDIOUS 2, AXRL, Mean as custard, RjwilmsiBot, Japheth the Warlock, Ripchip Bot, Salvio giuliano, Deagle AP, EmausBot, WikitanvirBot, RA0808, Jordan776, Wikipelli, P. S. F. Freitas, AvicBot, ZéroBot, Fingerginger1, Maxviwe, StringTheory11, H3llBot, Makecat, Wagino 20100516, L Kensington, Donner60, Whoop whoop pull up, JohnMCrain, Mjbmrbot, Special Cases, Washington Irving Esquire, ClueBot NG, Rich Smith, Jack Greenmaven, Hon-3s-T, Skoot13, Ethanpiot, Lanthanum-138, Widr, Lolm8, Bibcode Bot, Swamphlosion, Lowercase sigmabot, Gluonman, TCN7JM, Iankhou, Sandbh, MusikAnimal, Altaïr, WikisucksKNOBlegasses, VictorParker, Jimbo2440, Tycho Magnetic Anomaly-1, Softballbaby984, ThomasRules, BattyBot, Justincheng12345-bot, Abilanin, ChrisGualtieri, EuroCarGT, Dexbot, Webclient101, TwoTwoHello, King jakob c, RandomLittleHelper, Reatlas, Cteung, DavidLeighEllis, Ugog Nizdast, Ginsuloft, Noyster, DudeWithAFeud, Skr15081997, Matthewweber12, HotHabenero, Hotta stuffu, Sony Vark XIII, Monkbot, HiYahhFriend, ZYjacklin, Narky Blert, Jodihe93, Selimozd20, Anbgsm07, UZawMoeNaing, SandKitty256, Supdiop, KasparBot, Sat cheat, Lolhappyface and Anonymous: 594

- **Niels Bohr** *Source:* https://en.wikipedia.org/wiki/Niels_Bohr?oldid=687185810 *Contributors:* AxelBoldt, Magnus Manske, Trelvis, General Wesc, Ansible, The Anome, Berek, Tarquin, Gareth Owen, BenBaker, Andre Engels, Youssefsan, Danny, Deb, Roadrunner, Zoe, Heron, Rsabbatini, Fonzy, JDG, Edward, Michael Hardy, Tim Starling, David Martland, Nixdorf, Liftarn, Gabbe, Mic, Zanimum, Skysmith, Paul A, Looxix~enwiki, Ahoerstemeier, Samuelsen, Suisui, Angela, Darkwind, Ruhrjung, Harvester, Harry Potter, Schneelocke, EL Willy, Guaka, Wikiborg, Reddi, Lfh, The Anomebot, Wik, DJ Clayworth, Markhurd, Peregrine981, Tpbradbury, Maximus Rex, Furrykef, Jose Ramos, Jackson~enwiki, AnonMoos, Proteus, David.Monniaux, BenRG, Jeffq, Mjmcb1, Riddley, Robbot, Jwbrown77, Sanders muc, Twid, Calmypal, Academic Challenger, Flauto Dolce, Rorro, Blainster, Timrollpickering, Hadal, Saforrest, Wereon, Kent Wang, Lupo, Cyrius, Giftlite, DocWatson42, Lethe, Reub2000, MathKnight, Brian Kendig, Fastfission, LeYaYa, Fleminra, Curps, Michael Devore, Gamaliel, Bensaccount, Ssd, Andris, Xinoph, Richard cocks, Just Another Dan, Nlaporte, Chowbok, Utcursch, Alexf, Knutux, Quadell, Antandrus, The Singing Badger, PDH, Karol Langner, Rdsmith4, Balcer, Shturm, Icairns, Kaisersanders, Magnum1, Gcanyon, Thorwald, D6, TheBlueWizard, Jayjg, CALR, DanielCD, EugeneZelenko, Rich Farmbrough, Guanabot, Vsmith, Jpk, Aris Katsaris, Triskaideka, Samboy, Bender235, Friism, El C, Bluap, Kwamikagami, Momotaro, Gershwinrb, Pablo X, Jpgordon, Wee Jimmy, Bobo192, Viriditas, Elipongo, Jojit fb, Nk, MPerel, Merope, HasharBot~enwiki, Echobeats, Senor Purple, Alansohn, Gary, JYolkowski, Jhertel, Plumbago, Nwinther, ABCD, Monk127, Esrob, Batmanand, Ksnow, Wtmitchell, Velella, Hasdrubal~enwiki, CloudNine, Dirac1933, Redvers, HenryLi, Afshar, AnIco, Oleg Alexandrov, Mwalcoff, WilliamKF, Velho, Jeffrey O. Gustafson, OwenX, Mindmatrix, ScottDavis, Rianamit, Kzollman, WadeSimMiser, The Wordsmith, Dowew, MONGO, Mpatel, Jok2000, Twthmoses, WikiklrsC, Ardydavari, J M Rice, Atomicarchive, Banpei~enwiki, Pfalstad, Emerson7, Mandarax, Graham87, Magister Mathematicae, Ted Wilkes, Demonuk, Jclemens, Zoz, Sjakkalle, Rjwilmsi, Mayumashu, Koavf, Jake Wartenberg, Kinu, Jivecat, Vary, Tangotango, Pabix, Mike Peel, The wub, Bhadani, Olessi, Sango123, Oo64eva, MutterErde, Kasparov, Drrngrvy, FlaBot, Kristjan Wager, Ground Zero, Old Moonraker, Doc glasgow, Mathbot, JdforresterBot, Annacoder, Catsmeat, AJR, Kerowyn, RexNL, Gurch, Alphachimp, Srleffler, Physchim62, Snailwalker, Valentinian, Chobot, Sharkface217, Hall Monitor, Gwernol, The Rambling Man, YurikBot, Wavelength, RattusMaximus, RobotE, John Quincy Adding Machine, Sillybilly, Nobs01, SpuriousQ, Stephenb, Manop, CambridgeBayWeather, Rsrikanth05, Pseudomonas, Wimt, NawlinWiki, Hawkeye7, Wiki alf, Leutha, Aeusoes1, Mauimonica, Grafen, Jaxl, Dureo, Nick, Ruhrfisch, Pyroclastic, Bota47, AirLiner, DRosenbach, FF2010, Jeremyzone, J S Ayer, Emijrp, 2over0, Homagetocatalonia, Imaninjapirate, Nikkimaria, Trendall, Pb30, Petri Krohn, TBadger, CWenger, Kevin, JLaTondre, Mais oui!, T. Anthony, Easter Monkey, Curpsbot-unicodify, SorryGuy, NeilN, Phl, Kingbouk, GrinBot~enwiki, Sebbi, SmackBot, Espresso Addict, Monkeyblue, Smallest step, KnowledgeOfSelf, Scott.br, Royalguard11, Pgk, C.Fred, Thorseth, Yuyudevil, KocjoBot~enwiki, Big Adamsky, Hydkat, Gabrielleitao, Edgar181, PeeJay2K3, Ian Rose, Aksi great, Gilliam, Steverich, Hmains, Skizzik, Angelbo, Cabe6403, Poulsen, Mirokado, QEDquid, JRSP, Chris the speller, Master Jay, Keegan, Persian Poet Gal, MK8, Enkyklios, Jprg1966, EncMstr, Miquonranger03, Fluri, Iamakhilesh, DHN-bot~enwiki, Colonies Chris, Gracenotes, Can't sleep, clown will eat me, RyanEberhart, Janysc, Discharger12, Mhym, KerathFreeman, Addshore, Khoikhoi, Wen D House, ArtVandelay13, Khukri, Nakon, Shamir1, Mr.Morose, Dreadstar, Richard001, Marc-André Aßbrock, Lcarscad, Lacatosias, DMacks, Ultraexactzz, Harrias, BrotherFlounder, ElizabethFong, Bidabadi~enwiki, Sadi Carnot, RossF18, Ohconfucius, The undertow, SashatoBot, Lambiam, Nishkid64, Eliyak, Dr. Sunglasses, John, Jan.Smolik, Soumyasch, Shadowlynk, JorisvS, Hemmingsen, Minna Sora no Shita, NYCJosh, Zarniwoot, Tdudkowski, IronGargoyle, BadDude45, Bcem2, Stwalkerster, LuYiSi, Treznor, Mr Stephen, Samfreed, Waggers, Spiel496, Ryulong, JdH, Condem, MTSbot~enwiki, Keith-264, Dan Gluck, BananaFiend, Iridescent, EPO, Alessandro57, Joseph Solis in Australia, J Di, Twas Now, Newyorkbrad, Ryan4, Ewulp, Courcelles, Ziusudra, Dpeters11, Tawkerbot2, Ouishoebean, Axt, Nil pat13, Chris55, JForget, Paulmlieberman, Mattbr, Van helsing, BeenAroundAWhile, Marshall.frimoth, JohnCD, Stevo1000, Dgw, Tar-Meneldur, Art10, Myasuda, Gregbard, Funnyfarmofdoom, Slazenger, Cydebot, Kanags, Metacosm, Mato, Gogo Dodo, EdiOnjales, Hebrides, MWaller, Rracecarr, Studerby, Michael C Price, Carstensen, Christian75, DumbBOT, Phydend, ErrantX, SpK, Cielovista, Wexcan, Kirk Hilliard, Casliber, DavidSteinle, Malleus Fatuorum, Thijs!bot, Epbr123, Wikid77, Qwyrxian, Willworkforicecream, Kablammo, Ucanlookitup, Headbomb, West Brom 4ever, James086, Blacklake, X201, Martin Hedegaard, Miller17CU94, Bunzil, EdJohnston, Escarbot, Dzubint, Xionication, AntiVandalBot, RobotG, QuiteUnusual, Jeffreyge1, Jj137, North Shoreman, Vistor, Wahabijaz, Lklundin, Jrheilig, Res2216firestar, JAnDbot, Leuko, Husond, MER-C, Matthew Fennell, Andonic, Roleplayer, LittleOldMe, Acroterion, Bencherlite, Easchiff, Magioladitis, Connormah, WolfmanSF, Gurubanks, Bongwarrior, VoABot II, MartinDK, JNW, Mbc362, Molybdenum1, Avicennasis, Nick Cooper, KConWiki, Indon, Animum, Cgingold, Elentirmo, Dirac66, LookingGlass, Hamiltonstone, Allstarecho, Duendeverde, Chris G, DerHexer, Edward321, TheRanger, Edton, Seba5618, SquidSK, CharlesKiddell, MartinBot, BetBot~enwiki, NAHID, Rettetast, Anaxial, Bus stop, R'n'B, CommonsDelinker, AlexiusHoratius, Lahlahlove, J.delanoy, Captain panda, Elfelix, Eelric92, Tlim7882, Fowler&fowler, Nbauman, SureFire, SuperGirl, JoDonHo, TrollerS, Maurice Carbonaro, Extransit, Baztastik, Salih, PedEye1, McSly, SimulacrumDP, Notreallydavid, Jayden54, Tarotcards, Ipigott, Ziagard, Rocket71048576, NewEnglandYankee, In Transit, Cathyrne, Cycotic, Toon05, Xyl 54, MikkelHøgh, Bonadea, Arabianjew1, Varnent, Specter01010, Steel1943, CardinalDan, Sam Blacketer, VolkovBot, DOHC Holiday, Philip Trueman, TXiKiBoT, Oshwah, Sparkzy, GcSwRhIc, Sean D Martin, JayC, Qxz, Someguy1221, Cloudswrest, Johnred32, Steven J. Anderson, Martin451, Omcnew, Ripepette, DesmondW, Cremepuff222, Pishogue, Mr. Absurd, Duncan.Hull, Katimawan2005, Mr.Kennedy1, Y, Carinemily, Maarten van Emden, CoolKid1993, FKmailliW, RaseaC, Insanity Incarnate, W3areVenom, Sidhu Jyatha, AlleborgoBot, Symane, Jorbjorb, Emerson54, Haiviet~enwiki, Brettdog, Akonden, Engsys, Cj1340, Michellecrisp, SieBot, Froztbyte, Paul20070, Ttony21, Tresiden, YonaBot, BotMultichill, Sakkura, Sweet92, ConradMcShtoon, Plinkit, Cyberix, Dawn Bard, RJaguar3, Yintan, Utternutter, Happysailor, Likebox, Flyer22 Reborn, Tiptoety, Hxhbot, 2hiyup2, Arthur Smart, Sharktooth9000, Oxymoron83, Dtrain1121, Coleman b, KoshVorlon, Afernand74, OKBot, Kumioko (renamed), Svick, Vojvodaen, Coldcreation, Chain27, Mygerardromance, Pinkadelica, Denisarona, LarRan, Madmadabsrd, Hadseys, TSRL, TheCatalyst31, RS1900, Atif.t2, Gloss, Frostedpinkdoughnuts, Wokoslayi, Church, Sfan00 IMG, ClueBot, Binksternet, GorillaWarfare, Fyyer, The Thing That Should Not Be, All Hallow's Wraith, Postmortemjapan, Icarusgeek, Stogego, Humanist505, Jan1nad, MikeVitale, Saddhiyama, Drmies, Phatbuddy09, CounterVandalismBot, Ptrain, Big man110, Neverquick, Ishiho555, ChandlerMapBot, Jandew, Masterpiece2000, DragonBot, Djr32, Excirial, Alexbot, Samasamas1, JonnyLee, Garrettissupercool, Linesmodel, Winston365, Bchaosf, Muenda, Jimmy415, CAVincent, Cenarium, Arjayay, Razorflame, Hummer460, Saebjorn, Muro Bot, BOTarate, Inspector 34, Thingg, Cardinalem, Dank, Burner0718, Macderv15h, SoxBot III, Bletchley, XLinkBot, MessinaRagazza, Tarlneustaedter, Integralolrivative, KekeBookworm, MidwestGeek, SilvonenBot, NellieBly, Broncos5, Bolikesboys, Jbeans, TFBCT1, Candyland251, HexaChord, Sam grundy, Addbot, Proofreader77, 67fro8iu, Mortense, Thecarpy, Andrewsthistle, Manuel Trujillo Berges, Mcmuffinz, Atethnekos, Friginator, Number1scatterbrain, Blethering Scot, Ronhjones, Urbandweller, CanadianLinuxUser, Socersam627, Cst17, LaaknorBot, Lfackerson, Glane23, Favonian, AtheWeatherman, Feketekave, AgadaUrbanit, Numbo3-bot, Lilly1234~enwiki, Tide rolls, Lightbot, GJo, Zorrobot, ⴲԡ, Alfie66, Albeiror24, Luckas-bot, Yobot, Fraggle81, THEN WHO WAS PHONE?, SwisterTwister, Andreas Werle, Farlander, AnomieBOT, Fatal!ty, Torricelli01, CatState, 1exec1, Killiondude, Tucoxn, Dwayne, Piano non troppo, Djdanger95, AdjustShift, Particrashr69, JGleick, EryZ, RandomAct, Jalexsmith1991, Materialscientist, ImperatorExercitus, The High Fin Sperm Whale, Citation bot, RevelationDirect, Xqbot, Addihockey10, JimVC3, Davshul, Gap9551, GrouchoBot, Omnipaedista, Ten-pint, Anotherclown, RibotBOT, Acdc therealone, Webnetprof, Supergeek1694, Amaury, Nedim Ardoǧa, Kyoko Takeda, Sophus Bie, AustralianRupert, Astatine-210, Satyajit.das90, Colt .55, Žiedas, Tubbablub, Maxidater, Dougofborg, ⴲⴲ, Bonziboy21, VS6507, Scott A Herbert, BenzolBot, Grandiose, A little insignificant, Cannolis, Citation bot 1, Cubs197, I dream of horses, Haaqfun, Grammarspellchecker, Adlerbot, Plucas58, Calmer Waters, Skyerise, A8UDI, Fat&Happy, Fixer88, Foobarnix, Meaghan, Conqu2, Karategirl2011, PrinceRegentLuitpold, White Shadows, Cnwilliams, Gerda Arendt, Double sharp, Trappist the monk, Badger M., DixonDBot, Hickorybark, ItsZippy, Krede~enwiki, Oracleofottawa, Vrenator, Dgregory317, Monster Energy 69, Reach Out to the Truth, DARTH SIDIOUS 2, Dacman48, Whisky drinker, TjBot, Griffinsusername, Rahul93 reddy, DASHBot, EmausBot, WikitanvirBot, GA bot, Ken95, Immunize, 478jjjz, Nø, Sprout333, Racerx11, RA0808, Illogicalpie, Grumpy Gnome, Tommy2010, Caljomac97, Wikipelli, K6ka, MikeyMouse10, Hhhippo, Kkm010, ZéroBot, John Cline, Lemeza Kosugi, Access Denied, H3llBot, Demiurge1000, Rcsprinter123, JeanneMish, Danmuz, Vibratorhythm, Noodleki, Donner60, Flight714, Brigade Piron, Petrb, ClueBot NG, MelbourneStar, Bped1985, Movsesbot, O.Koslowski, Woogy31, ScottSteiner, Rezabot, Helpful Pixie Bot, Jfredber, Bibcode Bot, Lowercase sigmabot, BG19bot, Goodfood72398, Rushabh Khasnis, Viggoodin, Metal Velocidad, Kendall-K1, Mhakcm, Visciousz, Earlyap, Trevayne08, Soerfm, Breezylost, Aranea Mortem, Jubjub1277, Connorbishop, Hafniensis, Lastdodo1, Incogmax, Brad7777, DWDHMB12, Jhckragh, Tutelary, Ninmacer20, Christoskdimou, ChrisGualtieri, BijouTrouvaille, Ayurtefe, Postrally4, Giso6150, MrNiceGuy1113, Dexbot, Jjrecto, Mogism, JSydel, Agnostihuck, Lugia2453, Jamesx12345, RevMSWIE500, Urlan Wannop, Big Mike in KeNtucky, JustAMuggle, BruceJohnJennerLawso, Epicgenius, VoxelBot, Eyesnore, Bibliophilen, Everymorning, DavidLeighEllis, Crispulop, Monochrome Monitor, My name is not dave, Jeab1234, OccultZone, ChristopherBaras, Wesley696969, Wikiguy3000, Fre3 vodka, Doloreshughes, Iknowlotsandeverything, Volz ernest, Threnos, GriffinJones123, Spectroman~enwiki, Qxxxxq, KingofAwesome82, Chrisykimmy0116, Styletread, Nubbygarter, SlothyBau5, Thispageinanutshell1, Jonas Vinther, Monkbot, Choo choo987651, G12345678, HiYahhFriend, BrokenMirrors123, Ignoble7, Waldmirptuin, Jonarnold1985, EternalFloette, Peter238, Crystallizedcarbon, Albert Brady, Infernus 780, RWJord, Shchenzk, Ohfdgg, Johnny0151, Dizbigdick, El rabano volatil, Ellipapa, 3 of Diamonds, JJMC89, Fiber126, Csumstudent, SephardicScholar, NielsHenrikDavidBohr, Helenq12, Erfasser and Anonymous: 1215

- **Bohr model** *Source:* https://en.wikipedia.org/wiki/Bohr_model?oldid=686992786 *Contributors:* Damian Yerrick, AxelBoldt, Trelvis, Zundark, The Anome, Taw, Andre Engels, Arvindn, Stevertigo, D, JohnOwens, Michael Hardy, Tim Starling, Looxix~enwiki, Ellywa, Ahoerstemeier, Cyp, Glenn, Complex Analysis, Hashar, Charles Matthews, Tantalate, RickK, The Anomebot, Furrykef, Fibonacci, Omegatron, BenRG, Robbot, Vespristiano, Pmineault, Elysdir, Rsduhamel, Enochlau, Decumanus, Giftlite, BenFrantzDale, Lethe, MathKnight, Bensaccount, Dmmaus, Yath, Antandrus, Darksun, Grunt, Discospinster, Vsmith, ArnoldReinhold, Gianluigi, Mani1, Paul August, Kaszeta, El C, MrMarshmallow, CDN99, Army1987, Whosyourjudas, Smalljim, Viriditas, Cmdrjameson, R. S. Shaw, Evgeny, Matt McIrvin, MPerel, Nsaa, Jjron, Mdd, Jumbuck, Storm Rider, Alansohn, TheParanoidOne, Munchkinguy, Fritzpoll, PAR, Redfarmer, Snowolf, Velella, HenkvD, Sciurinæ, Bsadowski1, Linas, Mindmatrix, Benhocking, Mpatel, Mreult~enwiki, Schzmo, Terence, MFH, Ian**, SeventyThree, Christopher Thomas, Graham87, Galwhaa, DePiep, Mendaliv, Rjwilmsi, Vary, Salix alba, Tawker, Scorpiuss, Yamamoto Ichiro, Sanbeg, Nivix, Krackpipe, RexNL, Gurch, Goudzovski, Srleffler, Chobot, DVdm, Gwernol, YurikBot, Wolfmankurd, Postglock, JabberWok, Rsrikanth05, Spike Wilbury, Aeusoes1, Buster79, CAPS lOCK, Moe Epsilon, Zwobot, Dna-webmaster, Wknight94, Sperril, PTSE, Closedmouth, KGasso, Petri Krohn, Fram, HereToHelp, Spliffy, GrinBot~enwiki, SkerHawx, SmackBot, Thorseth, Blue520, Jacek Kendysz, WookieInHeat, Frymaster, Timotheus Canens, Gilliam, Chris the speller, Pieter Kuiper, OrangeDog, Complexica, DHN-bot~enwiki, Sbharris, Darth Panda, Can't sleep, clown will eat me, Wen D House, Nakon, Localzuk, GoldenBoar, Adam Schloss, DMacks, O RLY?, Ligulembot, Mion, Sadi Carnot, Kukini, Eliyak, John, Gobonobo, Tktktk, LestatdeLioncourt, Hemmingsen, 3897515, Goodnightmush, IronGargoyle, PseudoSudo, Smith609, Tasc, Beetstra, Martinp23, Domino42, Mets501, Doczilla, Ryulong, Jonhall, Lee Carre, Iridescent, Zootsuits, Joseph Solis in Australia, Morrowulf, Igoldste, Tony Fox, Beve, Blehfu, Courcelles, Profjohn, Tawkerbot2, Chetvorno, JForget, Frovingslosh, Olaf Davis, BeenAroundAWhile, CWY2190, Harej bot, Myasuda, Dept of Alchemy, Cydebot, WillowW, Gogo Dodo, Corpx, Xndr, Shirulashem, DumbBOT, Chrislk02, Sp, Pinky sl, FrancoGG, Epbr123, Wikid77, Goods21, Kablammo, Headbomb, John254, Martin Hedegaard, Pfranson, Dawnseeker2000, Mentifisto, AntiVandalBot, Majorly, Tlabshier, Spartaz, JAnDbot, Harryzilber, MER-C, Andonic, Belg4mit, Hut 8.5, Tstrobaugh, Bkpsusmitaa, R27182818, Ô, Acroterion, Casmith 789, Magioladitis, VoABot II, Wikidudeman, JamesBWatson, Dirac66, 28421u2232nfenfcenc, Schumi555, Cpl Syx, SlamDiego, Mikerobertsn, Vssun,

TheRanger, Robin S, Geboy, Kpxxbladexx415, Jemijohn, Mont95, ChemNerd, Slash, J.delanoy, Melamed katz, Uncle Dick, Jonpro, Cpiral, It Is Me Here, Bot-Schafter, AntiSpamBot, Andraaide, NewEnglandYankee, Pez2, Juliancolton, Copsi, Jarry1250, JavierMC, Dextercioby, Cuzkatzimhut, Hugo999, VolkovBot, Philip Trueman, TXiKiBoT, The Original Wildbear, Sarenne, Anonymous Dissident, Piperh, JhsBot, Leafyplant, Jackfork, LeaveSleaves, Itemirus, Thunderbird2, MrChupon, Murkee, DrJunge, EvilBunnyHead, SieBot, Servant Saber~enwiki, Coffee, Ivan Štambuk, GrooveDog, Keilana, Likebox, Flyer22 Reborn, Tiptoety, Grimey109, Oxymoron83, Antonio Lopez, Scorpion451, Jdaloner, Pac72, Lightmouse, Christovac, Nandobike, The Stickler, Mike2vil, Anchor Link Bot, Pinkadelica, Dolphin51, Elassint, ClueBot, Ferred, GorillaWarfare, The Thing That Should Not Be, ArdClose, Swedish fusilier, Lantay77, Neverquick, Auntof6, Djr32, Robert Skyhawk, Excirial, SpikeToronto, Willthedrill, Danmichaelo, Nengscoz416, Iohannes Animosus, Bite Size Monkeys, W.GUGLINSKI, Zerxan, Thehelpfulone, Kakofonous, Thingg, BlueDevil, DumZiBoT, Life of Riley, Jmanigold, Nukeh, Avoided, Skarebo, NellieBly, Mm40, ZooFari, MystBot, RyanCross, Some jerk on the Internet, DOI bot, WMdeMuynck, YURI2008, CanadianLinuxUser, Fluffernutter, Glane23, Bassbonerocks, Glass Sword, LinkFA-Bot, Hainetron, Apteva, Gail, Jarble, Alfie66, Jackelfive, Legobot, Luckas-bot, Yobot, Cflm001, IW.HG, TestEditBot, Tempodivalse, N1RK4UDSK714, AnomieBOT, DemocraticLuntz, Jim1138, Piano non troppo, Sz-iwbot, Materialscientist, The High Fin Sperm Whale, Citation bot, E2eamon, GB fan, ArthurBot, Ammubhave, Jonathan321, Jeffrey Mall, Coretheapple, Omnipaedista, Ilovenickjay, Tgervaisphd, FrescoBot, Qalander, Steve Quinn, Machine Elf 1735, Citation bot 1, Pinethicket, I dream of horses, Teamdojo, BRUTE, RandomStringOfCharacters, Ifritnile, Jordgette, Javierito92, Zink Dawg, Auscompgeek, Redskins247, DARTH SIDIOUS 2, WikitanvirBot, Nuujinn, Super48paul, RA0808, Solarra, Tommy2010, Wikipelli, K6ka, Hhhippo, JSquish, John Cline, Lateg, Quondum, Zloyvolsheb, QEDK, Wagino 20100516, Thine Antique Pen, Dilwala314, WikiPidi, Nick9876, RockMagnetist, Peter Karlsen, CharlotteMab, DASHBotAV, Mtlee7, Billyvespiethethird, Spicemix, Imuwithu2, Rocketrod1960, ClueBot NG, Pruegz778, 12nichja, Satellizer, A520, Widr, Shivsagardharam, Titodutta, Bibcode Bot, BG19bot, Wiki13, Shalom25, MusikAnimal, Eio, Klilidiplomus, Ihatecairns, Achowat, Riley Huntley, Samanthaclark11, Pratyya Ghosh, ChrisGualtieri, Martinkupilas, La marts boys, Bethechangeyouhopetosee, Sdk16420, Makecat-bot, Cerabot~enwiki, Lugia2453, 6033CloudyRainbowTrail, JustAMuggle, Reatlas, Epicgenius, Howicus, Melonkelon, Tentinator, Ihatepauldirac, Ihatedirac, MKCarriegirl, Ugog Nizdast, BruceBlaus, Mourici, Konveyor Belt, Adamharsh, Williamsmith29, Ugotpowned12, LucaMoro, Amortias, Joeyransom, Adam tunard creator of science, DallasSama, YashGarg10, Leawesomepotato, KINGSUFI, MKZG, Pazycraft and Anonymous: 718

- **Quantum chemistry** *Source:* https://en.wikipedia.org/wiki/Quantum_chemistry?oldid=684103433 *Contributors:* Michael Hardy, Looxix~enwiki, Milo~enwiki, Александър, Glenn, Timwi, 4lex, UninvitedCompany, Chuunen Baka, Gentgeen, Robbot, Gershom, Giftlite, Dratman, Zeimusu, Karol Langner, Nickptar, Edsanville, Cypa, Noisy, Guanabot, Vsmith, CDN99, Robotje, Cmdrjameson, La goutte de pluie, Alansohn, Keenan Pepper, Alex '05, Jheald, CloudNine, Capecodeph, Woohookitty, GregorB, Yurik, Rjwilmsi, Smoe, HappyCamper, Ligulem, Krash, Ian Pitchford, Ayla, Sunev, YurikBot, Okedem, Salsb, Shanel, Ithacagorges, Wiki alf, Vb, Nzzl, ChemGardener, Itub, SmackBot, M stone, Yamaguchi⍰⍰, Cool3, Hugo-cs, Kaliumfredrik, Bduke, Shalom Yechiel, BTDenyer, Minority2005, Sadi Carnot, Vgy7ujm, Zarniwoot, Ryulong, Bubbha, W.F.Galway, MC10, Rifleman 82, Christian75, DumbBOT, Thijs!bot, Headbomb, Martin Hedegaard, EdJohnston, Xebvor, Ratol, Koinut, Holdran, Jantop, Lijuni, Lampuchi, Perelaar, EmilyT, Acroterion, ChemNerd, CommonsDelinker, Maurice Carbonaro, Sidhekin, Terhorstj, Bob, Funandtrvl, A4bot, Rei-bot, Shonenknifefan1, SQL, BrianY, SHL-at-Sv, Euryalus, JerrySteal, Keilana, Vig vimarsh, ClueBot, Auntof6, Alexbot, Sun Creator, Muro Bot, DJ Sturm, Stickee, Avoided, Addbot, Lightbot, Meisfunny, Luckas-bot, Yobot, Evans1982, AnakngAraw, علم-حبوب, AnomieBOT, Ulric1313, Materialscientist, Citation bot, Bci2, Tasudrty, Haljolad, Voigfdsa, P99am, FrescoBot, Louperibot, Fygoat, 123Mike456Winston789, Danhe, EmausBot, RA0808, Rivanvx, Klbrain, H3llBot, AManWithNoPlan, TonyMath, Hpubliclibrary, Superdelocalizable, Mayur, Spaceboy909, ClueBot NG, Michael P. Barnett, Satellizer, Widr, Flomenbom, Helpful Pixie Bot, Electriccatfish2, PhnomPencil, Tautally, BattyBot, TwoTwoHello, SFK2, Joeinwiki, Darth Sitges, Elie.nasrallah, Aavika, DavidLeighEllis, The Herald, Cytokinetics, Internucleotide, KasparBot, Enricfs and Anonymous: 124

- **History of electrochemistry** *Source:* https://en.wikipedia.org/wiki/History_of_electrochemistry?oldid=671094387 *Contributors:* Samsara, Wjbeaty, ELApro, YUL89YYZ, Pschemp, Riana, SteinbDJ, Gene Nygaard, Carcharoth, HappyApple, Mandarax, Rjwilmsi, Nihiltres, Lmatt, SirGrant, RussBot, Nahallac Silverwinds, Jaxl, Nick, Retired username, Kkmurray, Tevildo, SmackBot, Hmains, Chris the speller, Renegade Lisp, Ske2, DMacks, Ligulembot, John, Dockingman, WMSwiki, Ssilvers, Gaviidae, CosineKitty, Arch dude, Linksfuss, Magioladitis, Doug Coldwell, Spellmaster, Edward321, Schmloof, STBot, R'n'B, ClueBot, Mild Bill Hiccup, Stephaninator, UB.Esser, Steelaway, CohesionBot, DumZiBoT, Bletchley, Avoided, Nicolae Coman, Addbot, Tassedethe, Luckas-bot, Yobot, AnomieBOT, Daniele Pugliesi, RibotBOT, Jorge333manrique, John of Reading, Mutley1989, SkateTier and Anonymous: 13

- **Timeline of chemistry** *Source:* https://en.wikipedia.org/wiki/Timeline_of_chemistry?oldid=683184685 *Contributors:* Rmhermen, Paul A, Stone, Graeme Bartlett, Icairns, Picapica, Rich Farmbrough, Circeus, Nsaa, Sleigh, Woohookitty, Carcharoth, Wikiklrsc, Macaddct1984, Ketiltrout, Rjwilmsi, Pjetter, The Rambling Man, Wolfmankurd, Fabartus, Sandstein, CWenger, ChemGardener, Itub, SmackBot, TestPilot, WilyD, Jagged 85, Jrockley, M stone, Bduke, Colonies Chris, Chtit draco, Abyssal, DMacks, Sadi Carnot, Nishkid64, Iridescent, CmdrObot, KyraVixen, Neelix, Myasuda, Ntsimp, Rifleman 82, Christian75, Thijs!bot, D.H, Billscottbob, Jayron32, WolfmanSF, VoABot II, KConWiki, David Eppstein, CommonsDelinker, Pomte, Maproom, Rjclaudio, Funandtrvl, Crohnie, Joseph A. Spadaro, OMCV, OTAVIO1981, Grrahnbahr, BrianGo28, Dabomb87, Tomasz Prochownik, ArepoEn, Tanvir Ahmmed, ClueBot, S Levchenkov, J8079s, 718 Bot, PixelBot, SchreiberBike, Publicanalyst, Camboxer, Addbot, DOI bot, CanadianLinuxUser, Tassedethe, דוד55, Teles, Luckas-bot, AnomieBOT, Ulric1313, Materialscientist, Citation bot, Xqbot, Iamion, J04n, Omnipaedista, Citation bot 1, RedBot, Manurup1997, WaitingForConnection, DARTH SIDIOUS 2, J36miles, EmausBot, Syncategoremata, John of Lancaster, Knight1993, BeGenderNeutral, H3llBot, Makecat, Music Sorter, Hccc, ClueBot NG, KLindblom, Krshwunk, Helpful Pixie Bot, Bibcode Bot, Car Henkel, Leonxlin, Solomon7968, Footy Kev81, Dexbot, DJRufflesLive22, Monkbot, SantiLak and Anonymous: 58

- **Timeline of chemical element discoveries** *Source:* https://en.wikipedia.org/wiki/Timeline_of_chemical_element_discoveries?oldid=686495166 *Contributors:* Tobias Hoevekamp, Magnus Manske, Derek Ross, Lee Daniel Crocker, Mav, Bryan Derksen, Olof, Tarquin, Malcolm Farmer, Eob, Rgamble, Rmhermen, Toby Bartels, William Avery, DavidLevinson, Zoe, Heron, Fonzy, Ewen, Tucci528, Edward, Michael Hardy, Axlrosen, TakuyaMurata, Ahoerstemeier, G~enwiki, Sugarfish, Jiang, [212], Gh, Raven in Orbit, Malbi, Timwi, Reddi, Stone, Fibonacci, Phoebe, Lord Emsworth, Bcorr, Gentgeen, Sappe, Zandperl, Chris 73, Psychonaut, Arkuat, Rursus, Centrx, Fastfission, Wwoods, Chameleon, Manuel Anastácio, Zeimusu, Yath, Ctachme, Icairns, Creidieki, Tsemii, Guanabot, FT2, Vsmith, Florian Blaschke, Mani1, Pavel Vozenilek, Uppland, Blade Hirato~enwiki, SpookyMulder, JoeSmack, Reinyday, Ctrl build, Como, Nsaa, Jakew, Anthony Appleyard, Keenan Pepper, Benjah-bmm27, Cloud Strife~enwiki, Bootstoots, Dave.Dunford, BDD, Nightstallion, Dismas, Marianika~enwiki, Zntrip, Megan1967, Linas, Georgia guy, Carcharoth, Benbest, Fbriere, Graham87, DePiep, Drbogdan, Rjwilmsi, Eoghanacht, Koavf, JanSuchy, Naraht, Mariocki, Nihiltres, RexNL, Goudzovski, Benlisquare, Gdrbot, YurikBot, Wavelength, Kwarizmi, Gaius Cornelius, DeadEyeArrow, Silverhill, Wknight94,

SamuelRiv, TheMadBaron, Closedmouth, Abune, Whobot, Meegs, Itub, SpLoT, SmackBot, Mira, Tarret, Ccalvin, Jagged 85, Renesis, Gilliam, Kurykh, WikiFlier, TheGeck0, Gracenotes, Can't sleep, clown will eat me, Skydiver, Runefurb, MrPMonday, RandomP, Koepsell, Sadi Carnot, Pilotguy, Lambiam, Perfectblue97, Smartyllama, Mgiganteus1, Olin, IronGargoyle, Alatius, Novangelis, JeffW, Newone, IvanLanin, Tawkerbot2, JRSpriggs, CmdrObot, Glenn4pr, Rifleman 82, Christian75, Headbomb, SGGH, Escarbot, Oreo Priest, AntiVandalBot, Nisselua, Tpth, Dylan Lake, Shift6, Dougher, Figma, Plantsurfer, Briancollins, Dricherby, LittleOldMe, Quantockgoblin, Hbent, Patstuart, Schmieder, R'n'B, EmleyMoor, Celephicus, DRKS, Collegebookworm, Moon Ranger, Warut, Bob, Kraniel, Sstrebel, JavierMC, Squids and Chips, Philip Trueman, Rei-bot, Kv75, CloakedHorror, Inx272, Rwell3471, Petergans, PlanetStar, Rfts, Nergaal, The sunder king, Mario Žamić, ClueBot, Surfeited, S Levchenkov, Jan1nad, Mild Bill Hiccup, J8079s, Excirial, Jusdafax, Eeekster, Estirabot, ZrikiSvargla, DumZiBoT, Drjezza, XLinkBot, Stickee, Avoided, Skarebo, Nsim, Sami Lab, Jamieb561, Roentgenium111, Lancshero, DOI bot, Guoguo12, Bezuidenhout, Ashanda, Icantouchmytoes, Tassedethe, Sanchitblazer, Legobot, Dor Cohen, Kilom691, Azylber, AnomieBOT, Piano non troppo, Materialscientist, Citation bot, Herr Mlinka, Coretheapple, Trongphu, Chris.urs-o, Spesh531, SD5, Riventree, Msary80, Trewal, Robo37, Trdsf, Citation bot 1, Intelligentsium, Redrose64, Wdcf, Double sharp, UTrunn, Armando-Martin, RjwilmsiBot, Skamecrazy123, EmausBot, John of Reading, Syncategoremata, GoingBatty, XinaNicole, Peterindelft, ZéroBot, BAICAN XXX, Josve05a, StringTheory11, H3llBot, Makecat, Kevjonesin, RockMagnetist, ClueBot NG, Gareth Griffith-Jones, Matt5595, Lanthanum-138, O.Koslowski, Rezabot, Helpful Pixie Bot, Bibcode Bot, Bths83Cu87Aiu06, MusikAnimal, Soerfm, Cengime, AlanPalgut, Siuenti, Dexbot, Cwobeel, Burzuchius, XXN, Jc86035, Limitderivative, Kevin12xd, Makecat (public), Monkbot, Hashimmmm, Hockey100050, Fattbutts and Anonymous: 221

27.9.2   Images

- **File:046CupolaSPietro.jpg** *Source:* https://upload.wikimedia.org/wikipedia/commons/5/5a/046CupolaSPietro.jpg *License:* CC BY-SA 3.0 *Contributors:* Own work *Original artist:* MarkusMark

- **File:1911_Solvay_conference.jpg** *Source:* https://upload.wikimedia.org/wikipedia/commons/c/ca/1911_Solvay_conference.jpg *License:* Public domain *Contributors:* Brussels, Belgium *Original artist:* Benjamin Couprie

- **File:AOs-1s-2pz.png** *Source:* https://upload.wikimedia.org/wikipedia/commons/8/86/AOs-1s-2pz.png *License:* Public domain *Contributors:* ? *Original artist:* ?

- **File:A_New_System_of_Chemical_Philosophy_fp.jpg** *Source:* https://upload.wikimedia.org/wikipedia/commons/9/97/A_New_System_of_Chemical_Philosophy_fp.jpg *License:* Public domain *Contributors:* En.wiki *Original artist:* haade

- **File:Affinity-table.jpg** *Source:* https://upload.wikimedia.org/wikipedia/commons/e/ee/Affinity-table.jpg *License:* Public domain *Contributors:* Table des differens rapports observes en chemie entre differentes substances; Memoires de l'Academie Royale des Sciences, pp. 202-212 *Original artist:* E.R. Geoffroy

- **File:Albert_Edelfelt_-_Louis_Pasteur_-_1885.jpg** *Source:* https://upload.wikimedia.org/wikipedia/commons/3/3c/Albert_Edelfelt_-_Louis_Pasteur_-_1885.jpg *License:* Public domain *Contributors:* Photograph originally posted on Flickr as Albert EDELFELT, Louis Pasteur, en 1885. Date of generation: 27 August 2009. Photographed by Ondra Havala. Modifications by the uploader: perspective corrected to fit a rectangle (the painting was possibly distorted during this operation), frame cropped out. *Original artist:* Albert Edelfelt

- **File:Alchemik_Sedziwoj_Matejko.JPG** *Source:* https://upload.wikimedia.org/wikipedia/commons/9/99/Alchemik_Sedziwoj_Matejko.JPG *License:* Public domain *Contributors:* www.pinakoteka.zascianek.pl *Original artist:* Jan Matejko

- **File:Alchemists_Workshop_detail_from_Title_Page_AQ24_(3).tif** *Source:* https://upload.wikimedia.org/wikipedia/commons/8/8d/Alchemists_Workshop_detail_from_Title_Page_AQ24_%283%29.tif *License:* Public domain *Contributors:* Chemical Heritage Foundation *Original artist:* Lazarus Ercker

- **File:Alchemy_of_Happiness.png** *Source:* https://upload.wikimedia.org/wikipedia/commons/0/03/Alchemy_of_Happiness.png *License:* Public domain *Contributors:*

- http://expositions.bnf.fr/splendeurs/notices/1-18.htm *Original artist:* Abū Hāmid al-Ghazzālī

- **File:Aluminium_sulfate.jpg** *Source:* https://upload.wikimedia.org/wikipedia/commons/d/d3/Aluminium_sulfate.jpg *License:* Public domain *Contributors:* ? *Original artist:* ?

- **File:Ambox_important.svg** *Source:* https://upload.wikimedia.org/wikipedia/commons/b/b4/Ambox_important.svg *License:* Public domain *Contributors:* Own work, based off of Image:Ambox scales.svg *Original artist:* Dsmurat (talk · contribs)

- **File:Antoine_lavoisier_color.jpg** *Source:* https://upload.wikimedia.org/wikipedia/commons/6/6c/Antoine_lavoisier_color.jpg *License:* Public domain *Contributors:* Courtesy of Chemical Achievers *Original artist:* Louis Jean Desire Delaistre, after Boilly

- **File:ArTube.jpg** *Source:* https://upload.wikimedia.org/wikipedia/commons/2/2f/ArTube.jpg *License:* CC BY-SA 2.5 *Contributors:* user-made *Original artist:* User:Pslawinski

- **File:Argon-glow.jpg** *Source:* https://upload.wikimedia.org/wikipedia/commons/5/53/Argon-glow.jpg *License:* CC BY 3.0 *Contributors:* http://images-of-elements.com/argon.php *Original artist:* Jurii

- **File:Argon_Spectrum.png** *Source:* https://upload.wikimedia.org/wikipedia/commons/3/37/Argon_Spectrum.png *License:* CC BY-SA 3.0 *Contributors:* Own work http://goiphone5.com/ *Original artist:* Abilanin

- **File:Argon_discharge_tube.jpg** *Source:* https://upload.wikimedia.org/wikipedia/commons/8/87/Argon_discharge_tube.jpg *License:* GFDL 1.2 *Contributors:* Own work *Original artist:* Alchemist-hp (<a href='//commons.wikimedia.org/wiki/User_talk:Alchemist-hp' title='User talk: Alchemist-hp'>talk</a>) (www.pse-mendelejew.de)

- **File:Aristoteles_Louvre.jpg** *Source:* https://upload.wikimedia.org/wikipedia/commons/a/a4/Aristoteles_Louvre.jpg *License:* CC BY-SA 2.5 *Contributors:* Eric Gaba (User:Sting), July 2005. *Original artist:* After Lysippos

- **File:Asterisks_one.svg** *Source:* https://upload.wikimedia.org/wikipedia/commons/4/49/Asterisks_one.svg *License:* CC BY-SA 3.0 *Contributors:* Own work *Original artist:* DePiep

- **File:Asterisks_one_(right).svg** *Source:* https://upload.wikimedia.org/wikipedia/commons/1/1c/Asterisks_one_%28right%29.svg *License:* CC BY-SA 3.0 *Contributors:* Own work *Original artist:* DePiep
- **File:Asterisks_two.svg** *Source:* https://upload.wikimedia.org/wikipedia/commons/3/3f/Asterisks_two.svg *License:* CC BY-SA 3.0 *Contributors:* Own work *Original artist:* DePiep
- **File:Avogadro_Amedeo.jpg** *Source:* https://upload.wikimedia.org/wikipedia/commons/3/3d/Avogadro_Amedeo.jpg *License:* Public domain *Contributors:* Edgar Fahs Smith collection *Original artist:* From a drawing by C. Sentier, executed in Torino at Litografia Doyen in 1856.
- **File:Axe_of_iron_from_Swedish_Iron_Age,_found_at_Gotland,_Sweden.jpg** *Source:* https://upload.wikimedia.org/wikipedia/commons/c/c4/Axe_of_iron_from_Swedish_Iron_Age%2C_found_at_Gotland%2C_Sweden.jpg *License:* Public domain *Contributors:* Nordisk familjebok (1910), vol.13, *Till art. Järnåldern. II* [1] *Original artist:* Nordisk familjebok
- **File:Bas_fourneau.png** *Source:* https://upload.wikimedia.org/wikipedia/commons/a/ac/Bas_fourneau.png *License:* Public domain *Contributors:* ? *Original artist:* ?
- **File:Bessemer_converter.jpg** *Source:* https://upload.wikimedia.org/wikipedia/commons/6/61/Bessemer_converter.jpg *License:* Public domain *Contributors:* ? *Original artist:* ?
- **File:Blausen_0342_ElectronEnergyLevels.png** *Source:* https://upload.wikimedia.org/wikipedia/commons/2/2c/Blausen_0342_ElectronEnergyLevels.png *License:* CC BY 3.0 *Contributors:* Own work *Original artist:* BruceBlaus
- **File:Bohr-atom-PAR.svg** *Source:* https://upload.wikimedia.org/wikipedia/commons/5/55/Bohr-atom-PAR.svg *License:* CC-BY-SA-3.0 *Contributors:* Transferred from en.wikipedia to Commons. *Original artist:* Original uplo:JabberWok]] at en.wikipedia
- **File:Bohr_atom_animation_2.gif** *Source:* https://upload.wikimedia.org/wikipedia/commons/1/17/Bohr_atom_animation_2.gif *License:* CC BY-SA 3.0 *Contributors:* Own work *Original artist:* Kurzondddd
- **File:Broglie_Big.jpg** *Source:* https://upload.wikimedia.org/wikipedia/commons/d/d2/Broglie_Big.jpg *License:* Public domain *Contributors:* http://www.physics.umd.edu/courses/Phys420/Spring2002/Parra_Spring2002/HTMPages/whoswho.htm *Original artist:* Unknown
- **File:Buckminsterfullerene-perspective-3D-balls.png** *Source:* https://upload.wikimedia.org/wikipedia/commons/0/0f/Buckminsterfullerene-perspective-3D-balls.png *License:* Public domain *Contributors:* Own work *Original artist:* Benjah-bmm27
- **File:Butan_Lewis.svg** *Source:* https://upload.wikimedia.org/wikipedia/commons/c/cb/Butan_Lewis.svg *License:* Public domain *Contributors:* Own work *Original artist:* NEUROtiker ⇌
- **File:C60a.png** *Source:* https://upload.wikimedia.org/wikipedia/commons/4/41/C60a.png *License:* CC-BY-SA-3.0 *Contributors:* Transferred from en.wikipedia to Commons. *Original artist:* The original uploader was Mstroeck at English Wikipedia Later versions were uploaded by Bryn C at en.wikipedia.
- **File:Carnot2.jpg** *Source:* https://upload.wikimedia.org/wikipedia/commons/e/ec/Carnot2.jpg *License:* Public domain *Contributors:* ? *Original artist:* ?
- **File:Carnot_engine_(hot_body_-_working_body_-_cold_body).jpg** *Source:* https://upload.wikimedia.org/wikipedia/commons/c/c7/Carnot_engine_%28hot_body_-_working_body_-_cold_body%29.jpg *License:* Public domain *Contributors:* Transferred from en.wikipedia; transferred to Commons by User:Burpelson AFB using CommonsHelper. *Original artist:* Libb Thims (talk). Original uploader was Libb Thims at en.wikipedia
- **File:Chemielabor_des_18._Jahrhunderts,_Naturhistorisches_Museum_Wien.jpg** *Source:* https://upload.wikimedia.org/wikipedia/commons/f/fa/Chemielabor_des_18._Jahrhunderts%2C_Naturhistorisches_Museum_Wien.jpg *License:* CC BY 3.0 *Contributors:* Own work *Original artist:* Sandstein
- **File:Chinese_Fining_and_Blast_Furnace.jpg** *Source:* https://upload.wikimedia.org/wikipedia/commons/8/8c/Chinese_Fining_and_Blast_Furnace.jpg *License:* Public domain *Contributors:* Transferred from en.wikipedia; Transfer was stated to be made by User:Rifleman_82. *Original artist:* Original uploader was PericlesofAthens at en.wikipedia
- **File:Coat_of_Arms_of_Niels_Bohr.svg** *Source:* https://upload.wikimedia.org/wikipedia/commons/d/da/Coat_of_Arms_of_Niels_Bohr.svg *License:* CC BY-SA 3.0 *Contributors:* Own work from File:Royal Coat of Arms of Denmark.svg (Collar of the Order of the Elephant) + File:Yin yang.svg. For images of his coat of arms as displayed at Frederiksborg Castle, Denmark, see: [1], [2], [3] *Original artist:* GJo
- **File:Commons-logo.svg** *Source:* https://upload.wikimedia.org/wikipedia/en/4/4a/Commons-logo.svg *License:* ? *Contributors:* ? *Original artist:* ?
- **File:Crookes_tube_two_views.jpg** *Source:* https://upload.wikimedia.org/wikipedia/commons/b/bf/Crookes_tube_two_views.jpg *License:* CC BY-SA 3.0 at *Contributors:* File:Crookes tube-not in use-lateral view-standing cross prPNr°07.jpg and File:Crookes tube-in use-lateral view-standing cross prPNr°11.jpg *Original artist:* D-Kuru
- **File:DNA_chemical_structure.svg** *Source:* https://upload.wikimedia.org/wikipedia/commons/e/e4/DNA_chemical_structure.svg *License:* CC-BY-SA-3.0 *Contributors:* <a href='//validator.w3.org/' data-x-rel='nofollow'><img alt='W3C' src='https://upload.wikimedia.org/wikipedia/commons/thumb/1/1a/Valid_SVG_1.1_%28green%29.svg/88px-Valid_SVG_1.1_%28green%29.svg.png' width='88' height='30' style='vertical-align: top' srcset='https://upload.wikimedia.org/wikipedia/commons/thumb/1/1a/Valid_SVG_1.1_%28green%29.svg/132px-Valid_SVG_1.1_%28green%29.svg.png 1.5x, https://upload.wikimedia.org/wikipedia/commons/thumb/1/1a/Valid_SVG_1.1_%28green%29.svg/176px-Valid_SVG_1.1_%28green%29.svg.png 2x' data-file-width='91' data-file-height='31' /></a>iThe source code of this SVG is <a data-x-rel='nofollow' class='external text' href='//validator.w3.org/check?uri=https%3A%2F%2Fcommons.wikimedia.org%2Fwiki%2FSpecial%3AFilepath%2FDNA_chemical_structure.svg,<span>,&,</span>,ss=1#source'>valid</a>. *Original artist:* Madprime (talk · contribs)
- **File:Dagger_India_Louvre_MR13434.jpg** *Source:* https://upload.wikimedia.org/wikipedia/commons/3/3b/Dagger_India_Louvre_MR13434.jpg *License:* Public domain *Contributors:* Own work *Original artist:* Marie-Lan Nguyen

- **File:Dalton_John_desk.jpg** *Source:* https://upload.wikimedia.org/wikipedia/commons/3/3f/Dalton_John_desk.jpg *License:* Public domain *Contributors:* Frontispiece of *John Dalton and the Rise of Modern Chemistry* by Henry Roscoe *Original artist:* Henry Roscoe (author), William Henry Worthington (engraver), and Joseph Allen (painter)

- **File:Daltons_symbols.gif** *Source:* https://upload.wikimedia.org/wikipedia/commons/3/39/Daltons_symbols.gif *License:* Public domain *Contributors:* ? *Original artist:* ?

- **File:Daniell-element---Elemento-Daniell.jpg** *Source:* https://upload.wikimedia.org/wikipedia/commons/f/fc/Daniell-element---Elemento-Daniell.jpg *License:* Public domain *Contributors:* Tratado elemental de física experimental y aplicada, y de meteorología... *Original artist:* A. Ganot (autor) ; J. Molau (traductor), J.M. Pérez (corrector), J. Canalejas (revisor).

- **File:David_-_Portrait_of_Monsieur_Lavoisier_and_His_Wife.jpg** *Source:* https://upload.wikimedia.org/wikipedia/commons/4/4e/David_-_Portrait_of_Monsieur_Lavoisier_and_His_Wife.jpg *License:* Public domain *Contributors:* Metropolitan Museum of Art, online database: entry 436106 *Original artist:* Jacques-Louis David

- **File:De_Re_Metallica_1556_p_357AQ20_(3).TIF** *Source:* https://upload.wikimedia.org/wikipedia/commons/1/10/De_Re_Metallica_1556_p_357AQ20_%283%29.TIF *License:* Public domain *Contributors:* Chemical Heritage Foundation *Original artist:* Chemical Heritage Foundation

- **File:Democritus2.jpg** *Source:* https://upload.wikimedia.org/wikipedia/commons/b/b9/Democritus2.jpg *License:* Public domain *Contributors:* ? *Original artist:* ?

- **File:Discovery_of_chemical_elements.svg** *Source:* https://upload.wikimedia.org/wikipedia/commons/3/3d/Discovery_of_chemical_elements.svg *License:* CC BY-SA 3.0 *Contributors:* Wikimedia Commons. *Original artist:* Sandbh

- **File:Dmitri_Ivanowitsh_Mendeleev.jpg** *Source:* https://upload.wikimedia.org/wikipedia/commons/b/b3/Dmitri_Ivanowitsh_Mendeleev.jpg *License:* Public domain *Contributors:* New York Public Library Archives *Original artist:* Historical and Public Figures Collection

- **File:Dmitry_Mendeleyev_Osnovy_Khimii_1869-1871_first_periodic_table.jpg** *Source:* https://upload.wikimedia.org/wikipedia/commons/1/1d/Dmitry_Mendeleyev_Osnovy_Khimii_1869-1871_first_periodic_table.jpg *License:* Public domain *Contributors:* Chemical Heritage Foundation *Original artist:* Dmitry Ivanovich Mendeleyev, 1834-1907

- **File:East&southern_africa_early_iron_age.png** *Source:* https://upload.wikimedia.org/wikipedia/commons/8/82/East%26southern_africa_early_iron_age.png *License:* Public domain *Contributors:* own painting in a PD map (File:BlankMap-World.png) *Original artist:* User:Ulamm

- **File:Edit-clear.svg** *Source:* https://upload.wikimedia.org/wikipedia/en/f/f2/Edit-clear.svg *License:* Public domain *Contributors:* The *Tango! Desktop Project.* *Original artist:*

  The people from the Tango! project. And according to the meta-data in the file, specifically: "Andreas Nilsson, and Jakub Steiner (although minimally)."

- **File:Eight_founding_schools.png** *Source:* https://upload.wikimedia.org/wikipedia/commons/8/85/Eight_founding_schools.png *License:* Public domain *Contributors:* Own work *Original artist:* Libb Thims

- **File:Electrolysis_Apparatus.png** *Source:* https://upload.wikimedia.org/wikipedia/commons/d/d1/Electrolysis_Apparatus.png *License:* CC BY-SA 3.0 *Contributors:* Own work *Original artist:* Ivan Akira

- **File:Electron_affinity_of_the_elements.svg** *Source:* https://upload.wikimedia.org/wikipedia/commons/6/6c/Electron_affinity_of_the_elements.svg *License:* CC BY-SA 3.0 *Contributors:* Based on Electron affinities of the elements 2.png by Sandbh. *Original artist:* DePiep

- **File:Electron_shell_010_Neon_-_no_label.svg** *Source:* https://upload.wikimedia.org/wikipedia/commons/3/3e/Electron_shell_010_Neon_-_no_label.svg *License:* CC BY-SA 2.0 uk *Contributors:* http://commons.wikimedia.org/wiki/Category:Electron_shell_diagrams (corresponding labeled version) *Original artist:* commons:User:Pumbaa (original work by commons:User:Greg Robson)

- **File:Elementspiral_(polyatomic).svg** *Source:* https://upload.wikimedia.org/wikipedia/commons/c/ce/Elementspiral_%28polyatomic%29.svg *License:* CC BY-SA 3.0 *Contributors:* Own work *Original artist:* DePiep

- **File:Empirical_atomic_radius_trends.png** *Source:* https://upload.wikimedia.org/wikipedia/commons/b/bc/Empirical_atomic_radius_trends.png *License:* GFDL *Contributors:* Own work *Original artist:* StringTheory11

- **File:Endohedral_fullerene.png** *Source:* https://upload.wikimedia.org/wikipedia/commons/e/e1/Endohedral_fullerene.png *License:* GFDL *Contributors:* Own work *Original artist:* Hajv01

- **File:Ernest_Rutherford_1908.jpg** *Source:* https://upload.wikimedia.org/wikipedia/commons/d/de/Ernest_Rutherford_1908.jpg *License:* Public domain *Contributors:* This image is available from the United States Library of Congress's Prints and Photographs division under the digital ID ggbain.03392.

  This tag does not indicate the copyright status of the attached work. A normal copyright tag is still required. See Commons:Licensing for more information. *Original artist:* Bain News Service, publisher

- **File:Erwin_Schrödinger_(1933).jpg** *Source:* https://upload.wikimedia.org/wikipedia/commons/2/2e/Erwin_Schr%C3%B6dinger_%281933%29.jpg *License:* Public domain *Contributors:* http://nobelprize.org/nobel_prizes/physics/laureates/1933/schrodinger-bio.html *Original artist:* Nobel foundation

- **File:Esoteric_Taijitu.svg** *Source:* https://upload.wikimedia.org/wikipedia/commons/2/21/Esoteric_Taijitu.svg *License:* Public domain *Contributors:* Own work *Original artist:* Kenny Shen

- **File:Ethanol-3D-balls.png** *Source:* https://upload.wikimedia.org/wikipedia/commons/b/b0/Ethanol-3D-balls.png *License:* Public domain *Contributors:* ? *Original artist:* ?

- **File:Evolution_of_atomic_models_infographic.svg** *Source:* https://upload.wikimedia.org/wikipedia/commons/8/8c/Evolution_of_atomic_models_infographic.svg *License:* CC BY 3.0 *Contributors:* Own work *Original artist:* Ville Takanen

- **File:First_Ionization_Energy.svg** *Source:* https://upload.wikimedia.org/wikipedia/commons/1/1d/First_Ionization_Energy.svg *License:* CC BY-SA 3.0 *Contributors:* http://commons.wikimedia.org/wiki/File:Erste_Ionisierungsenergie_PSE_color_coded.png *Original artist:* User: Sponk

- **File:Flag_of_Denmark.svg** *Source:* https://upload.wikimedia.org/wikipedia/commons/9/9c/Flag_of_Denmark.svg *License:* Public domain *Contributors:* Own work *Original artist:* User:Madden

- **File:Flag_of_India.svg** *Source:* https://upload.wikimedia.org/wikipedia/en/4/41/Flag_of_India.svg *License:* Public domain *Contributors:* ? *Original artist:* ?

- **File:Flag_of_Pakistan.svg** *Source:* https://upload.wikimedia.org/wikipedia/commons/3/32/Flag_of_Pakistan.svg *License:* Public domain *Contributors:* The drawing and the colors were based from flagspot.net. *Original artist:* User:Zscout370

- **File:Flag_of_Russia.svg** *Source:* https://upload.wikimedia.org/wikipedia/en/f/f3/Flag_of_Russia.svg *License:* PD *Contributors:* ? *Original artist:* ?

- **File:Folder_Hexagonal_Icon.svg** *Source:* https://upload.wikimedia.org/wikipedia/en/4/48/Folder_Hexagonal_Icon.svg *License:* Cc-by-sa-3.0 *Contributors:* ? *Original artist:* ?

- **File:Fotothek_df_n-08_0000320.jpg** *Source:* https://upload.wikimedia.org/wikipedia/commons/f/f8/Fotothek_df_n-08_0000320.jpg *License:* CC BY-SA 3.0 de *Contributors:* Deutsche Fotothek *Original artist:* Eugen Nosko

- **File:Fotothek_df_tg_0006097_Theosophie_\char"005E\relax{}_Alchemie.jpg** *Source:* https://upload.wikimedia.org/wikipedia/commons/6/6f/Fotothek_df_tg_0006097_Theosophie_%5E_Alchemie.jpg *License:* Public domain *Contributors:* Deutsche Fotothek *Original artist:* ?

- **File:Fotothek_df_tg_0007129_Theosophie_\char"005E\relax{}_Alchemie.jpg** *Source:* https://upload.wikimedia.org/wikipedia/commons/8/8f/Fotothek_df_tg_0007129_Theosophie_%5E_Alchemie.jpg *License:* Public domain *Contributors:* Deutsche Fotothek *Original artist:* ?

- **File:Galvani-frog-legs.PNG** *Source:* https://upload.wikimedia.org/wikipedia/en/a/ab/Galvani-frog-legs.PNG *License:* Public domain *Contributors:* ? *Original artist:* ?

- **File:Geber.jpg** *Source:* https://upload.wikimedia.org/wikipedia/commons/e/ea/Geber.jpg *License:* Public domain *Contributors:* http://histoirechimie.free.fr/Lien/Geber.jpg *Original artist:* ?

- **File:Geiger-Marsden_experiment_expectation_and_result.svg** *Source:* https://upload.wikimedia.org/wikipedia/commons/f/f9/Geiger-Marsden_experiment_expectation_and_result.svg *License:* CC BY-SA 3.0 *Contributors:* Own work *Original artist:* Kurzon

- **File:Georgius_Agricola.jpg** *Source:* https://upload.wikimedia.org/wikipedia/commons/6/63/Georgius_Agricola.jpg *License:* Public domain *Contributors:* http://kanitz.onlinehome.de/agricolagymnasium/agrigale.htm *Original artist:* {{creator:|Year = }}

- **File:Glenn_Seaborg_-_1964.jpg** *Source:* https://upload.wikimedia.org/wikipedia/commons/4/47/Glenn_Seaborg_-_1964.jpg *License:* Public domain *Contributors:* NAIL Control Number: NWDNS-326-COM-12 NARA (enter "Glenn Seaborg" in search form under Digital Copies tab) *Original artist:* Atomic Energy Commission. (1946 - 01/19/1975)

- **File:Goodyear-blimp.jpg** *Source:* https://upload.wikimedia.org/wikipedia/commons/2/2a/Goodyear-blimp.jpg *License:* Public domain *Contributors:* user-made *Original artist:* Derek Jensen (Tysto)

- **File:Guericke-electricaldevice.PNG** *Source:* https://upload.wikimedia.org/wikipedia/commons/3/32/Guericke-electricaldevice.PNG *License:* Public domain *Contributors:*

- Picture was obtained from http://www.corrosion-doctors.org/Biographies/GuerickeBio.htm *Original artist:* Original uploader was HappyApple at en.wikipedia

- **File:Hall-heroult-kk-2008-12-31.png** *Source:* https://upload.wikimedia.org/wikipedia/commons/2/24/Hall-heroult-kk-2008-12-31.png *License:* CC BY-SA 3.0 *Contributors:* Transferred from en.wikipedia to Commons by Vinhtantran using CommonsHelper. *Original artist:* Kashkhan at English Wikipedia

- **File:Hauksbee_Generator.JPG** *Source:* https://upload.wikimedia.org/wikipedia/commons/b/b6/Hauksbee_Generator.JPG *License:* Public domain *Contributors:* ? *Original artist:* ?

- **File:HeTube.jpg** *Source:* https://upload.wikimedia.org/wikipedia/commons/1/1f/HeTube.jpg *License:* CC BY-SA 2.5 *Contributors:* user-made *Original artist:* User:Pslawinski

- **File:Heinkel_He_111_during_the_Battle_of_Britain.jpg** *Source:* https://upload.wikimedia.org/wikipedia/commons/8/82/Heinkel_He_111_during_the_Battle_of_Britain.jpg *License:* Public domain *Contributors:* This is photograph MH6547 from the collections of the Imperial War Museums (collection no. 4700-05) *Original artist:* Unknown

- **File:Heisenbergbohr.jpg** *Source:* https://upload.wikimedia.org/wikipedia/commons/1/1a/Heisenbergbohr.jpg *License:* Public domain *Contributors:* http://www.fnal.gov/pub/inquiring/timeline/images/heisenbergbohr.jpg shown on http://www.fnal.gov/pub/inquiring/timeline/05.html *Original artist:* Fermilab, U.S. Department of Energy

- **File:Helium-glow.jpg** *Source:* https://upload.wikimedia.org/wikipedia/commons/0/00/Helium-glow.jpg *License:* CC BY 3.0 *Contributors:* http://images-of-elements.com/helium.php *Original artist:* Jurii

- **File:Helium_atom_QM.svg** *Source:* https://upload.wikimedia.org/wikipedia/commons/2/23/Helium_atom_QM.svg *License:* CC-BY-SA-3.0 *Contributors:* Own work *Original artist:* User:Yzmo

- **File:Helium_discharge_tube.jpg** *Source:* https://upload.wikimedia.org/wikipedia/commons/8/82/Helium_discharge_tube.jpg *License:* GFDL 1.2 *Contributors:* Own work *Original artist:* Alchemist-hp (<a href='//commons.wikimedia.org/wiki/User_talk:Alchemist-hp' title='User talk:Alchemist-hp'>talk</a>) (www.pse-mendelejew.de)

- **File:Helium_spectra.jpg** *Source:* https://upload.wikimedia.org/wikipedia/commons/c/c3/Helium_spectra.jpg *License:* Public domain *Contributors:* Transferred from en.wikipedia; transferred to Commons by User:João Sousa using CommonsHelper. *Original artist:* (teravolt (talk)). Original uploader was Teravolt at en.wikipedia

- **File:Helium_spectrum.jpg** *Source:* https://upload.wikimedia.org/wikipedia/commons/8/80/Helium_spectrum.jpg *License:* Public domain *Contributors:* http://imagine.gsfc.nasa.gov/docs/teachers/lessons/xray_spectra/worksheet-specgraph2-sol.html *Original artist:* NASA

- **File:Henning_brand.jpg** *Source:* https://upload.wikimedia.org/wikipedia/en/7/79/Henning_brand.jpg *License:* Public domain *Contributors:* ? *Original artist:* ?

- **File:Henry_Moseley.jpg** *Source:* https://upload.wikimedia.org/wikipedia/en/d/dd/Henry_Moseley.jpg *License:* PD-US *Contributors:* ? *Original artist:* ?

- **File:Hexahedron.jpg** *Source:* https://upload.wikimedia.org/wikipedia/commons/7/78/Hexahedron.jpg *License:* CC-BY-SA-3.0 *Contributors:* ? *Original artist:* ?

- **File:Heyrovského_polarograf_2.jpg** *Source:* https://upload.wikimedia.org/wikipedia/commons/1/17/Heyrovsk%C3%A9ho_polarograf_2.jpg *License:* CC BY-SA 3.0 *Contributors:* Own work *Original artist:* Lukáš Mižoch

- **File:Humphry_davy.jpg** *Source:* https://upload.wikimedia.org/wikipedia/commons/1/1f/Humphry_davy.jpg *License:* Public domain *Contributors:* ? *Original artist:* ?

- **File:HyderAli.jpg** *Source:* https://upload.wikimedia.org/wikipedia/commons/1/1c/HyderAli.jpg *License:* Public domain *Contributors:* Image: HyderAli.jpg *Original artist:* user:Anetode

- **File:Hydrochinon2.svg** *Source:* https://upload.wikimedia.org/wikipedia/commons/b/b2/Hydrochinon2.svg *License:* Public domain *Contributors:* Own work *Original artist:* NEUROtiker ⇌

- **File:Ice-calorimeter.jpg** *Source:* https://upload.wikimedia.org/wikipedia/commons/3/35/Ice-calorimeter.jpg *License:* Public domain *Contributors:* originally uploaded http://en.wikipedia.org/wiki/Image:Ice-calorimeter.jpg *Original artist:* Originally en:User:Sadi Carnot

- **File:Icosahedron.jpg** *Source:* https://upload.wikimedia.org/wikipedia/commons/e/eb/Icosahedron.jpg *License:* CC-BY-SA-3.0 *Contributors:* en:image:poly.pov *Original artist:* Created by en:User:Cyp and copied from the English Wikipedia.

- **File:Ionization_energies.png** *Source:* https://upload.wikimedia.org/wikipedia/commons/2/27/Ionization_energies.png *License:* Public domain *Contributors:* Self-made; based on data from: Martin, W. C.; Wiese, W. L. (1996) *Atomic, Molecular, & Optical Physics Handbook*, American Institute of Physics ISBN 156396242X *Original artist:* RJHall

- **File:Iron-Making.jpg** *Source:* https://upload.wikimedia.org/wikipedia/commons/5/57/Iron-Making.jpg *License:* Public domain *Contributors:* ? *Original artist:* ?

- **File:Isobutane_numbered_2D.svg** *Source:* https://upload.wikimedia.org/wikipedia/commons/a/a2/Isobutane_numbered_2D.svg *License:* Public domain *Contributors:* Own work *Original artist:* Rubber Duck (☮ • ✍)

- **File:JJ_Thomson_Cathode_Ray_2_explained.svg** *Source:* https://upload.wikimedia.org/wikipedia/commons/a/a3/JJ_Thomson_Cathode_Ray_2_explained.svg *License:* Public domain *Contributors:* Own work *Original artist:* Kurzon

- **File:JJ_Thomson_Crookes_Tube_Replica.jpg** *Source:* https://upload.wikimedia.org/wikipedia/commons/7/7d/JJ_Thomson_Crookes_Tube_Replica.jpg *License:* CC BY-SA 3.0 *Contributors:* Own work *Original artist:* Kurzon

- **File:Jabir_ibn_Hayyan.jpg** *Source:* https://upload.wikimedia.org/wikipedia/commons/0/04/Jabir_ibn_Hayyan.jpg *License:* Public domain *Contributors:*

    - Codici Ashburnhamiani 1166, Biblioteca Medicea Laurenziana, Florence

  Original artist: (uploaded by user Halfdan)

- **File:John_Dalton_by_Charles_Turner.jpg** *Source:* https://upload.wikimedia.org/wikipedia/commons/d/d4/John_Dalton_by_Charles_Turner.jpg *License:* Public domain *Contributors:* This image is available from the United States Library of Congress's Prints and Photographs division under the digital ID cph.3b12511.
  This tag does not indicate the copyright status of the attached work. A normal copyright tag is still required. See Commons:Licensing for more information. *Original artist:* Charles Turner

- **File:JosephWright-Alchemist-1.jpg** *Source:* https://upload.wikimedia.org/wikipedia/commons/c/c1/JosephWright-Alchemist-1.jpg *License:* Public domain *Contributors:* ? *Original artist:* ?

- **File:Joseph_louis_gay-lussac.jpg** *Source:* https://upload.wikimedia.org/wikipedia/commons/3/3a/Joseph_louis_gay-lussac.jpg *License:* Public domain *Contributors:* ? *Original artist:* ?

- **File:Josiah_Willard_Gibbs_-from_MMS-.jpg** *Source:* https://upload.wikimedia.org/wikipedia/commons/c/c7/Josiah_Willard_Gibbs_-from_MMS-.jpg *License:* Public domain *Contributors:* Frontispiece of *The Scientific Papers of J. Willard Gibbs*, in two volumes, eds. H. A. Bumstead and R. G. Van Name, (London and New York: Longmans, Green, and Co., 1906) *Original artist:* Unknown. Uploaded by Serge Lachinov (обработка для wiki)

- **File:Jöns_Jacob_Berzelius_from_Familj-Journalen1873.png** *Source:* https://upload.wikimedia.org/wikipedia/commons/b/b0/J%C3%B6ns_Jacob_Berzelius_from_Familj-Journalen1873.png *License:* Public domain *Contributors:* ? *Original artist:* ?

- **File:Kekule_acetic_acid_formulae.jpg** *Source:* https://upload.wikimedia.org/wikipedia/commons/f/f3/Kekule_acetic_acid_formulae.jpg *License:* Public domain *Contributors:* Lehrbuch der Organischen Chemie *Original artist:* A. Kekulé

- **File:Kepler-solar-system-2.gif** *Source:* https://upload.wikimedia.org/wikipedia/commons/1/1d/Kepler-solar-system-2.gif *License:* Public domain *Contributors:* ? *Original artist:* ?

- **File:Klechkovski_rule.svg** *Source:* https://upload.wikimedia.org/wikipedia/commons/9/95/Klechkovski_rule.svg *License:* CC-BY-SA-3.0 *Contributors:* No machine-readable source provided. Own work assumed (based on copyright claims). *Original artist:* No machine-readable author provided. Bono~commonswiki assumed (based on copyright claims).
- **File:Known-elements-1700-2000.png** *Source:* https://upload.wikimedia.org/wikipedia/commons/d/dc/Known-elements-1700-2000.png *License:* CC BY-SA 3.0 *Contributors:* Own work *Original artist:* Soerfm
- **File:KrTube.jpg** *Source:* https://upload.wikimedia.org/wikipedia/commons/e/e7/KrTube.jpg *License:* CC BY-SA 2.5 *Contributors:* user-made *Original artist:* User:Pslawinski
- **File:Krypton-glow.jpg** *Source:* https://upload.wikimedia.org/wikipedia/commons/9/9c/Krypton-glow.jpg *License:* CC BY 3.0 *Contributors:* http://images-of-elements.com/krypton.php *Original artist:* Jurii
- **File:Krypton_Spectrum.jpg** *Source:* https://upload.wikimedia.org/wikipedia/commons/a/a6/Krypton_Spectrum.jpg *License:* Public domain *Contributors:* en wikipedia ([1]) *Original artist:* Mrgoogfan
- **File:Krypton_discharge_tube.jpg** *Source:* https://upload.wikimedia.org/wikipedia/commons/5/50/Krypton_discharge_tube.jpg *License:* GFDL 1.2 *Contributors:* Own work *Original artist:* Alchemist-hp (<a href='//commons.wikimedia.org/wiki/User_talk:Alchemist-hp' title='User talk:Alchemist-hp'>talk</a>) (www.pse-mendelejew.de)
- **File:Lavoisier.jpg** *Source:* https://upload.wikimedia.org/wikipedia/commons/7/7e/Lavoisier.jpg *License:* Public domain *Contributors:* University of Texas , Austin *Original artist:* David
- **File:Lavoisier_-_Traité_élémentaire_de_chimie,_1789_-_3895821_F.tif** *Source:* https://upload.wikimedia.org/wikipedia/commons/2/27/Lavoisier_-_Trait%C3%A9_%C3%A9l%C3%A9mentaire_de_chimie%2C_1789_-_3895821_F.tif *License:* Public domain *Contributors:* Available in the digital library of the European Library of Information and Culture and uploaded in partnership (ID: '). *Original artist:* Lavoisier, Antoine Laurent
- **File:Libr0310.jpg** *Source:* https://upload.wikimedia.org/wikipedia/commons/2/24/Libr0310.jpg *License:* Public domain *Contributors:* ? *Original artist:* ?
- **File:Lomonosov_Chymiae_Physicae_1752.jpg** *Source:* https://upload.wikimedia.org/wikipedia/commons/3/30/Lomonosov_Chymiae_Physicae_1752.jpg *License:* CC-BY-SA-3.0 *Contributors:* Михаил Васильевич Ломоносов. 275 лет со дня рождения. Разрезной фотоальбом. М.: Планета. 1986 *Original artist:* Serge Lachinov (обработка для wiki)
- **File:Maquina_vapor_Watt_ETSIIM.jpg** *Source:* https://upload.wikimedia.org/wikipedia/commons/9/9e/Maquina_vapor_Watt_ETSIIM.jpg *License:* CC-BY-SA-3.0 *Contributors:* Enciclopedia Libre *Original artist:* Nicolás Pérez
- **File:Mariecurie.jpg** *Source:* https://upload.wikimedia.org/wikipedia/commons/d/d9/Mariecurie.jpg *License:* Public domain *Contributors:* http://www.mlahanas.de/Physics/Bios/MarieCurie.html *Original artist:* Unknown
- **File:Medeleeff_by_repin.jpg** *Source:* https://upload.wikimedia.org/wikipedia/commons/b/b3/Medeleeff_by_repin.jpg *License:* Public domain *Contributors:* http://www.picture.art-catalog.ru/picture.php?id_picture=4318 *Original artist:* Ilya Repin
- **File:Medieval-university.jpg** *Source:* https://upload.wikimedia.org/wikipedia/commons/a/ad/Medieval-university.jpg *License:* Public domain *Contributors:* Image from: Innozence IV, Apparatus super V libros Decretalium, 296ff., Paris, Bibliothèque de la Sorbonne, ms. 31, f. 278

This file from http://www.educ.fc.ul.pt/docentes/opombo/hfe/momentos/modelos/universidade.htm

Original artist: French school, 14th century

- **File:Mendeleev'{}s_1869_periodic_table.png** *Source:* https://upload.wikimedia.org/wikipedia/commons/b/bb/Mendeleev%27s_1869_periodic_table.png *License:* Public domain *Contributors:* Originally from en.wikipedia; description page is/was here. *Original artist:* Original uploader was Sadi Carnot at en.wikipedia
- **File:Mendeleev'{}s_periodic_table_(1869).svg** *Source:* https://upload.wikimedia.org/wikipedia/commons/4/46/Mendeleev%27s_periodic_table_%281869%29.svg *License:* Public domain *Contributors:* https://archive.org/stream/zeitschriftfrch12unkngoog#page/n414/mode/2up *Original artist:* Dimitri Mendeleev (in Zeitschrift für Chemie (1869))
- **File:Mendelejevs_periodiska_system_1871.png** *Source:* https://upload.wikimedia.org/wikipedia/commons/5/55/Mendelejevs_periodiska_system_1871.png *License:* Public domain *Contributors:* Källa:Dmitrij Ivanovitj Mendelejev (1834 - 1907). Originally from sv.wikipedia; description page is/was here. *Original artist:* Original uploader was Den fjättrade ankan at sv.wikipedia
- **File:Mendelejew_signature.jpg** *Source:* https://upload.wikimedia.org/wikipedia/commons/8/8a/Mendelejew_signature.jpg *License:* Public domain *Contributors:* http://www.prometeus.nsc.ru/archives/exhibit2/prmendel.ssi *Original artist:* ?
- **File:Mendeleyev_gold_Barry_Kent.JPG** *Source:* https://upload.wikimedia.org/wikipedia/commons/3/31/Mendeleyev_gold_Barry_Kent.JPG *License:* CC-BY-SA-3.0 *Contributors:* Own work *Original artist:* Robert Wielgórski (Barry Kent)
- **File:Metal_production_in_Ancient_Middle_East.svg** *Source:* https://upload.wikimedia.org/wikipedia/commons/0/0c/Metal_production_in_Ancient_Middle_East.svg *License:* CC BY-SA 3.0 *Contributors:*
- Regs_productoras_de_metales_en_la_Edad_Antigua_en_Oriente_Medio.svg*Original artist:* Regs_productoras_de_metales_en_la_Edad_Antig Crates
- **File:Millikan.jpg** *Source:* https://upload.wikimedia.org/wikipedia/commons/2/2f/Millikan.jpg *License:* Public domain *Contributors:* http://nobelprize.org/nobel_prizes/physics/laureates/1923/millikan-bio.html *Original artist:* Nobel foundation
- **File:Modern_3T_MRI.JPG** *Source:* https://upload.wikimedia.org/wikipedia/commons/b/bd/Modern_3T_MRI.JPG *License:* CC-BY-SA-3.0 *Contributors:* Photographed by User:KasugaHuang on Mar 27, 2006 at Tri-Service General Hospital, Taiwan. *Original artist:* User:KasugaHuang

- **File:Nagasakibomb.jpg** *Source:* https://upload.wikimedia.org/wikipedia/commons/e/e0/Nagasakibomb.jpg *License:* Public domain *Contributors:* http://www.archives.gov/research/military/ww2/photos/images/ww2-163.jpg National Archives image (208-N-43888) *Original artist:* Charles Levy from one of the B-29 Superfortresses used in the attack.

- **File:NatarajaMET.JPG** *Source:* https://upload.wikimedia.org/wikipedia/commons/1/1d/NatarajaMET.JPG *License:* CC BY-SA 2.5 *Contributors:* Originally from en.wikipedia; description page is (was) here
Original artist: User Kaysov on en.wikipedia

- **File:NeTube.jpg** *Source:* https://upload.wikimedia.org/wikipedia/commons/8/88/NeTube.jpg *License:* CC BY-SA 2.5 *Contributors:* usermade *Original artist:* User:Pslawinski

- **File:Neon-glow.jpg** *Source:* https://upload.wikimedia.org/wikipedia/commons/f/f8/Neon-glow.jpg *License:* CC BY 3.0 *Contributors:* http://images-of-elements.com/neon.php *Original artist:* Jurii

- **File:Neon_discharge_tube.jpg** *Source:* https://upload.wikimedia.org/wikipedia/commons/4/46/Neon_discharge_tube.jpg *License:* GFDL 1.2 *Contributors:* Own work *Original artist:* Alchemist-hp (<a href='//commons.wikimedia.org/wiki/User_talk:Alchemist-hp' title='User talk: Alchemist-hp'>talk</a>) (www.pse-mendelejew.de)

- **File:Neon_spectra.jpg** *Source:* https://upload.wikimedia.org/wikipedia/commons/9/99/Neon_spectra.jpg *License:* Public domain *Contributors:* Transferred from en.wikipedia
Original artist: Teravolt at en.wikipedia

- **File:Newlands_periodiska_system_1866.png** *Source:* https://upload.wikimedia.org/wikipedia/commons/e/e5/Newlands_periodiska_system_1866.png *License:* Public domain *Contributors:* John Alexander Reina Newlands (1838–1898) *Original artist:* John Alexander Reina Newlands

- **File:Niels_Bohr.jpg** *Source:* https://upload.wikimedia.org/wikipedia/commons/6/6d/Niels_Bohr.jpg *License:* Public domain *Contributors:* Niels Bohr's Nobel Prize biography, from 1922 *Original artist:* The American Institute of Physics credits the photo [1] to AB Lagrelius & Westphal, which is the Swedish company used by the Nobel Foundation for most photos of its book series *Les Prix Nobel*.

- **File:Niels_Bohr_Date_Unverified_LOC.jpg** *Source:* https://upload.wikimedia.org/wikipedia/commons/5/5e/Niels_Bohr_Date_Unverified_LOC.jpg *License:* Public domain *Contributors:* ? *Original artist:* ?

- **File:Niels_Bohr_Institute_1.jpg** *Source:* https://upload.wikimedia.org/wikipedia/commons/2/28/Niels_Bohr_Institute_1.jpg *License:* Public domain *Contributors:* ? *Original artist:* ?

- **File:Niels_Bohr_Signature.svg** *Source:* https://upload.wikimedia.org/wikipedia/commons/d/d3/Niels_Bohr_Signature.svg *License:* Public domain *Contributors:* Own work by uploader, traced in Adobe Illustrator from w:File:Niels Bohr-sig.jpg *Original artist:* Connormah, Niels Bohr

- **File:Niels_Bohr_and_Margrethe_engaged_1910.jpg** *Source:* https://upload.wikimedia.org/wikipedia/commons/a/a6/Niels_Bohr_and_Margrethe_engaged_1910.jpg *License:* Public domain *Contributors:* Niels Bohr Institute *Original artist:* Unknown

- **File:Nitrous-oxide-3D-balls.png** *Source:* https://upload.wikimedia.org/wikipedia/commons/9/93/Nitrous-oxide-3D-balls.png *License:* Public domain *Contributors:* Own work *Original artist:* Ben Mills

- **File:Nobel_Prize.png** *Source:* https://upload.wikimedia.org/wikipedia/en/e/ed/Nobel_Prize.png *License:* ? *Contributors:*
Derivative of File:NobelPrize.JPG *Original artist:*
Photograph: JonathunderMedal: Erik Lindberg (1873-1966)

- **File:Nuvola_apps_edu_science.svg** *Source:* https://upload.wikimedia.org/wikipedia/commons/5/59/Nuvola_apps_edu_science.svg *License:* LGPL *Contributors:* http://ftp.gnome.org/pub/GNOME/sources/gnome-themes-extras/0.9/gnome-themes-extras-0.9.0.tar.gz *Original artist:* David Vignoni / ICON KING

- **File:Nylon6_and_Nylon_66.png** *Source:* https://upload.wikimedia.org/wikipedia/commons/4/4d/Nylon6_and_Nylon_66.png *License:* CC-BY-SA-3.0 *Contributors:* Transferred from en.wikipedia *Original artist:* Michael Ströck (mstroeck) at en.wikipedia

- **File:Octahedron.svg** *Source:* https://upload.wikimedia.org/wikipedia/commons/0/07/Octahedron.svg *License:* CC-BY-SA-3.0 *Contributors:* Vectorisation of Image:Octahedron.jpg *Original artist:* User:Stannered

- **File:Office-book.svg** *Source:* https://upload.wikimedia.org/wikipedia/commons/a/a8/Office-book.svg *License:* Public domain *Contributors:* This and myself. *Original artist:* Chris Down/Tango project

- **File:P-Benzochinon.svg** *Source:* https://upload.wikimedia.org/wikipedia/commons/e/e4/P-Benzochinon.svg *License:* Public domain *Contributors:* Own work *Original artist:* NEUROtiker

- **File:P_vip.svg** *Source:* https://upload.wikimedia.org/wikipedia/en/6/69/P_vip.svg *License:* PD *Contributors:* ? *Original artist:* ?

- **File:Pauli.jpg** *Source:* https://upload.wikimedia.org/wikipedia/commons/4/43/Pauli.jpg *License:* Public domain *Contributors:* http://nobelprize.org/nobel_prizes/physics/laureates/1945/pauli-bio.html *Original artist:* Nobel foundation

- **File:People_icon.svg** *Source:* https://upload.wikimedia.org/wikipedia/commons/3/37/People_icon.svg *License:* CC0 *Contributors:* OpenClipart *Original artist:* OpenClipart

- **File:Periodic_Table_overview_(standard).svg** *Source:* https://upload.wikimedia.org/wikipedia/commons/d/d4/Periodic_Table_overview_%28standard%29.svg *License:* CC BY-SA 3.0 *Contributors:* Own work *Original artist:* DePiep

- **File:Periodic_Table_overview_(wide).svg** *Source:* https://upload.wikimedia.org/wikipedia/commons/f/f3/Periodic_Table_overview_%28wide%29.svg *License:* CC BY-SA 3.0 *Contributors:* Own work *Original artist:* DePiep

- **File:Periodic_table_(metals–metalloids–nonmetals,_32_columns).png** *Source:* https://upload.wikimedia.org/wikipedia/commons/7/78/Periodic_table_%28metals%E2%80%93metalloids%E2%80%93nonmetals%2C_32_columns%29.png *License:* CC BY-SA 4.0 *Contributors:* Own work *Original artist:* DePiep

- **File:Periodic_table_(polyatomic).svg** *Source:* https://upload.wikimedia.org/wikipedia/commons/9/98/Periodic_table_%28polyatomic%29. svg *License:* CC BY-SA 3.0 *Contributors:* Own work -- Actually "inspired by"/forked from earlier free versions on Wikipedia/Commons like this, but there is no option to note this in Upload. *Original artist:* DePiep

- **File:Periodic_table_blocks_spdf_(32_column).svg** *Source:* https://upload.wikimedia.org/wikipedia/commons/f/f2/Periodic_table_blocks_ spdf_%2832_column%29.svg *License:* CC BY-SA 3.0 *Contributors:* https://commons.wikimedia.org/wiki/File:Periodic_Table_2.svg *Original artist:* User:DePiep

- **File:Periodic_table_by_Mendeleev,_1869.svg** *Source:* https://upload.wikimedia.org/wikipedia/commons/c/ce/Periodic_table_by_ Mendeleev%2C_1869.svg *License:* Public domain *Contributors:* Own work *Original artist:* NikNaks

- **File:Periodic_table_by_Mendeleev,_1871.svg** *Source:* https://upload.wikimedia.org/wikipedia/commons/a/aa/Periodic_table_by_ Mendeleev%2C_1871.svg *License:* Public domain *Contributors:* Own work *Original artist:* NikNaks

- **File:Periodic_table_compilation.svg** *Source:* https://upload.wikimedia.org/wikipedia/commons/0/04/Periodic_table_compilation.svg *License:* CC BY 3.0 *Contributors:* a compilation of: files: Telluric screw of De Chancourtois.gif; A New System of Chemical Philosophy fp.jpg; Mendeleev law.jpg; Lavoisier Traité élémentaire de chimie p192.jpg; Periodic table large.svg *Original artist:* authors described in individual files

- **File:Periodic_table_monument.jpg** *Source:* https://upload.wikimedia.org/wikipedia/commons/0/00/Periodic_table_monument.jpg *License:* CC BY-SA 2.0 *Contributors:* http://www.flickr.com/photos/mmmdirt/279349599 *Original artist:* http://www.flickr.com/people/mmmdirt/

- **File:Periodic_trends.svg** *Source:* https://upload.wikimedia.org/wikipedia/commons/f/fe/Periodic_trends.svg *License:* CC0 *Contributors:* Own work *Original artist:* Mirek2

- **File:Periodic_variation_of_Pauling_electronegativities.png** *Source:* https://upload.wikimedia.org/wikipedia/commons/b/b4/Periodic_ variation_of_Pauling_electronegativities.png *License:* CC-BY-SA-3.0 *Contributors:* Own work *Original artist:* Physchim62

- **File:Pierre_Curie_by_Dujardin_c1906.jpg** *Source:* https://upload.wikimedia.org/wikipedia/commons/d/db/Pierre_Curie_by_Dujardin_ c1906.jpg *License:* Public domain *Contributors:* (Traité de radioactivité. ed.), Paris: Gauthier, 1910, http://openlibrary.org/books/ OL24142900M/Traité_de_radioactivité. *Original artist:* Dujardin

- **File:Pile_Leclanché.jpg** *Source:* https://upload.wikimedia.org/wikipedia/commons/3/32/Pile_Leclanch%C3%A9.jpg *License:* Public domain *Contributors:* Leçons de Physique ; Éditions Vuibert et Nony *Original artist:* Gillard

- **File:Portal-puzzle.svg** *Source:* https://upload.wikimedia.org/wikipedia/en/f/fd/Portal-puzzle.svg *License:* Public domain *Contributors:* ? *Original artist:* ?

- **File:Portrait_of_Albert_Einstein,_Niels_Bohr,_James_Franck_and_Rabi.jpg** *Source:* https://upload.wikimedia.org/wikipedia/ commons/9/99/Portrait_of_Albert_Einstein%2C_Niels_Bohr%2C_James_Franck_and_Rabi.jpg *License:* No restrictions *Contributors:* Portrait of Albert Einstein and Others (1879-1955), Physicist *Original artist:* Smithsonian Institution from United States

- **File:Priestley.jpg** *Source:* https://upload.wikimedia.org/wikipedia/commons/d/d5/Priestley.jpg *License:* Public domain *Contributors:* Scan of a print. Original is housed at the National Portrait Gallery, London. *Original artist:* Ellen Sharples (1769 - 1849)

- **File:Puddling_furnace.jpg** *Source:* https://upload.wikimedia.org/wikipedia/commons/5/51/Puddling_furnace.jpg *License:* Public domain *Contributors:* ? *Original artist:* ?

- **File:QtubIronPillar.JPG** *Source:* https://upload.wikimedia.org/wikipedia/commons/3/3f/QtubIronPillar.JPG *License:* Public domain *Contributors:* Original photograph *Original artist:* Photograph taken by Mark A. Wilson (Department of Geology, The College of Wooster). [1]

- **File:Question_book-new.svg** *Source:* https://upload.wikimedia.org/wikipedia/en/9/99/Question_book-new.svg *License:* Cc-by-sa-3.0 *Contributors:*
Created from scratch in Adobe Illustrator. Based on Image:Question book.png created by User:Equazcion *Original artist:* Tkgd2007

- **File:Radioactive.svg** *Source:* https://upload.wikimedia.org/wikipedia/commons/b/b5/Radioactive.svg *License:* Public domain *Contributors:* Created by Cary Bass using Adobe Illustrator on January 19, 2006. *Original artist:* Cary Bass

- **File:Raimundus_Lullus_alchemic_page.jpg** *Source:* https://upload.wikimedia.org/wikipedia/commons/4/40/Raimundus_Lullus_alchemic_ page.jpg *License:* Public domain *Contributors:* Book scan *Original artist:* Ramon Llull, 1232?−1316

- **File:Ritter_Wasserelektrolyse.jpg** *Source:* https://upload.wikimedia.org/wikipedia/commons/9/94/Ritter_Wasserelektrolyse.jpg *License:* Public domain *Contributors:* Original uploaded on de.wikipedia *Original artist:* Original uploaded by Jazi (Transfered by LeastCommonAncestor)

- **File:Robert-millikan2.jpg** *Source:* https://upload.wikimedia.org/wikipedia/commons/a/a6/Robert-millikan2.jpg *License:* Public domain *Contributors:* en-wiki *Original artist:* photograph by Clark Millikan

- **File:Robert_Boyle_0001.jpg** *Source:* https://upload.wikimedia.org/wikipedia/commons/b/b3/Robert_Boyle_0001.jpg *License:* Public domain *Contributors:* http://www.bbk.ac.uk/boyle/Issue4.html *Original artist:* Johann Kerseboom

- **File:Rutherford_gold_foil_experiment_results.svg** *Source:* https://upload.wikimedia.org/wikipedia/commons/c/c3/Rutherford_gold_foil_ experiment_results.svg *License:* Public domain *Contributors:* Own work *Original artist:* Drawn by User:Fastfission in Illustrator and Inkscape. --Fastfission 15:04, 14 April 2008 (UTC)

- **File:Sample_of_Urea.jpg** *Source:* https://upload.wikimedia.org/wikipedia/commons/7/70/Sample_of_Urea.jpg *License:* CC BY-SA 3.0 *Contributors:* Own work *Original artist:* LHcheM

- **File:SamudraguptaCoin.jpg** *Source:* https://upload.wikimedia.org/wikipedia/commons/f/fb/SamudraguptaCoin.jpg *License:* CC-BY-SA- 3.0 *Contributors:* self-made, photographed at the British Museum *Original artist:* PHGCOM

- **File:Savery-engine.jpg** *Source:* https://upload.wikimedia.org/wikipedia/commons/c/cc/Savery-engine.jpg *License:* Public domain *Contributors:* Image copy/pasted from http://www.humanthermodynamics.com/HT-history.html *Original artist:* Institute of Human Thermodynamics and IoHT Publishing Ltd.

- **File:Sceptical_chymist_1661_Boyle_Title_page_AQ18_(3).jpg** *Source:* https://upload.wikimedia.org/wikipedia/commons/d/db/Sceptical_chymist_1661_Boyle_Title_page_AQ18_%283%29.jpg *License:* Public domain *Contributors:* Chemical Heritage Foundation *Original artist:* Chemical Heritage Foundation

- **File:Schrodinger_Equation.png** *Source:* https://upload.wikimedia.org/wikipedia/en/5/53/Schrodinger_Equation.png *License:* PD *Contributors:* ? *Original artist:* ?

- **File:ShortPT20b.png** *Source:* https://upload.wikimedia.org/wikipedia/commons/c/c9/ShortPT20b.png *License:* CC BY-SA 3.0 *Contributors:* Own work *Original artist:* Sandbh

- **File:Socrates.png** *Source:* https://upload.wikimedia.org/wikipedia/commons/c/cd/Socrates.png *License:* Public domain *Contributors:* Transferred from en.wikipedia to Commons. *Original artist:* The original uploader was Magnus Manske at English Wikipedia Later versions were uploaded by Optimager at en.wikipedia.

- **File:Solvay_conference_1927.jpg** *Source:* https://upload.wikimedia.org/wikipedia/commons/6/6e/Solvay_conference_1927.jpg *License:* Public domain *Contributors:* http://w3.pppl.gov/ *Original artist:* Benjamin Couprie

- **File:Solvay_conference_1927_detail.jpg** *Source:* https://upload.wikimedia.org/wikipedia/commons/8/8c/Solvay_conference_1927_detail.jpg *License:* CC BY-SA 3.0 *Contributors:* Own work *Original artist:* Soerfm

- **File:Sommerfeld_ellipses.svg** *Source:* https://upload.wikimedia.org/wikipedia/commons/7/75/Sommerfeld_ellipses.svg *License:* Public domain *Contributors:* Own work *Original artist:* Pieter Kuiper

- **File:Speaker_Icon.svg** *Source:* https://upload.wikimedia.org/wikipedia/commons/2/21/Speaker_Icon.svg *License:* Public domain *Contributors:* ? *Original artist:* ?

- **File:Stylised_Lithium_Atom.svg** *Source:* https://upload.wikimedia.org/wikipedia/commons/e/e1/Stylised_Lithium_Atom.svg *License:* CC-BY-SA-3.0 *Contributors:* based off of Image:Stylised Lithium Atom.png by Halfdan. *Original artist:* SVG by Indolences. Recoloring and ironing out some glitches done by Rainer Klute.

- **File:Symbol_book_class2.svg** *Source:* https://upload.wikimedia.org/wikipedia/commons/8/89/Symbol_book_class2.svg *License:* CC BY-SA 2.5 *Contributors:* Mad by Lokal_Profil by combining: *Original artist:* Lokal_Profil

- **File:Symbol_list_class.svg** *Source:* https://upload.wikimedia.org/wikipedia/en/d/db/Symbol_list_class.svg *License:* Public domain *Contributors:* ? *Original artist:* ?

- **File:Symbol_question.svg** *Source:* https://upload.wikimedia.org/wikipedia/en/e/e0/Symbol_question.svg *License:* Public domain *Contributors:* ? *Original artist:* ?

- **File:System_boundary.svg** *Source:* https://upload.wikimedia.org/wikipedia/commons/b/b6/System_boundary.svg *License:* Public domain *Contributors:* en:Image:System-boundary.jpg *Original artist:* en:User:Wavesmikey, traced by User:Stannered

- **File:Tetrahedron.jpg** *Source:* https://upload.wikimedia.org/wikipedia/commons/8/83/Tetrahedron.jpg *License:* CC-BY-SA-3.0 *Contributors:* ? *Original artist:* ?

- **File:Text_document_with_red_question_mark.svg** *Source:* https://upload.wikimedia.org/wikipedia/commons/a/a4/Text_document_with_red_question_mark.svg *License:* Public domain *Contributors:* Created by bdesham with Inkscape; based upon Text-x-generic.svg from the Tango project. *Original artist:* Benjamin D. Esham (bdesham)

- **File:The_Defeat_of_Baz_Bahadur_of_Malwa_by_the_Mughal_Troops,_1561,_Akbarnama.jpg** *Source:* https://upload.wikimedia.org/wikipedia/commons/8/8d/The_Defeat_of_Baz_Bahadur_of_Malwa_by_the_Mughal_Troops%2C_1561%2C_Akbarnama.jpg *License:* Public domain *Contributors:* V& A Museum [1] *Original artist:* Jagan, Banwali

- **File:The_Sceptical_Chymist.jpg** *Source:* https://upload.wikimedia.org/wikipedia/commons/5/5d/The_Sceptical_Chymist.jpg *License:* Public domain *Contributors:* ? *Original artist:* ?

- **File:The_Shannon_Portrait_of_the_Hon_Robert_Boyle.jpg** *Source:* https://upload.wikimedia.org/wikipedia/commons/a/a3/The_Shannon_Portrait_of_the_Hon_Robert_Boyle.jpg *License:* Public domain *Contributors:* Chemical Heritage Foundation, Photograph by Will Brown. *Original artist:* Johann Kerseboom

- **File:Thermally_Agitated_Molecule.gif** *Source:* https://upload.wikimedia.org/wikipedia/commons/2/23/Thermally_Agitated_Molecule.gif *License:* CC-BY-SA-3.0 *Contributors:* http://en.wikipedia.org/wiki/Image:Thermally_Agitated_Molecule.gif *Original artist:* en:User:Greg L

- **File:Thermodynamics.png** *Source:* https://upload.wikimedia.org/wikipedia/commons/3/3d/Thermodynamics.png *License:* CC BY-SA 3.0 *Contributors:* Own work *Original artist:* Miketwardos

- **File:Transparent.gif** *Source:* https://upload.wikimedia.org/wikipedia/commons/c/ce/Transparent.gif *License:* Public domain *Contributors:* Own work *Original artist:* Edokter

- **File:Urea.png** *Source:* https://upload.wikimedia.org/wikipedia/commons/c/c0/Urea.png *License:* Public domain *Contributors:* user-created image *Original artist:* Ben Mills

- **File:Volta-and-napoleon.PNG** *Source:* https://upload.wikimedia.org/wikipedia/commons/2/28/Volta-and-napoleon.PNG *License:* Public domain *Contributors:* The image was obtained from a Czech website, available on http://www.quido.cz/objevy/monoclanek.htm *Original artist:* ?

- **File:VoltaBattery.JPG** *Source:* https://upload.wikimedia.org/wikipedia/commons/5/54/VoltaBattery.JPG *License:* CC-BY-SA-3.0 *Contributors:* Own work *Original artist:* GuidoB

- **File:Werner_Heisenberg_cropped.jpg** *Source:* https://upload.wikimedia.org/wikipedia/commons/1/1f/Werner_Heisenberg_cropped.jpg *License:* Public domain *Contributors:* http://nobelprize.org/nobel_prizes/physics/laureates/1932/heisenberg-bio.html or crop of File:Bundesarchiv Bild 183-1986-0310-501, Werner Heisenberg.jpg *Original artist:* o.Ang., possibly licensed to Nobel Foundation or Deutsches Bundesarchiv (German Federal Archive), Bild 183-1986-0310-501

- **File:Widmanstatten_IronMet.JPG** *Source:* https://upload.wikimedia.org/wikipedia/commons/9/9f/Widmanstatten_IronMet.JPG *License:* Public domain *Contributors:* ? *Original artist:* ?
- **File:Wikibooks-logo-en-noslogan.svg** *Source:* https://upload.wikimedia.org/wikipedia/commons/d/df/Wikibooks-logo-en-noslogan.svg *License:* CC BY-SA 3.0 *Contributors:* Own work *Original artist:* User:Bastique, User:Ramac et al.
- **File:Wikibooks-logo.svg** *Source:* https://upload.wikimedia.org/wikipedia/commons/f/fa/Wikibooks-logo.svg *License:* CC BY-SA 3.0 *Contributors:* Own work *Original artist:* User:Bastique, User:Ramac et al.
- **File:Wikidata-logo.svg** *Source:* https://upload.wikimedia.org/wikipedia/commons/f/ff/Wikidata-logo.svg *License:* Public domain *Contributors:* Own work *Original artist:* User:Planemad
- **File:Wikinews-logo.svg** *Source:* https://upload.wikimedia.org/wikipedia/commons/2/24/Wikinews-logo.svg *License:* CC BY-SA 3.0 *Contributors:* This is a cropped version of Image:Wikinews-logo-en.png. *Original artist:* Vectorized by Simon 01:05, 2 August 2006 (UTC) Updated by Time3000 17 April 2007 to use official Wikinews colours and appear correctly on dark backgrounds. Originally uploaded by Simon.
- **File:Wikiquote-logo.svg** *Source:* https://upload.wikimedia.org/wikipedia/commons/f/fa/Wikiquote-logo.svg *License:* Public domain *Contributors:* ? *Original artist:* ?
- **File:Wikisource-logo.svg** *Source:* https://upload.wikimedia.org/wikipedia/commons/4/4c/Wikisource-logo.svg *License:* CC BY-SA 3.0 *Contributors:* Rei-artur *Original artist:* Nicholas Moreau
- **File:Wikiversity-logo-Snorky.svg** *Source:* https://upload.wikimedia.org/wikipedia/commons/1/1b/Wikiversity-logo-en.svg *License:* CC BY-SA 3.0 *Contributors:* Own work *Original artist:* Snorky
- **File:Wikiversity-logo.svg** *Source:* https://upload.wikimedia.org/wikipedia/commons/9/91/Wikiversity-logo.svg *License:* CC BY-SA 3.0 *Contributors:* Snorky (optimized and cleaned up by verdy_p) *Original artist:* Snorky (optimized and cleaned up by verdy_p)
- **File:Wiktionary-logo-en.svg** *Source:* https://upload.wikimedia.org/wikipedia/commons/f/f8/Wiktionary-logo-en.svg *License:* Public domain *Contributors:* Vector version of Image:Wiktionary-logo-en.png. *Original artist:* Vectorized by Fvasconcellos (talk · contribs), based on original logo tossed together by Brion Vibber
- **File:Willamette_Meteorite_AMNH.jpg** *Source:* https://upload.wikimedia.org/wikipedia/commons/c/cc/Willamette_Meteorite_AMNH.jpg *License:* CC-BY-SA-3.0 *Contributors:* Own work *Original artist:* User:Dante Alighieri
- **File:William_Fettes_Douglas_-_The_Alchemist.jpg** *Source:* https://upload.wikimedia.org/wikipedia/commons/3/3a/William_Fettes_Douglas_-_The_Alchemist.jpg *License:* Public domain *Contributors:* ? *Original artist:* ?
- **File:XeF2.png** *Source:* https://upload.wikimedia.org/wikipedia/commons/2/27/XeF2.png *License:* Public domain *Contributors:* user-made *Original artist:* User:Smokefoot
- **File:XeTube.jpg** *Source:* https://upload.wikimedia.org/wikipedia/commons/5/59/XeTube.jpg *License:* CC BY-SA 2.5 *Contributors:* user-made *Original artist:* User:Pslawinski
- **File:Xenon-glow.jpg** *Source:* https://upload.wikimedia.org/wikipedia/commons/5/5d/Xenon-glow.jpg *License:* CC BY 3.0 *Contributors:* http://images-of-elements.com/xenon.php *Original artist:* Jurii
- **File:Xenon-tetrafluoride-3D-vdW.png** *Source:* https://upload.wikimedia.org/wikipedia/commons/e/e3/Xenon-tetrafluoride-3D-vdW.png *License:* Public domain *Contributors:* user-made *Original artist:* User:Benjah-bmm27
- **File:Xenon_Spectrum.jpg** *Source:* https://upload.wikimedia.org/wikipedia/commons/6/67/Xenon_Spectrum.jpg *License:* Public domain *Contributors:* Transferred from en.wikipedia; transferred to Commons by User:Homer Landskirty using CommonsHelper. *Original artist:* Teravolt (talk). Original uploader was Teravolt at en.wikipedia
- **File:Xenon_discharge_tube.jpg** *Source:* https://upload.wikimedia.org/wikipedia/commons/d/d7/Xenon_discharge_tube.jpg *License:* FAL *Contributors:* Own work *Original artist:* Alchemist-hp (<a href='//commons.wikimedia.org/wiki/User_talk:Alchemist-hp' title='User talk:Alchemist-hp'>talk</a>) (www.pse-mendelejew.de)
- **File:Xenon_short_arc_1.jpg** *Source:* https://upload.wikimedia.org/wikipedia/commons/9/9e/Xenon_short_arc_1.jpg *License:* CC BY 2.5 *Contributors:* user-made *Original artist:* provided that proper attribution of my copyright is made. - Atlant 19:15, 26 August 2005 (UTC)
- **File:Yuan_Dynasty_-_waterwheels_and_smelting.png** *Source:* https://upload.wikimedia.org/wikipedia/commons/7/7b/Yuan_Dynasty_-_waterwheels_and_smelting.png *License:* Public domain *Contributors:* http://www.waterhistory.org/histories/waterwheels/ *Original artist:* Wang Zhen
- **File:Zosimosapparat.jpg** *Source:* https://upload.wikimedia.org/wikipedia/commons/5/53/Zosimosapparat.jpg *License:* Public domain *Contributors:* Transferred from sv.wikipedia to Commons by natox. *Original artist:* The original uploader was Adragoor at Swedish Wikipedia
- **File:Дмитрий_Иванович_Менделеев_4.gif** *Source:* https://upload.wikimedia.org/wikipedia/commons/2/26/%D0%94%D0%BC%D0%B8%D1%82%D1%80%D0%B8%D0%B9_%D0%98%D0%B2%D0%B0%D0%BD%D0%BE%D0%B2%D0%B8%D1%87_%D0%9C%D0%B5%D0%BD%D0%B4%D0%B5%D0%BB%D0%B5%D0%B5%D0%B2_4.gif *License:* Public domain *Contributors:* Weeks, Mary Elvira (1933) *The Discovery of the Elements*, Easton, PA: Journal of Chemical Education, p. 208 ISBN: 0766138720. *Original artist:* Unknown

27.9.3 Content license

- Creative Commons Attribution-Share Alike 3.0

www.ingramcontent.com/pod-product-compliance
Lightning Source LLC
Chambersburg PA
CBHW080800180526
45168CB00006B/2276